Geometric Formulation of Classical and Quantum Mechanics

Geometric Formulation of Classical and Quantum Mechanics

Giovanni Giachetta
University of Camerino, Italy

Luigi Mangiarotti
University of Camerino, Italy

Gennadi Sardanashvily
Moscow State University, Russia

World Scientific

NEW JERSEY • LONDON • SINGAPORE • BEIJING • SHANGHAI • HONG KONG • TAIPEI • CHENNAI

Published by

World Scientific Publishing Co. Pte. Ltd.
5 Toh Tuck Link, Singapore 596224
USA office: 27 Warren Street, Suite 401-402, Hackensack, NJ 07601
UK office: 57 Shelton Street, Covent Garden, London WC2H 9HE

British Library Cataloguing-in-Publication Data
A catalogue record for this book is available from the British Library.

GEOMETRIC FORMULATION OF CLASSICAL AND QUANTUM MECHANICS

ISBN-13 978-981-4313-72-8
ISBN-10 981-4313-72-6

Printed in Singapore by World Scientific Printers.

Preface

Geometry of symplectic and Poisson manifolds is well known to provide the adequate mathematical formulation of autonomous Hamiltonian mechanics. The literature on this subject is extensive.

This book presents the advanced geometric formulation of classical and quantum non-relativistic mechanics in a general setting of time-dependent coordinate and reference frame transformations. This formulation of mechanics as like as that of classical field theory lies in the framework of general theory of dynamic systems, Lagrangian and Hamiltonian formalism on fibre bundles.

Non-autonomous dynamic systems, Newtonian systems, Lagrangian and Hamiltonian non-relativistic mechanics, relativistic mechanics, quantum non-autonomous mechanics are considered.

Classical non-relativistic mechanics is formulated as a particular field theory on smooth fibre bundles over the time axis $\mathbb{R}$. Quantum non-relativistic mechanics is phrased in the geometric terms of Banach and Hilbert bundles and connections on these bundles. A quantization scheme speaking this language is geometric quantization. Relativistic mechanics is adequately formulated as particular classical string theory of one-dimensional submanifolds.

The concept of a connection is the central link throughout the book. Connections on a configuration space of non-relativistic mechanics describe reference frames. Holonomic connections on a velocity space define non-relativistic dynamic equations. Hamiltonian connections in Hamiltonian non-relativistic mechanics define the Hamilton equations. Evolution of quantum systems is described in terms of algebraic connections. A connection on a prequantization bundle is the main ingredient in geometric quantization.

The book provides a detailed exposition of theory of partially integrable and superintegrable systems and their quantization, classical and quantum non-autonomous constraint systems, Lagrangian and Hamiltonian theory of Jacobi fields, classical and quantum mechanics with time-dependent parameters, the technique of non-adiabatic holonomy operators, formalism of instantwise quantization and quantization with respect to different reference frames.

Our book addresses to a wide audience of theoreticians and mathematicians of undergraduate, post-graduate and researcher levels. It aims to be a guide to advanced geometric methods in classical and quantum mechanics.

For the convenience of the reader, a few relevant mathematical topics are compiled in Appendixes, thus making our exposition self-contained.

Contents

Introduction

We address classical and quantum mechanics in a general setting of arbitrary time-dependent coordinate and reference frame transformations.

The technique of symplectic manifolds is well known to provide the adequate Hamiltonian formulation of autonomous mechanics [1; 104; 157]. Its familiar example is a mechanical system whose configuration space is a manifold M and whose phase space is the cotangent bundle T^*M of M provided with the canonical symplectic form

$$\Omega = dp_i \wedge dq^i, \tag{0.0.1}$$

written with respect to the holonomic coordinates $(q^i, p_i = \dot{q}_i)$ on T^*M. A Hamiltonian $\mathcal{H}$ of this mechanical system is defined as a real function on a phase space T^*M. Any autonomous Hamiltonian system locally is of this type.

However, this Hamiltonian formulation of autonomous mechanics is not extended to mechanics under time-dependent transformations because the symplectic form (0.0.1) fails to be invariant under these transformations. As a palliative variant, one develops time-dependent (non-autonomous) mechanics on a configuration space $Q = \mathbb{R} \times M$ where $\mathbb{R}$ is the time axis [37; 102]. Its phase space $\mathbb{R} \times T^*M$ is provided with the presymplectic form

$$\mathrm{pr}_2^*\Omega = dp_i \wedge dq^i \tag{0.0.2}$$

which is the pull-back of the canonical symplectic form Ω (0.0.1) on T^*M. A time-dependent Hamiltonian is defined as a function on this phase space. A problem is that the presymplectic form (0.0.2) also is broken by time-dependent transformations.

Throughout the book (except Chapter 10), we consider non-relativistic mechanics. Its configuration space is a fibre bundle $Q \to \mathbb{R}$ over the time axis $\mathbb{R}$ endowed with the standard Cartesian coordinate t possessing transition functions $t' = t$+const (this is not the case of time-reparametrized mechanics in Section 3.9). A velocity space of non-relativistic mechanics is the first order jet manifold J^1Q of sections of $Q \to \mathbb{R}$, and its phase space is the vertical cotangent bundle V^*Q of $Q \to \mathbb{R}$ endowed with the canonical Poisson structure [106; 139].

A fibre bundle $Q \to \mathbb{R}$ always is trivial. Its trivialization defines both an appropriate coordinate systems and a connection on this fibre bundle which is associated with a certain non-relativistic reference frame (Section 1.6). Formulated as theory on fibre bundles over $\mathbb{R}$, non-relativistic mechanics is covariant under gauge (atlas) transformations of these fibre bundles, i.e., the above mentioned time-dependent coordinate and reference frame transformations.

This formulation of non-relativistic mechanics is similar to that of classical field theory on fibre bundles over a smooth manifold X of dimension $n > 1$ [68]. A difference between mechanics and field theory however lies in the fact that all connections on fibre bundles over $X = \mathbb{R}$ are flat and, consequently, they are not dynamic variables. Therefore, this formulation of non-relativistic mechanics is covariant, but not invariant under time-dependent transformations.

Second order dynamic systems, Newtonian, Lagrangian and Hamiltonian mechanics are especially considered (Chapters 1–3).

Equations of motion of non-relativistic mechanics almost always are first and second order dynamic equations. Second order dynamic equations on a fibre bundle $Q \to \mathbb{R}$ are conventionally defined as the holonomic connections on the jet bundle $J^1Q \to \mathbb{R}$ (Section 1.4). These equations also are represented by connections on the jet bundle $J^1Q \to Q$ and, due to the canonical imbedding $J^1Q \to TQ$, they are proved to be equivalent to non-relativistic geodesic equations on the tangent bundle TQ of Q (Section 1.5). In Section 10.3, we compare non-relativistic geodesic equations and relativistic geodesic equations in relativistic mechanics. The notions of a free motion equation (Section 1.7.) and a relative acceleration (Section 1.8) are formulated in terms of connections on $J^1Q \to Q$ and $TQ \to Q$.

Generalizing the second Newton law, one introduces the notion of a Newtonian system characterized by a mass tensor (Section 1.9). If a mass tensor is non-degenerate, an equation of motion of a Newtonian system is equivalent to a dynamic equation. We also come to the definition of an external force.

Lagrangian non-relativistic mechanics is formulated in the framework of conventional Lagrangian formalism on fibre bundles [53; 68; 106] (Section 2.1). Its Lagrangian is defined as a density on the velocity space J^1Q, and the corresponding Lagrange equation is a second order differential equation on $Q \to \mathbb{R}$. Besides Lagrange equations, the Cartan and Hamilton–De Donder equations are considered in the framework of Lagrangian formalism. Note that the Cartan equation, but not the Lagrange one is associated to a Hamilton equation in Hamiltonian mechanics (Section 3.6). The relations between Lagrangian and Newtonian systems are established (Section 2.4). Lagrangian conservation laws are defined in accordance with the first Noether theorem (Section 2.5).

Hamiltonian mechanics on a phase space V^*Q is not familiar Poisson Hamiltonian theory on a Poisson manifold V^*Q because all Hamiltonian vector fields on V^*Q are vertical. Hamiltonian mechanics on V^*Q is formulated as particular (polysymplectic) Hamiltonian formalism on fibre bundles [53; 68; 106]. Its Hamiltonian is a section of the fibre bundle $T^*Q \to V^*Q$ (Section 3.3). The pull-back of the canonical Liouville form on T^*Q with respect to this section is a Hamiltonian one-form on V^*Q. The corresponding Hamiltonian connection on $V^*Q \to \mathbb{R}$ defines the first order Hamilton equation on V^*Q.

Furthermore, one can associate to any Hamiltonian system on V^*Q an autonomous symplectic Hamiltonian system on the cotangent bundle T^*Q such that the corresponding Hamilton equations on V^*Q and T^*Q are equivalent (Section 3.4). Moreover, the Hamilton equation on V^*Q also is equivalent to the Lagrange equation of a certain first order Lagrangian system on a configuration space V^*Q. As a consequence, Hamiltonian conservation laws can be formulated as the particular Lagrangian ones (Section 3.8).

Lagrangian and Hamiltonian formulations of mechanics fail to be equivalent, unless a Lagrangian is hyperregular. If a Lagrangian L on a velocity space J^1Q is hyperregular, one can associate to L an unique Hamiltonian form on a phase space V^*Q such that Lagrange equation on Q and the Hamilton equation on V^*Q are equivalent. In general, different Hamiltonian forms are associated to a non-regular Lagrangian. The comprehensive relations between Lagrangian and Hamiltonian systems can be established in the case of almost regular Lagrangians (Section 3.6).

In comparison with non-relativistic mechanics, if a configuration space of a mechanical system has no preferable fibration $Q \to \mathbb{R}$, we obtain a general formulation of relativistic mechanics, including Special Relativity on the Minkowski space $Q = \mathbb{R}^4$ (Chapter 10). A velocity space of relativistic mechanics is the first order jet manifold $J_1^1 Q$ of one-dimensional submanifolds of a configuration space Q [53; 139]. This notion of jets generalizes that of jets of sections of fibre bundles which is utilized in field theory and non-relativistic mechanics. The jet bundle $J_1^1 Q \to Q$ is projective, and one can think of its fibres as being spaces of three-velocities of a relativistic system. Four-velocities of a relativistic system are represented by elements of the tangent bundle TQ of a configuration space Q, while the cotangent bundle T^*Q, endowed with the canonical symplectic form, plays a role of the phase space of relativistic theory. As a result, Hamiltonian relativistic mechanics can be seen as a constraint Dirac system on the hyperboloids of relativistic momenta in the phase space T^*Q.

Note that the tangent bundle TQ of a configuration space Q plays a role of the space of four-velocities both in non-relativistic and relativistic mechanics. The difference is only that, given a fibration $Q \to \mathbb{R}$, the four-velocities of a non-relativistic system live in the subbundle (10.3.14) of TQ, whereas the four-velocities of a relativistic theory belong to the hyperboloids

$$g_{\mu\nu}\dot{q}^\mu \dot{q}^\nu = 1, \tag{0.0.3}$$

where g is an admissible pseudo-Riemannian metric in TQ. Moreover, as was mentioned above, both relativistic and non-relativistic equations of motion can be seen as geodesic equations on the tangent bundle TQ, but their solutions live in its different subbundles (0.0.3) and (10.3.14).

Quantum non-relativistic mechanics is phrased in the geometric terms of Banach and Hilbert manifolds and locally trivial Hilbert and C^*-algebra bundles (Chapter 4). A quantization scheme speaking this language is geometric quantization (Chapter 5).

Let us note that a definition of smooth Banach (and Hilbert) manifolds follows that of finite-dimensional smooth manifolds in general, but infinite-dimensional Banach manifolds are not locally compact, and they need not be paracompact [65; 100; 155]. It is essential that Hilbert manifolds satisfy the inverse function theorem and, therefore, locally trivial Hilbert bundles are defined. We restrict our consideration to Hilbert and C^*-algebra bundles over smooth finite-dimensional manifolds X, e.g., $X = \mathbb{R}$. Sections of such a Hilbert bundle make up a particular locally trivial continuous field of Hilbert spaces [33]. Conversely, one can think of any locally trivial continuous field of Hilbert spaces or C^*-algebras as being a module of sections of some topological fibre bundle. Given a Hilbert space E, let $B \subset B(E)$ be some C^*-algebra of bounded operators in E. The following fact reflects the non-equivalence of Schrödinger and Heisenberg quantum pictures. There is the obstruction to the existence of associated (topological) Hilbert and C^*-algebra bundles $\mathcal{E} \to X$ and $\mathcal{B} \to X$ with the typical fibres E and B, respectively. Firstly, transition functions of $\mathcal{E}$ define those of $\mathcal{B}$, but the latter need not be continuous, unless B is the algebra of compact operators in E. Secondly, transition functions of $\mathcal{B}$ need not give rise to transition functions of $\mathcal{E}$. This obstruction is characterized by the Dixmier–Douady class of $\mathcal{B}$ in the Čech cohomology group $H^3(X, \mathbb{Z})$ (Section 4.4).

One also meets a problem of the definition of connections on C^*-algebra bundles. It comes from the fact that a C^*-algebra need not admit non-zero bounded derivations. An unbounded derivation of a C^*-algebra A obeying certain conditions is an infinitesimal generator of a strongly (but not uniformly) continuous one-parameter group of automorphisms of A [18]. Therefore, one may introduce a connection on a C^*-algebra bundle in terms of parallel transport curves and operators, but not their infinitesimal generators [6]. Moreover, a representation of A does not imply necessarily a unitary representation of its strongly (not uniformly) continuous one-parameter group of automorphisms (Section 4.5). In contrast, connections on a Hilbert bundle over a smooth manifold can be defined both as particular first order differential operators on the module of its sections [65; 109] and a parallel displacement along paths lifted from the base [88].

The most of quantum models come from quantization of original classical systems. This is the case of *canonical quantization* which replaces the Poisson bracket $\{f, f'\}$ of smooth functions with the bracket $[\widehat{f}, \widehat{f}']$ of Hermitian operators in a Hilbert space such that *Dirac's condition*

$$[\widehat{f}, \widehat{f}'] = -i\widehat{\{f, f'\}} \tag{0.0.4}$$

holds. Canonical quantization of Hamiltonian non-relativistic mechanics on a configuration space $Q \to \mathbb{R}$ is geometric quantization [57; 65]. It takes the form of instantwise quantization phrased in the terms of Hilbert bundles over $\mathbb{R}$ (Section 5.4.3). This quantization depends on a reference frame, represented by a connection on a configuration space $Q \to \mathbb{R}$. Under quantization, this connection yields a connection on the quantum algebra of a phase space V^*Q. We obtain the relation between operators of energy with respect to different reference frames (Section 5.5).

The book provides a detailed exposition of a few important mechanical systems.

Chapter 6 is devoted to Hamiltonian systems with time-dependent constraints and their geometric quantization.

In Chapter 7, completely integrable, partially integrable and superintegrable Hamiltonian systems are described in a general setting of invariant submanifolds which need not be compact. In particular, this is the case of non-autonomous completely integrable and superintegrable systems. Geometric quantization of completely integrable and superintegrable Hamiltonian systems with respect to action-angle variables is considered. Using this quantization, the non-adiabatic holonomy operator is constructed in Section 9.6.

Given a mechanical system on a configuration space $Q \to \mathbb{R}$, its extension onto the vertical tangent bundle $VQ \to \mathbb{R}$ of $Q \to \mathbb{R}$ describes the Jacobi fields of the Lagrange and Hamilton equations (Chapter 8). In particular, we show that Jacobi fields of a completely integrable Hamiltonian system of m degrees of freedom make up an extended completely integrable system of $2m$ degrees of freedom, where m additional integrals of motion characterize a relative motion.

Chapter 9 addresses mechanical systems with time-dependent parameters. These parameters can be seen as sections of some smooth fibre bundle $\Sigma \to \mathbb{R}$ called the parameter bundle. Sections 9.1 and 9.2 are devoted to Lagrangian and Hamiltonian classical mechanics with parameters. In order to obtain the Lagrange and Hamilton equations, we treat parameters on the same level as dynamic variables. Geometric quantization of mechanical systems with time-dependent parameters is developed in Section 9.3. Berry's phase factor is a phenomenon peculiar to quantum systems depending on classical time-dependent parameters (Section 9.4). In Section 9.5, we study the Berry phase phenomena in completely integrable systems. The reason is that, being constant under an internal dynamic evolution, action variables of a completely integrable system are driven only by a perturbation holonomy operator without any adiabatic approximation

Let us note that, since time reparametrization is not considered, we believe that all quantities are physically dimensionless, but sometimes refer to the *universal unit system* where the velocity of light c and the Planck constant $\hbar$ are equal to 1, while the length unit is the Planck one

$$(G\hbar c^{-3})^{1/2} = G^{1/2} = 1,616 \cdot 10^{-33} cm,$$

where G is the Newtonian gravitational constant. Relative to the universal unit system, the physical dimension of the spatial and temporal Cartesian coordinates is [length], while the physical dimension of a mass is $[\text{length}]^{-1}$.

For the convenience of the reader, a few relevant mathematical topics are compiled in Chapter 11.

Chapter 1

Dynamic equations

Equations of motion of non-relativistic mechanics are first and second order differential equations on manifolds and fibre bundles over $\mathbb{R}$. Almost always, they are dynamic equations. Their solutions are called a *motion*.

This Chapter is devoted to theory of second order dynamic equations in non-relativistic mechanics, whose *configuration space* is a fibre bundle $Q \to \mathbb{R}$. They are defined as the holonomic connections on the jet bundle $J^1Q \to \mathbb{R}$ (Section 1.4). These equations are represented by connections on the jet bundle $J^1Q \to Q$. Due to the canonical imbedding $J^1Q \to TQ$ (1.1.6), they are proved equivalent to non-relativistic geodesic equations on the tangent bundle TQ of Q (Theorem 1.5.1). In Section 10.3, we compare non-relativistic geodesic equations and relativistic geodesic equations in relativistic mechanics. Any relativistic geodesic equation on the tangent bundle TQ defines the non-relativistic one, but the converse relitivization procedure is more intricate [106; 107; 109].

The notions of a non-relativistic reference frame, a relative velocity, a free motion equation and a relative acceleration are formulated in terms of connections on $Q \to \mathbb{R}$, $J^1Q \to Q$ and $TQ \to Q$.

Generalizing the second Newton law, we introduce the notion of a Newtonian system (Definition 1.9.1) characterized by a mass tensor. If a mass tensor is non-degenerate, an equation of motion of a Newtonian system is equivalent to a dynamic equation. The notion of an external force also is formulated.

1.1 Preliminary. Fibre bundles over $\mathbb{R}$

This section summarizes some peculiarities of fibre bundles over $\mathbb{R}$.

Let

$$\pi : Q \to \mathbb{R} \tag{1.1.1}$$

be a fibred manifold whose base is treated as a time axis. Throughout the book, the time axis $\mathbb{R}$ is parameterized by the Cartesian coordinate t with the transition functions $t' = t+$const. Of course, this is the case neither of relativistic mechanics (Chapter 10) nor the models with time reparametrization (Section 3.9). Relative to the Cartesian coordinate t, the time axis $\mathbb{R}$ is provided with the *standard vector field* ∂_t and the *standard one-form* dt which also is the volume element on $\mathbb{R}$. The symbol dt also stands for any pull-back of the standard one-form dt onto a fibre bundle over $\mathbb{R}$.

Remark 1.1.1. Point out one-to-one correspondence between the vector fields $f\partial_t$, the densities fdt and the real functions f on $\mathbb{R}$. Roughly speaking, we can neglect the contribution of $T\mathbb{R}$ and $T^*\mathbb{R}$ to some expressions (Remarks 1.1.3 and 1.9.1). However, one should be careful with such simplification in the framework of the universal unit system. For instance, coefficients f of densities fdt have the physical dimension $[\text{length}]^{-1}$, whereas functions f are physically dimensionless.

In order that the dynamics of a mechanical system can be defined at any instant $t \in \mathbb{R}$, we further assume that a fibred manifold $Q \to \mathbb{R}$ is a fibre bundle with a typical fibre M.

Remark 1.1.2. In accordance with Remark 11.4.1, a fibred manifold $Q \to \mathbb{R}$ is a fibre bundle if and only if it admits an Ehresmann connection Γ, i.e., the horizontal lift $\Gamma\partial_t$ onto Q of the standard vector field ∂_t on $\mathbb{R}$ is complete. By virtue of Theorem 11.2.5, any fibre bundle $Q \to \mathbb{R}$ is trivial. Its different trivializations

$$\psi : Q = \mathbb{R} \times M \tag{1.1.2}$$

differ from each other in fibrations $Q \to M$.

Given bundle coordinates (t, q^i) on the fibre bundle $Q \to \mathbb{R}$ (1.1.1), the first order jet manifold J^1Q of $Q \to \mathbb{R}$ is provided with the adapted coordinates (t, q^i, q^i_t) possessing transition functions (11.3.1) which read

$$q'^i_t = (\partial_t + q^j_t \partial_j)q'^i. \tag{1.1.3}$$

In non-relativistic mechanics on a configuration space $Q \to \mathbb{R}$, the jet manifold J^1Q plays a role of the *velocity space*.

Note that, if $Q = \mathbb{R} \times M$ coordinated by $(t, \overline{q}^i)$, there is the canonical isomorphism

$$J^1(\mathbb{R} \times M) = \mathbb{R} \times TM, \qquad \overline{q}^i_t = \dot{\overline{q}}^i, \tag{1.1.4}$$

that one can justify by inspection of the transition functions of the coordinates $\overline{q}^i_t$ and $\dot{\overline{q}}^i$ when transition functions of q^i are time-independent. Due to the isomorphism (1.1.4), every trivialization (1.1.2) yields the corresponding trivialization of the jet manifold

$$J^1Q = \mathbb{R} \times TM. \tag{1.1.5}$$

The canonical imbedding (11.3.5) of J^1Q takes the form

$$\lambda_{(1)} : J^1Q \ni (t, q^i, q^i_t) \to (t, q^i, \dot{t} = 1, \dot{q}^i = q^i_t) \in TQ, \tag{1.1.6}$$

$$\lambda_{(1)} = d_t = \partial_t + q^i_t \partial_i, \tag{1.1.7}$$

where by d_t is meant the total derivative. From now on, a jet manifold J^1Q is identified with its image in TQ. Using the morphism (1.1.6), one can define the contraction

$$J^1Q \underset{Q}{\times} T^*Q \underset{Q}{\to} Q \times \mathbb{R},$$

$$(q^i_t; \dot{\mathrm{t}}, \dot{q}_i) \to \lambda_{(1)} \rfloor (\dot{\mathrm{t}} dt + \dot{q}_i dq^i) = \dot{\mathrm{t}} + q^i_t \dot{q}_i, \tag{1.1.8}$$

where $(t, q^i, \dot{\mathrm{t}}, \dot{q}_i)$ are holonomic coordinates on the cotangent bundle T^*Q.

Remark 1.1.3. Following precisely the expression (11.3.5), one should write the morphism $\lambda_{(1)}$ (1.1.7) in the form

$$\lambda_{(1)} = dt \otimes (\partial_t + q^i_t \partial_i). \tag{1.1.9}$$

With respect to the universal unit system, the physical dimension of $\lambda_{(1)}$ (1.1.7) is $[\text{length}]^{-1}$, while $\lambda_{(1)}$ (1.1.9) is dimensionless.

A glance at the expression (1.1.6) shows that the affine jet bundle $J^1Q \to Q$ is modelled over the vertical tangent bundle VQ of a fibre bundle $Q \to \mathbb{R}$. As a consequence, there is the following canonical splitting (11.2.27) of the vertical tangent bundle $V_Q J^1Q$ of the affine jet bundle $J^1Q \to Q$:

$$\alpha : V_Q J^1Q = J^1Q \underset{Q}{\times} VQ, \qquad \alpha(\partial^t_i) = \partial_i, \tag{1.1.10}$$

together with the corresponding splitting of the vertical cotangent bundle $V^*_Q J^1Q$ of $J^1Q \to Q$:

$$\alpha^* : V^*_Q J^1Q = J^1Q \underset{Q}{\times} V^*Q, \qquad \alpha^*(\overline{d}q^i_t) = \overline{d}q^i, \tag{1.1.11}$$

where $\overline{d}q^i_t$ and $\overline{d}q^i$ are the holonomic bases for $V^*_Q J^1Q$ and V^*Q, respectively. Then the exact sequence (11.4.30) of vertical bundles over the composite fibre bundle

$$J^1Q \longrightarrow Q \longrightarrow \mathbb{R} \tag{1.1.12}$$

reads

$$\begin{array}{c} \xleftarrow{\qquad\qquad \alpha^{-1} \qquad\qquad} \\ 0 \longrightarrow V_Q J^1Q \xrightarrow{\ i\ } VJ^1Q \xrightarrow{\ \pi_V\ } J^1Q \underset{Q}{\times} VQ \longrightarrow 0. \end{array}$$

Hence, we obtain the following linear endomorphism over J^1Q of the vertical tangent bundle VJ^1Q of the jet bundle $J^1Q \to \mathbb{R}$:

$$\begin{aligned} &\widehat{v} = i \circ \alpha^{-1} \circ \pi_V : VJ^1Q \to VJ^1Q, \\ &\widehat{v}(\partial_i) = \partial^t_i, \qquad \widehat{v}(\partial^t_i) = 0. \end{aligned} \tag{1.1.13}$$

This endomorphism obeys the nilpotency rule

$$\widehat{v} \circ \widehat{v} = 0. \tag{1.1.14}$$

Combining the canonical horizontal splitting (11.2.27), the corresponding epimorphism

$$\begin{aligned} &\mathrm{pr}_2 : J^1Q \underset{Q}{\times} TQ \to J^1Q \underset{Q}{\times} VQ = V_Q J^1Q, \\ &\partial_t \to -q^i_t \partial^t_i, \qquad \partial_i \to \partial^t_i, \end{aligned}$$

and the monomorphism $VJ^1Q \to TJ^1Q$, one can extend the endomorphism (1.1.13) to the tangent bundle TJ^1Q:

$$\begin{aligned} &\widehat{v} : TJ^1Q \to TJ^1Q, \\ &\widehat{v}(\partial_t) = -q^i_t \partial^t_i, \qquad \widehat{v}(\partial_i) = \partial^t_i, \qquad \widehat{v}(\partial^t_i) = 0. \end{aligned} \tag{1.1.15}$$

This is called the *vertical endomorphism*. It inherits the nilpotency property (1.1.14). The transpose of the vertical endomorphism $\widehat{v}$ (1.1.15) is

$$\begin{aligned} &\widehat{v}^* : T^*J^1Q \to T^*J^1Q, \\ &\widehat{v}^*(dt) = 0, \qquad \widehat{v}^*(dq^i) = 0, \qquad \widehat{v}^*(dq^i_t) = \theta^i, \end{aligned} \tag{1.1.16}$$

where $\theta^i = dq^i - q^i_t dt$ are the contact forms (11.3.6). The nilpotency rule $\widehat{v}^* \circ \widehat{v}^* = 0$ also is fulfilled. The homomorphisms $\widehat{v}$ and $\widehat{v}^*$ are associated with the tangent-valued one-form $\widehat{v} = \theta^i \otimes \partial^t_i$ in accordance with the relations (11.2.52) – (11.2.53).

In view of the morphism $\lambda_{(1)}$ (1.1.6), any connection

$$\Gamma = dt \otimes (\partial_t + \Gamma^i \partial_i) \tag{1.1.17}$$

on a fibre bundle $Q \to \mathbb{R}$ can be identified with a nowhere vanishing *horizontal vector field*

$$\Gamma = \partial_t + \Gamma^i \partial_i \tag{1.1.18}$$

on Q which is the horizontal lift $\Gamma\partial_t$ (11.4.3) of the standard vector field ∂_t on $\mathbb{R}$ by means of the connection (1.1.17). Conversely, any vector field Γ on Q such that $dt\rfloor\Gamma = 1$ defines a connection on $Q \to \mathbb{R}$. Therefore, the connections (1.1.17) further are identified with the vector fields (1.1.18). The integral curves of the vector field (1.1.18) coincide with the integral sections for the connection (1.1.17).

Connections on a fibre bundle $Q \to \mathbb{R}$ constitute an affine space modelled over the vector space of vertical vector fields on $Q \to \mathbb{R}$. Accordingly, the covariant differential (11.4.8), associated with a connection Γ on $Q \to \mathbb{R}$, takes its values into the vertical tangent bundle VQ of $Q \to \mathbb{R}$:

$$D^\Gamma : J^1Q \underset{Q}{\to} VQ, \qquad \dot{q}^i \circ D^\Gamma = q^i_t - \Gamma^i. \tag{1.1.19}$$

A connection Γ on a fibre bundle $Q \to \mathbb{R}$ is obviously flat. It yields a horizontal distribution on Q. The integral manifolds of this distribution are integral curves of the vector field (1.1.18) which are transversal to fibres of a fibre bundle $Q \to \mathbb{R}$.

Theorem 1.1.1. *By virtue of Theorem 11.4.1, every connection Γ on a fibre bundle $Q \to \mathbb{R}$ defines an atlas of local constant trivializations of $Q \to \mathbb{R}$ such that the associated bundle coordinates (t, q^i) on Q possess the transition functions $q^i \to q'^i(q^j)$ independent of t, and*

$$\Gamma = \partial_t \tag{1.1.20}$$

with respect to these coordinates. Conversely, every atlas of local constant trivializations of the fibre bundle $Q \to \mathbb{R}$ determines a connection on $Q \to \mathbb{R}$ which is equal to (1.1.20) relative to this atlas.

A connection Γ on a fibre bundle $Q \to \mathbb{R}$ is said to be *complete* if the horizontal vector field (1.1.18) is complete. In accordance with Remark 11.4.1, a connection on a fibre bundle $Q \to \mathbb{R}$ is complete if and only if it is an Ehresmann connection. The following holds [106].

Theorem 1.1.2. *Every trivialization of a fibre bundle $Q \to \mathbb{R}$ yields a complete connection on this fibre bundle. Conversely, every complete connection Γ on $Q \to \mathbb{R}$ defines its trivialization (1.1.2) such that the horizontal* *vector field (1.1.18) equals ∂_t relative to the bundle coordinates associated with this trivialization.*

Let J^1J^1Q be the repeated jet manifold of a fibre bundle $Q \to \mathbb{R}$ provided with the adapted coordinates $(t, q^i, q^i_t, \widehat{q}^i_t, q^i_{tt})$ possessing transition functions

$$\begin{aligned} &q'^i_t = d_t q'^i, \qquad \widehat{q}'^i_t = \widehat{d}_t q'^i, \qquad q'^i_{tt} = \widehat{d}_t q'^i_t, \\ &d_t = \partial_t + q^j_t \partial_j + q^j_{tt} \partial^t_j, \qquad \widehat{d}_t = \partial_t + \widehat{q}^j_t \partial_j + q^j_{tt} \partial^t_j. \end{aligned}$$

There is the canonical isomorphism k between the affine fibrations π_{11} (11.3.10) and $J^1\pi^1_0$ (11.3.11) of J^1J^1Q over J^1Q, i.e.,

$$\pi_{11} \circ k = J^1_0 \pi_{01}, \qquad k \circ k = \operatorname{Id} J^1 J^1 Q,$$

where

$$q^i_t \circ k = \widehat{q}^i_t, \qquad \widehat{q}^i_t \circ k = q^i_t, \qquad q^i_{tt} \circ k = q^i_{tt}. \tag{1.1.21}$$

In particular, the affine bundle π_{11} (11.3.10) is modelled over the vertical tangent bundle VJ^1Q of $J^1Q \to \mathbb{R}$ which is canonically isomorphic to the underlying vector bundle $J^1VQ \to J^1Q$ of the affine bundle $J^1\pi^1_0$ (11.3.11).

For a fibre bundle $Q \to \mathbb{R}$, the sesquiholonomic jet manifold $\widehat{J}^2Q$ coincides with the second order jet manifold J^2Q coordinated by $(t, q^i, q^i_t, q^i_{tt})$, possessing transition functions

$$q'^i_t = d_t q'^i, \qquad q'^i_{tt} = d_t q'^i_t. \tag{1.1.22}$$

The affine bundle $J^2Q \to J^1Q$ is modelled over the vertical tangent bundle

$$V_Q J^1 Q = J^1 Q \underset{Q}{\times} VQ \to J^1 Q$$

of the affine jet bundle $J^1Q \to Q$. There are the imbeddings

$$J^2Q \xrightarrow{\lambda_{(2)}} TJ^1Q \xrightarrow{T\lambda_{(1)}} V_Q TQ = T^2Q \subset TTQ,$$

$$\lambda_{(2)} : (t, q^i, q^i_t, q^i_{tt}) \to (t, q^i, q^i_t, \dot{t} = 1, \dot{q}^i = q^i_t, \dot{q}^i_t = q^i_{tt}), \tag{1.1.23}$$

$$\begin{aligned} T\lambda_{(1)} \circ \lambda_{(2)} &: (t, q^i, q^i_t, q^i_{tt}) \\ &\to (t, q^i, \dot{t} = \dot{\mathrm{t}} = 1, \dot{q}^i = \dot{\mathrm{q}}^i = q^i_t, \ddot{t} = 0, \ddot{q}^i = q^i_{tt}), \end{aligned} \tag{1.1.24}$$

where $(t, q^i, \dot{t}, \dot{q}^i, \dot{\mathrm{t}}, \dot{\mathrm{q}}^i, \ddot{t}, \ddot{q}^i)$ are the coordinates on the double tangent bundle TTQ and $T^2Q \subset TTQ$ is *second tangent bundle* the *second tangent bundle* given by the coordinate relation $\dot{t} = \dot{\mathrm{t}}$.

Due to the morphism (1.1.23), any connection ξ on the jet bundle $J^1Q \to \mathbb{R}$ (defined as a section of the affine bundle π_{11} (11.3.10)) is represented by a horizontal vector field on J^1Q such that $\xi \rfloor dt = 1$.

A connection Γ (1.1.18) on a fibre bundle $Q \to \mathbb{R}$ has the jet prolongation to the section $J^1\Gamma$ of the affine bundle $J^1\pi_0^1$. By virtue of the isomorphism k (1.1.21), every connection Γ on $Q \to \mathbb{R}$ gives rise to the connection

$$J\Gamma = k \circ J^1\Gamma : J^1Q \to J^1J^1Q,$$
$$J\Gamma = \partial_t + \Gamma^i\partial_i + d_t\Gamma^i\partial_i^t, \tag{1.1.25}$$

on the jet bundle $J^1Q \to \mathbb{R}$.

A connection on the jet bundle $J^1Q \to \mathbb{R}$ is said to be *holonomic* if it is a section

$$\xi = dt \otimes (\partial_t + q_t^i\partial_i + \xi^i\partial_i^t)$$

of the holonomic subbundle $J^2Q \to J^1Q$ of $J^1J^1Q \to J^1Q$. In view of the morphism (1.1.23), a holonomic connection is represented by a horizontal vector field

$$\xi = \partial_t + q_t^i\partial_i + \xi^i\partial_i^t \tag{1.1.26}$$

on J^1Q. Conversely, every vector field ξ on J^1Q such that

$$dt\rfloor\xi = 1, \qquad \widehat{v}(\xi) = 0,$$

where $\widehat{v}$ is the vertical endomorphism (1.1.15), is a holonomic connection on the jet bundle $J^1Q \to \mathbb{R}$.

Holonomic connections (1.1.26) make up an affine space modelled over the linear space of vertical vector fields on the affine jet bundle $J^1Q \to Q$, i.e., which live in V_QJ^1Q.

A holonomic connection ξ defines the corresponding covariant differential (1.1.19) on the jet manifold J^1Q:

$$D^\xi : J^1J^1Q \underset{J^1Q}{\longrightarrow} V_QJ^1Q \subset VJ^1Q,$$

$$\dot q^i \circ D^\xi = 0, \qquad \dot q_t^i \circ D^\xi = q_{tt}^i - \xi^i,$$

which takes its values into the vertical tangent bundle V_QJ^1Q of the jet bundle $J^1Q \to Q$. Then by virtue of Theorem 11.3.1, any integral section $\overline{c} : () \to J^1Q$ for a holonomic connection ξ is holonomic, i.e., $\overline{c} = \dot c$ where c is a curve in Q.

1.2 Autonomous dynamic equations

Let us start with dynamic equations on a manifold. From the physical viewpoint, they are treated as *autonomous dynamic equations* in autonomous non-relativistic mechanics.

Let Z, $\dim Z > 1$, be a smooth manifold coordinated by (z^λ).

Definition 1.2.1. Let u be a vector field u on Z. A closed subbundle $u(Z)$ of the tangent bundle TZ given by the coordinate relations

$$\dot{z}^\lambda = u^\lambda(z) \tag{1.2.1}$$

is said to be an *autonomous dynamic equation!first order* on a manifold Z. This is a system of first order differential equations on a fibre bundle $\mathbb{R} \times Z \to \mathbb{R}$ in accordance with Definition 11.3.5.

By a *solution* of the autonomous first order dynamic equation (1.2.1) is meant an integral curve of the vector field u.

Definition 1.2.2. An *autonomous dynamic equation!second order* on a manifold Z is defined as an autonomous first order dynamic equation on the tangent bundle TZ which is associated with a *holonomic vector field*

$$\Xi = \dot{z}^\lambda \partial_\lambda + \Xi^\lambda(z^\mu, \dot{z}^\mu)\dot{\partial}_\lambda \tag{1.2.2}$$

on TZ. This vector field, by definition, obeys the condition

$$J(\Xi) = u_{TZ},$$

where J is the endomorphism (11.2.55) and u_{TZ} is the Liouville vector field (11.2.34) on TZ.

The holonomic vector field (1.2.2) also is called the autonomous second order dynamic equation.

Let the double tangent bundle TTZ be provided with coordinates $(z^\lambda, \dot{z}^\lambda, \dot{\mathrm{z}}^\lambda, \ddot{z}^\lambda)$. With respect to these coordinates, an autonomous second order dynamic equation defined by the holonomic vector field Ξ (1.2.2) reads

$$\dot{\mathrm{z}}^\lambda = \dot{z}^\lambda, \qquad \ddot{z}^\lambda = \Xi^\lambda(z^\mu, \dot{z}^\mu). \tag{1.2.3}$$

By a *solution* of the second order dynamic equation (1.2.3) is meant a curve $c : (,) \to Z$ in a manifold Z whose tangent prolongation $\dot{c} : (,) \to TZ$ is an integral curve of the holonomic vector field Ξ or, equivalently, whose second order tangent prolongation $\ddot{c}$ lives in the subbundle (1.2.3). It satisfies an autonomous second order differential equation

$$\ddot{c}^\lambda(t) = \Xi^\lambda(c^\mu(t), \dot{c}^\mu(t)).$$

Remark 1.2.1. In fact, the autonomous second order dynamic equation (1.2.3) is a closed subbundle

$$\ddot{z}^\lambda = \Xi^\lambda(z^\mu, \dot{z}^\mu) \tag{1.2.4}$$

of the second tangent bundle $T^2Z \subset TTZ$.

Autonomous second order dynamic equations on a manifold Z are exemplified by geodesic equations on the tangent bundle TZ.

Given a connection

$$K = dz^\mu \otimes (\partial_\mu + K^\nu_\mu \dot{\partial}_\nu) \tag{1.2.5}$$

on the tangent bundle $TZ \to Z$, let

$$\widehat{K} : TZ \underset{Z}{\times} TZ \to TTZ \tag{1.2.6}$$

be the corresponding linear bundle morphism over TZ which splits the exact sequence (11.2.20):

$$0 \longrightarrow VTZ \longrightarrow TTZ \longrightarrow TZ \underset{Z}{\times} TZ \longrightarrow 0.$$

Note that, in contrast with K (11.4.20), the connection K (1.2.5) need not be linear.

Definition 1.2.3. A *geodesic equation* on TZ with respect to the connection K (1.2.5) is defined as the range

$$\dot{z}^\lambda = \dot{z}^\lambda, \qquad \ddot{z}^\mu = K^\mu_\nu \dot{z}^\nu \tag{1.2.7}$$

in $T^2Z \subset TTZ$ of the morphism (1.2.6) restricted to the diagonal $TZ \subset TZ \times TZ$.

By a *solution* of a geodesic equation on TZ is meant a *geodesic curve* c in Z whose tangent prolongation $\dot{c}$ is an integral section (a *geodesic vector field*) over $c \subset Z$ for a connection K.

It is readily observed that the range (1.2.7) of the morphism $\widehat{K}$ (1.2.6) is a holonomic vector field

$$\widehat{K}(TZ) = \dot{z}^\lambda \partial_\lambda + K^\mu_\nu \dot{z}^\nu \dot{\partial}_\mu \tag{1.2.8}$$

on TZ whose integral curve is a geodesic vector field. It follows that any geodesic equation (1.2.6) on TZ is an autonomous second order dynamic equation on Z. The converse is not true in general. Nevertheless, there is the following theorem [118].

Theorem 1.2.1. *Every autonomous second order dynamic equation (1.2.3) on a manifold Z defines a connection K_Ξ on the tangent bundle $TZ \to Z$ whose components are*

$$K^\mu_\nu = \frac{1}{2}\dot{\partial}_\nu \Xi^\mu. \tag{1.2.9}$$

However, the autonomous second order dynamic equation (1.2.3) fails to be a geodesic equation with respect to the connection (1.2.9) in general. In particular, the geodesic equation (1.2.7) with respect to a connection K determines the connection (1.2.9) on $TZ \to Z$ which does not necessarily coincide with K.

Theorem 1.2.2. *An autonomous second order dynamic equation Ξ on Z is a geodesic equation for the connection (1.2.9) if and only if Ξ is a spray, i.e.,*

$$[u_{TZ}, \Xi] = \Xi,$$

where u_{TZ} is the Liouville vector field (11.2.34) on TZ, i.e.,

$$\Xi^i = a_{ij}(q^k)\dot{q}^i\dot{q}^j$$

and the connection K (1.2.9) is linear.

1.3 Dynamic equations

Let $Q \to X$ (1.1.1) be a configuration space of non-relativistic mechanics. Refereing to Definition 11.3.5 of a differential equation on a fibre bundle, one defines a *dynamic equation* on $Q \to \mathbb{R}$ as a differential equation which is algebraically solved for the highest order derivatives.

Definition 1.3.1. Let Γ (1.1.18) be a connection on a fibre bundle $Y \to \mathbb{R}$. The corresponding covariant differential D^Γ (1.1.19) is a first order differential operator on Y. Its kernel, given by the coordinate equation

$$q^i_t = \Gamma^i(t, q^i), \tag{1.3.1}$$

is a closed subbundle of the jet bundle $J^1Y \to \mathbb{R}$. By virtue of Definition 11.3.5, it is a first order differential equation on a fibre bundle $Y \to \mathbb{R}$ called the *first order dynamic equation* on $Y \to \mathbb{R}$.

Due to the canonical imbedding $J^1Q \to TQ$ (1.1.6), the equation (1.3.1) is equivalent to the autonomous first order dynamic equation

$$\dot{t} = 1, \qquad \dot{q}^i = \Gamma^i(t, q^i) \tag{1.3.2}$$

on a manifold Y (Definition 1.2.2). It is defined by the vector field (1.1.18). *Solutions* of the first order dynamic equation (1.3.1) are integral sections for a connection Γ.

Definition 1.3.2. Let us consider the first order dynamic equation (1.3.1) on the jet bundle $J^1Q \to \mathbb{R}$, which is associated with a holonomic connection ξ (1.1.26) on $J^1Q \to \mathbb{R}$. This is a closed subbundle of the second order jet bundle $J^2Q \to \mathbb{R}$ given by the coordinate relations

$$q^i_{tt} = \xi^i(t, q^j, q^j_t). \tag{1.3.3}$$

Consequently, it is a second order differential equation on a fibre bundle $Q \to \mathbb{R}$ in accordance with Definition 11.3.5. This equation is called a *second order dynamic equation*. The corresponding horizontal vector field ξ (1.1.26) also is termed the second order dynamic equation.

The second order dynamic equation (1.3.3) possesses the coordinate transformation law

$$q'^i_{tt} = \xi'^i, \qquad \xi'^i = (\xi^j\partial_j + q^j_t q^k_t \partial_j\partial_k + 2q^j_t\partial_j\partial_t + \partial^2_t)q'^i(t, q^j), \tag{1.3.4}$$

derived from the formula (1.1.22).

A *solution* of the second order dynamic equation (1.3.3) is a curve c in Q whose second order jet prolongation $\ddot{c}$ lives in (1.3.3). Any integral section $\overline{c}$ for the holonomic connection ξ obviously is the jet prolongation $\dot{c}$ of a solution c of the second order dynamic equation (1.3.3), i.e.,

$$\ddot{c}^i = \xi^i \circ \dot{c}, \tag{1.3.5}$$

and *vice versa*.

Remark 1.3.1. By very definition, the second order dynamic equation (1.3.3) on a fibre bundle $Q \to \mathbb{R}$ is equivalent to the system of first order differential equations

$$\widehat{q}^i_t = q^i_t, \qquad q^i_{tt} = \xi^i(t, q^j, q^j_t), \tag{1.3.6}$$

on the jet bundle $J^1Q \to \mathbb{R}$. Any solution $\overline{c}$ of these equations takes its values into J^2Q and, by virtue of Theorem 11.3.1, is holonomic, i.e., $\overline{c} = \dot{c}$. The equations (1.3.3) and (1.3.6) are therefore equivalent. The equation (1.3.6) is said to be the *first order reduction of the second order dynamic equation* (1.3.3).

A second order dynamic equation ξ on a fibre bundle $Q \to \mathbb{R}$ is said to be *conservative* if there exist a trivialization (1.1.2) of Q and the corresponding trivialization (1.1.5) of J^1Q such that the vector field ξ (1.1.26) on J^1Q is projectable onto TM. Then this projection

$$\Xi_\xi = \dot{q}^i\partial_i + \xi^i(q^j, \dot{q}^j)\dot{\partial}_i$$

is an autonomous second order dynamic equation on the typical fibre M of $Q \to \mathbb{R}$ in accordance with Definition 1.2.2. Its solution is seen as a section of the fibre bundle $\mathbb{R} \times M \to \mathbb{R}$ (1.1.2). Conversely, every autonomous second order dynamic equation Ξ (1.2.2) on a manifold M can be seen as a conservative second order dynamic equation

$$\xi_\Xi = \partial_t + \dot{q}^i \partial_i + \Xi^i \dot{\partial}_i \tag{1.3.7}$$

on the fibre bundle $\mathbb{R} \times M \to \mathbb{R}$ in accordance with the isomorphism (1.1.5).

The following theorem holds [106].

Theorem 1.3.1. *Any second order dynamic equation ξ (1.3.3) on a fibre bundle $Q \to \mathbb{R}$ is equivalent to an autonomous second order dynamic equation Ξ on a manifold Q which makes the diagram*

$$\begin{array}{ccc} J^2Q & \longrightarrow & T^2Q \\ {\scriptstyle \xi}\uparrow & & \uparrow{\scriptstyle \Xi} \\ J^1Q & \overset{\lambda_{(1)}}{\longrightarrow} & TQ \end{array}$$

commutative and obeys the relations

$$\xi^i = \Xi^i(t, q^j, \dot{t} = 1, \dot{q}^j = q^j_t), \qquad \Xi^t = 0.$$

Accordingly, the second order dynamic equation (1.3.3) is written in the form

$$q^i_{tt} = \Xi^i \,|_{\dot{t}=1, \dot{q}^j = q^j_t},$$

which is equivalent to the autonomous second order dynamic equation

$$\ddot{t} = 0, \qquad \dot{t} = 1, \qquad \ddot{q}^i = \Xi^i, \tag{1.3.8}$$

on Q.

1.4 Dynamic connections

In order to say something more, let us consider the relationship between the holonomic connections on the jet bundle $J^1Q \to \mathbb{R}$ and the connections on the affine jet bundle $J^1Q \to Q$ (see Propositions 1.4.1 and 1.4.2 below).

By $J^1_Q J^1 Q$ throughout is meant the first order jet manifold of the affine jet bundle $J^1Q \to Q$. The adapted coordinates on $J^1_Q J^1 Q$ are $(q^\lambda, q^i_t, q^i_{\lambda t})$, where we use the compact notation $\lambda = (0, i)$, $q^0 = t$. Let

$$\gamma : J^1Q \to J^1_Q J^1 Q$$

be a connection on the affine jet bundle $J^1Q \to Q$. It takes the coordinate form

$$\gamma = dq^\lambda \otimes (\partial_\lambda + \gamma^i_\lambda \partial^t_i), \tag{1.4.1}$$

together with the coordinate transformation law

$$\gamma'^i_\lambda = (\partial_j q'^i \gamma^j_\mu + \partial_\mu q'^i_t)\frac{\partial q^\mu}{\partial q'^\lambda}. \tag{1.4.2}$$

Remark 1.4.1. In view of the canonical splitting (1.1.10), the curvature (11.4.13) of the connection γ (1.4.1) reads

$$R : J^1Q \to \overset{2}{\wedge} T^*Q \underset{J^1Q}{\otimes} VQ,$$

$$R = \frac{1}{2} R^i_{\lambda\mu} dq^\lambda \wedge dq^\mu \otimes \partial_i = \left(\frac{1}{2} R^i_{kj} dq^k \wedge dq^j + R^i_{0j} dt \wedge dq^j\right) \otimes \partial_i,$$

$$R^i_{\lambda\mu} = \partial_\lambda \gamma^i_\mu - \partial_\mu \gamma^i_\lambda + \gamma^j_\lambda \partial_j \gamma^i_\mu - \gamma^j_\mu \partial_j \gamma^i_\lambda. \tag{1.4.3}$$

Using the contraction (1.1.8), we obtain the soldering form

$$\lambda_{(1)} \rfloor R = [(R^i_{kj} q^k_t + R^i_{0j}) dq^j - R^i_{0j} q^j_t dt] \otimes \partial_i$$

on the affine jet bundle $J^1Q \to Q$. Its image by the canonical projection $T^*Q \to V^*Q$ (2.2.5) is the tensor field

$$\overline{R} : J^1Q \to V^*Q \underset{Q}{\otimes} VQ, \qquad \overline{R} = (R^i_{kj} q^k_t + R^i_{0j}) \overline{d} q^j \otimes \partial_i, \tag{1.4.4}$$

and then we come to the scalar field

$$\widetilde{R} : J^1Q \to \mathbb{R}, \qquad \widetilde{R} = R^i_{ki} q^k_t + R^i_{0i}, \tag{1.4.5}$$

on the jet manifold J^1Q.

Proposition 1.4.1. *Any connection γ (1.4.1) on the affine jet bundle $J^1Q \to Q$ defines the holonomic connection*

$$\xi_\gamma = \rho \circ \gamma : J^1Q \to J^1_Q J^1Q \to J^2Q, \tag{1.4.6}$$

$$\xi_\gamma = \partial_t + q^i_t \partial_i + (\gamma^i_0 + q^j_t \gamma^i_j) \partial^t_i,$$

on the jet bundle $J^1Q \to \mathbb{R}$.

Proof. Let us consider the composite fibre bundle (1.1.12) and the morphism ρ (11.4.25) which reads

$$\rho : J^1_Q J^1Q \ni (q^\lambda, q^i_t, q^i_{\lambda t}) \tag{1.4.7}$$

$$\to (q^\lambda, q^i_t, \widehat{q}^i_t = q^i_t, q^i_{tt} = q^i_{0t} + q^j_t q^i_{jt}) \in J^2Q.$$

A connection γ (1.4.1) and the morphism ρ (1.4.7) combine into the desired holonomic connection ξ_γ (1.4.6) on the jet bundle $J^1Q \to \mathbb{R}$. □

It follows that every connection γ (1.4.1) on the affine jet bundle $J^1Q \to Q$ yields the second order dynamic equation

$$q^i_{tt} = \gamma^i_0 + q^j_t \gamma^i_j \tag{1.4.8}$$

on the configuration bundle $Q \to \mathbb{R}$. This is precisely the restriction to J^2Q of the kernel Ker $\widetilde{D}^\gamma$ of the vertical covariant differential $\widetilde{D}^\gamma$ (11.4.36) defined by the connection γ:

$$\widetilde{D}^\gamma : J^1J^1Q \to V_QJ^1Q, \qquad \dot{q}^i_t \circ \widetilde{D}^\gamma = q^i_{tt} - \gamma^i_0 - q^j_t \gamma^i_j. \tag{1.4.9}$$

Therefore, connections on the jet bundle $J^1Q \to Q$ are called the *dynamic connections*. The corresponding equation (1.3.5) can be written in the form

$$\ddot{c}^i = \rho \circ \gamma \circ \dot{c},$$

where ρ is the morphism (1.4.7).

Of course, different dynamic connections can lead to the same second order dynamic equation (1.4.8).

Proposition 1.4.2. *Any holonomic connection ξ (1.1.26) on the jet bundle $J^1Q \to \mathbb{R}$ yields the dynamic connection*

$$\gamma_\xi = dt \otimes \left[\partial_t + (\xi^i - \frac{1}{2}q^j_t \partial^t_j \xi^i)\partial^t_i\right] + dq^j \otimes \left[\partial_j + \frac{1}{2}\partial^t_j \xi^i \partial^t_i\right] \tag{1.4.10}$$

on the affine jet bundle $J^1Q \to Q$ [106; 109].

It is readily observed that the dynamic connection γ_ξ (1.4.10), defined by a second order dynamic equation, possesses the property

$$\gamma^k_i = \partial^t_i \gamma^k_0 + q^j_t \partial^t_i \gamma^k_j, \tag{1.4.11}$$

which implies the relation

$$\partial^t_j \gamma^k_i = \partial^t_i \gamma^k_j.$$

Therefore, a dynamic connection γ, obeying the condition (1.4.11), is said to be *symmetric*. The *torsion of a dynamic connection* γ is defined as the tensor field

$$T : J^1Q \to V^*Q \underset{Q}{\otimes} VQ,$$

$$T = T^k_i \overline{d}q^i \otimes \partial_k, \qquad T^k_i = \gamma^k_i - \partial^t_i \gamma^k_0 - q^j_t \partial^t_i \gamma^k_j. \tag{1.4.12}$$

It follows at once that a dynamic connection is symmetric if and only if its torsion vanishes.

Let γ be the dynamic connection (1.4.1) and ξ_γ the corresponding second order dynamic equation (1.4.6). Then the dynamic connection (1.4.10) associated with the second order dynamic equation ξ_γ takes the form

$$\gamma_{\xi_\gamma}{}_i^k = \frac{1}{2}(\gamma_i^k + \partial_i^t\gamma_0^k + q_t^j\partial_i^t\gamma_j^k), \qquad \gamma_{\xi_\gamma}{}_0^k = \gamma_0^k + q_t^j\gamma_j^k - q_t^i\gamma_{\xi_\gamma}{}_i^k.$$

It is readily observed that $\gamma = \gamma_{\xi_\gamma}$ if and only if the torsion T (1.4.12) of the dynamic connection γ vanishes.

Example 1.4.1. Since a jet bundle $J^1Q \to Q$ is affine, it admits an affine connection

$$\gamma = dq^\lambda \otimes [\partial_\lambda + (\gamma_{\lambda 0}^i(q^\mu) + \gamma_{\lambda j}^i(q^\mu)q_t^j)\partial_i^t]. \tag{1.4.13}$$

This connection is symmetric if and only if $\gamma_{\lambda\mu}^i = \gamma_{\mu\lambda}^i$. One can easily justify that an affine dynamic connection generates a quadratic second order dynamic equation, and *vice versa*. Nevertheless, a non-affine dynamic connection, whose symmetric part is affine, also defines a quadratic second order dynamic equation. The affine connection (1.4.13) on an affine jet bundle $J^1Q \to Q$ yields the linear connection

$$\overline{\gamma} = dq^\lambda \otimes [\partial_\lambda + \gamma_{\lambda j}^i(q^\mu)\dot{q}_t^j\dot{\partial}_i]$$

on the vertical tangent bundle $VQ \to Q$.

Using the notion of a dynamic connection, we can modify Theorem 1.2.1 as follows. Let Ξ be an autonomous second order dynamic equation on a manifold M, and let ξ_Ξ (1.3.7) be the corresponding conservative second order dynamic equation on the bundle $\mathbb{R} \times M \to \mathbb{R}$. The latter yields the dynamic connection γ (1.4.10) on a fibre bundle

$$\mathbb{R} \times TM \to \mathbb{R} \times M.$$

Its components γ_j^i are exactly those of the connection (1.2.9) on the tangent bundle $TM \to M$ in Theorem 1.2.1, while γ_0^i make up a vertical vector field

$$e = \gamma_0^i\dot{\partial}_i = \left(\Xi^i - \frac{1}{2}\dot{q}^j\dot{\partial}_j\Xi^i\right)\dot{\partial}_i \tag{1.4.14}$$

on $TM \to M$. Thus, we have shown the following.

Proposition 1.4.3. *Every autonomous second order dynamic equation Ξ (1.2.3) on a manifold M admits the decomposition*

$$\Xi^i = K_j^i\dot{q}^j + e^i$$

where K is the connection (1.2.9) on the tangent bundle $TM \to M$, and e is the vertical vector field (1.4.14) on $TM \to M$.

1.5 Non-relativistic geodesic equations

In this Section, we aim to show that every second order dynamic equation on a configuration bundle $Q \to \mathbb{R}$ is equivalent to a geodesic equation on the tangent bundle $TQ \to Q$ [56; 107].

We start with the relation between the dynamic connections γ on the affine jet bundle $J^1Q \to Q$ and the connections

$$K = dq^\lambda \otimes (\partial_\lambda + K^\mu_\lambda \dot{\partial}_\mu) \tag{1.5.1}$$

on the tangent bundle $TQ \to Q$ of the configuration space Q. Note that they need not be linear. We follow the compact notation (11.2.30).

Let us consider the diagram

$$\begin{array}{ccc} J^1_Q J^1 Q & \overset{J^1\lambda_{(1)}}{\longrightarrow} & J^1_Q TQ \\ {\scriptstyle\gamma}\Big\uparrow & & \Big\uparrow{\scriptstyle K} \\ J^1 Q & \overset{\lambda_{(1)}}{\longrightarrow} & TQ \end{array} \tag{1.5.2}$$

where $J^1_Q TQ$ is the first order jet manifold of the tangent bundle $TQ \to Q$, coordinated by

$$(t, q^i, \dot{t}, \dot{q}^i, (\dot{t})_\mu, (\dot{q}^i)_\mu).$$

The jet prolongation over Q of the canonical imbedding $\lambda_{(1)}$ (1.1.6) reads

$$J^1\lambda_{(1)} : (t, q^i, q^i_t, q^i_{\mu t}) \to (t, q^i, \dot{t} = 1, \dot{q}^i = q^i_t, (\dot{t})_\mu = 0, (\dot{q}^i)_\mu = q^i_{\mu t}).$$

Then we have

$$J^1\lambda_{(1)} \circ \gamma : (t, q^i, q^i_t) \to (t, q^i, \dot{t} = 1, \dot{q}^i = q^i_t, (\dot{t})_\mu = 0, (\dot{q}^i)_\mu = \gamma^i_\mu),$$
$$K \circ \lambda_{(1)} : (t, q^i, q^i_t) \to (t, q^i, \dot{t} = 1, \dot{q}^i = q^i_t, (\dot{t})_\mu = K^0_\mu, (\dot{q}^i)_\mu = K^i_\mu).$$

It follows that the diagram (1.5.2) can be commutative only if the components K^0_μ of the connection K (1.5.1) on the tangent bundle $TQ \to Q$ vanish.

Since the transition functions $t \to t'$ are independent of q^i, a connection

$$\widetilde{K} = dq^\lambda \otimes (\partial_\lambda + K^i_\lambda \dot{\partial}_i) \tag{1.5.3}$$

with $K^0_\mu = 0$ may exist on the tangent bundle $TQ \to Q$ in accordance with the transformation law

$$K'^i_\lambda = (\partial_j q'^i K^j_\mu + \partial_\mu \dot{q}'^i)\frac{\partial q^\mu}{\partial q'^\lambda}. \tag{1.5.4}$$

Now the diagram (1.5.2) becomes commutative if the connections γ and $\widetilde{K}$ fulfill the relation

$$\gamma^i_\mu = K^i_\mu \circ \lambda_{(1)} = K^i_\mu(t, q^i, \dot{t} = 1, \dot{q}^i = q^i_t). \tag{1.5.5}$$

It is easily seen that this relation holds globally because the substitution of $\dot{q}^i = q^i_t$ in (1.5.4) restates the transformation law (1.4.2) of a connection on the affine jet bundle $J^1Q \to Q$. In accordance with the relation (1.5.5), the desired connection $\widetilde{K}$ is an extension of the section $J^1\lambda \circ \gamma$ of the affine jet bundle $J^1_Q TQ \to TQ$ over the closed submanifold $J^1Q \subset TQ$ to a global section. Such an extension always exists by virtue of Theorem 11.2.2, but it is not unique. Thus, we have proved the following.

Proposition 1.5.1. *In accordance with the relation (1.5.5), every second order dynamic equation on a configuration bundle $Q \to \mathbb{R}$ can be written in the form*

$$q^i_{tt} = K^i_0 \circ \lambda_{(1)} + q^j_t K^i_j \circ \lambda_{(1)}, \tag{1.5.6}$$

where $\widetilde{K}$ is the connection (1.5.3) on the tangent bundle $TQ \to Q$. Conversely, each connection $\widetilde{K}$ (1.5.3) on $TQ \to Q$ defines the dynamic connection γ (1.5.5) on the affine jet bundle $J^1Q \to Q$ and the second order dynamic equation (1.5.6) on a configuration bundle $Q \to \mathbb{R}$.

Then we come to the following theorem.

Theorem 1.5.1. *Every second order dynamic equation (1.3.3) on a configuration bundle $Q \to \mathbb{R}$ is equivalent to the geodesic equation*

$$\ddot{q}^0 = 0, \qquad \dot{q}^0 = 1,$$
$$\ddot{q}^i = K^i_\lambda(q^\mu, \dot{q}^\mu)\dot{q}^\lambda, \tag{1.5.7}$$

on the tangent bundle TQ relative to the connection $\widetilde{K}$ (1.5.3) with the components $K^0_\lambda = 0$ and K^i_λ (1.5.5). We call this equation the non-relativistic geodesic equation *Its solution is a geodesic curve in Q which also obeys the second order dynamic equation (1.5.6), and vice versa.*

In accordance with this theorem, the autonomous second order equation (1.3.8) in Theorem 1.3.1 can be chosen as a non-relativistic geodesic equation. It should be emphasized that, written relative to the bundle coordinates (t, q^i), the non-relativistic geodesic equation (1.5.7) and the connection $\widetilde{K}$ (1.5.5) are well defined with respect to any coordinates on Q.

From the physical viewpoint, the most relevant second order dynamic equations are the quadratic ones

$$\xi^i = a^i_{jk}(q^\mu)q^j_t q^k_t + b^i_j(q^\mu)q^j_t + f^i(q^\mu). \tag{1.5.8}$$

This property is global due to the transformation law (1.3.4). Then one can use the following two facts.

Proposition 1.5.2. *There is one-to-one correspondence between the affine connections γ on the affine jet bundle $J^1Q \to Q$ and the linear connections K (1.5.3) on the tangent bundle $TQ \to Q$.*

Proof. This correspondence is given by the relation (1.5.5), written in the form

$$\begin{aligned}\gamma^i_\mu &= \gamma^i_{\mu 0} + \gamma^i_{\mu j}q^j_t = K_\mu{}^i{}_0(q^\nu)\dot{t} + K_\mu{}^i{}_j(q^\nu)\dot{q}^j|_{\dot{t}=1,\dot{q}^i=q^i_t} \\ &= K_\mu{}^i{}_0(q^\nu) + K_\mu{}^i{}_j(q^\nu)q^j_t,\end{aligned}$$

i.e., $\gamma^i_{\mu\lambda} = K_\mu{}^i{}_\lambda$. □

In particular, if an affine dynamic connection γ is symmetric, so is the corresponding linear connection K.

Corollary 1.5.1. *Every quadratic second order dynamic equation (1.5.8) on a configuration bundle $Q \to \mathbb{R}$ of non-relativistic mechanics is equivalent to the non-relativistic geodesic equation*

$$\begin{aligned}&\ddot{q}^0 = 0, \qquad \dot{q}^0 = 1, \\ &\ddot{q}^i = a^i_{jk}(q^\mu)\dot{q}^j\dot{q}^k + b^i_j(q^\mu)\dot{q}^j\dot{q}^0 + f^i(q^\mu)\dot{q}^0\dot{q}^0\end{aligned} \tag{1.5.9}$$

on the tangent bundle TQ with respect to the symmetric linear connection $\widetilde{K}$ (1.5.3):

$$K_\lambda{}^0{}_\nu = 0, \quad K_0{}^i{}_0 = f^i, \quad K_0{}^i{}_j = \frac{1}{2}b^i_j, \quad K_k{}^i{}_j = a^i_{kj}, \tag{1.5.10}$$

on the tangent bundle $TQ \to Q$.

The geodesic equation (1.5.9), however, is not unique for the second order dynamic equation (1.5.8).

Proposition 1.5.3. *Any quadratic second order dynamic equation (1.5.8), being equivalent to a non-relativistic geodesic equation with respect to the symmetric linear connection $\widetilde{K}$ (1.5.10), also is equivalent to the geodesic equation with respect to an affine connection K' on $TQ \to Q$ which differs from $\widetilde{K}$ (1.5.10) in a soldering form σ on $TQ \to Q$ with the components*

$$\sigma^0_\lambda = 0, \qquad \sigma^i_k = h^i_k + (s-1)h^i_k\dot{q}^0, \qquad \sigma^i_0 = -sh^i_k\dot{q}^k - h^i_0\dot{q}^0 + h^i_0,$$

where s and h^i_λ are local functions on Q.

Proposition 1.5.3 also can be deduced from the following lemma.

Lemma 1.5.1. *Every affine vertical vector field*

$$\sigma = [f^i(q^\mu) + b^i_j(q^\mu)q^j_t]\partial^0_i \tag{1.5.11}$$

on the affine jet bundle $J^1Q \to Q$ is extended to the soldering form

$$\sigma = (f^i dq^0 + b^i_k dq^k) \otimes \dot\partial_i \tag{1.5.12}$$

on the tangent bundle $TQ \to Q$.

Proof. Similarly to Proposition 1.5.2, one can show that there is one-to-one correspondence between the $V_Q J^1Q$-valued affine vector fields (1.5.11) on the jet manifold J^1Q and the linear vertical vector fields

$$\overline{\sigma} = [b^i_j(q^\mu)\dot q^j + f^i(q^\mu)\dot q^0]\dot\partial_i$$

on the tangent bundle TQ. This linear vertical vector field determines the desired soldering form (1.5.12). □

In Section 10.3, we use Theorem 1.5.1, Corollary 1.5.1 and Proposition 1.5.3 in order to study the relationship between non-relativistic and relativistic equations of motion [56].

Now let us extend our inspection of dynamic equations to connections on the tangent bundle $TM \to M$ of the typical fibre M of a configuration bundle $Q \to \mathbb{R}$. In this case, the relationship fails to be canonical, but depends on a trivialization (1.1.2) of $Q \to \mathbb{R}$.

Given such a trivialization, let $(t, \overline{q}^i)$ be the associated coordinates on Q, where $\overline{q}^i$ are coordinates on M with transition functions independent of t. The corresponding trivialization (1.1.5) of $J^1Q \to \mathbb{R}$ takes place in the coordinates $(t, \overline{q}^i, \dot{\overline{q}}^i)$, where $\dot{\overline{q}}^i$ are coordinates on TM. With respect to these coordinates, the transformation law (1.4.2) of a dynamic connection γ on the affine jet bundle $J^1Q \to Q$ reads

$$\gamma'^i_0 = \frac{\partial \overline{q}'^i}{\partial \overline{q}^j}\gamma^j_0 \qquad \gamma'^i_k = \left(\frac{\partial \overline{q}'^i}{\partial \overline{q}^j}\gamma^j_n + \frac{\partial \dot{\overline{q}}'^i}{\partial \overline{q}^n}\right)\frac{\partial \overline{q}^n}{\partial \overline{q}'^k}.$$

It follows that, given a trivialization of $Q \to \mathbb{R}$, a connection γ on $J^1Q \to Q$ defines the time-dependent vertical vector field

$$\gamma^i_0(t, \overline{q}^j, \dot{\overline{q}}^j)\frac{\partial}{\partial \dot{\overline{q}}^i} : \mathbb{R} \times TM \to VTM$$

and the time-dependent connection

$$d\overline{q}^k \otimes \left(\frac{\partial}{\partial \overline{q}^k} + \gamma^i_k(t, \overline{q}^j, \dot{\overline{q}}^j)\frac{\partial}{\partial \dot{\overline{q}}^i}\right) : \mathbb{R} \times TM \to J^1TM \subset TTM \tag{1.5.13}$$

on the tangent bundle $TM \to M$.

Conversely, let us consider a connection

$$\overline{K} = d\overline{q}^k \otimes \left(\frac{\partial}{\partial \overline{q}^k} + \overline{K}^i_k(\overline{q}^j, \dot{\overline{q}}^j) \frac{\partial}{\partial \dot{\overline{q}}^i} \right)$$

on the tangent bundle $TM \to M$. Given the above-mentioned trivialization of the configuration bundle $Q \to \mathbb{R}$, the connection $\overline{K}$ defines the connection $\widetilde{K}$ (1.5.3) with the components

$$K^i_0 = 0, \qquad K^i_k = \overline{K}^i_k,$$

on the tangent bundle $TQ \to Q$. The corresponding dynamic connection γ on the affine jet bundle $J^1Q \to Q$ reads

$$\gamma^i_0 = 0, \qquad \gamma^i_k = \overline{K}^i_k. \tag{1.5.14}$$

Using the transformation law (1.4.2), one can extend the expression (1.5.14) to arbitrary bundle coordinates (t, q^i) on the configuration space Q as follows:

$$\gamma^i_k = \left[\frac{\partial q^i}{\partial \overline{q}^j} \overline{K}^j_n(\overline{q}^j(q^r), \dot{\overline{q}}^j(q^r, q^r_t)) + \frac{\partial^2 q^i}{\partial \overline{q}^n \partial \overline{q}^j} \dot{\overline{q}}^j + \frac{\partial \Gamma^i}{\partial \overline{q}^n} \right] \partial_k \overline{q}^n, \tag{1.5.15}$$
$$\gamma^i_0 = \partial_t \Gamma^i + \partial_j \Gamma^i q^j_t - \gamma^i_k \Gamma^k,$$

where

$$\Gamma^i = \partial_t q^i(t, \overline{q}^j)$$

is the connection on $Q \to \mathbb{R}$, corresponding to a given trivialization of Q, i.e., $\Gamma^i = 0$ relative to $(t, \overline{q}^i)$. The second order dynamic equation on Q defined by the dynamic connection (1.5.15) takes the form

$$q^i_{tt} = \partial_t \Gamma^i + q^j_t \partial_j \Gamma^i + \gamma^i_k (q^k_t - \Gamma^k). \tag{1.5.16}$$

By construction, it is a conservative second order dynamic equation. Thus, we have proved the following.

Proposition 1.5.4. *Any connection $\overline{K}$ on the typical fibre M of a configuration bundle $Q \to \mathbb{R}$ yields a conservative second order dynamic equation (1.5.16) on Q.*

1.6 Reference frames

From the physical viewpoint, a reference frame in non-relativistic mechanics determines a tangent vector at each point of a configuration space Q, which characterizes the velocity of an *observer* at this point. This speculation leads to the following mathematical definition of a reference frame in non-relativistic mechanics [106; 112; 139].

Definition 1.6.1. A non-relativistic *reference frame* is a connection Γ on a configuration space $Q \to \mathbb{R}$.

By virtue of this definition, one can think of the horizontal vector field (1.1.18) associated with a connection Γ on $Q \to \mathbb{R}$ as being a family of observers, while the corresponding covariant differential (1.1.19):

$$\dot{q}^i_\Gamma = D^\Gamma(q^i_t) = q^i_t - \Gamma^i,$$

determines the *relative velocity* with respect to a reference frame Γ. Accordingly, q^i_t are regarded as the *absolute velocities*.

In particular, given a motion $c : \mathbb{R} \to Q$, its covariant derivative $\nabla^\Gamma c$ (11.4.9) with respect to a connection Γ is a velocity of this motion relative to a reference frame Γ. For instance, if c is an integral section for a connection Γ, a velocity of the motion c relative to a reference frame Γ is equal to 0. Conversely, every motion $c : \mathbb{R} \to Q$ defines a reference frame Γ_c such that a velocity of c relative to Γ_c vanishes. This reference frame Γ_c is an extension of a section $c(\mathbb{R}) \to J^1Q$ of an affine jet bundle $J^1Q \to Q$ over the closed submanifold $c(\mathbb{R}) \in Q$ to a global section in accordance with Theorem 11.2.2.

Remark 1.6.1. Bearing in mind time reparametrization, one should define relative velocities as elements of $VQ \otimes_Q T^*\mathbb{R}$. They as like as absolute velocities possess the physical dimension $[q] - 1$.

By virtue of Theorem 1.1.1, any reference frame Γ on a configuration bundle $Q \to \mathbb{R}$ is associated with an atlas of local constant trivializations, and *vice versa.* A connection Γ takes the form $\Gamma = \partial_t$ (1.1.20) with respect to the corresponding coordinates $(t, \overline{q}^i)$, whose transition functions $\overline{q}^i \to \overline{q}'^i$ are independent of time. One can think of these coordinates as also being a reference frame, corresponding to the connection (1.1.20). They are called the *adapted coordinates* to a reference frame Γ. Thus, we come to the following definition, equivalent to Definition 1.6.1.

Definition 1.6.2. In non-relativistic mechanics, a reference frame is an atlas of local constant trivializations of a configuration bundle $Q \to \mathbb{R}$.

In particular, with respect to the coordinates $\overline{q}^i$ adapted to a reference frame Γ, the velocities relative to this reference frame coincide with the absolute ones

$$D^\Gamma(\overline{q}^i_t) = \dot{\overline{q}}^i_\Gamma = \overline{q}^i_t.$$

Remark 1.6.2. By analogy with gauge field theory, we agree to call transformations of bundle atlases of a fibre bundle $Q \to \mathbb{R}$ the *gauge transformations.* To be precise, one should call them passive gauge transformations, while by active gauge transformations are meant automorphisms of a fibre bundle. In non-relativistic mechanics, gauge transformations also are reference frame transformations in accordance with Theorem 1.1.1. An object on a fibre bundle is said to be *gauge covariant* or, simply, *covariant* if its definition is atlas independent. It is called *gauge invariant* if its form is maintained under atlas transformations.

A reference frame is said to be *complete* if the associated connection Γ is complete. By virtue of Proposition 1.1.2, every complete reference frame defines a trivialization of a bundle $Q \to \mathbb{R}$, and *vice versa.*

Remark 1.6.3. Given a reference frame Γ, one should solve the equations

$$\Gamma^i(t, q^j(t, \overline{q}^a)) = \frac{\partial q^i(t, \overline{q}^a)}{\partial t}, \tag{1.6.1}$$

$$\frac{\partial \overline{q}^a(t, q^j)}{\partial q^i}\Gamma^i(t, q^j) + \frac{\partial \overline{q}^a(t, q^j)}{\partial t} = 0 \tag{1.6.2}$$

in order to find the coordinates $(t, \overline{q}^a)$ adapted to Γ. Let (t, q^a_1) and (t, q^i_2) be the adapted coordinates for reference frames Γ_1 and Γ_2, respectively. In accordance with the equality (1.6.2), the components Γ^i_1 of the connection Γ_1 with respect to the coordinates (t, q^i_2) and the components Γ^a_2 of the connection Γ_2 with respect to the coordinates (t, q^a_1) fulfill the relation

$$\frac{\partial q^a_1}{\partial q^i_2}\Gamma^i_1 + \Gamma^a_2 = 0.$$

Using the relations (1.6.1) – (1.6.2), one can rewrite the coordinate transformation law (1.3.4) of second order dynamic equations as follows. Let

$$\overline{q}^a_{tt} = \overline{\xi}^a \tag{1.6.3}$$

be a second order dynamic equation on a configuration space Q written with respect to a reference frame $(t, \overline{q}^n)$. Then, relative to arbitrary bundle coordinates (t, q^i) on $Q \to \mathbb{R}$, the second order dynamic equation (1.6.3) takes the form

$$q^i_{tt} = d_t\Gamma^i + \partial_j\Gamma^i(q^j_t - \Gamma^j) - \frac{\partial q^i}{\partial \overline{q}^a}\frac{\partial \overline{q}^a}{\partial q^j \partial q^k}(q^j_t - \Gamma^j)(q^k_t - \Gamma^k) + \frac{\partial q^i}{\partial \overline{q}^a}\overline{\xi}^a, \quad (1.6.4)$$

where Γ is a connection corresponding to the reference frame $(t, \overline{q}^n)$. The second order dynamic equation (1.6.4) can be expressed in the relative velocities $\dot{q}^i_\Gamma = q^i_t - \Gamma^i$ with respect to the initial reference frame $(t, \overline{q}^a)$. We have

$$d_t\dot{q}^i_\Gamma = \partial_j\Gamma^i\dot{q}^j_\Gamma - \frac{\partial q^i}{\partial \overline{q}^a}\frac{\partial \overline{q}^a}{\partial q^j \partial q^k}\dot{q}^j_\Gamma\dot{q}^k_\Gamma + \frac{\partial q^i}{\partial \overline{q}^a}\overline{\xi}^a(t, q^j, \dot{q}^j_\Gamma). \quad (1.6.5)$$

Accordingly, any second order dynamic equation (1.3.3) can be expressed in the relative velocities $\dot{q}^i_\Gamma = q^i_t - \Gamma^i$ with respect to an arbitrary reference frame Γ as follows:

$$d_t\dot{q}^i_\Gamma = (\xi - J\Gamma)^i_t = \xi^i - d_t\Gamma, \quad (1.6.6)$$

where $J\Gamma$ is the prolongation (1.1.25) of a connection Γ onto the jet bundle $J^1Q \to \mathbb{R}$.

For instance, let us consider the following particular reference frame Γ for a second order dynamic equation ξ. The covariant derivative of a reference frame Γ with respect to the corresponding dynamic connection γ_ξ (1.4.10) reads

$$\nabla^\gamma\Gamma = Q \to T^*Q \times V_Q J^1Q, \quad (1.6.7)$$
$$\nabla^\gamma\Gamma = \nabla^\gamma_\lambda\Gamma^k dq^\lambda \otimes \partial_k, \qquad \nabla^\gamma_\lambda\Gamma^k = \partial_\lambda\Gamma^k - \gamma^k_\lambda \circ \Gamma.$$

A connection Γ is called a *geodesic reference frame* for the second order dynamic equation ξ if

$$\Gamma\rfloor\nabla^\gamma\Gamma = \Gamma^\lambda(\partial_\lambda\Gamma^k - \gamma^k_\lambda \circ \Gamma) = (d_t\Gamma^i - \xi^i \circ \Gamma)\partial_i = 0. \quad (1.6.8)$$

Proposition 1.6.1. *Integral sections c for a reference frame Γ are solutions of a second order dynamic equation ξ if and only if Γ is a geodesic reference frame for ξ.*

Proof. The proof follows at once from substitution of the equality (1.6.8) in the second order dynamic equation (1.6.6). □

Remark 1.6.4. The left- and right-hand sides of the equation (1.6.6) separately are not well-behaved objects. This equation is brought into the covariant form (1.8.6).

Reference frames play a prominent role in many constructions of non-relativistic mechanics. They enable us to write the covariant forms: (1.8.5) – (1.8.6) of dynamic equations, (2.3.5) of quadratic Lagrangians and (3.3.17) of Hamiltonians of non-relativistic mechanics.

With a reference frame, we obtain the converse of Theorem 1.5.1.

Theorem 1.6.1. *Given a reference frame* Γ, *any connection* K *(1.5.1) on the tangent bundle* $TQ \to Q$ *defines a second order dynamic equation*

$$\xi^i = (K^i_\lambda - \Gamma^i K^0_\lambda)\dot{q}^\lambda \mid_{\dot{q}^0=1,\dot{q}^j=q^j_t} .$$

This theorem is a corollary of Proposition 1.5.1 and the following lemma.

Lemma 1.6.1. *Given a connection* Γ *on a fibre bundle* $Q \to \mathbb{R}$ *and a connection* K *on the tangent bundle* $TQ \to Q$, *there is the connection* $\widetilde{K}$ *on* $TQ \to Q$ *with the components*

$$\widetilde{K}^0_\lambda = 0, \qquad \widetilde{K}^i_\lambda = K^i_\lambda - \Gamma^i K^0_\lambda.$$

1.7 Free motion equations

Let us point out the following interesting class of second order dynamic equations which we agree to call the free motion equations.

Definition 1.7.1. We say that the second order dynamic equation (1.3.3) is a *free motion equation* if there exists a reference frame $(t, \overline{q}^i)$ on the configuration space Q such that this equation reads

$$\overline{q}^i_{tt} = 0. \tag{1.7.1}$$

With respect to arbitrary bundle coordinates (t, q^i), a free motion equation takes the form

$$q^i_{tt} = d_t\Gamma^i + \partial_j\Gamma^i(q^j_t - \Gamma^j) - \frac{\partial q^i}{\partial \overline{q}^m}\frac{\partial \overline{q}^m}{\partial q^j \partial q^k}(q^j_t - \Gamma^j)(q^k_t - \Gamma^k), \tag{1.7.2}$$

where $\Gamma^i = \partial_t q^i(t, \overline{q}^j)$ is the connection associated with the initial frame $(t, \overline{q}^i)$ (cf. (1.6.4)). One can think of the right-hand side of the equation (1.7.2) as being the general coordinate expression for an *inertial force* in non-relativistic mechanics. The corresponding dynamic connection γ_ξ on the affine jet bundle $J^1Q \to Q$ reads

$$\gamma^i_k = \partial_k\Gamma^i - \frac{\partial q^i}{\partial \overline{q}^m}\frac{\partial \overline{q}^m}{\partial q^j \partial q^k}(q^j_t - \Gamma^j), \tag{1.7.3}$$
$$\gamma^i_0 = \partial_t\Gamma^i + \partial_j\Gamma^i q^j_t - \gamma^i_k\Gamma^k.$$

It is affine. By virtue of Proposition 1.5.2, this dynamic connection defines a linear connection K on the tangent bundle $TQ \to Q$, whose curvature necessarily vanishes. Thus, we come to the following criterion of a second order dynamic equation to be a free motion equation.

Proposition 1.7.1. *If ξ is a free motion equation on a configuration space Q, it is quadratic, and the corresponding symmetric linear connection (1.5.10) on the tangent bundle $TQ \to Q$ is a curvature-free connection.*

This criterion is not a sufficient condition because it may happen that the components of a curvature-free symmetric linear connection on $TQ \to Q$ vanish with respect to the coordinates on Q which are not compatible with a fibration $Q \to \mathbb{R}$.

The similar criterion involves the curvature of a dynamic connection (1.7.3) of a free motion equation.

Proposition 1.7.2. *If ξ is a free motion equation, then the curvature R (1.4.3) of the corresponding dynamic connection γ_ξ is equal to 0, and so are the tensor field $\overline{R}$ (1.4.4) and the scalar field $\widetilde{R}$ (1.4.5).*

Proposition 1.7.2 also fails to be a sufficient condition. If the curvature R (1.4.3) of a dynamic connection γ_ξ vanishes, it may happen that components of γ_ξ are equal to zero with respect to non-holonomic bundle coordinates on an affine jet bundle $J^1Q \to Q$.

Nevertheless, we can formulate the necessary and sufficient condition of the existence of a free motion equation on a configuration space Q.

Proposition 1.7.3. *A free motion equation on a fibre bundle $Q \to \mathbb{R}$ exists if and only if a typical fibre M of Q admits a curvature-free symmetric linear connection.*

Proof. Let a free motion equation take the form (1.7.1) with respect to some atlas of local constant trivializations of a fibre bundle $Q \to \mathbb{R}$. By virtue of Proposition 1.4.2, there exists an affine dynamic connection γ on the affine jet bundle $J^1Q \to Q$ whose components relative to this atlas are equal to 0. Given a trivialization chart of this atlas, the connection γ defines the curvature-free symmetric linear connection (1.5.13) on M. The converse statement follows at once from Proposition 1.5.4. □

Corollary 1.7.1. *A free motion equation on a fibre bundle $Q \to \mathbb{R}$ exists if and only if a typical fibre M of Q and, consequently, Q itself are locally affine manifolds, i.e., toroidal cylinders (see Section 11.4.3).*

The free motion equation (1.7.2) is simplified if the coordinate transition functions $\overline{q}^i \to q^i$ are affine in coordinates $\overline{q}^i$. Then we have

$$q^i_{tt} = \partial_t \Gamma^i - \Gamma^j \partial_j \Gamma^i + 2q^j_t \partial_j \Gamma^i. \tag{1.7.4}$$

Example 1.7.1. Let us consider a free motion on a plane $\mathbb{R}^2$. The corresponding configuration bundle is $\mathbb{R}^3 \to \mathbb{R}$ coordinated by $(t, \overline{\mathbf{r}})$. The dynamic equation of this motion is

$$\overline{\mathbf{r}}_{tt} = 0. \tag{1.7.5}$$

Let us choose the *rotatory reference frame* with the adapted coordinates

$$\mathbf{r} = A\overline{\mathbf{r}}, \qquad A = \begin{pmatrix} \cos\omega t & -\sin\omega t \\ \sin\omega t & \cos\omega t \end{pmatrix}. \tag{1.7.6}$$

Relative to these coordinates, a connection Γ corresponding to the initial reference frame reads

$$\Gamma = \partial_t \mathbf{r} = \partial_t A \cdot A^{-1}\mathbf{r}.$$

Then the free motion equation (1.7.5) with respect to the rotatory reference frame (1.7.6) takes the familiar form

$$\mathbf{r}_{tt} = \omega^2 \mathbf{r} + 2\begin{pmatrix} 0 & -1 \\ 1 & 0 \end{pmatrix} \mathbf{r}_t. \tag{1.7.7}$$

The first term in the right-hand side of the equation (1.7.7) is the *centrifugal force* $-\Gamma^j \partial_j \Gamma^i$, while the second one is the *Coriolis force* $2q^j_t \partial_j \Gamma^i$.

The following lemma shows that the free motion equation (1.7.4) is affine in the coordinates q^i and q^i_t [106].

Lemma 1.7.1. *Let $(t, \overline{q}^a)$ be a reference frame on a configuration bundle $Q \to \mathbb{R}$ and Γ the corresponding connection. Components Γ^i of this connection with respect to another coordinate system (t, q^i) are affine functions in the coordinates q^i if and only if the transition functions between the coordinates $\overline{q}^a$ and q^i are affine.*

One can easily find the geodesic reference frames for the free motion equation

$$q^i_{tt} = 0. \tag{1.7.8}$$

They are $\Gamma^i = v^i = \text{const.}$ By virtue of Lemma 1.7.1, these reference frames define the adapted coordinates

$$\overline{q}^i = k^i_j q^j - v^i t - a^i, \quad k^i_j = \text{const.}, \quad v^i = \text{const.}, \quad a^i = \text{const.} \tag{1.7.9}$$

The equation (1.7.8) obviously keeps its free motion form under the transformations (1.7.9) between the geodesic reference frames. Geodesic reference frames for a free motion equation are called *inertial*.

1.8 Relative acceleration

In comparison with the notion of a relative velocity, the definition of a relative acceleration is more intricate.

To consider a relative acceleration with respect to a reference frame Γ, one should prolong a connection Γ on a configuration space $Q \to \mathbb{R}$ to a holonomic connection ξ_Γ on the jet bundle $J^1Q \to \mathbb{R}$. Note that the jet prolongation $J\Gamma$ (1.1.25) of Γ onto $J^1Q \to \mathbb{R}$ is not holonomic. We can construct the desired prolongation by means of a dynamic connection γ on an affine jet bundle $J^1Q \to Q$.

Lemma 1.8.1. *Let us consider the composite bundle (1.1.12). Given a reference frame Γ on $Q \to \mathbb{R}$ and a dynamic connections γ on $J^1Q \to Q$, there exists a dynamic connection $\widetilde{\gamma}$ on $J^1Q \to Q$ with the components*

$$\widetilde{\gamma}^i_k = \gamma^i_k, \qquad \widetilde{\gamma}^i_0 = d_t\Gamma^i - \gamma^i_k\Gamma^k. \tag{1.8.1}$$

Proof. Combining a connection Γ on $Q \to \mathbb{R}$ and a connection γ on $J^1Q \to Q$ gives the composite connection (11.4.29) on $J^1Q \to \mathbb{R}$ which reads

$$B = dt \otimes (\partial_t + \Gamma^i\partial_i + (\gamma^i_k\Gamma^k + \gamma^i_0)\partial^t_i).$$

Let $J\Gamma$ be the jet prolongation (1.1.25) of a connection Γ on $J^1Q \to \mathbb{R}$. Then the difference

$$J\Gamma - B = dt \otimes (d_t\Gamma^i - \gamma^i_k\Gamma^k - \gamma^i_0)\partial^t_i$$

is a V_QJ^1Q-valued soldering form on the jet bundle $J^1Q \to \mathbb{R}$, which also is a soldering form on the affine jet bundle $J^1Q \to Q$. The desired connection (1.8.1) is

$$\widetilde{\gamma} = \gamma + J\Gamma - B = dt \otimes (\partial_t + (d_t\Gamma^i - \gamma^i_k\Gamma^k)\partial^t_i) + dq^k \otimes (\partial_k + \gamma^i_k\partial^t_i). \qquad \square$$

Now, we construct a certain soldering form on an affine jet bundle $J^1Q \to Q$ and add it to this connection. Let us apply the canonical projection $T^*Q \to V^*Q$ and then the imbedding $\Gamma : V^*Q \to T^*Q$ to the covariant derivative (1.6.7) of the reference frame Γ with respect to the dynamic connection γ. We obtain the V_QJ^1Q-valued one-form

$$\sigma = [-\Gamma^i(\partial_i\Gamma^k - \gamma^k_i \circ \Gamma)dt + (\partial_i\Gamma^k - \gamma^k_i \circ \Gamma)dq^i] \otimes \partial^t_k$$

on Q whose pull-back onto J^1Q is a desired soldering form. The sum

$$\gamma_\Gamma = \widetilde{\gamma} + \sigma,$$

called the *frame connection*, reads

$$\gamma_\Gamma{}^i_0 = d_t\Gamma^i - \gamma^i_k\Gamma^k - \Gamma^k(\partial_k\Gamma^i - \gamma^i_k \circ \Gamma), \qquad (1.8.2)$$
$$\gamma_\Gamma{}^i_k = \gamma^i_k + \partial_k\Gamma^i - \gamma^i_k \circ \Gamma.$$

This connection yields the desired holonomic connection

$$\xi^i_\Gamma = d_t\Gamma^i + (\partial_k\Gamma^i + \gamma^i_k - \gamma^i_k \circ \Gamma)(q^k_t - \Gamma^k)$$

on the jet bundle $J^1Q \to \mathbb{R}$.

Let ξ be a second order dynamic equation and $\gamma = \gamma_\xi$ the connection (1.4.10) associated with ξ. Then one can think of the vertical vector field

$$a_\Gamma = \xi - \xi_\Gamma = (\xi^i - \xi^i_\Gamma)\partial^t_i \qquad (1.8.3)$$

on the affine jet bundle $J^1Q \to Q$ as being a *relative acceleration* with respect to the reference frame Γ in comparison with the *absolute acceleration* ξ.

For instance, let us consider a reference frame Γ which is geodesic for the second order dynamic equation ξ, i.e., the relation (1.6.8) holds. Then the relative acceleration of a motion c with respect to a reference frame Γ is

$$(\xi - \xi_\Gamma) \circ \Gamma = 0.$$

Let ξ now be an arbitrary second order dynamic equation, written with respect to coordinates (t, q^i) adapted to a reference frame Γ, i.e., $\Gamma^i = 0$. In these coordinates, the relative acceleration with respect to a reference frame Γ is

$$a^i_\Gamma = \xi^i(t, q^j, q^j_t) - \frac{1}{2}q^k_t(\partial_k\xi^i - \partial_k\xi^i\,|_{q^j_t=0}). \qquad (1.8.4)$$

Given another bundle coordinates (t, q'^i) on $Q \to \mathbb{R}$, this dynamic equation takes the form (1.6.5), while the relative acceleration (1.8.4) with respect to a reference frame Γ reads

$$a'^i_\Gamma = \partial_j q'^i a^j_\Gamma.$$

Then we can write the second order dynamic equation (1.3.3) in the form which is covariant under coordinate transformations:

$$\widetilde{D}_{\gamma_\Gamma} q^i_t = d_t q^i_t - \xi^i_\Gamma = a_\Gamma, \qquad (1.8.5)$$

where $\widetilde{D}_{\gamma_\Gamma}$ is the vertical covariant differential (1.4.9) with respect to the frame connection γ_Γ (1.8.2) on an affine jet bundle $J^1Q \to Q$.

In particular, if ξ is a free motion equation which takes the form (1.7.1) with respect to a reference frame Γ, then

$$\widetilde{D}_{\gamma_\Gamma} q^i_t = 0$$

relative to arbitrary bundle coordinates on the configuration bundle $Q \to \mathbb{R}$.

The left-hand side of the second order dynamic equation (1.8.5) also can be expressed in the relative velocities such that this dynamic equation takes the form

$$d_t \dot q^i_\Gamma - \gamma_\Gamma{}^i_k \dot q^k_\Gamma = a_\Gamma \tag{1.8.6}$$

which is the covariant form of the equation (1.6.6).

The concept of a relative acceleration is understood better when we deal with a quadratic second order dynamic equation ξ, and the corresponding dynamic connection γ is affine.

Lemma 1.8.2. *If a dynamic connection γ is affine, i.e.,*

$$\gamma^i_\lambda = \gamma^i_{\lambda 0} + \gamma^i_{\lambda k} q^k_t,$$

so is a frame connection γ_Γ for any frame Γ.

Proof. The proof follows from direct computation. We have

$$\gamma_\Gamma{}^i_0 = \partial_t \Gamma^i + (\partial_j \Gamma^i - \gamma^i_{kj}\Gamma^k)(q^j_t - \Gamma^j),$$
$$\gamma_\Gamma{}^i_k = \partial_k \Gamma^i + \gamma^i_{kj}(q^j_t - \Gamma^j)$$

or

$$\begin{aligned}
&\gamma_\Gamma{}^i_{jk} = \gamma^i_{jk}, \\
&\gamma_\Gamma{}^i_{0k} = \partial_k \Gamma^i - \gamma^i_{jk}\Gamma^j, \qquad \gamma_\Gamma{}^i_{k0} = \partial_k \Gamma^i - \gamma^i_{kj}\Gamma^j, \\
&\gamma_\Gamma{}^i_{00} = \partial_t \Gamma^i - \Gamma^j \partial_j \Gamma^i + \gamma^i_{jk}\Gamma^j \Gamma^k.
\end{aligned} \tag{1.8.7}$$

□

In particular, we obtain

$$\gamma_\Gamma{}^i_{jk} = \gamma^i_{jk}, \qquad \gamma_\Gamma{}^i_{0k} = \gamma_\Gamma{}^i_{k0} = \gamma_\Gamma{}^i_{00} = 0$$

relative to the coordinates adapted to a reference frame Γ.

A glance at the expression (1.8.7) shows that, if a dynamic connection γ is symmetric, so is a frame connection γ_Γ.

Corollary 1.8.1. *If a second order dynamic equation ξ is quadratic, the relative acceleration a_Γ (1.8.3) is always affine, and it admits the decomposition*

$$a^i_\Gamma = -(\Gamma^\lambda \nabla^\gamma_\lambda \Gamma^i + 2\dot q^\lambda_\Gamma \nabla^\gamma_\lambda \Gamma^i), \tag{1.8.8}$$

where $\gamma = \gamma_\xi$ is the dynamic connection (1.4.10), and

$$\dot q^\lambda_\Gamma = q^\lambda_t - \Gamma^\lambda, \qquad q^0_t = 1, \qquad \Gamma^0 = 1$$

is the relative velocity with respect to the reference frame Γ.

Note that the splitting (1.8.8) gives a *generalized Coriolis theorem*. In particular, the well-known analogy between inertial and electromagnetic forces is restated. Corollary 1.8.1 shows that this analogy can be extended to an arbitrary quadratic dynamic equation.

1.9 Newtonian systems

Equations of motion of non-relativistic mechanics need not be exactly dynamic equations. For instance, the second Newton law of point mechanics contains a mass. The notion of a Newtonian system generalizes the second Newton law as follows.

Let m be a *fibre metric* (bilinear form) in the vertical tangent bundle $V_QJ^1Q \to J^1Q$ of $J^1Q \to Q$. It reads

$$m : J^1Q \to \overset{2}{\underset{J^1Q}{\vee}} V_Q^*J^1Q, \qquad m = \frac{1}{2}m_{ij}\overline{d}q_t^i \vee \overline{d}q_t^j, \tag{1.9.1}$$

where $\overline{d}q_t^i$ are the holonomic bases for the vertical cotangent bundle $V_Q^*J^1Q$ of $J^1Q \to Q$. It defines the map

$$\widehat{m} : V_QJ^1Q \to V_Q^*J^1Q.$$

Definition 1.9.1. Let $Q \to \mathbb{R}$ be a fibre bundle together with:

(i) a fibre metric m (1.9.1) satisfying the symmetry condition

$$\partial_k^t m_{ij} = \partial_j^t m_{ik}, \tag{1.9.2}$$

(ii) a holonomic connection ξ (1.1.26) on a jet bundle $J^1Q \to \mathbb{R}$ related to the fibre metric m by the compatibility condition

$$\xi\rfloor dm_{ij} + \frac{1}{2}m_{ik}\partial_j^t\xi^k + m_{jk}\partial_i^t\xi^k = 0. \tag{1.9.3}$$

A triple (Q, m, ξ) is called the *Newtonian system*.

We agree to call a metric m in Definition 1.9.1 the *mass tensor* of a Newtonian system (Q, m, ξ). The equation of motion of this Newtonian system is defined to be

$$\widehat{m}(D^\xi) = 0, \qquad m_{ik}(q_{tt}^k - \xi^k) = 0. \tag{1.9.4}$$

Due to the conditions (1.9.2) and (1.9.3), it is brought into the form

$$d_t(m_{ik}q_t^k) - m_{ik}\xi^k = 0.$$

Therefore, one can think of this equation as being a generalization of the *second Newton law*.

If a mass tensor m (1.9.1) is non-degenerate, the equation of motion (1.9.4) is equivalent to the second order dynamic equation

$$D^\xi = 0, \qquad q^k_{tt} - \xi^k = 0.$$

Because of the canonical vertical splitting (1.1.11), the mass tensor (1.9.1) also is a map

$$m : J^1Q \to \overset{2}{\underset{J^1Q}{\vee}} V^*Q, \qquad m = \frac{1}{2} m_{ij} dq^i \vee dq^j. \tag{1.9.5}$$

Remark 1.9.1. To be precise, one should define a mass tensor as a map

$$m : J^1Q \to \overset{2}{\underset{J^1Q}{\vee}} V^*_Q J^1Q \underset{J^1Q}{\otimes} T^*\mathbb{R},$$

but we follow Remark 1.1.1, without considering time reparametrization. In the universal unit system, a mass tensor m is of physical dimension $-2[q]+1$. For instance, the physical dimension of a mass tensor of a point mass with respect to Cartesian coordinates q^i is [length]$^{-1}$, while that with respect to the angle coordinates is [length].

A Newtonian system (Q, m, ξ) is said to be *standard*, if its mass tensor m is the pull-back onto $V_Q J^1Q$ of a fibre metric

$$m : Q \to \overset{2}{\underset{Q}{\vee}} V^*Q \tag{1.9.6}$$

in the vertical tangent bundle $VQ \to Q$ in accordance with the isomorphisms (1.1.10) and (1.1.11), i.e., m is independent of the velocity coordinates q^i_t.

Given a mass tensor, one can introduce the notion of an external force.

Definition 1.9.2. An *external force* is defined as a section of the vertical cotangent bundle $V^*_Q J^1Q \to J^1Q$. Let us also bear in mind the isomorphism (1.1.11).

It should be emphasized that there are no canonical isomorphisms between the vertical cotangent bundle $V^*_Q J^1Q$ and the vertical tangent bundle $V_Q J^1Q$ of J^1Q. One must therefore distinguish forces and accelerations which are related by means of a mass tensor (see Remark 1.9.2 below).

Let $(Q, \widehat{m}, \xi)$ be a Newtonian system and f an external force. Then

$$\xi^i_f = \xi^i + (m^{-1})^{ik} f_k \tag{1.9.7}$$

is a dynamic equation, but the triple (Q, m, ξ_f) is not a Newtonian system in general. As it follows from a direct computation, if and only if an external force possesses the property

$$\partial^t_i f_j + \partial^t_j f_i = 0, \tag{1.9.8}$$

then ξ_f (1.9.7) fulfills the compatibility condition (1.9.3), and $(Q, \widehat{m}, \xi_f)$ also is a Newtonian system.

Example 1.9.1. For instance, the *Lorentz force*

$$f_i = eF_{\lambda i}q_t^\lambda, \qquad q_t^0 = 1, \tag{1.9.9}$$

where

$$F_{\lambda\mu} = \partial_\lambda A_\mu - \partial_\mu A_\lambda \tag{1.9.10}$$

is the electromagnetic strength, obeys the condition (1.9.8). Note that the Lorentz force (1.9.9) as like as other forces can be expressed in the relative velocities $\dot{q}_\Gamma$ with respect to an arbitrary reference frame Γ:

$$f_i = e\frac{\partial\overline{q}^j}{\partial q^i}\left(\frac{\partial\overline{q}^n}{\partial q^k}\overline{F}_{nj}\dot{q}_\Gamma^k + \overline{F}_{0j}\right),$$

where $\overline{q}$ are the coordinates adapted to a reference frame Γ, and $\overline{F}$ is an electromagnetic strength, written with respect to these coordinates.

Remark 1.9.2. The contribution of an external force f to a second order dynamic equation

$$q_{tt}^i - \xi^i = (m^{-1})^{ik} f_k$$

of a Newtonian system obviously depends on a mass tensor. It should be emphasized that, besides external forces, we have a *universal force* which is a holonomic connection

$$\xi^i = K_{\mu\lambda}^i q_t^\mu q_t^\lambda, \qquad q_t^0 = 1,$$

associated with the symmetric linear connection $\widetilde{K}$ (1.5.3) on the tangent bundle $TQ \to Q$. From the physical viewpoint, this is a non-relativistic *gravitational force*, including an inertial force, whose contribution to a second order dynamic equation is independent of a mass tensor.

1.10 Integrals of motion

Let an equation of motion of a mechanical system on a fibre bundle $Y \to \mathbb{R}$ be described by an r-order differential equation $\mathfrak{E}$ given by a closed subbundle of the jet bundle $J^rY \to \mathbb{R}$ in accordance with Definition 11.3.5.

Definition 1.10.1. An *integral of motion* of this mechanical system is defined as a $(k < r)$-order differential operator Φ on Y such that $\mathfrak{E}$ belongs to the kernel of an r-order jet prolongation of the differential operator $d_t\Phi$, i.e.,

$$J^{r-k-1}(d_t\Phi)|_{\mathfrak{E}} = J^{r-k}\Phi|_{\mathfrak{E}} = 0. \tag{1.10.1}$$

It follows that an integral of motion Φ is constant on solutions s of a differential equation $\mathfrak{E}$, i.e., there is the *differential conservation law*

$$(J^k s)^*\Phi = \text{const.}, \qquad (J^{k+1}s)^* d_t\Phi = 0. \tag{1.10.2}$$

We agree to write the condition (1.10.1) as the *weak equality*

$$J^{r-k-1}(d_t\Phi) \approx 0, \tag{1.10.3}$$

which holds *on-shell*, i.e., on solutions of a differential equation $\mathfrak{E}$ by the formula (1.10.2).

In non-relativistic mechanics (without time-reparametrization), we can restrict our consideration to integrals of motion Φ which are functions on J^kY. As was mentioned above, equations of motion of non-relativistic mechanics mainly are of first or second order. Accordingly, their integrals of motion are functions on Y or J^kY. In this case, the corresponding weak equality (1.10.1) takes the form

$$d_t\Phi \approx 0 \tag{1.10.4}$$

of a *weak conservation law* or, simply, a *conservation law*.

Different integrals of motion need not be independent. Let integrals of motion $\Phi_1, \ldots, \Phi_m$ of a mechanical system on Y be functions on J^kY. They are called *independent* if

$$d\Phi_1 \wedge \cdots \wedge d\Phi_m \neq 0 \tag{1.10.5}$$

everywhere on J^kY. In this case, any motion J^ks of this mechanical system lies in the common level surfaces of functions $\Phi_1, \ldots, \Phi_m$ which bring J^kY into a fibred manifold.

Integrals of motion can come from symmetries. This is the case of Lagrangian and Hamiltonian mechanics (Sections 2.5 and 3.8).

Definition 1.10.2. Let an equation of motion of a mechanical system be an r-order differential equation $\mathfrak{E} \subset J^rY$. Its *infinitesimal symmetry* (or, simply, a *symmetry*) is defined as a vector field on J^rY whose restriction to $\mathfrak{E}$ is tangent to $\mathfrak{E}$.

For instance, let us consider first order dynamic equations.

Proposition 1.10.1. *Let $\mathfrak{E}$ be the autonomous first order dynamic equation (1.2.1) given by a vector field u on a manifold Z. A vector field ϑ on Z is its symmetry if and only if $[u, \vartheta] \approx 0$.*

Proof. The first order dynamic equation (1.2.1) is a subbundle of TZ. The functorial lift of ϑ into TZ is (11.2.31). Then the condition of Definition (1.10.2) leads to a desired weak equality. □

One can show that a smooth real function F on a manifold Z is an integral of motion of the autonomous first order dynamic equation (1.2.1) (i.e., it is constant on solutions of this equation) if and only if its Lie derivative along u vanishes:

$$\mathbf{L}_u F = u^\lambda \partial_\lambda \Phi = 0. \tag{1.10.6}$$

Proposition 1.10.2. *Let $\mathfrak{E}$ be the first order dynamic equation (1.3.1) given by a connection Γ (1.1.18) on a fibre bundle $Y \to \mathbb{R}$. Then a vector field ϑ on Y is its symmetry if and only if $[\Gamma, \vartheta] \approx 0$.*

Proof. The first order dynamic equation (1.3.1) on a fibre bundle $Y \to \mathbb{R}$ is equivalent to the autonomous first order dynamic equation (1.3.2) given by the vector field Γ (1.1.18) on a manifold Y. Then the result is a corollary of Proposition 1.10.1. □

A smooth real function Φ on Y is an integral of motion of the first order dynamic equation (1.3.1) in accordance with the equality (1.10.4) if and only if

$$\mathbf{L}_\Gamma \Phi = (\partial_t + \Gamma^i \partial_i)\Phi = 0. \tag{1.10.7}$$

Following Definition 1.10.2, let us introduce the notion of a symmetry of differential operators in the following relevant case. Let us consider an r-order differential operator on a fibre bundle $Y \to \mathbb{R}$ which is represented by an exterior form $\mathcal{E}$ on J^rY (Definition 11.3.4). Let its kernel $\mathrm{Ker}\,\mathcal{E}$ be an r-order differential equation on $Y \to \mathbb{R}$.

Proposition 1.10.3. *It is readily justified that a vector field ϑ on J^rY is a symmetry of the equation $\mathrm{Ker}\,\mathcal{E}$ in accordance with Definition 1.10.2 if and only if*

$$\mathbf{L}_\vartheta \mathcal{E} \approx 0. \tag{1.10.8}$$

Motivated by Proposition 1.10.3, we come to the following.

Definition 1.10.3. Let $\mathcal{E}$ be the above mentioned differential operator. A vector field ϑ on J^rY is called a *symmetry* of a differential operator $\mathcal{E}$ if the Lie derivative $\mathbf{L}_\vartheta \mathcal{E}$ vanishes.

By virtue of Proposition 1.10.3, a symmetry of a differential operator $\mathcal{E}$ also is a symmetry of the differential equation $\operatorname{Ker}\mathcal{E}$.

Note that there exist integrals of motion which are not associated with symmetries of an equation of motion (see Example 2.5.4 below).

Chapter 2

Lagrangian mechanics

Lagrangian non-relativistic mechanics on a velocity space is formulated in the framework of Lagrangian formalism on fibre bundles [53; 68; 106]. This formulation is based on the variational bicomplex and the first variational formula, without appealing to the variational principle. Besides Lagrange equations, the Cartan and Hamilton–De Donder equations are considered in the framework of Lagrangian formalism. Note that the Cartan equation, but not the Lagrange one is associated to the Hamilton equation (Section 3.6). The relations between Lagrangian and Newtonian systems are investigated. Lagrangian conservation laws are defined by means of the first Noether theorem.

2.1 Lagrangian formalism on $Q \to \mathbb{R}$

Let $\pi : Q \to \mathbb{R}$ be a fibre bundle (1.1.1). The finite order jet manifolds J^kQ of $Q \to \mathbb{R}$ form the inverse sequence

$$Q \overset{\pi^1_0}{\longleftarrow} J^1Q \longleftarrow \cdots J^{r-1}Q \overset{\pi^r_{r-1}}{\longleftarrow} J^rQ \longleftarrow \cdots, \tag{2.1.1}$$

where π^r_{r-1} are affine bundles. Its projective limit $J^\infty Q$ is a paracompact Fréchet manifold. One can think of its elements as being infinite order jets of sections of $Q \to \mathbb{R}$ identified by their Taylor series at points of $\mathbb{R}$. Therefore, $J^\infty Q$ is called the *infinite order jet manifold*. A bundle coordinate atlas (t, q^i) of $Q \to \mathbb{R}$ provides $J^\infty Q$ with the manifold coordinate atlas

$$(t, q^i, q^i_t, q^i_{tt}, \ldots), \qquad {q'}^i_{t\Lambda} = d_t {q'}^i_\Lambda, \tag{2.1.2}$$

where $\Lambda = (t \cdots t)$ denotes a multi-index of length $|\Lambda|$ and

$$d_t = \partial_t + q^i_t \partial_i + q^i_{tt} \partial^t_i + \cdots + q^i_{t\Lambda} \partial^\Lambda_i + \cdots$$

is the *total derivative*.

Let $\mathcal{O}^*_r = \mathcal{O}^*(J^rQ)$ be a graded differential algebra of exterior forms on a jet manifold J^rQ. The inverse sequence (2.1.1) of jet manifolds yields the direct sequence of differential graded algebras $\mathcal{O}^*_r$:

$$\mathcal{O}^*(Q) \xrightarrow{\pi^{1\,*}_0} \mathcal{O}^*_1 \longrightarrow \cdots \mathcal{O}^*_{r-1} \xrightarrow{\pi^{r\,\,*}_{r-1}} \mathcal{O}^*_r \longrightarrow \cdots, \tag{2.1.3}$$

where $\pi^r_{r-1}{}^*$ are the pull-back monomorphisms. Its direct limit

$$\mathcal{O}^*_\infty Q = \overrightarrow{\lim}\, \mathcal{O}^*_r \tag{2.1.4}$$

(or, simply, $\mathcal{O}^*_\infty$) consists of all exterior forms on finite order jet manifolds modulo the pull-back identification. In particular, $\mathcal{O}^0_\infty$ is the ring of all smooth functions on finite order jet manifolds. The $\mathcal{O}^*_\infty$ (2.1.4) is a differential graded algebra which inherits the operations of the exterior differential d and exterior product $\wedge$ of exterior algebras $\mathcal{O}^*_r$.

Theorem 2.1.1. *The cohomology $H^*(\mathcal{O}^*_\infty)$ of the de Rham complex*

$$0 \longrightarrow \mathbb{R} \longrightarrow \mathcal{O}^0_\infty \xrightarrow{d} \mathcal{O}^1_\infty \xrightarrow{d} \cdots \tag{2.1.5}$$

of the differential graded algebra $\mathcal{O}^*_\infty$ equals the de Rham cohomology $H^*_{\rm DR}(Q)$ of a fibre bundle Q [68].

Corollary 2.1.1. *Since Q (1.1.1) is a trivial fibre bundle over $\mathbb{R}$, the de Rham cohomology $H^*_{\rm DR}(Q)$ of Q equals the de Rham cohomology of its typical fibre M in accordance with the well-known Künneth formula. Therefore, the cohomology $H^*(\mathcal{O}^*_\infty)$ of the de Rham complex (2.1.5) equals the de Rham cohomology $H^*_{\rm DR}(M)$ of M.*

Since elements of the differential graded algebra $\mathcal{O}^*_\infty$ (2.1.4) are exterior forms on finite order jet manifolds, this $\mathcal{O}^0_\infty$-algebra is locally generated by the horizontal form dt and contact one-forms

$$\theta^i_\Lambda = dq^i_\Lambda - q^i_{t\Lambda}dt.$$

Moreover, there is the canonical decomposition

$$\mathcal{O}^*_\infty = \oplus \mathcal{O}^{k,m}_\infty, \qquad m = 0, 1,$$

of $\mathcal{O}^*_\infty$ into $\mathcal{O}^0_\infty$-modules $\mathcal{O}^{k,m}_\infty$ of k-contact and $(m = 0, 1)$-horizontal forms together with the corresponding projectors

$$h_k : \mathcal{O}^*_\infty \to \mathcal{O}^{k,*}_\infty, \qquad h^m : \mathcal{O}^*_\infty \to \mathcal{O}^{*,m}_\infty.$$

Accordingly, the exterior differential on $\mathcal{O}^*_\infty$ is decomposed into the sum $d = d_V + d_H$ of the *vertical differential*

$$d_V : \mathcal{O}^{k,m}_\infty \to \mathcal{O}^{k+1,m}_\infty, \qquad d_V \circ h^m = h^m \circ d \circ h^m,$$
$$d_V(\phi) = \theta^i_\Lambda \wedge \partial^\Lambda_i \phi, \qquad \phi \in \mathcal{O}^*_\infty,$$

and the *total differential*

$$d_H : \mathcal{O}^{k,m}_\infty \to \mathcal{O}^{k,m+1}_\infty, \quad d_H \circ h_k = h_k \circ d \circ h_k, \quad d_H \circ h_0 = h_0 \circ d,$$
$$d_H(\phi) = dt \wedge d_t\phi, \qquad \phi \in \mathcal{O}^*_\infty. \tag{2.1.6}$$

These differentials obey the nilpotent conditions

$$d_H \circ d_H = 0, \qquad d_V \circ d_V = 0, \qquad d_H \circ d_V + d_V \circ d_H = 0,$$

and make $\mathcal{O}^{*,*}_\infty$ into a bicomplex.

One introduces the following two additional operators acting on $\mathcal{O}^{*,n}_\infty$.

(i) There exists an $\mathbb{R}$-module endomorphism

$$\varrho = \sum_{k>0} \frac{1}{k} \overline{\varrho} \circ h_k \circ h^1 : \mathcal{O}^{*>0,1}_\infty \to \mathcal{O}^{*>0,1}_\infty, \tag{2.1.7}$$
$$\overline{\varrho}(\phi) = \sum_{0 \leq |\Lambda|} (-1)^{|\Lambda|} \theta^i \wedge [d_\Lambda(\partial^\Lambda_i \rfloor \phi)], \qquad \phi \in \mathcal{O}^{>0,1}_\infty,$$

possessing the following properties.

Lemma 2.1.1. *For any $\phi \in \mathcal{O}^{>0,1}_\infty$, the form $\phi - \varrho(\phi)$ is locally d_H-exact on each coordinate chart (2.1.2). The operator ϱ obeys the relation*

$$(\varrho \circ d_H)(\psi) = 0, \qquad \psi \in \mathcal{O}^{>0,0}_\infty. \tag{2.1.8}$$

It follows from Lemma 2.1.1 that ϱ (2.1.7) is a projector, i.e., $\varrho \circ \varrho = \varrho$.

(ii) One defines the variational operator

$$\delta = \varrho \circ d : \mathcal{O}^{*,1}_\infty \to \mathcal{O}^{*+1,1}_\infty. \tag{2.1.9}$$

Lemma 2.1.2. *The variational operator δ (2.1.9) is nilpotent, i.e., $\delta \circ \delta = 0$, and it obeys the relation*

$$\delta \circ \varrho = \delta. \tag{2.1.10}$$

With operators ϱ (2.1.7) and δ (2.1.9), the bicomplex $\mathcal{O}^{*,*}$ is brought into the *variational bicomplex*. Let us denote $\mathbf{E}_k = \varrho(\mathcal{O}^{k,1}_\infty)$. We have

$$\begin{array}{ccccccc}
 & & \vdots & & \vdots & & \vdots \\
 & & d_V \uparrow & & d_V \uparrow & & -\delta \uparrow \\
0 \to & & \mathcal{O}^{1,0}_\infty & \overset{d_H}{\longrightarrow} & \mathcal{O}^{1,1}_\infty & \overset{\varrho}{\longrightarrow} & \mathbf{E}_1 \to 0 \\
 & & d_V \uparrow & & d_V \uparrow & & -\delta \uparrow \\
0 \to \mathbb{R} \to & & \mathcal{O}^0_\infty & \overset{d_H}{\longrightarrow} & \mathcal{O}^{0,1}_\infty & \equiv & \mathcal{O}^{0,1}_\infty
\end{array} \tag{2.1.11}$$

This variational bicomplex possesses the following cohomology [68].

Theorem 2.1.2. *The bottom row and the last column of the variational bicomplex (2.1.11) make up the variational complex*

$$0 \to \mathbb{R} \to \mathcal{O}^0_\infty \xrightarrow{d_H} \mathcal{O}^{0,1}_\infty \xrightarrow{\delta} \mathbf{E}_1 \xrightarrow{\delta} \mathbf{E}_2 \longrightarrow \cdots . \tag{2.1.12}$$

Its cohomology is isomorphic to the de Rham cohomology of a fibre bundle Q and, consequently, the de Rham cohomology of its typical fibre M (Corollary 2.1.1).

Theorem 2.1.3. *The rows of contact forms of the variational bicomplex (2.1.11) are exact sequences.*

Note that Theorem 2.1.3 gives something more. Due to the relations (2.1.6) and (2.1.10), we have the cochain morphism

$$\begin{array}{ccccccccc}
\mathcal{O}^0_\infty & \xrightarrow{d} & \mathcal{O}^1_\infty & \xrightarrow{d} & \mathcal{O}^2_\infty & \xrightarrow{d} & \mathcal{O}^3_\infty & \to & \cdots \\
{\scriptstyle h_0}\downarrow & & {\scriptstyle h_0}\downarrow & & {\scriptstyle \varrho}\downarrow & & {\scriptstyle \varrho}\downarrow & & \\
\mathcal{O}^{0,0}_\infty & \xrightarrow{d_H} & \mathcal{O}^{0,1}_\infty & \xrightarrow{\delta} & \mathbf{E}_1 & \xrightarrow{\delta} & \mathbf{E}_2 & \longrightarrow & \cdots
\end{array}$$

of the de Rham complex (2.1.5) of the differential graded algebra $\mathcal{O}^*_\infty$ to its variational complex (2.1.12). By virtue of Theorems 2.1.1 and 2.1.2, the corresponding homomorphism of their cohomology groups is an isomorphism. A consequence of this fact is the following.

Theorem 2.1.4. *Any δ-closed form $\phi \in \mathcal{O}^{k,1}$, $k = 0, 1$, is split into the sum*

$$\phi = h_0\sigma + d_H\xi, \qquad k = 0, \qquad \xi \in \mathcal{O}^{0,0}_\infty, \tag{2.1.13}$$

$$\phi = \varrho(\sigma) + \delta(\xi), \qquad k = 1, \qquad \xi \in \mathcal{O}^{0,1}_\infty, \tag{2.1.14}$$

where σ is a closed $(1 + k)$-form on Q.

In Lagrangian formalism on a fibre bundle $Q \to \mathbb{R}$, a finite order *Lagrangian* and its *Lagrange operator* are defined as elements

$$L = \mathcal{L}dt \in \mathcal{O}^{0,1}_\infty, \tag{2.1.15}$$

$$\mathcal{E}_L = \delta L = \mathcal{E}_i\theta^i \wedge dt \in \mathbf{E}_1, \tag{2.1.16}$$

$$\mathcal{E}_i = \sum_{0\leq|\Lambda|} (-1)^{|\Lambda|} d_\Lambda(\partial^\Lambda_i \mathcal{L}), \tag{2.1.17}$$

of the variational complex (2.1.12). Components $\mathcal{E}_i$ (2.1.17) of the Lagrange operator (2.1.16) are called the *variational derivatives*. Elements of $\mathbf{E}_1$ are called the *Lagrange-type operators*.

We agree to call a pair $(\mathcal{O}^*_\infty, L)$ the *Lagrangian system*.

Corollary 2.1.2. *A finite order Lagrangian L (2.1.15) is variationally trivial, i.e., $\delta(L) = 0$ if and only if*

$$L = h_0\sigma + d_H\xi, \qquad \xi \in \mathcal{O}^{0,0}_\infty, \tag{2.1.18}$$

where σ is a closed one-form on Q.

Corollary 2.1.3. *A finite order Lagrange-type operator $\mathcal{E} \in \mathbf{E}_1$ satisfies the Helmholtz condition $\delta(\mathcal{E}) = 0$ if and only if*

$$\mathcal{E} = \delta L + \varrho(\sigma), \qquad L \in \mathcal{O}^{0,1}_\infty, \tag{2.1.19}$$

where σ is a closed two-form on Q.

Given a Lagrangian L (2.1.15) and its Lagrange operator δL (2.1.16), the kernel $\operatorname{Ker}\delta L \subset J^{2r}Q$ of δL is called the *Lagrange equation*. It is locally given by the equalities

$$\mathcal{E}_i = \sum_{0\leq|\Lambda|} (-1)^{|\Lambda|} d_\Lambda(\partial_i^\Lambda \mathcal{L}) = 0. \tag{2.1.20}$$

However, it may happen that the Lagrange equation is not a differential equation in accordance with Definition 11.3.2 because $\operatorname{Ker}\delta L$ need not be a closed subbundle of $J^{2r}Q \to \mathbb{R}$.

Example 2.1.1. Let $Q = \mathbb{R}^2 \to \mathbb{R}$ be a configuration space, coordinated by (t, q). The corresponding velocity phase space J^1Q is equipped with the adapted coordinates (t, q, q_t). The Lagrangian

$$L = \frac{1}{2}q^2 q_t^2 dt$$

on J^1Q leads to the Lagrange operator

$$\mathcal{E}_L = [qq_t^2 - d_t(q^2q_t)]\overline{d}q \wedge dt$$

whose kernel is not a submanifold at the point $q = 0$.

Theorem 2.1.5. *Owing to the exactness of the row of one-contact forms of the variational bicomplex (2.1.11) at the term $\mathcal{O}^{1,1}_\infty$, there is the decomposition*

$$dL = \delta L - d_H\mathfrak{L}, \tag{2.1.21}$$

where a one-form $\mathfrak{L}$ is a Lepage equivalent *of a Lagrangian L [68].*

Let us restrict our consideration to first order Lagrangian theory on a fibre bundle $Q \to \mathbb{R}$. This is the case of Lagrangian non-relativistic mechanics.

A *first order Lagrangian* is defined as a density

$$L = \mathcal{L}dt, \qquad \mathcal{L} : J^1Q \to \mathbb{R}, \tag{2.1.22}$$

on a velocity space J^1Q. The corresponding *second-order Lagrange operator* (2.1.16) reads

$$\delta L = (\partial_i \mathcal{L} - d_t \partial_i^t \mathcal{L})\theta^i \wedge dt. \tag{2.1.23}$$

Let us further use the notation

$$\pi_i = \partial_i^t \mathcal{L}, \qquad \pi_{ji} = \partial_j^t \partial_i^t \mathcal{L}. \tag{2.1.24}$$

The kernel $\operatorname{Ker} \delta L \subset J^2Q$ of the Lagrange operator defines the *second order Lagrange equation*

$$(\partial_i - d_t \partial_i^t)\mathcal{L} = 0. \tag{2.1.25}$$

Its *solutions* are (local) sections c of the fibre bundle $Q \to \mathbb{R}$ whose second order jet prolongations $\ddot{c}$ live in (2.1.25). They obey the equations

$$\partial_i \mathcal{L} \circ \dot{c} - \frac{d}{dt}(\pi_i \circ \dot{c}) = 0. \tag{2.1.26}$$

Definition 2.1.1. Given a Lagrangian L, a holonomic connection

$$\xi_L = \partial_t + q_t^i \partial_i + \xi^i \partial_i^t$$

on the jet bundle $J^1Q \to \mathbb{R}$ is said to be the *Lagrangian connection* if it takes its values into the kernel of the Lagrange operator δL, i.e., if it satisfies the relation

$$\partial_i \mathcal{L} - \partial_t \pi_i - q_t^j \partial_j \pi_i - \xi^j \pi_{ji} = 0. \tag{2.1.27}$$

A Lagrangian connection need not be unique.

Let us bring the relation (2.1.27) into the form

$$\partial_i \mathcal{L} - d_t \pi_i + (q_{tt}^j - \xi^j)\pi_{ji} = 0. \tag{2.1.28}$$

If a Lagrangian connection ξ_L exists, it defines the second order dynamic equation

$$q_{tt}^i = \xi_L^i \tag{2.1.29}$$

on $Q \to \mathbb{R}$, whose solutions also are solutions of the Lagrange equation (2.1.25) by virtue of the relation (2.1.28). Conversely, since the jet bundle $J^2Q \to J^1Q$ is affine, every solution c of the Lagrange equation also is an integral section for a holonomic connection ξ, which is a global extension of the local section $J^1c(\mathbb{R}) \to J^2c(\mathbb{R})$ of this jet bundle over the closed imbedded submanifold $J^1c(\mathbb{R}) \subset J^1Q$. Hence, every solution of the Lagrange equation also is a solution of some second order dynamic equation, but it is not necessarily a Lagrangian connection.

Every first order Lagrangian L (2.1.22) yields the bundle morphism

$$\widehat{L} : J^1Q \underset{Q}{\longrightarrow} V^*Q, \qquad p_i \circ \widehat{L} = \pi_i, \tag{2.1.30}$$

where (t, q^i, p_i) are holonomic coordinates on the vertical cotangent bundle V^*Q of $Q \to \mathbb{R}$. This morphism is called the *Legendre map*, and

$$\pi_\Pi : V^*Q \to Q, \tag{2.1.31}$$

is called the *Legendre bundle*. As was mentioned above, the vertical cotangent bundle V^*Q plays a role of the phase space of non-relativistic mechanics on a configuration space $Q \to \mathbb{R}$. The range $N_L = \widehat{L}(J^1Q)$ of the Legendre map (2.1.30) is called the *Lagrangian constraint space*.

Definition 2.1.2. A Lagrangian L is said to be:

• *hyperregular* if the Legendre map $\widehat{L}$ is a diffeomorphism;

• *regular* if $\widehat{L}$ is a local diffeomorphism, i.e., $\det(\pi_{ij}) \neq 0$;

• *semiregular* if the inverse image $\widehat{L}^{-1}(p)$ of any point $p \in N_L$ is a connected submanifold of J^1Q;

• *almost regular* if a Lagrangian constraint space N_L is a closed imbedded subbundle $i_N : N_L \to V^*Q$ of the Legendre bundle $V^*Q \to Q$ and the Legendre map

$$\widehat{L} : J^1Q \to N_L \tag{2.1.32}$$

is a fibred manifold with connected fibres (i.e., a Lagrangian is semiregular).

Remark 2.1.1. A glance at the equation (2.1.27) shows that a regular Lagrangian L admits a unique Lagrangian connection

$$\xi_L^j = (\pi^{-1})^{ij}(-\partial_i\mathcal{L} + \partial_t\pi_i + q_t^k\partial_k\pi_i). \tag{2.1.33}$$

In this case, the Lagrange equation (2.1.25) for L is equivalent to the second order dynamic equation associated to the Lagrangian connection (2.1.33).

2.2 Cartan and Hamilton–De Donder equations

Given a first order Lagrangian L, its Lepage equivalent $\mathfrak{L}$ in the decomposition (2.1.21) is the *Poincaré–Cartan form*

$$H_L = \pi_i dq^i - (\pi_i q_t^i - \mathcal{L})dt \tag{2.2.1}$$

(see the notation (2.1.24)). This form takes its values into the subbundle $J^1Q \underset{Q}{\times} T^*Q$ of T^*J^1Q. Hence, we have a morphism

$$\widehat{H}_L : J^1Q \to T^*Q, \tag{2.2.2}$$

whose range

$$Z_L = \widehat{H}_L(J^1Q) \tag{2.2.3}$$

is an imbedded subbundle $i_L : Z_L \to T^*Q$ of the cotangent bundle T^*Q. One calls $\widehat{H}_L$ the *homogeneous Legendre map* and T^*Q the *homogeneous Legendre bundle*. Let (t, q^i, p_0, p_i) denote the holonomic coordinates of T^*Q possessing transition functions

$$p'_i = \frac{\partial q^j}{\partial q'^i} p_j, \qquad p'_0 = \left(p_0 + \frac{\partial q^j}{\partial t} p_j\right). \tag{2.2.4}$$

With respect to these coordinates, the morphism $\widehat{H}_L$ (2.2.2) reads

$$(p_0, p_i) \circ \widehat{H}_L = (\mathcal{L} - q^i_t \pi_i, \pi_i).$$

A glance at the transition functions (2.2.4) shows that T^*Q is a one-dimensional affine bundle

$$\zeta : T^*Q \to V^*Q \tag{2.2.5}$$

over the vertical cotangent bundle V^*Q (cf. (11.2.19)). Moreover, the Legendre map $\widehat{L}$ (2.1.30) is exactly the composition of morphisms

$$\widehat{L} = \zeta \circ H_L : J^1Q \underset{Q}{\to} V^*Q. \tag{2.2.6}$$

It is readily observed that the Poincaré–Cartan form H_L (2.2.1) also is the Poincaré–Cartan form $H_L = H_{\widetilde{L}}$ of the first order Lagrangian

$$\widetilde{L} = \widehat{h}_0(H_L) = (\mathcal{L} + (q^i_{(t)} - q^i_t)\pi_i)dt, \qquad \widehat{h}_0(dq^i) = q^i_{(t)}dt, \tag{2.2.7}$$

on the repeated jet manifold J^1J^1Y [53; 68]. The Lagrange operator for $\widetilde{L}$ reads (called the *Lagrange–Cartan operator*)

$$\delta\widetilde{L} = [(\partial_i\mathcal{L} - \widehat{d}_t\pi_i + \partial_i\pi_j(q^j_{(t)} - q^j_t))dq^i + \partial^t_i\pi_j(q^j_{(t)} - q^j_t)dq^i_t] \wedge dt. \tag{2.2.8}$$

Its kernel $\operatorname{Ker}\delta\overline{L} \subset J^1J^1Q$ defines the *Cartan equation*

$$\partial^t_i\pi_j(q^j_{(t)} - q^j_t) = 0, \tag{2.2.9}$$

$$\partial_i\mathcal{L} - \widehat{d}_t\pi_i + \partial_i\pi_j(q^j_{(t)} - q^j_t) = 0 \tag{2.2.10}$$

on J^1Q. Since $\delta\widetilde{L}|_{J^2Q} = \delta L$, the Cartan equation (2.2.9) – (2.2.10) is equivalent to the Lagrange equation (2.1.25) on integrable sections of $J^1Q \to X$.

It is readily observed that these equations are equivalent if a Lagrangian L is regular.

The Cartan equation (2.2.9) – (2.2.10) on sections $\overline{c}:\mathbb{R}\to J^1Q$ is equivalent to the relation

$$\overline{c}^*(u\rfloor dH_L)=0, \tag{2.2.11}$$

which is assumed to hold for all vertical vector fields u on $J^1Q\to\mathbb{R}$.

The cotangent bundle T^*Q admits the *Liouville form*

$$\Xi=p_0dt+p_idq^i. \tag{2.2.12}$$

Accordingly, its imbedded subbundle Z_L (2.2.3) is provided with the pull-back *De Donder form* $\Xi_L=i_L^*\Xi$. There is the equality

$$H_L=\widehat{H}_L^*\Xi_L=\widehat{H}_L^*(i_L^*\Xi). \tag{2.2.13}$$

By analogy with the Cartan equation (2.2.11), the *Hamilton–De Donder equation* for sections $\overline{r}$ of $Z_L\to\mathbb{R}$ is written as

$$\overline{r}^*(u\rfloor d\Xi_L)=0, \tag{2.2.14}$$

where u is an arbitrary vertical vector field on $Z_L\to\mathbb{R}$.

Theorem 2.2.1. *Let the homogeneous Legendre map $\widehat{H}_L$ be a submersion. Then a section $\overline{c}$ of $J^1Q\to\mathbb{R}$ is a solution of the Cartan equation (2.2.11) if and only if $\widehat{H}_L\circ\overline{c}$ is a solution of the Hamilton–De Donder equation (2.2.14), i.e., the Cartan and Hamilton–De Donder equations are quasi-equivalent [68; 74].*

Remark 2.2.1. As was mentioned above, the vertical cotangent bundle V^*Q plays a role of the phase space of non-relativistic mechanics on a configuration space Q. Accordingly, the cotangent bundle T^*Q is its homogeneous phase space (Section 3.3).

2.3 Quadratic Lagrangians

Quadratic Lagrangians provide the most physically relevant case of non-relativistic mechanical systems.

Given a configuration bundle $Q\to\mathbb{R}$, let us consider a *quadratic Lagrangian*

$$L=\left(\frac{1}{2}a_{ij}q_t^iq_t^j+b_iq_t^i+c\right)dt, \tag{2.3.1}$$

where a, b and c are local functions on Q. This property is global due to the transformation law of the velocity coordinates q^i_t. The associated Legendre map reads

$$p_i \circ \widehat{L} = a_{ij} q^j_t + b_i. \tag{2.3.2}$$

Lemma 2.3.1. *The Lagrangian (2.3.1) is semiregular.*

Proof. For any point p of the Lagrangian constraint space N_L (2.3.2), the system of linear algebraic equations (2.3.2) for q^i_t has solutions which make up an affine space modelled over the linear space of solutions of the homogeneous linear algebraic equations

$$0 = a_{ij} \dot{q}^j,$$

where $\dot{q}^j$ are the holonomic coordinates on a vertical tangent bundle VQ. This affine space is obviously connected. □

Let us assume that the Lagrangian L (2.3.1) is almost regular, i.e., the matrix a_{ij} is of constant rank.

The Legendre map (2.3.2) is an affine morphism over Q. It defines the corresponding linear morphism

$$\overline{L} : VQ \underset{Q}{\to} V^*Q, \qquad p_i \circ \overline{L} = a_{ij} \dot{q}^j,$$

whose range $\overline{N}$ is a linear subbundle of the Legendre bundle $V^*Q \to Q$. Accordingly, the Lagrangian constraint space N_L, given by the equations (2.3.2), is an affine subbundle $N_L \to Q$, modelled over $\overline{N}$, of the Legendre bundle $V^*Q \to Q$. Hence, the fibre bundle $N_L \to Q$ has a global section. For the sake of simplicity, let us assume that this is the canonical zero section $\widehat{0}(Q)$ of $V^*Q \to Q$. Then $\overline{N} = N_L$.

The kernel

$$\operatorname{Ker} \widehat{L} = \widehat{L}^{-1}(\widehat{0}(Q))$$

of the Legendre map is an affine subbundle of the affine jet bundle $J^1Q \to Q$, which is modelled over the vector bundle

$$\operatorname{Ker} \overline{L} = \overline{L}^{-1}(\widehat{0}(Q)) \subset VQ.$$

Then there exists a connection

$$\Gamma : Q \to \operatorname{Ker} \widehat{L}, \tag{2.3.3}$$

$$a_{ij}\Gamma^j + b_i = 0, \tag{2.3.4}$$

on the configuration bundle $Q \to \mathbb{R}$, which takes its values into $\operatorname{Ker} \widehat{L}$. It is called the *Lagrangian frame connection*.

Thus, any quadratic Lagrangian defines a reference frame given by some Lagrangian frame connection (2.3.3). It is called the *Lagrangian reference frame.*

With a Lagrangian frame connection, the quadratic Lagrangian (2.3.1) can be brought into the covariant form

$$L = \left(\frac{1}{2}a_{ij}(q_t^i - \Gamma^i)(q_t^j - \Gamma^j) + c'\right) dt, \qquad (2.3.5)$$

i.e., it factorizes trough relative velocities $\dot{q}_\Gamma^i = q_t^i - \Gamma^i$ with respect to the Lagrangian reference frame (2.3.3).

For instance, if the quadratic Lagrangian (2.3.1) is regular, there is a unique solution (2.3.3) of the algebraic equations (2.3.4). Thus, the regular Lagrangian admits a unique Lagrangian frame connection and a Lagrangian reference frame.

The matrix a in the Lagrangian L (2.3.1) can be seen as a degenerate fibre metric of constant rank in $VQ \to Q$. Then the following holds.

Lemma 2.3.2. *Given a k-dimensional vector bundle $E \to Z$, let a be a section of rank r of the tensor bundle $\overset{2}{\vee} E^* \to Z$. There is a splitting*

$$E = \operatorname{Ker} a \underset{Z}{\oplus} E', \qquad (2.3.6)$$

where $E' = E/\operatorname{Ker} a$ is the quotient bundle, and a is a non-degenerate fibre metric in E'.

Proof. Since a exists, the structure group $GL(k, \mathbb{R})$ of the vector bundle $E \to Z$ is reducible to the subgroup $GL(r, k-r; \mathbb{R})$ of general linear transformations of $\mathbb{R}^k$ which keep its r-dimensional subspace, and to its subgroup $GL(r, \mathbb{R}) \times GL(k-r, \mathbb{R})$. □

Theorem 2.3.1. *Given an almost regular quadratic Lagrangian L, there exists a linear map*

$$\sigma : V^*Q \to VQ, \qquad \dot{q}^i \circ \sigma = \sigma^{ij} p_j, \qquad (2.3.7)$$

over Q such that

$$\widehat{L} \circ \sigma \circ i_N = i_N.$$

Proof. The map (2.3.7) is a solution of the algebraic equations

$$a_{ij}\sigma^{jk}a_{kb} = a_{ib}. \qquad (2.3.8)$$

By virtue of Lemma 2.3.2, there exist the bundle slitting

$$VQ = \operatorname{Ker} a \underset{Q}{\oplus} E' \qquad (2.3.9)$$

and a (non-holonomic) atlas of this bundle such that transition functions of $\operatorname{Ker} a$ and E' are independent. Since a is a non-degenerate fibre metric in E', there exists an atlas of E' such that a is brought into a diagonal matrix with non-vanishing components a_{AA}. Due to the splitting (2.3.9), we have the corresponding bundle splitting

$$V^*Q = (\operatorname{Ker} a)^* \underset{Q}{\oplus} \operatorname{Im} a. \tag{2.3.10}$$

Then a desired map σ is represented by the direct sum $\sigma_1 \oplus \sigma_0$ of an arbitrary section σ_1 of the bundle

$$\overset{2}{\vee}(\operatorname{Ker} a^*) \to Q$$

and a section σ_0 of the bundle $\overset{2}{\vee} E' \to Q$, which has non-vanishing components $\sigma^{AA} = (a_{AA})^{-1}$ with respect to the above mentioned atlas of E'. Moreover, σ satisfies the particular relations

$$\sigma_0 = \sigma_0 \circ \overline{L} \circ \sigma_0, \quad a \circ \sigma_1 = 0, \quad \sigma_1 \circ a = 0. \tag{2.3.11}$$

□

Remark 2.3.1. Using the relations (2.3.11), one can write the above assumption, that the Lagrangian constraint space $N_L \to Q$ admits a global zero section, in the form

$$b_i = a_{ij}\sigma_0^{jk} b_k. \tag{2.3.12}$$

If the quadratic Lagrangian (2.3.1) is regular, the map (2.3.7) is uniquely defined by the equation (2.3.8).

With the relations (2.3.7), (2.3.8) and (2.3.12), we obtain the splitting

$$J^1Q = \mathcal{S}(J^1Q) \underset{Q}{\oplus} \mathcal{F}(J^1Q) = \operatorname{Ker} \widehat{L} \underset{Q}{\oplus} \operatorname{Im}(\sigma_0 \circ \widehat{L}), \tag{2.3.13}$$

$$\begin{aligned} q_t^i &= \mathcal{S}^i + \mathcal{F}^i \\ &= [q_t^i - \sigma_0^{ik}(a_{kj}q_t^j + b_k)] + [\sigma_0^{ik}(a_{kj}q_t^j + b_k)]. \end{aligned} \tag{2.3.14}$$

It is readily observed that, with respect to the coordinates $\mathcal{S}^i$ and $\mathcal{F}^i$ (2.3.14), the Lagrangian (2.3.1) reads

$$\mathcal{L} = \frac{1}{2} a_{ij}\mathcal{F}^i\mathcal{F}^j + c', \tag{2.3.15}$$

where

$$\mathcal{F}^i = \sigma_0^{ik} a_{kj}(q_t^j - \Gamma^j) \tag{2.3.16}$$

for some Lagrangian reference frame Γ (2.3.3) on $Y \to X$.

Example 2.3.1. Let us consider a regular quadratic Lagrangian

$$L = \left[\frac{1}{2}m_{ij}(q^\mu)q_t^i q_t^j + k_i(q^\mu)q_t^i + \phi(q^\mu)\right] dt, \tag{2.3.17}$$

where m_{ij} is a non-degenerate positive-definite fibre metric in the vertical tangent bundle $VQ \to Q$. The corresponding Lagrange equation takes the form

$$q_{tt}^i = -(m^{-1})^{ik}\{_{\lambda k\nu}\}q_t^\lambda q_t^\nu, \qquad q_t^0 = 1, \tag{2.3.18}$$

where

$$\{_{\lambda\mu\nu}\} = -\frac{1}{2}(\partial_\lambda g_{\mu\nu} + \partial_\nu g_{\mu\lambda} - \partial_\mu g_{\lambda\nu})$$

are the Christoffel symbols of the metric

$$g_{00} = -2\phi, \qquad g_{0i} = -k_i, \qquad g_{ij} = -m_{ij} \tag{2.3.19}$$

on the tangent bundle TQ. Let us assume that this metric is non-degenerate. By virtue of Corollary 1.5.1, the second order dynamic equation (2.3.18) is equivalent to the non-relativistic geodesic equation (1.5.9) on the tangent bundle TQ which reads

$$\ddot{q}^0 = 0, \qquad \dot{q}^0 = 1, \qquad \ddot{q}^i = \{_\lambda{}^i{}_\nu\}\dot{q}^\lambda\dot{q}^\nu - g^{i0}\{_{\lambda 0\nu}\}\dot{q}^\lambda\dot{q}^\nu. \tag{2.3.20}$$

Let us now bring the Lagrangian (2.3.17) into the form (2.3.5):

$$\mathcal{L} = \left[\frac{1}{2}m_{ij}(q^\mu)(q_t^i - \Gamma^i)(q_t^j - \Gamma^j) + \phi'(q^\mu)\right] dt, \tag{2.3.21}$$

where Γ is a Lagrangian frame connection on $Q \to \mathbb{R}$. This connection defines an atlas of local constant trivializations of a fibre bundle $Q \to \mathbb{R}$ and the corresponding coordinates $(t, \overline{q}^i)$ on Q such that the transition functions $\overline{q}^i \to \overline{q}'^i$ are independent of t, and $\Gamma^i = 0$ with respect to $(t, \overline{q}^i)$. In these coordinates, the Lagrangian (2.3.21) reads

$$L = \left[\frac{1}{2}\overline{m}_{ij}\overline{q}_t^i\overline{q}_t^j + \phi'(q^\nu(\overline{q}^\mu))\right] dt. \tag{2.3.22}$$

Let us assume that ϕ' is a nowhere vanishing function on Q. Then the Lagrange equation (2.3.18) takes the form

$$\overline{q}_{tt}^i = \{_\lambda{}^i{}_\nu\}\overline{q}_t^\lambda\overline{q}_t^\nu, \qquad \overline{q}_t^0 = 1,$$

where $\{_\lambda{}^i{}_\nu\}$ are the Christoffel symbols of the metric (2.3.19), whose components with respect to the coordinates $(t, \overline{q}^i)$ read

$$g_{00} = -2\phi', \qquad g_{0i} = 0, \qquad g_{ij} = -\overline{m}_{ij}. \tag{2.3.23}$$

The corresponding non-relativistic geodesic equation (1.5.9) on the tangent bundle TQ reads

$$\ddot{\overline{q}}^0 = 0, \qquad \dot{\overline{q}}^0 = 1,$$
$$\ddot{\overline{q}}^i = \{_\lambda{}^i{}_\nu\}\dot{\overline{q}}^\lambda\dot{\overline{q}}^\nu. \tag{2.3.24}$$

Its spatial part (2.3.24) is exactly the spatial part of a geodesic equation with respect to the Levi–Civita connection for the metric (2.3.23) on TQ.

2.4 Lagrangian and Newtonian systems

Let L be a Lagrangian on a velocity space J^1Q and $\widehat{L}$ the Legendre map (2.1.30). Due to the vertical splitting (11.2.27) of VV^*Q, the vertical tangent map $V\widehat{L}$ to $\widehat{L}$ reads

$$V\widehat{L} : V_Q J^1Q \to V^*Q \underset{Q}{\times} V^*Q.$$

It yields the linear bundle morphism

$$\widehat{m} = (\mathrm{Id}_{J^1Q}, \mathrm{pr}_2 \circ V\widehat{L}) : V_Q J^1Q \to V_Q^* J^1Q, \quad \widehat{m} : \partial_i^t \to \pi_{ij}\overline{d}q_t^j, \tag{2.4.1}$$

and consequently a fibre metric

$$m : J^1Q \to \overset{2}{\underset{J^1Q}{\vee}} V_Q^* J^1Q$$

in the vertical tangent bundle $V_Q J^1Q \to J^1Q$. This fibre metric m obviously satisfies the symmetry condition (1.9.2).

Let a Lagrangian L be regular. Then the fibre metric m (2.4.1) is non-degenerate. In accordance with Remark 2.1.1, if a Lagrangian L is regular, there exists a unique Lagrangian connection ξ_L for L which obeys the equality

$$m_{ik}\xi_L^k + \partial_t\pi_i + \partial_j\pi_i q_t^j - \partial_i\mathcal{L} = 0. \tag{2.4.2}$$

The derivation of this equality with respect to q_t^j results in the relation (1.9.3). Thus, any regular Lagrangian L defines a Newtonian system characterized by the mass tensor $m_{ij} = \pi_{ij}$.

Remark 2.4.1. Any fibre metric m in $VQ \to Q$ can be seen as a mass metric of a standard Newtonian system, given by the Lagrangian

$$\mathcal{L} = \frac{1}{2} m_{ij}(q^\mu)(q_t^i - \Gamma^i)(q_t^j - \Gamma^j), \tag{2.4.3}$$

where Γ is a reference frame. If $\widehat{m}$ is positive-definite, one can think of the Lagrangian (2.4.3) as being a kinetic energy with respect to the reference frame Γ.

Now let us investigate the conditions for a Newtonian system to be the Lagrangian one.

The equation (1.9.4) is the kernel of the second order differential Lagrange type operator

$$\mathcal{E} : J^2Q \to V^*Q, \qquad \mathcal{E} = m_{ik}(\xi^k - q_{tt}^k)\theta^i \wedge dt. \tag{2.4.4}$$

A glance at the variational complex (2.1.12) shows that this operator is a Lagrange operator of some Lagrangian only if it obeys the Helmholtz condition

$$\delta(\mathcal{E}_i\theta^i\wedge dt) = [(2\partial_j - d_t\partial_j^t + d_t^2\partial_j^{tt})\mathcal{E}_i\theta^j\wedge\theta^i$$
$$+ (\partial_j^t\mathcal{E}_i + \partial_i^t\mathcal{E}_j - 2d_t\partial_j^{tt}\mathcal{E}_i)\theta_t^i\wedge\theta^j + (\partial_j^{tt}\mathcal{E}_i - \partial_i^{tt}\mathcal{E}_j)\theta_{tt}^j\wedge\theta^i]\wedge dt = 0.$$

This condition falls into the equalities

$$\partial_j\mathcal{E}_i - \partial_i\mathcal{E}_j + \frac{1}{2}d_t(\partial_i^t\mathcal{E}_j - \partial_j^t\mathcal{E}_i) = 0, \qquad (2.4.5)$$
$$\partial_j^t\mathcal{E}_i + \partial_i^t\mathcal{E}_j - 2d_t\partial_j^{tt}\mathcal{E}_i = 0, \qquad (2.4.6)$$
$$\partial_j^{tt}\mathcal{E}_i - \partial_i^{tt}\mathcal{E}_j = 0. \qquad (2.4.7)$$

It is readily observed, that the condition (2.4.7) is satisfied since the mass tensor is symmetric. The condition (2.4.6) holds due to the equality (1.9.3) and the property (1.9.2). Thus, it is necessary to verify the condition (2.4.5) for a Newtonian system to be a Lagrangian one. If this condition holds, the operator $\mathcal{E}$ (2.4.4) takes the form (2.1.19) in accordance with Corollary 2.1.3. If the second de Rham cohomology of Q (or, equivalently, M) vanishes, this operator is a Lagrange operator.

Example 2.4.1. Let ξ be a free motion equation which takes the form (1.7.8) with respect to a reference frame $(t,\overline{q}^i)$, and let m be a mass tensor which depends only on the velocity coordinates $\overline{q}^i_t$. Such a mass tensor may exist in accordance with affine coordinate transformations (1.7.9) which maintain the equation (1.7.8). Then ξ and $\widehat{m}$ make up a Newtonian system. This system is a Lagrangian one if m is constant with respect to the above-mentioned reference frame $(t,\overline{q}^i)$. Relative to arbitrary coordinates on a configuration space Q, the corresponding Lagrangian takes the form (2.4.3), where Γ is the connection associated with the reference frame $(t,\overline{q}^i)$.

Example 2.4.2. Let us consider a one-dimensional motion of a point mass m_0 subject to *friction*. It is described by the equation

$$m_0q_{tt} = -kq_t, \qquad k>0, \qquad (2.4.8)$$

on the configuration space $\mathbb{R}^2\to\mathbb{R}$ coordinated by (t,q). This mechanical system is characterized by the mass function $m=m_0$ and the holonomic connection

$$\xi = \partial_t + q_t\partial_q - \frac{k}{m}q_t\partial_q^t, \qquad (2.4.9)$$

but it is neither a Newtonian nor a Lagrangian system. The conditions (2.4.5) and (2.4.7) are satisfied for an arbitrary mass function $m(t, q, q_t)$, whereas the conditions (1.9.3) and (2.4.6) take the form

$$-kq_t\partial_q^t m - km + \partial_t m + q_t\partial_q m = 0. \tag{2.4.10}$$

The mass function $m =$ const. fails to satisfy this relation. Nevertheless, the equation (2.4.10) has a solution

$$m = m_0 \exp\left[\frac{k}{m_0}t\right]. \tag{2.4.11}$$

The mechanical system characterized by the mass function (2.4.11) and the holonomic connection (2.4.9) is both a Newtonian and Lagrangian system with the *Havas Lagrangian*

$$\mathcal{L} = \frac{1}{2}m_0 \exp\left[\frac{k}{m_0}t\right] q_t^2 \tag{2.4.12}$$

[133]. The corresponding Lagrange equation is equivalent to the equation of motion (2.4.8).

In conclusion, let us mention mechanical systems whose motion equations are Lagrange equations plus additional non-Lagrangian external forces. They read

$$(\partial_i - d_t\partial_i^t)\mathcal{L} + f_i(t, q^j, q_t^j) = 0. \tag{2.4.13}$$

Let a Lagrangian system be the Newtonian one, and let an external force f satisfy the condition (1.9.8). Then the equation (2.4.13) describe a Newtonian system.

2.5 Lagrangian conservation laws

In Lagrangian mechanics, integrals of motion come from variational symmetries of a Lagrangian (Theorem 2.5.3) in accordance with the first Noether theorem (Theorem 2.5.2). However, not all integrals of motion are of this type (Example 2.5.4).

2.5.1 *Generalized vector fields*

Given a Lagrangian system $(\mathcal{O}^*_\infty, L)$ on a fibre bundle $Q \to \mathbb{R}$, its *infinitesimal transformations* are defined to be contact derivations of the real ring $\mathcal{O}^0_\infty$ [64; 68].

Let us consider the $\mathcal{O}^0_\infty$-module $\mathfrak{d}\mathcal{O}^0_\infty$ of derivations of the real ring $\mathcal{O}^0_\infty$. This module is isomorphic to the $\mathcal{O}^0_\infty$-dual $(\mathcal{O}^1_\infty)^*$ of the module of one-forms $\mathcal{O}^1_\infty$. Let $\vartheta\rfloor\phi$, $\vartheta \in \mathfrak{d}\mathcal{O}^0_\infty$, $\phi \in \mathcal{O}^1_\infty$, be the corresponding interior product. Extended to a differential graded algebra $\mathcal{O}^*_\infty$, it obeys the rule (11.2.48).

Restricted to the coordinate chart (2.1.2), any derivation of a real ring $\mathcal{O}^0_\infty$ takes the coordinate form

$$\vartheta = \vartheta^t\partial_t + \vartheta^i\partial_i + \sum_{0<|\Lambda|}\vartheta^i_\Lambda\partial^\Lambda_i,$$

where

$$\partial^\Lambda_i(q^j_\Sigma) = \partial^\Lambda_i\rfloor dq^j_\Sigma = \delta^j_i\delta^\Lambda_\Sigma.$$

Not concerned with time-reparametrization, we restrict our consideration to derivations

$$\vartheta = u^t\partial_t + \vartheta^i\partial_i + \sum_{0<|\Lambda|}\vartheta^i_\Lambda\partial^\Lambda_i, \qquad u^t = 0, 1. \tag{2.5.1}$$

Their coefficients ϑ^i, ϑ^i_Λ possess the transformation law

$$\vartheta'^i = \frac{\partial q'^i}{\partial q^j}\vartheta^j + \frac{\partial q'^i}{\partial t}u^t, \qquad \vartheta'^i_\Lambda = \sum_{|\Sigma|\leq|\Lambda|}\frac{\partial q'^i_\Lambda}{\partial q^j_\Sigma}\vartheta^j_\Sigma + \frac{\partial q'^i_\Lambda}{\partial t}u^t.$$

Any derivation ϑ (2.5.1) of a ring $\mathcal{O}^0_\infty$ yields a derivation (a Lie derivative) $\mathbf{L}_\vartheta$ of a differential graded algebra $\mathcal{O}^*_\infty$ which obeys the relations (11.2.49) – (11.2.50).

A derivation $\vartheta \in \mathfrak{d}\mathcal{O}^0_\infty$ (2.5.1) is called *contact* if the Lie derivative $\mathbf{L}_\vartheta$ preserves an ideal of contact forms of a differential graded algebra $\mathcal{O}^*_\infty$, i.e., the Lie derivative $\mathbf{L}_\vartheta$ of a contact form is a contact form.

Lemma 2.5.1. *A derivation ϑ (2.5.1) is contact if and only if it takes the form*

$$\vartheta = u^t\partial_t + u^i\partial_i + \sum_{0<|\Lambda|}[d_\Lambda(u^i - q^i_t u^t) + q^i_{t\Lambda}u^t]\partial^\Lambda_i. \tag{2.5.2}$$

A glance at the expression (2.5.2) enables one to regard a contact derivation ϑ as an infinite order jet prolongation $\vartheta = J^\infty u$ of its restriction

$$u = u^t\partial_t + u^i(t, q^i, q^i_\Lambda)\partial_i, \qquad u^t = 0, 1, \tag{2.5.3}$$

to a ring $C^\infty(Q)$. Since coefficients u^i of u (2.5.3) generally depend on jet coordinates q^i_Λ, $0 < |\Lambda| \leq r$, one calls u (2.5.3) the *generalized vector field*. It can be represented as a section of the pull-back bundle

$$J^rQ \underset{Q}{\times} TQ \to J^rQ.$$

In particular, let u (2.5.3) be a vector field

$$u = u^t\partial_t + u^i(t, q^i)\partial_i, \qquad u^t = 0, 1, \tag{2.5.4}$$

on a configuration space $Q \to \mathbb{R}$. One can think of this vector field as being an infinitesimal generator of a local one-parameter group of local automorphisms of a fibre bundle $Q \to \mathbb{R}$. If $u^t = 0$, the vertical vector field (2.5.4) is an infinitesimal generator of a local one-parameter group of local vertical automorphisms of $Q \to \mathbb{R}$. If $u^t = 1$, the vector field u (2.5.4) is projected onto the standard vector field ∂_t on a base $\mathbb{R}$ which is an infinitesimal generator of a group of translations of $\mathbb{R}$.

Any contact derivation ϑ (2.5.2) admits the horizontal splitting

$$\vartheta = \vartheta_H + \vartheta_V = u^t d_t + \left[u^i_V\partial_i + \sum_{0<|\Lambda|} d_\Lambda u^i_V \partial_i^\Lambda\right], \tag{2.5.5}$$

$$u = u_H + u_V = u^t(\partial_t + q^i_t\partial_i) + (u^i - q^i_t u^t)\partial_i. \tag{2.5.6}$$

Lemma 2.5.2. *Any vertical contact derivation*

$$\vartheta = u^i\partial_i + \sum_{0<|\Lambda|} d_\Lambda u^i \partial_i^\Lambda \tag{2.5.7}$$

obeys the relations

$$\vartheta\rfloor d_H\phi = -d_H(\vartheta\rfloor\phi), \qquad \mathbf{L}_\vartheta(d_H\phi) = d_H(\mathbf{L}_\vartheta\phi), \qquad \phi \in \mathcal{O}^*_\infty. \tag{2.5.8}$$

We restrict our consideration to first order Lagrangian mechanics. In this case, contact derivations (2.1.1) can be reduced to the first order jet prolongation

$$\vartheta = J^1u = u^t\partial_t + u^i\partial_i + d_t u^i \partial_i^t \tag{2.5.9}$$

of the generalized vector fields u (2.5.3).

2.5.2 *First Noether theorem*

Let L be a Lagrangian (2.1.22) on a velocity space J^1Q. Let us consider its Lie derivative $\mathbf{L}_\vartheta L$ along the contact derivation ϑ (2.5.9).

Theorem 2.5.1. *The Lie derivative $\mathbf{L}_\vartheta L$ fulfils the first variational formula*

$$\mathbf{L}_{J^1u}L = u_V\rfloor\delta L + d_H(u\rfloor H_L), \tag{2.5.10}$$

where $\mathfrak{L} = H_L$ is the Poincaré–Cartan form (2.2.1). Its coordinate expression reads

$$[u^t\partial_t + u^i\partial_i + d_t u^i\partial_i^t]\mathcal{L} = (u^i - q^i_t u^t)\mathcal{E}_i + d_t[\pi_i(u^i - u^t q^i_t) + u^t\mathcal{L}]. \tag{2.5.11}$$

Proof. The formula (2.5.10) results from the decomposition (2.1.21) and the relations (2.5.8) [68]. □

The generalized vector field u (2.5.3) is said to be the *variational symmetry* of a Lagrangian L if the Lie derivative $\mathbf{L}_{J^1u}L$ is d_H-exact, i.e.,

$$\mathbf{L}_{J^1u}L = d_H\sigma. \tag{2.5.12}$$

Variational symmetries of L constitute a real vector space which we denote $\mathcal{G}_L$.

Proposition 2.5.1. *A glance at the first variational formula (2.5.11) shows that a generalized vector field u is a variational symmetry if and only if the exterior form*

$$u_V\rfloor\delta L = (u^i - q^i_t u^t)\mathcal{E}_i dt \tag{2.5.13}$$

is d_H-exact.

Proposition 2.5.2. *The generalized vector field u (2.5.3) is a variational symmetry of a Lagrangian L if and only if its vertical part u_V (2.5.6) also is a variational symmetry.*

Proof. A direct computation shows that

$$\mathbf{L}_{J^1u}L = \mathbf{L}_{J^1u_V}L + d_H(u^t\mathcal{L}). \tag{2.5.14}$$

□

A corollary of the first variational formula (2.5.10) is the *first Noether theorem.*

Theorem 2.5.2. *If a contact derivation ϑ (2.5.2) is a variational symmetry (2.5.12) of a Lagrangian L, the first variational formula (2.5.10) restricted to the kernel of the Lagrange operator* $\operatorname{Ker}\delta L$ *yields a weak conservation law*

$$0 \approx d_H(u\rfloor H_L - \sigma), \tag{2.5.15}$$

$$0 \approx d_t(\pi_i(u^i - u^t q^i_t) + u^t\mathcal{L} - \sigma), \tag{2.5.16}$$

of the generalized symmetry current

$$\mathfrak{T}_u = u\rfloor H_L - \sigma = \pi_i(u^i - u^t q^i_t) + u^t\mathcal{L} - \sigma \tag{2.5.17}$$

along a generalized vector field u. The generalized symmetry current (2.5.17) obviously is defined with the accuracy of a constant summand.

The weak conservation law (2.5.15) on the shell $\delta L = 0$ is called the *Lagrangian conservation law*. It leads to the differential conservation law (1.10.2):

$$0 = \frac{d}{dt}[\mathfrak{T}_u \circ J^{r+1}c],$$

on solutions c of the Lagrange equation (2.1.26).

Proposition 2.5.3. *Let u be a variational symmetry of a Lagrangian L. By virtue of Proposition 2.5.2, its vertical part u_V is so. It follows from the equality (2.5.14) that the conserved generalized symmetry current $\mathfrak{T}_u$ (2.5.17) along u equals that $\mathfrak{T}_{u_V}$ along u_V.*

A glance at the conservation law (2.5.16) shows the following.

Theorem 2.5.3. *If a variational symmetry u is a generalized vector field independent of higher order jets q^i_Λ, $|\Lambda| > 1$, the conserved generalized current $\mathfrak{T}_u$ (2.5.17) along u plays a role of an integral of motion.*

Therefore, we further restrict our consideration to variational symmetries at most of first jet order for the purpose of obtaining integrals of motion. However, it may happen that a Lagrangian system possesses integrals of motion which do not come from variational symmetries (Example 2.5.4).

A variational symmetry u of a Lagrangian L is called its *exact symmetry* if

$$\mathbf{L}_{J^1u} L = 0. \tag{2.5.18}$$

In this case, the first variational formula (2.5.10) takes the form

$$0 = u_V \rfloor \delta L + d_H(u \rfloor H_L). \tag{2.5.19}$$

It leads to the weak conservation law (2.5.15):

$$0 \approx d_t \mathfrak{T}_u, \tag{2.5.20}$$

of the *symmetry current*

$$\mathfrak{T}_u = u \rfloor H_L = \pi_i(u^i - u^t q^i_t) + u^t \mathcal{L} \tag{2.5.21}$$

along a generalized vector field u.

Remark 2.5.1. In accordance with the standard terminology, if variational and exact symmetries are generalized vector fields (2.5.3), they are called *generalized symmetries* [21; 42; 87; 124]. Accordingly, by variational and exact symmetries one means only vector fields u (2.5.4) on Q. We agree to call them *classical symmetries*. Classical exact symmetries are symmetries of a Lagrangian, and they are named the *Lagrangian symmetries*.

Remark 2.5.2. Given a Lagrangian L, let $\widetilde{L}$ be its partner (2.2.7) on the repeated jet manifold J^1J^1Q. Since H_L (2.2.1) is the Poincaré–Cartan both for L and $\widetilde{L}$, a Lagrangian $\widetilde{L}$ does not lead to new conserved currents.

Remark 2.5.3. Let us describe the relation between symmetries of a Lagrangian and and symmetries of the corresponding Lagrange equation. Let u be the vector field (2.5.4) and

$$J^2u = u^t\partial_t + u^i\partial_i + d_tu^i\partial_i^t + d_{tt}u^i\partial_i^{tt}$$

its second order jet prolongation. Given a Lagrangian L on J^1Q, the relation

$$\mathbf{L}_{J^2u}\delta L = \delta(\mathbf{L}_{J^1u}L) \tag{2.5.22}$$

holds [53; 124]. Note that this equality need not be true in the case of a generalized vector field u. A vector field u is called the *local variational symmetry* of a Lagrangian L if the Lie derivative $\mathbf{L}_{J^1u}L$ of L along u is variationally trivial, i.e.,

$$\delta(\mathbf{L}_{J^1u}L) = 0.$$

Then it follows from the equality (2.5.22) that a local (classical) variational symmetry of L also is a symmetry of the Lagrange operator δL, i.e.,

$$\mathbf{L}_{J^2u}\delta L = 0,$$

and *vice versa*. Consequently, any local classical variational symmetry u of a Lagrangian L is a symmetry of the Lagrange equation (2.1.25) in accordance with Proposition 1.10.3. By virtue of Theorem 2.1.2, any local classical variational symmetry is a classical variational symmetry if a typical fibre M of Q is simply connected.

Remark 2.5.4. The first variational formula (2.5.10) also can be utilized when a Lagrangian possesses symmetries, but an equation of motion is the sum (2.4.13) of a Lagrange equation and an additional non-Lagrangian external force. Let us substitute $\mathcal{E}_i = -f_i$ from this equality in the first variational formula (2.5.10), and let us assume that the Lie derivative of a Lagrangian L along a vector field u vanishes. Then we have the transformation law

$$(u^i - q_t^i)f_i \approx d_t\mathfrak{T}_u \tag{2.5.23}$$

of the symmetry current $\mathfrak{T}_u$ (2.5.21).

2.5.3 *Noether conservation laws*

It is readily observed that the first variational formula (2.5.11) is linear in a generalized vector field u. Therefore, one can consider superposition of the identities (2.5.11) for different generalized vector fields.

For instance, if u and u' are generalized vector fields (2.5.3), projected onto the standard vector field ∂_t on $\mathbb{R}$, the difference of the corresponding identities (2.5.11) results in the first variational formula (2.5.11) for the vertical generalized vector field $u - u'$.

Conversely, every generalized vector field u (2.5.4), projected onto ∂_t, can be written as the sum

$$u = \Gamma + v \tag{2.5.24}$$

of some reference frame

$$\Gamma = \partial_t + \Gamma^i \partial_i \tag{2.5.25}$$

and a vertical generalized vector field v on Q.

It follows that the first variational formula (2.5.11) for the generalized vector field u (2.5.4) can be represented as a superposition of those for a reference frame Γ (2.5.25) and a vertical generalized vector field v.

If $u = v$ is a vertical generalized vector field, the first variational formula (2.5.11) reads

$$(v^i \partial_i + d_t v^i \partial_i^t)\mathcal{L} = v^i \mathcal{E}_i + d_t(\pi_i v^i).$$

If v is an exact symmetry of L, we obtain from (2.5.20) the weak conservation law

$$0 \approx d_t(\pi_i v^i). \tag{2.5.26}$$

By analogy with field theory [68], it is called the *Noether conservation law* of the Noether current

$$\mathfrak{T}_v = \pi_i v^i. \tag{2.5.27}$$

If a generalized vector field v is independent of higher order jets q^i_Λ, $|\Lambda| > 1$, the Noether current (2.5.27) is an integral of motion by virtue of Theorem 2.5.3.

Example 2.5.1. Let us assume that, given a trivialization $Q = \mathbb{R} \times M$ in bundle coordinates (t, q^i), a Lagrangian L is independent of a coordinate q^1. Then the Lie derivative of L along the vertical vector field $v = \partial_1$ equals zero, and we have the conserved Noether current (2.5.27) which reduces to the momentum $\mathfrak{T}_v = \pi_1$. With respect to arbitrary bundle coordinates (t, q'^i), this conserved Noether current takes the form

$$\mathfrak{T}_v = \frac{\partial q'^i}{\partial q^1}\pi'_i.$$

It is an integral of motion.

Example 2.5.2. Let us consider a free motion on a configuration space Q. It is described by a Lagrangian

$$L = \left(\frac{1}{2}\overline{m}_{ij}\overline{q}^i_t\overline{q}^j_t\right) dt, \qquad \overline{m}_{ij} = \text{const.}, \tag{2.5.28}$$

written with respect to a reference frame $(t, \overline{q}^i)$ such that the free motion dynamic equation takes the form (1.7.1). As it follows from Example 2.5.1, this Lagrangian admits $\dim Q - 1$ independent integrals of motion π_i.

Example 2.5.3. Let us consider a point mass in the presence of a central potential. Its configuration space is

$$Q = \mathbb{R} \times \mathbb{R}^3 \to \mathbb{R} \tag{2.5.29}$$

endowed with the Cartesian coordinates (t, q^i). A Lagrangian of this mechanical system reads

$$\mathcal{L} = \frac{1}{2}\left(\sum_i (q^i_t)^2\right) - V(r), \qquad r = \left(\sum_i (q^i)^2\right)^{1/2}. \tag{2.5.30}$$

The vector fields

$$v^a_b = q^a\partial_b - q^b\partial_a \tag{2.5.31}$$

are infinitesimal generators of the group $SO(3)$ acting on $\mathbb{R}^3$. Their jet prolongation (2.5.9) reads

$$J^1 v^a_b = q^a\partial_b - q^b\partial_a + q^a_t\partial^t_b - q^b_t\partial^t_a. \tag{2.5.32}$$

It is readily observed that vector fields (2.5.31) are symmetries of the Lagrangian (2.5.30). The corresponding conserved Noether currents (2.5.27) are *orbital momenta*

$$M^a_b = \mathfrak{T}^b_a = (q^a\pi_b - q^b\pi_a) = q^a q^b_t - q^b q^a_t. \tag{2.5.33}$$

They are integrals of motion, which however fail to be independent.

Example 2.5.4. Let us consider the Lagrangian system in Example 2.5.3 where

$$V(r) = -\frac{1}{r} \tag{2.5.34}$$

is the *Kepler potential.* This Lagrangian system possesses the integrals of motion

$$A^a = \sum_b (q^a q_t^b - q^b q_t^a) q_t^b - \frac{q^a}{r}, \tag{2.5.35}$$

besides the orbital momenta (2.5.33). They are components of the *Rung–Lenz vector.* There is no Lagrangian symmetry whose generalized symmetry currents are A^a (2.5.35).

2.5.4 *Energy conservation laws*

In the case of a reference frame Γ (2.5.25), where $u^t = 1$, the first variational formula (2.5.11) reads

$$(\partial_t + \Gamma^i \partial_i + d_t \Gamma^i \partial_i^t)\mathcal{L} = (\Gamma^i - q_t^i)\mathcal{E}_i - d_t(\pi_i(q_t^i - \Gamma^i) - \mathcal{L}), \tag{2.5.36}$$

where

$$E_\Gamma = -\mathfrak{T}_\Gamma = \pi_i(q_t^i - \Gamma^i) - \mathcal{L} \tag{2.5.37}$$

is the *energy function relative to a reference frame* Γ [36; 106; 139].

With respect to the coordinates adapted to a reference frame Γ, the first variational formula (2.5.36) takes the form

$$\partial_t \mathcal{L} = (\Gamma^i - q_t^i)\mathcal{E}_i - d_t(\pi_i q_t^i - \mathcal{L}), \tag{2.5.38}$$

and E_Γ (2.5.37) coincides with the *canonical energy function*

$$E_L = \pi_i q_t^i - \mathcal{L}.$$

A glance at the expression (2.5.38) shows that the vector field Γ (2.5.25) is an exact symmetry of a Lagrangian L if and only if, written with respect to coordinates adapted to Γ, this Lagrangian is independent on the time t. In this case, the energy function E_Γ (2.5.38) relative to a reference frame Γ is conserved:

$$0 \approx -d_t E_\Gamma. \tag{2.5.39}$$

It is an integral of motion in accordance with Theorem 2.5.3.

Example 2.5.5. Let us consider a free motion on a configuration space Q described by the Lagrangian (2.5.28) written with respect to a reference frame $(t, \overline{q}^i)$ such that the free motion dynamic equation takes the form (1.7.1). Let Γ be the associated connection. Then the conserved energy function E_Γ (2.5.37) relative to this reference frame Γ is precisely the kinetic energy of this free motion. With respect to arbitrary bundle coordinates (t, q^i) on Q, it takes the form

$$E_\Gamma = \pi_i(q^i_t - \Gamma^i) - \mathcal{L} = \frac{1}{2} m_{ij}(t, q^k)(q^i_t - \Gamma^i)(q^j_t - \Gamma^j).$$

Example 2.5.6. Let us consider a one-dimensional motion of a point mass m_0 subject to friction on the configuration space $\mathbb{R}^2 \to \mathbb{R}$, coordinated by (t, q) (Example 2.4.2). It is described by the dynamic equation (2.4.8) which is the Lagrange equation for the Lagrangian L (2.4.12). It is readily observed that the Lie derivative of this Lagrangian along the vector field

$$\Gamma = \partial_t - \frac{1}{2} \frac{k}{m_0} q \partial_q \tag{2.5.40}$$

vanishes. Consequently, we have the conserved energy function (2.5.37) with respect to the reference frame Γ (2.5.40). This energy function reads

$$E_\Gamma = \frac{1}{2} m_0 \exp\left[\frac{k}{m_0} t\right] q_t \left(q_t + \frac{k}{m_0} q\right) = \frac{1}{2} m \dot{q}^2_\Gamma - \frac{mk^2}{8m_0^2} q^2,$$

where m is the mass function (2.4.11).

Since any generalized vector field u (2.5.3) can be represented as the sum (2.5.24) of a reference frame Γ (2.5.25) and a vertical generalized vector field v, the symmetry current (2.5.21) along the generalized vector field u (2.5.4) is the difference

$$\mathfrak{T}_u = \mathfrak{T}_v - E_\Gamma$$

of the Noether current $\mathfrak{T}_v$ (2.5.27) along the vertical generalized vector field v and the energy function E_Γ (2.5.37) relative to a reference frame Γ [36; 139]. Conversely, energy functions relative to different reference frames Γ and Γ' differ from each other in the Noether current along the vertical vector field $\Gamma' - \Gamma$:

$$E_\Gamma - E_{\Gamma'} = \mathfrak{T}_{\Gamma - \Gamma'}.$$

One can regard this vector field $\Gamma' - \Gamma$ as the *relative velocity* of a reference frame Γ' with respect to Γ.

2.6 Gauge symmetries

Treating gauge symmetries of Lagrangian field theory, one is traditionally based on an example of the Yang–Mills gauge theory of principal connections on a principal bundle. This notion of gauge symmetries is generalized to Lagrangian theory on an arbitrary fibre bundle [67; 68], including non-relativistic mechanics on a fibre bundle $Q \to \mathbb{R}$.

Definition 2.6.1. Let $E \to \mathbb{R}$ be a vector bundle and $E(\mathbb{R})$ the $C^\infty(\mathbb{R})$ module of sections χ of $E \to \mathbb{R}$. Let ζ be a linear differential operator on $E(\mathbb{R})$ taking its values into the vector space $\mathcal{G}_L$ of variational symmetries of a Lagrangian L (see Definition 11.5.1). Elements

$$u_\zeta = \zeta(\chi) \tag{2.6.1}$$

of $\operatorname{Im}\zeta$ are called the *gauge symmetry* of a Lagrangian L parameterized by sections χ of $E \to \mathbb{R}$. These sections are called the *gauge parameters*.

Remark 2.6.1. The differential operator ζ in Definition 2.6.1 takes its values into the vector space $\mathcal{G}_L$ as a subspace of the $C^\infty(\mathbb{R})$-module $\mathfrak{d}\mathcal{O}^0_\infty$, but it sends the $C^\infty(\mathbb{R})$-module $E(\mathbb{R})$ into the real vector space $\mathcal{G}_L \subset \mathfrak{d}\mathcal{O}^0_\infty$.

Equivalently, the gauge symmetry (2.6.1) is given by a section $\widetilde{\zeta}$ of the fibre bundle

$$(J^rQ \underset{Q}{\times} J^mE) \underset{Q}{\times} TQ \to J^rQ \underset{Q}{\times} J^mE$$

(see Definition 11.3.3) such that

$$u_\zeta = \zeta(\chi) = \widetilde{\zeta} \circ \chi$$

for any section χ of $E \to \mathbb{R}$. Hence, it is a generalized vector field u_ζ on the product $Q \times E$ represented by a section of the pull-back bundle

$$J^k(Q \underset{\mathbb{R}}{\times} E) \underset{Q}{\times} T(Q \underset{\mathbb{R}}{\times} E) \to J^k(Q \underset{\mathbb{R}}{\times} E), \qquad k = \max(r, m),$$

which lives in

$$TQ \subset T(Q \times E).$$

This generalized vector field yields the contact derivation $J^\infty u_\zeta$ (2.5.2) of the real ring $\mathcal{O}^0_\infty[Q \times E]$ which obeys the following condition.

Condition 2.6.1. Given a Lagrangian

$$L \in \mathcal{O}^{0,n}_\infty E \subset \mathcal{O}^{0,n}_\infty[Q \times E],$$

let us consider its Lie derivative

$$\mathbf{L}_{J^\infty u_\zeta} L = J^\infty u_\zeta \rfloor dL + d(J^\infty u_\zeta \rfloor L), \tag{2.6.2}$$

where d is the exterior differential of $\mathcal{O}^*_\infty[Q \times E]$. Then for any section χ of $E \to \mathbb{R}$, the pull-back $\chi^* \mathbf{L}_{J^\infty u_\zeta} L$ is d_H-exact.

It follows at once from the first variational formula (2.5.10) for the Lie derivative (2.6.2) that Condition 2.6.1 holds only if u_ζ is projected onto a generalized vector field on Q and, in this case, if and only if the density $(u_\zeta)_V \rfloor \mathcal{E}$ is d_H-exact (Proposition 2.5.1). Thus, we come to the following equivalent definition of gauge symmetries.

Definition 2.6.2. Let $E \to \mathbb{R}$ be a vector bundle. A gauge symmetry of a Lagrangian L parameterized by sections χ of $E \to \mathbb{R}$ is defined as a contact derivation $\vartheta = J^\infty u$ of the real ring $\mathcal{O}^0_\infty[Q \times E]$ such that:

(i) it vanishes on the subring $\mathcal{O}^0_\infty E$,

(ii) the generalized vector field u is linear in coordinates χ^a_Λ on $J^\infty E$, and it is projected onto a generalized vector field on Q, i.e., it takes the form

$$u = \partial_t + \left(\sum_{0 \leq |\Lambda| \leq m} u^{i\Lambda}_a (t, q^j_\Sigma) \chi^a_\Lambda \right) \partial_i, \tag{2.6.3}$$

(iii) the vertical part of u (2.6.3) obeys the equality

$$u_V \rfloor \delta L = d_H \sigma. \tag{2.6.4}$$

For the sake of convenience, the generalized vector field (2.6.3) also is called the gauge symmetry. In accordance with Proposition 2.5.2, the u (2.6.3) is a gauge symmetry if and only if its vertical part is so. Owing to this fact and Proposition 2.5.3, we can restrict our consideration to vertical gauge symmetries

$$u = \left(\sum_{0 \leq |\Lambda| \leq m} u^{i\Lambda}_a (t, q^j_\Sigma) \chi^a_\Lambda \right) \partial_i. \tag{2.6.5}$$

Gauge symmetries possess the following particular properties.

(i) Let $E' \to \mathbb{R}$ be another vector bundle and ζ' a linear $E(\mathbb{R})$-valued differential operator on a $C^\infty(\mathbb{R})$-module $E'(\mathbb{R})$ of sections of $E' \to \mathbb{R}$. Then

$$u_{\zeta'(\chi')} = (\zeta \circ \zeta')(\chi')$$

also is a gauge symmetry of L parameterized by sections χ' of $E' \to \mathbb{R}$. It factorizes through the gauge symmetry u_ζ (2.6.1).

(ii) Given a gauge symmetry, the corresponding conserved symmetry current $\mathfrak{T}_u$ (2.5.17) vanishes on-shell (Theorem 2.6.2 below).

(iii) The *second Noether theorem* associates to a gauge symmetry of a Lagrangian L the Noether identities of its Lagrange operator δL.

Theorem 2.6.1. *Let u (2.6.5) be a gauge symmetry of a Lagrangian L, then its Lagrange operator δL obeys the Noether identities (2.6.6).*

Proof. The density (2.6.4) is variationally trivial and, therefore, its variational derivatives with respect to variables χ^a vanish, i.e.,

$$\mathcal{E}_a = \sum_{0\leq|\Lambda|} (-1)^{|\Lambda|} d_\Lambda (u_a^{i\Lambda}\mathcal{E}_i) = 0. \tag{2.6.6}$$

These are the Noether identities for the Lagrange operator δL [68]. □

For instance, if the gauge symmetry u (2.6.3) is of second jet order in gauge parameters, i.e.,

$$u = (u_a^i \chi^a + u_a^{it}\chi_t^a + u_a^{itt}\chi_{tt}^a)\partial_i, \tag{2.6.7}$$

the corresponding Noether identities (2.6.6) take the form

$$u_a^i\mathcal{E}_i - d_t(u_a^{it}\mathcal{E}_i) + d_{tt}(u_a^{itt}\mathcal{E}_i) = 0. \tag{2.6.8}$$

If a Lagrangian L admits a gauge symmetry u (2.6.5), i.e., $\mathbf{L}_{J^1u}L = \sigma$, the weak conservation law (2.5.16) of the corresponding generalized symmetry current $\mathfrak{T}_u$ (2.5.17) holds. We call it the *gauge conservation law*. Because gauge symmetries depend on derivatives of gauge parameters, all gauge conservation laws in first order Lagrangian mechanics possess the following peculiarity.

Theorem 2.6.2. *If u (2.6.5) is a gauge symmetry of a first order Lagrangian L, the corresponding conserved generalized symmetry current $\mathfrak{T}_u$ (2.5.17) vanishes on-shell, i.e., $\mathfrak{T}_u \approx 0$.*

Proof. Let a gauge symmetry u be at most of jet order N in gauge parameters. Then the corresponding generalized symmetry current $\mathfrak{T}_u$ is decomposed into the sum

$$\mathfrak{T}_u = \sum_{1<|\Lambda|\leq N} J_a^\Lambda \chi_\Lambda^a + J_a^t\chi_t^a + J_a\chi^a. \tag{2.6.9}$$

The first variational formula (2.5.11) takes the form

$$0 = \left[\sum_{|\Lambda|=1}^{N} u^{i}_{V}{}^{\Lambda}_{a}\chi^a_\Lambda + u^i_{V a}\chi^a\right]\mathcal{E}_i + d_t\left(\sum_{|\Lambda|=1}^{N} J^\Lambda_a\chi^a_\Lambda + J_a\chi^a\right).$$

It falls into the set of equalities for each $\chi^a_{t\Lambda}$, χ^a_Λ, $|\Lambda| = 1,\ldots,N$, and χ^a as follows:

$$0 = J^\Lambda_a, \qquad |\Lambda| = N, \tag{2.6.10}$$

$$0 = u^{i}_{V}{}^{t\Lambda}_{a}\mathcal{E}_i + J^\Lambda_a + d_tJ^{t\Lambda}_a, \qquad 1 \leq |\Lambda| < N, \tag{2.6.11}$$

$$0 = u^{i}_{V}{}^{t}_{a}\mathcal{E}_i + J_a + d_tJ^t_a, \tag{2.6.12}$$

$$0 = u^i_{V a}\mathcal{E}_i + d_tJ_a. \tag{2.6.13}$$

With the equalities (2.6.10) – (2.6.12), the decomposition (2.6.9) takes the form

$$\begin{aligned}\mathfrak{T}_u = &- \sum_{1<|\Lambda|<N} [(u^{i}_{V}{}^{t\Lambda}_{a}\mathcal{E}_i + d_tJ^{t\Lambda}_a]\chi^a_\Lambda \\ &- (u^{i}_{V}{}^{tt}_{a}\mathcal{E}_i + d_tJ^{tt}_a)\chi^a_t - (u^{i}_{V}{}^{t}_{a}\mathcal{E}_i + d_tJ^t_a)\chi^a.\end{aligned}$$

A direct computation leads to the expression

$$\begin{aligned}\mathfrak{T}_u = &- \left(\sum_{1\geq|\Lambda|<N} u^{i}_{V}{}^{t\Lambda}_{a}\chi^a_\Lambda + u^{i}_{V}{}^{t}_{a}\chi^a\right)\mathcal{E}_i \\ &- \left(\sum_{1\geq|\Lambda|<N} d_tJ^{t\Lambda}\chi^a_\Lambda + d_tJ^t_a\chi^a\right).\end{aligned} \tag{2.6.14}$$

The first summand of this expression vanishes on-shell. Its second one contains the terms d_tJ^Λ, $|\Lambda| = 1,\ldots,N$. By virtue of the equalities (2.6.11), every d_tJ^Λ, $|\Lambda| < N$, is expressed in the terms vanishing on-shell and the term $d_td_tJ^{t\Lambda}$. Iterating the procedure and bearing in mind the equality (2.6.10), one can easily show that the second summand of the expression (2.6.14) also vanishes on-shell. Thus, the generalized symmetry current $\mathfrak{T}_u$ vanishes on-shell. □

Note that the statement of Theorem 2.6.2 is a particular case of the fact that symmetry currents of gauge symmetries in field theory are reduced to a superpotential [68; 143].

Chapter 3

Hamiltonian mechanics

As was mentioned above, a phase space of non-relativistic mechanics is the vertical cotangent bundle V^*Q of its configuration space $Q \to \mathbb{R}$. This phase space is provided with the canonical Poisson structure (3.3.7). However, Hamiltonian mechanics on a phase space V^*Q is not familiar Poisson Hamiltonian theory on a Poisson manifold V^*Q (Section 3.2) because all Hamiltonian vector fields on V^*Q are vertical. Hamiltonian non-relativistic mechanics on V^*Q is formulated as particular (polysymplectic) Hamiltonian formalism on fibre bundles [53; 68; 106]. Its Hamiltonian is a section of the fibre bundle $T^*Q \to V^*Q$ (2.2.5). The pull-back of the canonical Liouville form (2.2.12) on T^*Q with respect to this section is a Hamiltonian one-form on V^*Q. The corresponding Hamiltonian connection (3.3.21) on $V^*Q \to \mathbb{R}$ defines a first order Hamilton equation on V^*Q.

Note that one can associate to any Hamiltonian non-relativistic system on V^*Q an autonomous symplectic Hamiltonian system on the cotangent bundle T^*Q such that the corresponding Hamilton equations on V^*Q and T^*Q are equivalent (Section 3.4). Moreover, a Hamilton equation on V^*Q also is equivalent to the Lagrange equation of a certain first order Lagrangian on a configuration space V^*Q (Section 3.5).

Lagrangian and Hamiltonian formulations of mechanics fail to be equivalent, unless a Lagrangian is hyperregular. The comprehensive relations between Lagrangian and Hamiltonian systems can be established in the case of almost regular Lagrangians (Section 3.6).

3.1 Geometry of Poisson manifolds

Throughout the book, all Poisson manifolds are assumed to be regular. We start with symplectic manifolds which are non-degenerate Poisson manifolds.

3.1.1 Symplectic manifolds

Let Z be a smooth manifold. Any exterior two-form Ω on Z yields the linear bundle morphism

$$\Omega^\flat : TZ \underset{Z}{\to} T^*Z, \qquad \Omega^\flat : v \to -v\rfloor\Omega(z), \quad v \in T_zZ, \quad z \in Z. \tag{3.1.1}$$

One says that a two-form Ω is of *rank* r if the morphism (3.1.1) has a rank r. The *kernel* $\operatorname{Ker}\Omega$ of Ω is defined as the kernel of the morphism (3.1.1). If Ω is of constant rank, its kernel is a subbundle of the tangent bundle TZ. In particular, $\operatorname{Ker}\Omega$ contains the canonical zero section $\widehat{0}$ of $TZ \to Z$. If $\operatorname{Ker}\Omega = \widehat{0}$ (one customarily writes $\operatorname{Ker}\Omega = 0$), a two-form Ω is said to be *non-degenerate*. It is called an *almost symplectic form*. Equipped with such a form, a manifold Z becomes an *almost symplectic* manifold. It is never odd-dimensional. Unless otherwise stated, we put $\dim Z = 2m$.

A closed almost symplectic form is called *symplectic*. Accordingly, a manifold equipped with a symplectic form is a *symplectic manifold*. A symplectic manifold (Z,Ω) is orientable. It is usually oriented so that $\overset{m}{\wedge}\Omega$ is a volume form on Z, i.e., it defines a positive measure on Z.

A manifold morphism ζ of a symplectic manifold (Z,Ω) to a symplectic manifold (Z',Ω') is called a *symplectic morphism* if $\Omega = \zeta^*\Omega'$. Any symplectic morphism is an immersion. A symplectic isomorphism is sometimes called a *symplectomorphism* [104].

A vector field u on a symplectic manifold (Z,Ω) is an infinitesimal generator of a local one-parameter group of symplectic local automorphisms if and only if the Lie derivative $\mathbf{L}_u\Omega$ vanishes. It is called the *canonical vector field*. A canonical vector field u on a symplectic manifold (Z,Ω) is said to be *Hamiltonian* if the closed one-form $u\rfloor\Omega$ is exact. Any smooth function $f \in C^\infty(Z)$ on Z defines a unique Hamiltonian vector field ϑ_f, called the *Hamiltonian vector field of a function* f such that

$$\vartheta_f\rfloor\Omega = -df, \qquad \vartheta_f = \Omega^\sharp(df), \tag{3.1.2}$$

where $\Omega^\sharp$ is the inverse isomorphism to $\Omega^\flat$ (3.1.1).

Remark 3.1.1. There is another convention [1], where a Hamiltonian vector field differs in the minus sign from (3.1.2).

Example 3.1.1. Given an m-dimensional manifold M coordinated by (q^i), let

$$\pi_{*M} : T^*M \to M$$

be its cotangent bundle equipped with the holonomic coordinates $(q^i, p_i = \dot{q}_i)$. It is endowed with the *canonical Liouville form*

$$\Xi = p_i dq^i$$

and the *canonical symplectic form*

$$\Omega_T = d\Xi = dp_i \wedge dq^i. \tag{3.1.3}$$

Their coordinate expressions are maintained under holonomic coordinate transformations. The Hamiltonian vector field ϑ_f (3.1.2) with respect to the canonical symplectic form (3.1.3) reads

$$\vartheta_f = \partial^i f \partial_i - \partial_i f \partial^i. \tag{3.1.4}$$

Of course, Ω_T (3.1.3) is not a unique symplectic form on the cotangent bundle T^*M. Given a closed two-form ϕ on a manifold M and its pull-back $\pi^*_{*M}\phi$ onto T^*M, the form

$$\Omega_\phi = \Omega + \pi^*_{*M}\phi \tag{3.1.5}$$

also is a symplectic form on T^*M.

The canonical symplectic form (3.1.3) plays a prominent role in symplectic geometry in view of the classical *Darboux theorem* [104].

Theorem 3.1.1. *Each point of a symplectic manifold (Z, Ω) has an open neighborhood equipped with coordinates (q^i, p_i), called canonical or* Darboux *coordinates, such that Ω takes the coordinate form (3.1.3).*

One defines the following special submanifolds of a symplectic manifold.

Let $i_N : N \to Z$ be a submanifold of a symplectic manifold (Z, Ω). The subset

$$\mathrm{Orth}_\Omega TN = \bigcup_{z \in N} \{v \in T_z Z : v \rfloor u \rfloor \Omega = 0, \ u \in T_z N\} \tag{3.1.6}$$

of $TZ|_N$ is called *orthogonal* to TN relative to the symplectic form Ω or, simply, the *Ω-orthogonal space to TN*. There are the following bijections

$$\mathrm{Orth}_\Omega(\mathrm{Orth}_\Omega TN) = TN \subset TZ|_N,$$
$$\Omega^\flat(\mathrm{Orth}_\Omega TN) = \mathrm{Ann}\, TN \subset T^*Z|_N,$$
$$\mathrm{Ann}\,(\mathrm{Orth}_\Omega TN) = \Omega^\flat(TN) \subset T^*Z|_N.$$

If N_1 and N_2 are two submanifolds of Z, then $TN_1 \subset TN_2$ implies

$$\mathrm{Orth}_\Omega TN_1 \supset \mathrm{Orth}_\Omega TN_2$$

over $N_1 \cap N_2$, and *vice versa*. We also have

$$\mathrm{Orth}_\Omega(TN_1 \cap TN_2) = \mathrm{Orth}_\Omega TN_1|_{N_1\cap N_2} + \mathrm{Orth}_\Omega TN_2|_{N_1\cap N_2},$$
$$TN \cap \mathrm{Orth}_\Omega TN = \mathrm{Orth}_\Omega(\mathrm{Orth}_\Omega TN + TN).$$

It should be emphasized that

$$TN \cap \mathrm{Orth}_\Omega TN \neq 0, \qquad TZ|_N \neq TN + \mathrm{Orth}_\Omega TN,$$

in general. The first set is exactly the kernel of the pull-back $\Omega_N = i_N^*\Omega$ of the symplectic form Ω onto a submanifold N.

As was mentioned above, one considers the following special types of submanifolds of a symplectic manifold such that this pull-back Ω_N is of constant rank. A submanifold N of Z is said to be:

- *coisotropic* if $\mathrm{Orth}_\Omega TN \subseteq TN$, $\dim N \geq m$;
- *symplectic* if Ω_N is a symplectic form on N;
- *isotropic* if $TN \subseteq \mathrm{Orth}_\Omega TN$, $\dim N \leq m$;
- *Lagrangian* if N is both coisotropic and isotropic, i.e., $\mathrm{Orth}_\Omega N = TN$, $\dim N = m$.

Clearly, $\Omega_N = 0$ if N is isotropic. A one-dimensional submanifold always is isotropic, while that of codimension 1 is coisotropic.

3.1.2 *Presymplectic manifolds*

A two-form ω on a manifold Z is said to be *presymplectic* if it is closed, but not necessarily non-degenerate. A manifold equipped with a presymplectic form is called *presymplectic*.

Example 3.1.2. Let (Z, Ω) be a symplectic manifold and $i_N : N \to Z$ its coisotropic submanifold. Then $i_N^*\Omega$ is a presymplectic form on N.

The kernel $\mathrm{Ker}\,\omega$ of a presymplectic form ω of constant rank is an involutive distribution, called the *characteristic distribution* [104]. It defines the *characteristic foliation* of a presymplectic manifold (Z, ω). The pull-back of the presymplectic form ω onto any leaf of this foliation equals zero.

The notion of a Hamiltonian vector field on a symplectic manifold is extended in a straightforward manner to a presymplectic manifold. However, a function on a presymplectic manifold need not admit an associated Hamiltonian vector field.

Any presymplectic form has a *symplectic realization*, i.e., can be represented as the pull-back of a symplectic form. Indeed, a presymplectic form ω on a manifold Z is the pull-back

$$\omega = \widehat{0}^*\Omega_\omega = \widehat{0}^*(\Omega + \pi_{*Z}^*\omega)$$

of the symplectic form Ω_ω (3.1.5) on the cotangent bundle T^*Z of Z by its zero section $\widehat{0}$. It is easily justified that the zero section $\widehat{0}(Z) \subset T^*Z$ is a coisotropic submanifold with respect to the symplectic form Ω_ω on T^*Z. Therefore, the morphism $\widehat{0}$ of the presymplectic manifold (Z, ω) into the symplectic manifold (T^*Z, Ω_ω) exemplifies the *coisotropic imbedding*. This construction can be refined as follows.

If a presymplectic form is of constant rank, it admits the following symplectic realization [72].

Proposition 3.1.1. *Given a presymplectic manifold (Z, ω) where ω is of constant rank, there exists a symplectic form on a tubular neighborhood of the zero section $\widehat{0}$ of the dual bundle $(\operatorname{Ker}\omega)^*$ to the characteristic distribution $\operatorname{Ker}\omega \to Z$ such that (Z, ω) can be coisotropically imbedded onto $\widehat{0}(Z)$.*

If the characteristic foliation of a presymplectic form is simple, there is another important variant of symplectic realization, namely, along the leaves of this foliation [73].

Proposition 3.1.2. *Let a presymplectic form ω on a manifold Z be of constant rank, and let its characteristic foliation be simple, i.e., a fibred manifold $\pi : Z \to P$. Then the base P of this fibred manifold is equipped with a symplectic form Ω such that ω is the pull-back of Ω by π.*

3.1.3 *Poisson manifolds*

A *Poisson bracket* on the ring $C^\infty(Z)$ of smooth real functions on a manifold Z (or a *Poisson structure* on Z) is defined as an $\mathbb{R}$-bilinear map

$$C^\infty(Z) \times C^\infty(Z) \ni (f, g) \to \{f, g\} \in C^\infty(Z)$$

which satisfies the following conditions:

- $\{g, f\} = -\{f, g\}$;
- $\{f, \{g, h\}\} + \{g, \{h, f\}\} + \{h, \{f, g\}\} = 0$, called the *Jacobi identity*;
- $\{h, fg\} = \{h, f\}g + f\{h, g\}$.

A manifold Z endowed with a Poisson structure is called a *Poisson manifold*. A Poisson bracket makes $C^\infty(Z)$ into a real Lie algebra, called the *Poisson algebra*. A Poisson structure is characterized by a particular bivector field as follows.

Theorem 3.1.2. *Every Poisson bracket on a manifold Z is uniquely* *defined as*

$$\{f, f'\} = w(df, df') = w^{\mu\nu}\partial_\mu f \partial_\nu f' \tag{3.1.7}$$

by a bivector field w whose Schouten–Nijenhuis bracket $[w, w]_{\rm SN}$ vanishes. It is called a Poisson bivector field [157].

Example 3.1.3. Any manifold admits the *zero Poisson structure* characterized by the zero Poisson bivector field $w = 0$.

Example 3.1.4. Let vector fields u and v on a manifold Z mutually commute. Then $u \wedge v$ is a Poisson bivector field.

A function $f \in C^\infty(Z)$ is called the *Casimir function* of a Poisson structure on X if its Poisson bracket with any function on Z vanishes. Casimir functions form a real ring $\mathcal{C}(Z)$. Obviously, the Poisson algebra $C^\infty(X)$ also is a Lie $\mathcal{C}(Z)$-algebra.

Any bivector field w on a manifold Z yields a linear bundle morphism

$$w^\sharp : T^*Z \underset{Z}{\to} TZ, \qquad w^\sharp : \alpha \to -w(z)\lfloor\alpha, \qquad \alpha \in T^*_z Z. \tag{3.1.8}$$

One says that w is of *rank* r if the morphism (3.1.8) is of this rank. If a Poisson bivector field is of constant rank, the Poisson structure is called *regular*. Throughout the book, only regular Poisson structures are considered. A Poisson structure determined by a Poisson bivector field w is said to be *non-degenerate* if w is of maximal rank.

Remark 3.1.2. The morphism (3.1.8) is naturally generalized to the homomorphism of graded commutative algebras $\mathcal{O}^*(Z) \to \mathcal{T}_*(Z)$ in accordance with the relation

$$w^\sharp(\phi)(\sigma_1, \ldots, \sigma_r) = (-1)^r \phi(w^\sharp(\sigma_1), \ldots, w^\sharp(\sigma_r)),$$
$$\phi \in \mathcal{O}^r(Z), \qquad \sigma_i \in \mathcal{O}^1(Z).$$

It is an isomorphism if the bivector field w is non-degenerate.

There is one-to-one correspondence $\Omega_w \leftrightarrow w_\Omega$ between the almost symplectic forms and the non-degenerate bivector fields which is given by the equalities

$$w_\Omega(\phi, \sigma) = \Omega_w(w^\sharp_\Omega(\phi), w^\sharp_\Omega(\sigma)), \qquad \phi, \sigma \in \mathcal{O}^1(Z), \tag{3.1.9}$$

$$\Omega_w(\vartheta, \nu) = w_\Omega(\Omega^\flat_w(\vartheta), \Omega^\flat_w(\nu)), \qquad \vartheta, \nu \in \mathcal{T}(Z), \tag{3.1.10}$$

where the morphisms $w^\sharp_\Omega$ (3.1.8) and $\Omega^\flat_w$ (3.1.1) are mutually inverse, i.e.,

$$w^\sharp_\Omega = \Omega^\sharp_w, \qquad w^{\alpha\nu}_\Omega \Omega_{w\alpha\beta} = \delta^\nu_\beta.$$

Furthermore, one can show that there is one-to-one correspondence between the symplectic forms and the non-degenerate Poisson bivector fields. However, this correspondence is not preserved under manifold morphisms in general.

Namely, let (Z_1, w_1) and (Z_2, w_2) be Poisson manifolds. A manifold morphism $\varrho : Z_1 \to Z_2$ is said to be a *Poisson morphism* if

$$\{f \circ \varrho, f' \circ \varrho\}_1 = \{f, f'\}_2 \circ \varrho, \qquad f, f' \in C^\infty(Z_2),$$

or, equivalently, if

$$w_2 = T\varrho \circ w_1,$$

where $T\varrho$ is the tangent map to ϱ. Herewith, the rank of w_1 is superior or equal to that of w_2. Therefore, there are no pull-back and push-forward operations of Poisson structures in general. Nevertheless, let us mention the following construction [157].

Theorem 3.1.3. *Let (Z, w) be a Poisson manifold and $\pi : Z \to Y$ a fibration such that, for every pair of functions (f, g) on Y and for each point $y \in Y$, the restriction of the function $\{\pi^* f, \pi^* g\}$ to the fibre $\pi^{-1}(y)$ is constant, i.e., $\{\pi^* f, \pi^* g\}$ is the pull-back onto Z of some function on Y. Then there exists a coinduced Poisson structure w' on Y for which π is a Poisson morphism.*

Example 3.1.5. The direct product $Z \times Z'$ of Poisson manifolds (Z, w) and (Z', w') can be endowed with the *product of Poisson structures*, given by the bivector field $w + w'$ such that the surjections pr_1 and pr_2 are Poisson morphisms.

Example 3.1.6. Let (Z_1, Ω_1) and (Z_2, Ω_2) be symplectic manifolds equipped with the associated non-degenerate Poisson structures w_1 and w_2. If $\dim Z_1 > \dim Z_2$, a Poisson morphism $\varrho : Z_1 \to Z_2$ need not be a symplectic one, i.e., $w_2 = T\varrho \circ w_1$ and $\Omega_1 \neq \varrho^* \Omega_2$.

A vector field u on a Poisson manifold (Z, w) is an infinitesimal generator of a local one-parameter group of Poisson automorphisms if and only if the Lie derivative

$$\mathbf{L}_u w = [u, w]_{\mathrm{SN}} \tag{3.1.11}$$

vanishes. It is called the *canonical vector field* for the Poisson structure w. In particular, for any real smooth function f on a Poisson manifold (Z, w), let us put

$$\vartheta_f = w^\sharp(df) = -w\lfloor df = w^{\mu\nu}\partial_\mu f \partial_\nu. \tag{3.1.12}$$

It is a canonical vector field, called the *Hamiltonian vector field of a function* f with respect to the Poisson structure w. Hamiltonian vector fields fulfil the relations

$$\{f,g\} = \vartheta_f \rfloor dg, \tag{3.1.13}$$

$$[\vartheta_f, \vartheta_g] = \vartheta_{\{f,g\}}, \qquad f,g \in C^\infty(Z). \tag{3.1.14}$$

For instance, the Hamiltonian vector field ϑ_f (3.1.2) of a function f on a symplectic manifold (Z,Ω) coincides with that (3.1.12) with respect to the corresponding Poisson structure w_Ω. The Poisson bracket defined by a symplectic form Ω reads

$$\{f,g\} = \vartheta_g \rfloor \vartheta_f \rfloor \Omega.$$

Since a Poisson manifold (Z,w) is assumed to be regular, the range $\mathbf{T} = w^\sharp(T^*Z)$ of the morphism (3.1.8) is a subbundle of TZ called the *characteristic distribution* on (Z,w). It is spanned by Hamiltonian vector fields, and it is involutive by virtue of the relation (3.1.14). It follows that a Poisson manifold Z admits local adapted coordinates in Theorem 11.2.14. Moreover, one can choose particular adapted coordinates which bring a Poisson structure into the following canonical form [157].

Theorem 3.1.4. *For any point z of a Poisson manifold (Z,w), there exist coordinates*

$$(z^1,\ldots,z^{k-2m},q^1,\ldots,q^m,p_1,\ldots,p_m) \tag{3.1.15}$$

on a neighborhood of z such that

$$w = \frac{\partial}{\partial p_i} \wedge \frac{\partial}{\partial q^i}, \qquad \{f,g\} = \frac{\partial f}{\partial p_i}\frac{\partial g}{\partial q^i} - \frac{\partial f}{\partial q^i}\frac{\partial g}{\partial p_i}. \tag{3.1.16}$$

The coordinates (3.1.15) are called the *canonical* or *Darboux coordinates* for the Poisson structure w. The Hamiltonian vector field of a function f written in this coordinates is

$$\vartheta_f = \partial^i f \partial_i - \partial_i f \partial^i.$$

Of course, the canonical coordinates for a symplectic form Ω in Theorem 3.1.1 also are canonical coordinates in Theorem 3.1.4 for the corresponding non-degenerate Poisson bivector field w, i.e.,

$$\Omega = dp_i \wedge dq^i, \qquad w = \partial^i \wedge \partial_i.$$

With respect to these coordinates, the mutually inverse bundle isomorphisms $\Omega^\flat$ (3.1.1) and $w^\sharp$ (3.1.8) read

$$\Omega^\flat : v^i\partial_i + v_i\partial^i \to -v_i dq^i + v^i dp_i,$$

$$w^\sharp : v_i dq^i + v^i dp_i \to v^i\partial_i - v_i\partial^i.$$

Given a Poisson manifold (Z, w) and its characteristic distribution $\mathbf{T}$, the above mentioned notions of coisotropic and Lagrangian submanifolds of a symplectic manifold are generalized to a Poisson manifold as follows. A submanifold N of a Poisson manifold is said to be:

- *coisotropic* if $w^\sharp(\operatorname{Ann} TN) \subseteq TN$,
- *Lagrangian* if $w^\sharp(\operatorname{Ann} TN) = TN \cap \mathbf{T}$.

Integral manifolds of the characteristic distribution $\mathbf{T}$ of a Poisson manifold (Z, w) constitute a (regular) foliation $\mathcal{F}$ of Z whose tangent bundle $T\mathcal{F}$ is $\mathbf{T}$. It is called the *characteristic foliation of a Poisson manifold*. By the very definition of the characteristic distribution $\mathbf{T} = T\mathcal{F}$, the Poisson bivector field w is subordinate to $\overset{2}{\wedge} T\mathcal{F}$. Therefore, its restriction $w|_F$ to any leaf F of $\mathcal{F}$ is a non-degenerate Poisson bivector field on F. It provides F with a non-degenerate Poisson structure $\{,\}_F$ and, consequently, a symplectic structure. Clearly, the local Darboux coordinates for the Poisson structure w in Theorem 3.1.4 also are the local adapted coordinates

$$(z^1, \ldots, z^{k-2m}, z^i = q^i, z^{m+i} = p_i), \qquad i = 1, \ldots, m,$$

(11.2.65) for the characteristic foliation $\mathcal{F}$, and the symplectic structures along its leaves read

$$\Omega_F = dp_i \wedge dq^i.$$

Remark 3.1.3. Provided with this symplectic structure, the leaves of the characteristic foliation of a Poisson manifold Z are assembled into a symplectic foliation of Z. Moreover, there is one-to-one correspondence between the symplectic foliations of a manifold Z and the Poisson structures on Z (Section 3.1.5).

Since any foliation is locally simple, a local structure of an arbitrary Poisson manifold reduces to the following [157; 163].

Theorem 3.1.5. *Each point of a Poisson manifold has an open neighborhood which is Poisson equivalent to the product of a manifold with the zero Poisson structure and a symplectic manifold.*

Let (Z, w) be a Poisson manifold. By its *symplectic realization* is meant a symplectic manifold (Z', Ω) together with a Poisson morphism $Z' \to Z$ which is a surjective submersion.

Theorem 3.1.6. *Each point of a Poisson manifold has an open neighborhood which is realizable by a symplectic manifold.*

Proof. In local Darboux coordinates, this symplectic realization is described as follows. The Poisson structure given by the Poisson bracket (3.1.16) with respect to the canonical coordinates is coinduced from the symplectic structure given by the symplectic form

$$\Omega = dp_i \wedge dq^i + d\overline{z}_\lambda \wedge dz^\lambda$$

with respect to the coordinates

$$(z^1, \ldots, z^{k-2m}, \overline{z}_1, \ldots, \overline{z}_{k-2m}, q^1, \ldots, q^m, p_1, \ldots, p_m)$$

by the surjection

$$(z^\lambda, \overline{z}_\lambda, q^i, p_i) \to (z^\lambda, q^i, p_i). \qquad \square$$

Remark 3.1.4. It follows from Theorem 3.1.5 that each point of a Poisson manifold has an open neighborhood which is a presymplectic manifold with respect to the presymplectic form

$$\Omega = dp_i \wedge dq^i,$$

written relative to the local Darboux coordinates (z^λ, q^i, p_i). Moreover, let the direct product in Theorem 3.1.5 be global, i.e., a Poisson manifold (Z, w) is the Poisson product $Z = P \times Y$ of a symplectic manifold (P, Ω) and a manifold Y with the zero Poisson structure. Then Z is provided with the presymplectic form $\mathrm{pr}_1^*\Omega$. Conversely, let the characteristic foliation $\pi : Z \to P$ of a presymplectic form ω on a manifold Z in Proposition 3.1.2 be a trivial bundle $Z = P \times Y$. Then Z is a Poisson manifold given by the Poisson product of the symplectic manifold (P, Ω) and Y equipped with the zero Poisson structure.

3.1.4 *Lichnerowicz–Poisson cohomology*

Given a Poisson manifold (Z, w), let us introduce the operator

$$\widehat{w} : \mathcal{T}_r(Z) \to \mathcal{T}_{r+1}(Z), \qquad \widehat{w}(\vartheta) = -[w, \vartheta], \qquad \vartheta \in \mathcal{T}_*(Z), \tag{3.1.17}$$

on the graded commutative algebra $\mathcal{T}_*(Z)$ of multivector fields on Z, where $[.,.]$ is the Schouten–Nijenhuis bracket. This operator is nilpotent and obeys the rule

$$\widehat{w}(\vartheta \wedge \upsilon) = \widehat{w}(\vartheta) \wedge \upsilon + (-1)^{|\vartheta|}\vartheta \wedge \widehat{w}(\upsilon). \tag{3.1.18}$$

Called the *contravariant exterior differential* [157], it makes $\mathcal{T}_*(Z)$ into a differential algebra. Its de Rham complex is the *Lichnerowicz–Poisson complex*

$$\begin{aligned} 0 \to S_{\mathcal{F}}(Z) \longrightarrow C^\infty(Z) \overset{\widehat{w}}{\longrightarrow} \mathcal{T}_1(Z) \overset{\widehat{w}}{\longrightarrow} \cdots \\ \mathcal{T}_{r-1}(Z) \overset{\widehat{w}}{\longrightarrow} \mathcal{T}_r(Z) \overset{\widehat{w}}{\longrightarrow} \cdots, \end{aligned} \tag{3.1.19}$$

where $S_{\mathcal{F}}(Z)$ denotes the center of a Poisson algebra $C^\infty(Z)$. Accordingly, the cohomology $H^*_{\rm LP}(Z,w)$ of this complex is called the *Lichnerowicz–Poisson cohomology* (henceforth the *LP cohomology*) of a Poisson manifold.

Example 3.1.7. If $f \in \mathcal{T}_0(Z) = C^\infty(Z)$ is a function,

$$-\widehat{w}(f) = [w,f] = \vartheta_f$$

is its Hamiltonian vector field. Hence, the LP cohomology group $H^0_{\rm LP}(Z,w)$ coincides with the center $S_{\mathcal{F}}$ of the Poisson algebra $C^\infty(Z)$. The first LP cohomology group $H^1_{\rm LP}(Z,w)$ is the space of canonical vector fields u for the Poisson bivector field w (i.e., $\mathbf{L}_u w = -\widehat{w}(u) = 0$) modulo Hamiltonian vector fields $-\widehat{w}(f)$, $f \in C^\infty(Z)$. The second LP cohomology group $H^2_{\rm LP}(Z,w)$ contains an element $[w]$ whose representative is the Poisson bivector field w. We have $[w]=0$ if there is a vector field u on Z such that

$$w = \widehat{w}(u) = -\mathbf{L}_u w.$$

If $[w]=0$, a Poisson manifold (Z,w) is called *exact* or *homogeneous*.

The contravariant exterior differential $\widehat{w}$ is related to the exterior differential by means of the formula

$$\widehat{w}(w^\sharp(\phi)) = -w^\sharp(d\phi), \qquad \phi \in \mathcal{O}^*(Z).$$

This formula shows that $w^\sharp$ is a cochain homomorphism of the de Rham complex $(\mathcal{O}^*(Z),d)$ of exterior forms on Z to the Lichnerowicz–Poisson complex $(\mathcal{T}_*,-\widehat{w})$ (3.1.19). It yields the homomorphism

$$[w^\sharp] : H^*_{\rm DR}(Z) \to H^*_{\rm LP}(Z,w) \tag{3.1.20}$$

of the de Rham cohomology to the LP cohomology.

3.1.5 *Symplectic foliations*

There is above-mentioned one-to-one correspondence between the symplectic foliations of a manifold Z and the Poisson structures on Z. We start with some basic facts on geometry and cohomology of foliations.

Let $\mathcal{F}$ be a (regular) foliation of a k-dimensional manifold Z provided with the adapted coordinate atlas (11.2.65). The real Lie algebra $\mathcal{T}_1(\mathcal{F})$ of global sections of the tangent bundle $T\mathcal{F} \to Z$ to $\mathcal{F}$ is a $C^\infty(Z)$-submodule of the derivation module of the $\mathbb{R}$-ring $C^\infty(Z)$ of smooth real functions on Z. Its kernel $S_{\mathcal{F}}(Z) \subset C^\infty(Z)$ consists of functions constant on leaves of

$\mathcal{F}$. Therefore, $\mathcal{T}_1(\mathcal{F})$ is the Lie $S_{\mathcal{F}}(Z)$-algebra of derivations of $C^\infty(Z)$, regarded as a $S_{\mathcal{F}}(Z)$-ring. Then one can introduce the *leafwise differential calculus* [58; 65] as the Chevalley–Eilenberg differential calculus over the $S_{\mathcal{F}}(Z)$-ring $C^\infty(Z)$. It is defined as a subcomplex

$$0 \to S_{\mathcal{F}}(Z) \longrightarrow C^\infty(Z) \xrightarrow{\widetilde{d}} \mathfrak{F}^1(Z) \cdots \xrightarrow{\widetilde{d}} \mathfrak{F}^{\dim \mathcal{F}}(Z) \to 0 \qquad (3.1.21)$$

of the Chevalley–Eilenberg complex of the Lie $S_{\mathcal{F}}(Z)$-algebra $\mathcal{T}_1(\mathcal{F})$ with coefficients in $C^\infty(Z)$ which consists of $C^\infty(Z)$-multilinear skew-symmetric maps

$$\overset{r}{\times} \mathcal{T}_1(\mathcal{F}) \to C^\infty(Z), \qquad r = 1, \ldots, \dim \mathcal{F}.$$

These maps are global sections of exterior products $\overset{r}{\wedge} T\mathcal{F}^*$ of the dual $T\mathcal{F}^* \to Z$ of $T\mathcal{F} \to Z$. They are called the *leafwise forms* on a foliated manifold $(Z, \mathcal{F})$, and are given by the coordinate expression

$$\phi = \frac{1}{r!}\phi_{i_1\ldots i_r}\widetilde{d}z^{i_1} \wedge \cdots \wedge \widetilde{d}z^{i_r},$$

where $\{\widetilde{d}z^i\}$ are the duals of the holonomic fibre bases $\{\partial_i\}$ for $T\mathcal{F}$. Then one can think of the Chevalley–Eilenberg coboundary operator

$$\widetilde{d}\phi = \widetilde{d}z^k \wedge \partial_k \phi = \frac{1}{r!}\partial_k \phi_{i_1\ldots i_r}\widetilde{d}z^k \wedge \widetilde{d}z^{i_1} \wedge \cdots \wedge \widetilde{d}z^{i_r}$$

as being the *leafwise exterior differential*. Accordingly, the complex (3.1.21) is called the *leafwise de Rham complex* (or the *tangential de Rham complex*). This is the complex $(\mathcal{A}^{0,*}, d_f)$ in [155]. Its cohomology $H^*_{\mathcal{F}}(Z)$, called the *leafwise de Rham cohomology*, equals the cohomology $H^*(Z; S_{\mathcal{F}})$ of Z with coefficients in the sheaf $S_{\mathcal{F}}$ of germs of elements of $S_{\mathcal{F}}(Z)$ [119]. We aim to relate the leafwise de Rham cohomology $H^*_{\mathcal{F}}(Z)$ with the de Rham cohomology $H^*_{\rm DR}(Z)$ of Z and the LP cohomology $H^*_{\rm LP}(Z, w)$ [58].

Let us consider the exact sequence (11.2.67) of vector bundles over Z. Since it admits a splitting, the epimorphism $i^*_{\mathcal{F}}$ yields that of the algebra $\mathcal{O}^*(Z)$ of exterior forms on Z to the algebra $\mathfrak{F}^*(Z)$ of leafwise forms. It obeys the condition $i^*_{\mathcal{F}} \circ d = \widetilde{d} \circ i^*_{\mathcal{F}}$, and provides the cochain morphism

$$\begin{aligned} &i^*_{\mathcal{F}} : (\mathbb{R}, \mathcal{O}^*(Z), d) \to (S_{\mathcal{F}}(Z), \mathcal{F}^*(Z), \widetilde{d}), \\ &dz^\lambda \to 0, \quad dz^i \to \widetilde{d}z^i, \end{aligned} \qquad (3.1.22)$$

of the de Rham complex of Z to the leafwise de Rham complex (3.1.21) and the corresponding homomorphism

$$[i^*_{\mathcal{F}}]^* : H^*_{\rm DR}(Z) \to H^*_{\mathcal{F}}(Z) \qquad (3.1.23)$$

of the de Rham cohomology of Z to the leafwise one. Let us note that $[i^*_{\mathcal{F}}]^{r>0}$ need not be epimorphisms [155].

Given a leaf $i_F : F \to Z$ of $\mathcal{F}$, we have the pull-back homomorphism

$$(\mathbb{R}, \mathcal{O}^*(Z), d) \to (\mathbb{R}, \mathcal{O}^*(F), d) \tag{3.1.24}$$

of the de Rham complex of Z to that of F and the corresponding homomorphism of the de Rham cohomology groups

$$H^*_{\rm DR}(Z) \to H^*_{\rm DR}(F). \tag{3.1.25}$$

Proposition 3.1.3. *The homomorphisms (3.1.24) – (3.1.25) factorize through the homomorphisms (3.1.22) – (3.1.23) [65].*

Let us turn now to symplectic foliations. Let $\mathcal{F}$ be an even dimensional foliation of a manifold Z. A $\widetilde{d}$-closed non-degenerate leafwise two-form $\Omega_{\mathcal{F}}$ on a foliated manifold $(Z, \mathcal{F})$ is called *symplectic*. Its pull-back $i^*_F\Omega_{\mathcal{F}}$ onto each leaf F of $\mathcal{F}$ is a symplectic form on F. A foliation $\mathcal{F}$ provided with a symplectic leafwise form $\Omega_{\mathcal{F}}$ is called the *symplectic foliation*.

If a symplectic leafwise form $\Omega_{\mathcal{F}}$ exists, it yields the bundle isomorphism

$$\Omega^\flat_{\mathcal{F}} : T\mathcal{F} \underset{Z}{\to} T\mathcal{F}^*, \qquad \Omega^\flat_{\mathcal{F}} : v \to -v\rfloor\Omega_{\mathcal{F}}(z), \qquad v \in T_z\mathcal{F}. \tag{3.1.26}$$

The inverse isomorphism $\Omega^\sharp_{\mathcal{F}}$ determines the bivector field

$$w_\Omega(\alpha, \beta) = \Omega_{\mathcal{F}}(\Omega^\sharp_{\mathcal{F}}(i^*_{\mathcal{F}}\alpha), \Omega^\sharp_{\mathcal{F}}(i^*_{\mathcal{F}}\beta)), \qquad \alpha, \beta \in T^*_zZ, \quad z \in Z, \tag{3.1.27}$$

on Z subordinate to $\overset{2}{\wedge} T\mathcal{F}$. It is a Poisson bivector field (see the relation (3.1.34) below). The corresponding Poisson bracket reads

$$\{f, f'\}_{\mathcal{F}} = \vartheta_f\rfloor\widetilde{d}f', \qquad \vartheta_f\rfloor\Omega_{\mathcal{F}} = -\widetilde{d}f, \qquad \vartheta_f = \Omega^\sharp_{\mathcal{F}}(\widetilde{d}f). \tag{3.1.28}$$

Its kernel is $S_{\mathcal{F}}(Z)$.

Conversely, let (Z, w) be a Poisson manifold and $\mathcal{F}$ its characteristic foliation. Since $\operatorname{Ann} T\mathcal{F} \subset T^*Z$ is precisely the kernel of a Poisson bivector field w, the bundle homomorphism

$$w^\sharp : T^*Z \underset{Z}{\to} TZ$$

factorizes in a unique fashion

$$w^\sharp : T^*Z \xrightarrow[Z]{i^*_{\mathcal{F}}} T\mathcal{F}^* \xrightarrow[Z]{w^\sharp_{\mathcal{F}}} T\mathcal{F} \xrightarrow[Z]{i_{\mathcal{F}}} TZ \tag{3.1.29}$$

through the bundle isomorphism

$$w^\sharp_{\mathcal{F}} : T\mathcal{F}^* \underset{Z}{\to} T\mathcal{F}, \qquad w^\sharp_{\mathcal{F}} : \alpha \to -w(z)\lfloor\alpha, \qquad \alpha \in T_z\mathcal{F}^*. \tag{3.1.30}$$

The inverse isomorphism $w^\flat_{\mathcal{F}}$ yields the symplectic leafwise form

$$\Omega_{\mathcal{F}}(v,v') = w(w^\flat_{\mathcal{F}}(v), w^\flat_{\mathcal{F}}(v')), \qquad v,v' \in T_z\mathcal{F}, \qquad z \in Z. \tag{3.1.31}$$

The formulas (3.1.27) and (3.1.31) establish the above mentioned equivalence between the Poisson structures on a manifold Z and its symplectic foliations, though this equivalence need not be preserved under morphisms.

Let us consider the Lichnerowicz–Poisson complex 3.1.19. We have the cochain morphism

$$\begin{aligned}
&w^\sharp : (\mathbb{R}, \mathcal{O}^*(Z), d) \to (S_{\mathcal{F}}(Z), \mathcal{T}_*(Z), -\widehat{w}), \qquad (3.1.32)\\
&w^\sharp(\phi)(\sigma_1, \ldots, \sigma_r) = (-1)^r \phi(w^\sharp(\sigma_1), \ldots, w^\sharp(\sigma_r)), \quad \sigma_i \in \mathcal{O}^1(Z),\\
&\widehat{w} \circ w^\sharp = -w^\sharp \circ d,
\end{aligned}$$

of the de Rham complex to the Lichnerowicz–Poisson one and the corresponding homomorphism (3.1.20) of the de Rham cohomology of Z to the LP cohomology of the complex (3.1.31) [157].

Proposition 3.1.4. *The cochain morphism $w^\sharp$ (3.1.32) factorizes through the leafwise complex (3.1.21) and, accordingly, the cohomology homomorphism $[w^\sharp]$ (3.1.20) does through the leafwise cohomology*

$$H^*_{\rm DR}(Z) \xrightarrow{[i^*_{\mathcal{F}}]} H^*_{\mathcal{F}}(Z) \longrightarrow H^*_{\rm LP}(Z,w). \tag{3.1.33}$$

Proof. Let $\mathcal{T}_*(\mathcal{F}) \subset \mathcal{T}_*(Z)$ denote the graded commutative subalgebra of multivector fields on Z subordinate to $T\mathcal{F}$, where $\mathcal{T}_0(\mathcal{F}) = C^\infty(Z)$. Clearly, $(S_{\mathcal{F}}(Z), \mathcal{T}_*(\mathcal{F}), \widehat{w})$ is a subcomplex of the Lichnerowicz–Poisson complex (3.1.19). Since

$$\widehat{w} \circ \Omega^\sharp_{\mathcal{F}} = -\Omega^\sharp_{\mathcal{F}} \circ \widetilde{d}, \tag{3.1.34}$$

the bundle isomorphism $w^\sharp_{\mathcal{F}} = \Omega^\sharp_{\mathcal{F}}$ (3.1.30) yields the cochain isomorphism

$$\Omega^\sharp_{\mathcal{F}} : (S_{\mathcal{F}}(Z), \mathfrak{F}^*(Z), \widetilde{d}) \to (S_{\mathcal{F}}(Z), \mathcal{T}_*(\mathcal{F}), -\widehat{w})$$

of the leafwise de Rham complex (3.1.21) to the subcomplex $(\mathcal{T}_*(\mathcal{F}), \widehat{w})$ of the Lichnerowicz–Poisson complex (3.1.19). Then the composition

$$i_{\mathcal{F}} \circ \Omega^\sharp_{\mathcal{F}} : (S_{\mathcal{F}}(Z), \mathfrak{F}^*(Z), \widetilde{d}) \to (S_{\mathcal{F}}(Z), \mathcal{T}_*(Z), -\widehat{w}) \tag{3.1.35}$$

is a cochain monomorphism of the leafwise de Rham complex to the LP one (3.1.19). In view of the factorization (3.1.29), the cochain morphism (3.1.32) factorizes through the cochain morphisms (3.1.22) and (3.1.35). Accordingly, the cohomology homomorphism $[w^\sharp]$ (3.1.20) factorizes through the cohomology homomorphisms $[i^*_{\mathcal{F}}]$ (3.1.23) and

$$[i_{\mathcal{F}} \circ \Omega^\sharp_{\mathcal{F}}] : H^*_{\mathcal{F}}(Z) \to H^*_{\rm LP}(Z,w). \tag{3.1.36}$$

□

3.1.6 *Group action on Poisson manifolds*

By G throughout is meant a real connected Lie group, $\mathfrak{g}$ is its right Lie algebra, and $\mathfrak{g}^*$ is the Lie coalgebra (see Section 11.2.9). We start with the symplectic case [1; 104].

Let a Lie group G act on a symplectic manifold (Z, Ω) on the left by symplectomorphisms. Such an action of G is called *symplectic*. Since G is connected, its action on a manifold Z is symplectic if and only if the homomorphism $\varepsilon \to \xi_\varepsilon$, $\varepsilon \in \mathfrak{g}$, (11.2.69) of the Lie algebra $\mathfrak{g}$ to the Lie algebra $\mathcal{T}_1(Z)$ of vector fields on Z is carried out by canonical vector fields for the symplectic form Ω on Z. If all these vector fields are Hamiltonian, the action of G on Z is called a *Hamiltonian action*. One can show that, in this case, ξ_ε, $\varepsilon \in \mathfrak{g}$, are Hamiltonian vector fields of functions on Z of the following particular type.

Proposition 3.1.5. *An action of a Lie group G on a symplectic manifold Z is Hamiltonian if and only if there exists a mapping*

$$\widehat{J} : Z \to \mathfrak{g}^*, \tag{3.1.37}$$

called the momentum mapping, such that

$$\xi_\varepsilon \rfloor \Omega = -dJ_\varepsilon, \qquad J_\varepsilon(z) = \langle \widehat{J}(z), \varepsilon \rangle, \qquad \varepsilon \in \mathfrak{g}. \tag{3.1.38}$$

The momentum mapping (3.1.37) is defined up to a constant map. Indeed, if $\widehat{J}$ and $\widehat{J}'$ are different momentum mappings for the same symplectic action of G on Z, then

$$d(\langle \widehat{J}(z) - \widehat{J}'(z), \varepsilon \rangle) = 0, \qquad \varepsilon \in \mathfrak{g}.$$

A symplectic manifold provided with a Hamiltonian action of a Lie group is called the *Hamiltonian manifold*.

Given $g \in G$, let us us consider the difference

$$\sigma(g) = \widehat{J}(gz) - \mathrm{Ad}^* g(\widehat{J}(z)), \tag{3.1.39}$$

where $\mathrm{Ad}^* g$ is the coadjoint representation (11.2.72) on Γ^*. One can show (see, e.g., [1]) that the difference (3.1.39) is constant on a symplectic manifold Z and that it fulfils the equality

$$\sigma(gg') = \sigma(g) + \mathrm{Ad}^* g(\sigma(g')). \tag{3.1.40}$$

This equality (3.1.40) is a one-cocycle of cohomology $H^*(G; \mathfrak{g}^*)$ of the group G with coefficients in the Lie coalgebra $\mathfrak{g}^*$ [65; 105]. This cocycle is a coboundary if there exists an element $\mu \in \mathfrak{g}^*$ such that

$$\sigma(g) = \mu - \mathrm{Ad}^* g(\mu). \tag{3.1.41}$$

Let $\widehat{J}'$ be another momentum mapping associated to the same Hamiltonian action of G on Z. Since the difference $\widehat{J} - \widehat{J}'$ is constant on Z, then the difference of the corresponding cocycles $\sigma - \sigma'$ is the coboundary (3.1.41) where $\mu = \widehat{J} - \widehat{J}'$. Thus, a Hamiltonian action of a Lie group G on a symplectic manifold (Z, Ω) defines a cohomology class $[\sigma] \in H^1(G; \mathfrak{g}^*)$ of G.

A momentum mapping $\widehat{J}$ is called *equivariant* if $\sigma(g) = 0$, $g \in G$. It defines the zero cohomology class of the group G.

Example 3.1.8. Let a symplectic form on Z be exact, i.e., $\Omega = d\theta$, and let θ be G-invariant, i.e.,

$$\mathbf{L}_{\xi_\varepsilon}\theta = d(\xi_\varepsilon\rfloor\theta) + \xi_\varepsilon\rfloor\Omega = 0, \qquad \varepsilon \in \mathfrak{g}.$$

Then the momentum mapping $\widehat{J}$ (3.1.37) can be given by the relation

$$\langle \widehat{J}(z), \varepsilon\rangle = (\xi_\varepsilon\rfloor\theta)(z).$$

It is equivariant. In accordance with the relation (11.2.72), it suffices to show that

$$J_\varepsilon(gz) = J_{\mathrm{Ad}\, g^{-1}(\varepsilon)}(z), \qquad (\xi_\varepsilon\rfloor\theta)(gz) = (\xi_{\mathrm{Ad}\, g^{-1}(\varepsilon)}\rfloor\theta)(z).$$

This holds by virtue of the relation (11.2.70). For instance, let T^*Q be a symplectic manifold equipped with the canonical symplectic form Ω_T (3.1.3). Let a left action of a Lie group G on Q have the infinitesimal generators $\tau_m = \varepsilon^i_m(q)\partial_i$. The canonical lift of this action onto T^*Q has the infinitesimal generators (11.2.29):

$$\xi_m = \widetilde{\tau}_m = v e^i_m\partial_i - p_j\partial_i\varepsilon^j_m\partial^i, \tag{3.1.42}$$

and preserves the canonical Liouville form θ on T^*Q. The ξ_m (3.1.42) are Hamiltonian vector fields of the functions $J_m = \varepsilon^i_m(q)p_i$, determined by the equivariant momentum mapping $\widehat{J} = \varepsilon^i_m(q)p_i\varepsilon^m$.

Now a desired Poisson bracket of functions J_ε (3.1.38) is established as follows.

Theorem 3.1.7. *A momentum mapping $\widehat{J}$ associated to a symplectic action of a Lie group G on a symplectic manifold Z obeys the relation*

$$\{J_\varepsilon, J_{\varepsilon'}\} = J_{[\varepsilon,\varepsilon']} - \langle T_e\sigma(\varepsilon'), \varepsilon\rangle \tag{3.1.43}$$

(see, e.g., [1] where the left Lie algebra is utilized and Hamiltonian vector fields differ in the minus sign from those here).

In the case of an equivariant momentum mapping, the relation (3.1.43) leads to a homomorphism

$$\{J_\varepsilon, J_{\varepsilon'}\} = J_{[\varepsilon,\varepsilon']} \tag{3.1.44}$$

of the Lie algebra $\mathfrak{g}$ to the Poisson algebra of functions on a symplectic manifold Z (cf. Proposition 3.1.6 below).

Now let a Lie group G act on a Poisson manifold (Z, w) on the left by Poisson automorphism. This is a *Poisson action*. Since G is connected, its action on a manifold Z is a Poisson action if and only if the homomorphism $\varepsilon \to \xi_\varepsilon$, $\varepsilon \in \mathfrak{g}$, (11.2.69) of the Lie algebra $\mathfrak{g}$ to the Lie algebra $\mathcal{T}_1(Z)$ of vector fields on Z is carried out by canonical vector fields for the Poisson bivector field w, i.e., the condition (3.1.11) holds. The equivalent conditions are

$$\begin{aligned}
&\xi_\varepsilon(\{f,g\}) = \{\xi_\varepsilon(f), g\} + \{f, \xi_\varepsilon(g)\}, \qquad f,g \in C^\infty(Z),\\
&\xi_\varepsilon(\{f,g\}) = [\xi_\varepsilon, \vartheta_f](g) - [\xi_\varepsilon, \vartheta_g](f),\\
&[\xi_\varepsilon, \vartheta_f] = \vartheta_{\xi_\varepsilon(f)},
\end{aligned}$$

where ϑ_f is the Hamiltonian vector field (3.1.12) of a function f.

A Hamiltonian action of G on a Poisson manifold Z is defined similarly to that on a symplectic manifold. Its infinitesimal generators are tangent to leaves of the symplectic foliation of Z, and there is a Hamiltonian action of G on every symplectic leaf. Proposition 3.1.5 together with the notions of a momentum mapping and an equivariant momentum mapping also are extended to a Poisson action. However, the difference σ (3.1.39) is constant only on leaves of the symplectic foliation of Z in general. At the same time, one can say something more on an equivariant momentum mapping (that also is valid for a symplectic action) [157].

Proposition 3.1.6. *An equivariant momentum mapping $\widehat{J}$ (3.1.37) is a Poisson morphism to the Lie coalgebra $\mathfrak{g}^*$, provided with the Lie–Poisson structure (11.2.73).*

3.2 Autonomous Hamiltonian systems

This Section addresses autonomous Hamiltonian systems on Poisson, symplectic and presymplectic manifolds.

3.2.1 *Poisson Hamiltonian systems*

Given a Poisson manifold (Z, w), a *Poisson Hamiltonian system* $(w, \mathcal{H})$ on Z for a *Hamiltonian* $\mathcal{H} \in C^\infty(Z)$ with respect to a Poisson structure w is defined as the set

$$S_{\mathcal{H}} = \bigcup_{z\in Z} \{v \in T_zZ : v - w^\sharp(d\mathcal{H})(z) = 0\}. \tag{3.2.1}$$

By a *solution* of this Hamiltonian system is meant a vector field ϑ on Z which takes its values into $TN \cap S_{\mathcal{H}}$. Clearly, the Poisson Hamiltonian system (3.2.1) has a unique solution which is the Hamiltonian vector field

$$\vartheta_{\mathcal{H}} = w^\sharp(d\mathcal{H}) \tag{3.2.2}$$

of $\mathcal{H}$. Hence, $S_{\mathcal{H}}$ (3.2.1) is an autonomous first order dynamic equation (Definition 1.2.1), called the *Hamilton equation* for the Hamiltonian $\mathcal{H}$ with respect to the Poisson structure w.

Relative to local canonical coordinates (z^λ, q^i, p_i) (3.1.15) for the Poisson structure w on Z and the corresponding holonomic coordinates $(z^\lambda, q^i, p_i, \dot z^\lambda, \dot q^i, \dot p_i)$ on TZ, the Hamilton equation (3.2.1) and the Hamiltonian vector field (3.2.2) take the form

$$\dot q^i = \partial^i\mathcal{H}, \qquad \dot p_i = -\partial_i\mathcal{H}, \qquad \dot z^\lambda = 0, \tag{3.2.3}$$

$$\vartheta_{\mathcal{H}} = \partial^i\mathcal{H}\partial_i - \partial_i\mathcal{H}\partial^i. \tag{3.2.4}$$

Solutions of the Hamilton equation (3.2.3) are integral curves of the Hamiltonian vector field (3.2.4).

Let $(Z, w, \mathcal{H})$ be a Poisson Hamiltonian system. Its *integral of motion* is a smooth function F on Z whose Lie derivative

$$\mathbf{L}_{\vartheta_{\mathcal{H}}} F = \{\mathcal{H}, F\} \tag{3.2.5}$$

along the Hamiltonian vector field $\vartheta_{\mathcal{H}}$ (3.2.4) vanishes in accordance with the equality (1.10.6). The equality (3.2.5) is called the *evolution equation.*

It is readily observed that the Poisson bracket $\{F, F'\}$ of any two integrals of motion F and F' also is an integral of motion. Consequently, the integrals of motion of a Poisson Hamiltonian system constitute a real Lie algebra.

Since

$$\vartheta_{\{\mathcal{H},F\}} = [\vartheta_{\mathcal{H}}, \vartheta_F], \qquad \{\mathcal{H}, F\} = -\mathbf{L}_{\vartheta_F}\mathcal{H},$$

the Hamiltonian vector field ϑ_F of any integral of motion F of a Poisson Hamiltonian system is a symmetry both of the Hamilton equation (3.2.3) (Proposition 1.10.3) and a Hamiltonian $\mathcal{H}$ (Definition 1.10.3).

3.2.2 *Symplectic Hamiltonian systems*

Let (Z,Ω) be a symplectic manifold. The notion of a symplectic Hamiltonian system is a repetition of the Poisson one, but all expressions are rewritten in terms of a symplectic form Ω as follows.

A *symplectic Hamiltonian system* $(\Omega,\mathcal{H})$ on a manifold Z for a Hamiltonian $\mathcal{H}$ with respect to a symplectic structure Ω is the set

$$S_{\mathcal{H}} = \bigcup_{z\in Z}\{v\in T_zZ\,:\, v\rfloor\Omega + d\mathcal{H}(z) = 0\}. \tag{3.2.6}$$

As in the general case of Poisson Hamiltonian systems, the symplectic one $(\Omega,\mathcal{H})$ has a unique solution which is the Hamiltonian vector field

$$\vartheta_{\mathcal{H}}\rfloor\Omega = -d\mathcal{H} \tag{3.2.7}$$

of $\mathcal{H}$. Hence, $S_{\mathcal{H}}$ (3.2.6) is an autonomous first order dynamic equation, called the *Hamilton equation* for the Hamiltonian $\mathcal{H}$ with respect to the symplectic structure Ω. Relative to the local canonical coordinates (q^i,p_i) for the symplectic structure Ω, the Hamilton equation (3.2.6) and the Hamiltonian vector field (3.2.7) read

$$\dot q^i = \partial^i\mathcal{H}, \qquad \dot p_i = -\partial_i\mathcal{H}, \tag{3.2.8}$$

$$\vartheta_{\mathcal{H}} = \partial^i\mathcal{H}\partial_i - \partial_i\mathcal{H}\partial^i. \tag{3.2.9}$$

Integrals of motion of a symplectic Hamiltonian system are defined just as those of a Poisson Hamiltonian system.

3.2.3 *Presymplectic Hamiltonian systems*

The notion of a Hamiltonian system is naturally extended to presymplectic manifolds [70; 106]. Given a presymplectic manifold (Z,Ω), a *presymplectic Hamiltonian system* for a Hamiltonian $\mathcal{H}\in C^\infty(Z)$ is the set

$$S_{\mathcal{H}} = \bigcup_{z\in Z}\{v\in T_zZ\,:\, v\rfloor\Omega + d\mathcal{H}(z) = 0\}. \tag{3.2.10}$$

A solution of this Hamiltonian system is a Hamiltonian vector field $\vartheta_{\mathcal{H}}$ of $\mathcal{H}$. The necessary and sufficient conditions of its existence are the following [70; 106].

Proposition 3.2.1. *The equation*

$$v\rfloor\Omega + d\mathcal{H}(z) = 0, \qquad v\in T_zZ, \tag{3.2.11}$$

has a solution only at points of the set

$$N_2 = \{z\in Z\,:\, \mathrm{Ker}\,_z\Omega\subset \mathrm{Ker}\,_z d\mathcal{H}\}. \tag{3.2.12}$$

Proof. It is readily observed that the fibre (3.2.11) of the set $S_{\mathcal{H}}$ (3.2.10) over $z \in Z$ is an affine space modelled over the fibre

$$\operatorname{Ker}_z\Omega = \{v \in T_zZ \,:\, v\rfloor\Omega = 0\}$$

of the kernel of the presymplectic form Ω. Let a vector $v \in T_zZ$ satisfy the equation (3.2.11). Then the contraction of the right-hand side of this equation with an arbitrary element $u \in \operatorname{Ker}_z\Omega$ leads to the equality $u\rfloor d\mathcal{H}(z) = 0$. In order to prove the converse, it suffices to show that $d\mathcal{H}(z) \in \operatorname{Im}\Omega^\flat$. This inclusion results from the injections

$$d\mathcal{H}(z) \in \operatorname{Ann}(\operatorname{Ker} d\mathcal{H}(z)) \subset \operatorname{Ann}(\operatorname{Ker}_z\Omega) = \operatorname{Im}\Omega^\flat.$$

□

Let us suppose that a presymplectic form Ω is of constant rank and that the set N_2 (3.2.12) is a submanifold of Z, but not necessarily connected. Then $\operatorname{Ker}\Omega$ is a closed vector subbundle of the tangent bundle TZ, while $S_{\mathcal{H}}|_{N_2}$ is an affine bundle over N_2. The latter has a section over N_2, but this section need not live in TN_2, i.e., it is not necessarily a vector field on the submanifold N_2. Then one aims to find a submanifold $N \subset N_2 \subset Z$ such that

$$S_{\mathcal{H}}|_N \cap T_zN \neq \emptyset, \qquad z \in N,$$

or, equivalently,

$$d\mathcal{H}(z) \in \Omega^\flat(TN), \qquad z \in N.$$

If such a submanifold exists, it may be obtained by means of the following *constraint algorithm* . Let us consider the overlap $S_{\mathcal{H}}|_{N_2} \cap TN_2$ and its projection to Z. We obtain the subset

$$N_3 = \pi_Z(S_{\mathcal{H}}|_{N_2} \cap TN_2) \subset Z.$$

If N_3 is a submanifold, let us consider the overlap $S_{\mathcal{H}}|_{N_3} \cap TN_3$. Its projection to Z gives a subset $N_4 \subset Z$, and so on. Since a manifold Z is finite-dimensional, the procedure is stopped after a finite number of steps by one of the following results.

• There is a number $i \geq 2$ such that a set N_i is empty. This means that a presymplectic Hamiltonian system has no solution.

• A set N_i, $i \geq 2$, fails to be a submanifold. It follows that a solution need not exist at each point of N_i.

• If $N_{i+1} = N_i$ for some $i \geq 2$, this is a desired submanifold N. A local solution of the presymplectic Hamiltonian system (3.2.10) exists around each point of N. If $\Omega^\flat|_{TN}$ is of constant rank, there is a global solution on N.

Sections of the vector bundle $\mathrm{Ker}\,\Omega \to Z$ are sometimes called *gauge fields* in order to emphasize that, being solutions of the presymplectic Hamiltonian system $(\Omega, 0)$ for the zero Hamiltonian, they do not contribute to a physical state, and are responsible for a certain *gauge freedom* [152]. At the same time, there are physically interesting presymplectic Hamiltonian systems, e.g., in relativistic mechanics when a Hamiltonian is equal to zero (Section 10.5). In this case, $\mathrm{Ker}\,d\mathcal{H} = TZ$ and the Hamiltonian system (3.2.11) has a solution everywhere on a manifold Z.

The above mentioned gauge freedom also is related to the pull-back construction in Proposition 3.1.2. Let a presymplectic form Ω on a manifold Z be of constant rank and let its characteristic foliation be simple, i.e., a fibred manifold $\pi : Z \to P$. Then Ω is the pull-back $\pi^*\Omega_P$ of a certain symplectic form Ω_P on P. Let a Hamiltonian $\mathcal{H}$ also be the pull-back $\pi^*\mathcal{H}_P$ of a function $\mathcal{H}_P$ on P. Then we have

$$\mathrm{Ker}\,\Omega = VN \subset \mathrm{Ker}\,d\mathcal{H},$$

and the presymplectic Hamiltonian system $(\Omega, \mathcal{H})$ has a solution everywhere on a manifold Z. Any such a solution $\vartheta_{\mathcal{H}}$ is projected onto a unique solution of the symplectic Hamiltonian system $(\Omega_P, \mathcal{H}_P)$ on the manifold P, while gauge fields are vertical vector fields on the fibred manifold $Z \to P$.

3.3 Hamiltonian formalism on $Q \to \mathbb{R}$

As was mentioned above, a *phase space* of non-relativistic mechanics on a configuration space $Q \to \mathbb{R}$ is the vertical cotangent bundle (2.1.31):

$$V^*Q \xrightarrow{\pi_\Pi} Q \xrightarrow{\pi} \mathbb{R},$$

of $Q \to \mathbb{R}$ equipped with the holonomic coordinates $(t, q^i, p_i = \dot{q}_i)$ with respect to the fibre bases $\{\overline{d}q^i\}$ for the bundle $V^*Q \to Q$ [106; 139].

Remark 3.3.1. A generic phase space of Hamiltonian mechanics is a fibre bundle $\Pi \to \mathbb{R}$ endowed with a regular Poisson structure whose characteristic distribution belongs to the vertical tangent bundle $V\Pi$ of $\Pi \to \mathbb{R}$ [81]. It can be seen locally as the Poisson product over $\mathbb{R}$ of a fibre bundle $V^*Q \to \mathbb{R}$ and a fibre bundle over $\mathbb{R}$, equipped with the zero Poisson structure.

The cotangent bundle T^*Q of the configuration space Q is endowed with the holonomic coordinates (t, q^i, p_0, p_i), possessing the transition functions

(2.2.4). It admits the Liouville form Ξ (2.2.12), the symplectic form

$$\Omega_T = d\Xi = dp_0 \wedge dt + dp_i \wedge dq^i, \tag{3.3.1}$$

and the corresponding Poisson bracket

$$\{f,g\}_T = \partial^0 f \partial_t g - \partial^0 g \partial_t f + \partial^i f \partial_i g - \partial^i g \partial_i f, \quad f,g \in C^\infty(T^*Q). \tag{3.3.2}$$

Provided with the structures (3.3.1) – (3.3.2), the cotangent bundle T^*Q of Q plays a role of the *homogeneous phase space* of Hamiltonian non-relativistic mechanics.

There is the canonical one-dimensional affine bundle (2.2.5):

$$\zeta : T^*Q \to V^*Q. \tag{3.3.3}$$

A glance at the transformation law (2.2.4) shows that it is a trivial affine bundle. Indeed, given a global section h of ζ, one can equip T^*Q with the global fibre coordinate

$$I_0 = p_0 - h, \qquad I_0 \circ h = 0, \tag{3.3.4}$$

possessing the identity transition functions. With respect to the coordinates

$$(t, q^i, I_0, p_i), \qquad i = 1, \ldots, m, \tag{3.3.5}$$

the fibration (3.3.3) reads

$$\zeta : \mathbb{R} \times V^*Q \ni (t, q^i, I_0, p_i) \to (t, q^i, p_i) \in V^*Q. \tag{3.3.6}$$

Let us consider the subring of $C^\infty(T^*Q)$ which comprises the pull-back $\zeta^* f$ onto T^*Q of functions f on the vertical cotangent bundle V^*Q by the fibration ζ (3.3.3). This subring is closed under the Poisson bracket (3.3.2). Then by virtue of Theorem 3.1.3, there exists the degenerate coinduced Poisson structure

$$\{f,g\}_V = \partial^i f \partial_i g - \partial^i g \partial_i f, \qquad f,g \in C^\infty(V^*Q), \tag{3.3.7}$$

on a phase space V^*Q such that

$$\zeta^*\{f,g\}_V = \{\zeta^* f, \zeta^* g\}_T. \tag{3.3.8}$$

The holonomic coordinates on V^*Q are canonical for the Poisson structure (3.3.7).

With respect to the Poisson bracket (3.3.7), the Hamiltonian vector fields of functions on V^*Q read

$$\vartheta_f = \partial^i f \partial_i - \partial_i f \partial^i, \qquad f \in C^\infty(V^*Q), \tag{3.3.9}$$

$$[\vartheta_f, \vartheta_{f'}] = \vartheta_{\{f,f'\}_V}. \tag{3.3.10}$$

They are vertical vector fields on $V^*Q \to \mathbb{R}$. Accordingly, the characteristic distribution of the Poisson structure (3.3.7) is the vertical tangent bundle $VV^*Q \subset TV^*Q$ of a fibre bundle $V^*Q \to \mathbb{R}$. The corresponding symplectic foliation on the phase space V^*Q coincides with the fibration $V^*Q \to \mathbb{R}$.

It is readily observed that the ring $\mathcal{C}(V^*Q)$ of Casimir functions on a Poisson manifold V^*Q consists of the pull-back onto V^*Q of functions on $\mathbb{R}$. Therefore, the Poisson algebra $C^\infty(V^*Q)$ is a Lie $C^\infty(\mathbb{R})$-algebra.

Remark 3.3.2. The Poisson structure (3.3.7) can be introduced in a different way [106; 139]. Given any section h of the fibre bundle (3.3.3), let us consider the pull-back forms

$$\begin{aligned} \mathbf{\Theta} &= h^*(\Xi \wedge dt) = p_i dq^i \wedge dt, \\ \mathbf{\Omega} &= h^*(d\Xi \wedge dt) = dp_i \wedge dq^i \wedge dt \end{aligned} \tag{3.3.11}$$

on V^*Q. They are independent of the choice of h. With $\mathbf{\Omega}$ (3.3.11), the Hamiltonian vector field ϑ_f (3.3.9) for a function f on V^*Q is given by the relation

$$\vartheta_f \rfloor \mathbf{\Omega} = -df \wedge dt,$$

while the Poisson bracket (3.3.7) is written as

$$\{f, g\}_V dt = \vartheta_g \rfloor \vartheta_f \rfloor \mathbf{\Omega}.$$

Moreover, one can show that a projectable vector field ϑ on V^*Q such that $\vartheta\rfloor dt$ =const. is a canonical vector field for the Poisson structure (3.3.7) if and only if

$$\mathbf{L}_\vartheta \mathbf{\Omega} = d(\vartheta \rfloor \mathbf{\Omega}) = 0. \tag{3.3.12}$$

In contrast with autonomous Hamiltonian mechanics, the Poisson structure (3.3.7) fails to provide any dynamic equation on a fibre bundle $V^*Q \to \mathbb{R}$ because Hamiltonian vector fields (3.3.9) of functions on V^*Q are vertical vector fields, but not connections on $V^*Q \to \mathbb{R}$ (see Definition 1.3.1). Hamiltonian dynamics on V^*Q is described as a particular Hamiltonian dynamics on fibre bundles [68; 106; 139].

A *Hamiltonian* on a phase space $V^*Q \to \mathbb{R}$ of non-relativistic mechanics is defined as a global section

$$h : V^*Q \to T^*Q, \qquad p_0 \circ h = \mathcal{H}(t, q^j, p_j), \tag{3.3.13}$$

of the affine bundle ζ (3.3.3). Given the Liouville form Ξ (2.2.12) on T^*Q, this section yields the pull-back *Hamiltonian form*

$$H = (-h)^*\Xi = p_k dq^k - \mathcal{H} dt \tag{3.3.14}$$

on V^*Q. This is the well-known *invariant of Poincaré–Cartan* [4].

It should be emphasized that, in contrast with a Hamiltonian in autonomous mechanics, the Hamiltonian $\mathcal{H}$ (3.3.13) is not a function on V^*Q, but it obeys the transformation law

$$\mathcal{H}'(t,q'^i,p'_i) = \mathcal{H}(t,q^i,p_i) + p'_i\partial_t q'^i. \tag{3.3.15}$$

Remark 3.3.3. Any connection Γ (1.1.18) on a configuration bundle $Q \to \mathbb{R}$ defines the global section $h_\Gamma = p_i\Gamma^i$ (3.3.13) of the affine bundle ζ (3.3.3) and the corresponding Hamiltonian form

$$H_\Gamma = p_k dq^k - \mathcal{H}_\Gamma dt = p_k dq^k - p_i\Gamma^i dt. \tag{3.3.16}$$

Furthermore, given a connection Γ, any Hamiltonian form (3.3.14) admits the splitting

$$H = H_\Gamma - \mathcal{E}_\Gamma dt, \tag{3.3.17}$$

where

$$\mathcal{E}_\Gamma = \mathcal{H} - \mathcal{H}_\Gamma = \mathcal{H} - p_i\Gamma^i \tag{3.3.18}$$

is a function on V^*Q. It is called the *Hamiltonian function* relative to a reference frame Γ. With respect to the coordinates adapted to a reference frame Γ, we have $\mathcal{E}_\Gamma = \mathcal{H}$. Given different reference frames Γ and Γ', the decomposition (3.3.17) leads at once to the relation

$$\mathcal{E}_{\Gamma'} = \mathcal{E}_\Gamma + \mathcal{H}_\Gamma - \mathcal{H}_{\Gamma'} = \mathcal{E}_\Gamma + (\Gamma^i - \Gamma'^i)p_i \tag{3.3.19}$$

between the Hamiltonian functions with respect to different reference frames.

Given a Hamiltonian form H (3.3.14), there exists a unique horizontal vector field (1.1.18):

$$\gamma_H = \partial_t - \gamma^i\partial_i - \gamma_i\partial^i,$$

on V^*Q (i.e., a connection on $V^*Q \to \mathbb{R}$) such that

$$\gamma_H \rfloor dH = 0. \tag{3.3.20}$$

This vector field, called the *Hamilton vector field*, reads

$$\gamma_H = \partial_t + \partial^k\mathcal{H}\partial_k - \partial_k\mathcal{H}\partial^k. \tag{3.3.21}$$

In a different way (Remark 3.3.2), the Hamilton vector field γ_H is defined by the relation

$$\gamma_H \rfloor \mathbf{\Omega} = dH.$$

Consequently, it is canonical for the Poisson structure $\{,\}_V$ (3.3.7). This vector field yields the first order dynamic *Hamilton equation*

$$q_t^k = \partial^k \mathcal{H}, \tag{3.3.22}$$

$$p_{tk} = -\partial_k \mathcal{H} \tag{3.3.23}$$

on $V^*Q \to \mathbb{R}$ (Definition 1.3.1), where $(t, q^k, p_k, q_t^k, \dot{p}_{tk})$ are the adapted coordinates on the first order jet manifold J^1V^*Q of $V^*Q \to \mathbb{R}$.

Due to the canonical imbedding $J^1V^*Q \to TV^*Q$ (1.1.6), the Hamilton equation (3.3.22) – (3.3.23) is equivalent to the autonomous first order dynamic equation

$$\dot{t} = 1, \qquad \dot{q}^i = \partial^i \mathcal{H}, \qquad \dot{p}_i = -\partial_i \mathcal{H} \tag{3.3.24}$$

on a manifold V^*Q (Definition 1.2.1).

A *solution* of the Hamilton equation (3.3.22) – (3.3.23) is an integral section r for the connection γ_H.

Remark 3.3.4. Similarly to the Cartan equation (2.2.11), the Hamilton equation (3.3.22) – (3.3.23) is equivalent to the condition

$$r^*(u \rfloor dH) = 0 \tag{3.3.25}$$

for any vertical vector field u on $V^*Q \to \mathbb{R}$.

We agree to call (V^*Q, H) the *Hamiltonian system* of $k = \dim Q - 1$ degrees of freedom.

In order to describe evolution of a Hamiltonian system at any instant, the Hamilton vector field γ_H (3.3.21) is assumed to be complete, i.e., it is an Ehresmann connection (Remark 1.1.2). In this case, the Hamilton equation (3.3.22) – (3.3.23) admits a unique global solution through each point of the phase space V^*Q. By virtue of Theorem 1.1.2, there exists a trivialization of a fibre bundle $V^*Q \to \mathbb{R}$ (not necessarily compatible with its fibration $V^*Q \to Q$) such that

$$\gamma_H = \partial_t, \qquad H = \overline{p}_i d\overline{q}^i \tag{3.3.26}$$

with respect to the associated coordinates $(t, \overline{q}^i, \overline{p}_i)$. A direct computation shows that the Hamilton vector field γ_H (3.3.21) satisfies the relation (3.3.12) and, consequently, it is an infinitesimal generator of a one-parameter group of automorphisms of the Poisson manifold $(V^*Q, \{,\}_V)$. Then one can show that $(t, \overline{q}^i, \overline{p}_i)$ are canonical coordinates for the Poisson manifold $(V^*Q, \{,\}_V)$ [106], i.e.,

$$w = \frac{\partial}{\partial \overline{p}_i} \wedge \frac{\partial}{\partial \overline{q}^i}.$$

Since $\mathcal{H} = 0$, the Hamilton equation (3.3.22) – (3.3.23) in these coordinates takes the form

$$\overline{q}^i_t = 0, \qquad \overline{p}_{ti} = 0,$$

i.e., $(t, \overline{q}^i, \overline{p}_i)$ are the *initial data coordinates.*

3.4 Homogeneous Hamiltonian formalism

As was mentioned above, one can associate to any Hamiltonian system on a phase space V^*Q an equivalent autonomous symplectic Hamiltonian system on the cotangent bundle T^*Q (Theorem 3.4.1).

Given a Hamiltonian system (V^*Q, H), its Hamiltonian $\mathcal{H}$ (3.3.13) defines the function

$$\mathcal{H}^* = \partial_t \rfloor (\Xi - \zeta^*(-h)^*\Xi)) = p_0 + h = p_0 + \mathcal{H} \tag{3.4.1}$$

on T^*Q. Let us regard $\mathcal{H}^*$ (3.4.1) as a Hamiltonian of an autonomous Hamiltonian system on the symplectic manifold (T^*Q, Ω_T). The corresponding autonomous Hamilton equation on T^*Q takes the form

$$\dot{t} = 1, \qquad \dot{p}_0 = -\partial_t \mathcal{H}, \qquad \dot{q}^i = \partial^i \mathcal{H}, \qquad \dot{p}_i = -\partial_i \mathcal{H}. \tag{3.4.2}$$

Remark 3.4.1. Let us note that the splitting $\mathcal{H}^* = p_0 + \mathcal{H}$ (3.4.1) is ill defined. At the same time, any reference frame Γ yields the decomposition

$$\mathcal{H}^* = (p_0 + \mathcal{H}_\Gamma) + (\mathcal{H} - \mathcal{H}_\Gamma) = \mathcal{H}^*_\Gamma + \mathcal{E}_\Gamma, \tag{3.4.3}$$

where $\mathcal{H}_\Gamma$ is the Hamiltonian (3.3.16) and $\mathcal{E}_\Gamma$ (3.3.18) is the Hamiltonian function relative to a reference frame Γ.

The Hamiltonian vector field $\vartheta_{\mathcal{H}^*}$ of $\mathcal{H}^*$ (3.4.1) on T^*Q is

$$\vartheta_{\mathcal{H}^*} = \partial_t - \partial_t \mathcal{H} \partial^0 + \partial^i \mathcal{H} \partial_i - \partial_i \mathcal{H} \partial^i. \tag{3.4.4}$$

Written relative to the coordinates (3.3.5), this vector field reads

$$\vartheta_{\mathcal{H}^*} = \partial_t + \partial^i \mathcal{H} \partial_i - \partial_i \mathcal{H} \partial^i. \tag{3.4.5}$$

It is identically projected onto the Hamilton vector field γ_H (3.3.21) on V^*Q such that

$$\zeta^*(\mathbf{L}_{\gamma_H} f) = \{\mathcal{H}^*, \zeta^* f\}_T, \qquad f \in C^\infty(V^*Q). \tag{3.4.6}$$

Therefore, the Hamilton equation (3.3.22) – (3.3.23) is equivalent to the autonomous Hamilton equation (3.4.2).

Obviously, the Hamiltonian vector field $\vartheta_{\mathcal{H}^*}$ (3.4.5) is complete if the Hamilton vector field γ_H (3.3.21) is complete.

Thus, the following has been proved [31; 65; 108].

Theorem 3.4.1. *A Hamiltonian system (V^*Q, H) of k degrees of freedom is equivalent to an autonomous Hamiltonian system $(T^*Q, \mathcal{H}^*)$ of $k+1$ degrees of freedom on a symplectic manifold (T^*Q, Ω_T) whose Hamiltonian is the function $\mathcal{H}^*$ (3.4.1).*

We agree to call $(T^*Q, \mathcal{H}^*)$ the *homogeneous Hamiltonian system* and $\mathcal{H}^*$ (3.4.1) the *homogeneous Hamiltonian.*

3.5 Lagrangian form of Hamiltonian formalism

It is readily observed that the Hamiltonian form H (3.3.14) is the Poincaré–Cartan form of the Lagrangian

$$L_H = h_0(H) = (p_i q_t^i - \mathcal{H})dt \tag{3.5.1}$$

on the jet manifold J^1V^*Q of $V^*Q \to \mathbb{R}$ [109; 110; 139].

Remark 3.5.1. In fact, the Lagrangian (3.5.1) is the pull-back onto J^1V^*Q of the form L_H on the product $V^*Q \times_Q J^1Q$.

The Lagrange operator (2.1.16) associated to the Lagrangian L_H reads

$$\mathcal{E}_H = \delta L_H = [(q_t^i - \partial^i\mathcal{H})dp_i - (p_{ti} + \partial_i\mathcal{H})dq^i] \wedge dt. \tag{3.5.2}$$

The corresponding Lagrange equation (2.1.20) is of first order, and it coincides with the Hamilton equation (3.3.22) – (3.3.23) on J^1V^*Q.

Due to this fact, the Lagrangian L_H (3.5.1) plays a prominent role in Hamiltonian non-relativistic mechanics.

In particular, let u (2.5.4) be a vector field on a configuration space Q. Its functorial lift (11.2.32) onto the cotangent bundle T^*Q is

$$\widetilde{u} = u^t\partial_t + u^i\partial_i - p_j\partial_i u^j\partial^i. \tag{3.5.3}$$

This vector field is identically projected onto a vector field, also given by the expression (3.5.3), on the phase space V^*Q as a base of the trivial fibre bundle (3.3.3). Then we have the equality

$$\mathbf{L}_{\widetilde{u}}H = \mathbf{L}_{J^1\widetilde{u}}L_H = (-u^t\partial_t\mathcal{H} + p_i\partial_t u^i - u^i\partial_i\mathcal{H} + p_i\partial_j u^i\partial^j\mathcal{H})dt. \tag{3.5.4}$$

This equality enables us to study conservation laws in Hamiltonian mechanics similarly to those in Lagrangian mechanics (Section 3.8).

3.6 Associated Lagrangian and Hamiltonian systems

As was mentioned above, Lagrangian and Hamiltonian formulations of mechanics fail to be equivalent. For instance, there exist physically interesting systems whose phase spaces fail to be the cotangent bundles of configuration spaces, and they do not admit any Lagrangian description [149]. The comprehensive relations between Lagrangian and Hamiltonian systems can be established in the case of almost regular Lagrangians [106; 108; 139]. This is a particular case of the relations between Lagrangian and Hamiltonian theories on fibre bundles [55; 68].

In order to compare Lagrangian and Hamiltonian formalisms, we are based on the facts that:

(i) every first order Lagrangian L (2.1.15) on a velocity space J^1Q induces the Legendre map (2.1.30) of this velocity space to a phase space V^*Q;

(ii) every Hamiltonian form H (3.3.14) on a phase space V^*Q yields the *Hamiltonian map*

$$\widehat{H} : V^*Q \underset{Q}{\longrightarrow} J^1Q, \qquad q^i_t \circ \widehat{H} = \partial^i \mathcal{H} \tag{3.6.1}$$

of this phase space to a velocity space J^1Q.

Remark 3.6.1. A Hamiltonian form H is called *regular* if the Hamiltonian map $\widehat{H}$ (3.6.1) is regular, i.e., a local diffeomorphism.

Remark 3.6.2. It is readily observed that a section r of a fibre bundle $V^*Q \to \mathbb{R}$ is a solution of the Hamilton equation (3.3.22) – (3.3.23) for the Hamiltonian form H if and only if it obeys the equality

$$J^1(\pi_\Pi \circ r) = \widehat{H} \circ r, \tag{3.6.2}$$

where $\pi_\Pi : V^*Q \to Q$.

Given a Lagrangian L, the Hamiltonian form H (3.3.14) is said to be *associated* with L if H satisfies the relations

$$\widehat{L} \circ \widehat{H} \circ \widehat{L} = \widehat{L}, \tag{3.6.3}$$

$$\widehat{H}^* L_H = \widehat{H}^* L, \tag{3.6.4}$$

where L_H is the Lagrangian (3.5.1).

A glance at the equality (3.6.3) shows that $\widehat{L} \circ \widehat{H}$ is the projector of V^*Q onto the Lagrangian constraint space N_L which is given by the coordinate conditions

$$p_i = \pi_i(t, q^j, \partial^j \mathcal{H}(t, q^j, p_j)). \tag{3.6.5}$$

The relation (3.6.4) takes the coordinate form

$$\mathcal{H} = p_i\partial^i\mathcal{H} - \mathcal{L}(t, q^j, \partial^j\mathcal{H}). \tag{3.6.6}$$

Acting on this equality by the exterior differential, we obtain the relations

$$\partial_t\mathcal{H}(p) = -(\partial_t\mathcal{L}) \circ \widehat{H}(p), \qquad p \in N_L,$$

$$\partial_i\mathcal{H}(p) = -(\partial_i\mathcal{L}) \circ \widehat{H}(p), \qquad p \in N_L, \tag{3.6.7}$$

$$(p_i - (\partial_i\mathcal{L})(t, q^j, \partial^j\mathcal{H}))\partial^i\partial^a\mathcal{H} = 0. \tag{3.6.8}$$

The relation (3.6.8) shows that an L-associated Hamiltonian form H is not regular outside the Lagrangian constraint space N_L.

For instance, let L be a hyperregular Lagrangian, i.e., the Legendre map $\widehat{L}$ (2.1.30) is a diffeomorphism. It follows from the relation (3.6.3) that, in this case, $\widehat{H} = \widehat{L}^{-1}$. Then the relation (3.6.6) takes the form

$$\mathcal{H} = p_i\widehat{L}^{-1i} - \mathcal{L}(t, q^j, \widehat{L}^{-1j}). \tag{3.6.9}$$

It defines a unique Hamiltonian form associated with a hyperregular Lagrangian. Let s be a solution of the Lagrange equation (2.1.25) for a Lagrangian L. A direct computation shows that $\widehat{L} \circ J^1s$ is a solution of the Hamilton equation (3.3.22) – (3.3.23) for the Hamiltonian form H (3.6.9). Conversely, if r is a solution of the Hamilton equation (3.3.22) – (3.3.23) for the Hamiltonian form H (3.6.9), then $s = \pi_\Pi \circ r$ is a solution of the Lagrange equation (2.1.25) for L (see the equality (3.6.2)). It follows that, in the case of hyperregular Lagrangians, Hamiltonian formalism is equivalent to Lagrangian one.

If a Lagrangian is not regular, an associated Hamiltonian form need not exist.

Example 3.6.1. Let Q be a fibre bundle $\mathbb{R}^2 \to \mathbb{R}$ with coordinates (t, q). Its jet manifold $J^1Q = \mathbb{R}^3$ and its Legendre bundle $V^*Q = \mathbb{R}^3$ are coordinated by (t, q, q_t) and (t, q, p), respectively. Let us put

$$L = \exp(q_t)dt. \tag{3.6.10}$$

This Lagrangian is regular, but not hyperregular. The corresponding Legendre map reads

$$p \circ \widehat{L} = \exp q_t.$$

It follows that the Lagrangian constraint space N_L is given by the coordinate relation $p > 0$. This is an open subbundle of the Legendre bundle, and $\widehat{L}$ is a diffeomorphism of J^1Q onto N_L. Hence, there is a unique Hamiltonian form

$$H = pdq - p(\ln p - 1)dt$$

on N_L which is associated with the Lagrangian (3.6.10). This Hamiltonian form however is not smoothly extended to V^*Q.

A Hamiltonian form is called *weakly associated* with a Lagrangian L if the condition (3.6.4) (namely, the condition (3.6.8) holds on the Lagrangian constraint space N_L.

For instance, any Hamiltonian form is weakly associated with the Lagrangian $L = 0$, while the associated Hamiltonian forms are only H_Γ (3.3.16).

A hyperregular Lagrangian L has a unique weakly associated Hamiltonian form (3.6.9) which also is L-associated. In the case of a regular Lagrangian L, the Lagrangian constraint space N_L is an open subbundle of the vector Legendre bundle $V^*Q \to Q$. If $N_L \neq V^*Q$, a weakly associated Hamiltonian form fails to be defined everywhere on V^*Q in general. At the same time, N_L itself can be provided with the pull-back symplectic structure with respect to the imbedding $N_L \to V^*Q$, so that one may consider Hamiltonian forms on N_L.

Note that, in contrast with associated Hamiltonian forms, a weakly associated Hamiltonian form may be regular.

In order to say something more, let us restrict our consideration to almost regular Lagrangians L (Definition 2.1.2) [106; 108; 139].

Lemma 3.6.1. *The Poincaré–Cartan form H_L (2.2.1) of an almost regular Lagrangian L is constant on the inverse image $\widehat{L}^{-1}(z)$ of any point $z \in N_L$.*

A corollary of Lemma 3.6.1 is the following.

Theorem 3.6.1. *All Hamiltonian forms weakly associated with an almost regular Lagrangian L coincide with each other on the Lagrangian constraint space N_L, and the Poincaré–Cartan form H_L (2.2.1) of L is the pull-back*

$$H_L = \widehat{L}^* H, \qquad \pi_i q^i_t - \mathcal{L} = \mathcal{H}(t, q^j, \pi_j), \tag{3.6.11}$$

of such a Hamiltonian form H.

It follows that, given Hamiltonian forms H and H' weakly associated with an almost regular Lagrangian L, their difference is a density

$$H' - H = (\mathcal{H} - \mathcal{H}')dt$$

vanishing on the Lagrangian constraint space N_L. However, $\widehat{H}|_{N_L} \neq \widehat{H}'|_{N_L}$ in general. Therefore, the Hamilton equations for H and H' do not necessarily coincide on the Lagrangian constraint space N_L.

Theorem 3.6.1 enables us to relate the Lagrange equation for an almost regular Lagrangian L with the Hamilton equation for Hamiltonian forms weakly associated to L.

Theorem 3.6.2. *Let a section r of $V^*Q \to \mathbb{R}$ be a solution of the Hamilton equation (3.3.22) – (3.3.23) for a Hamiltonian form H weakly associated with an almost regular Lagrangian L. If r lives in the Lagrangian constraint space N_L, the section $s = \pi \circ r$ of $\pi : Q \to \mathbb{R}$ satisfies the Lagrange equation (2.1.25), while $\overline{s} = \widehat{H} \circ r$ obeys the Cartan equation (2.2.9) – (2.2.10).*

The proof is based on the relation

$$\widetilde{L} = (J^1\widehat{L})^* L_H,$$

where $\widetilde{L}$ is the Lagrangian (2.2.7), while L_H is the Lagrangian (3.5.1). This relation is derived from the equality (3.6.11). The converse assertion is more intricate.

Theorem 3.6.3. *Given an almost regular Lagrangian L, let a section $\overline{s}$ of the jet bundle $J^1Q \to \mathbb{R}$ be a solution of the Cartan equation (2.2.9) – (2.2.10). Let H be a Hamiltonian form weakly associated with L, and let H satisfy the relation*

$$\widehat{H} \circ \widehat{L} \circ \overline{s} = J^1 s, \tag{3.6.12}$$

where s is the projection of $\overline{s}$ onto Q. Then the section $r = \widehat{L} \circ \overline{s}$ of a fibre bundle $V^*Q \to \mathbb{R}$ is a solution of the Hamilton equation (3.3.22) – (3.3.23) for H.

We say that a set of Hamiltonian forms H weakly associated with an almost regular Lagrangian L is *complete* if, for each solution s of the Lagrange equation, there exists a solution r of the Hamilton equation for a Hamiltonian form H from this set such that $s = \pi_\Pi \circ r$. By virtue of Theorem 3.6.3, a set of weakly associated Hamiltonian forms is complete if, for every solution s of the Lagrange equation for L, there exists a Hamiltonian form H from this set which fulfills the relation (3.6.12) where $\overline{s} = J^1 s$, i.e.,

$$\widehat{H} \circ \widehat{L} \circ J^1 s = J^1 s. \tag{3.6.13}$$

In the case of almost regular Lagrangians, one can formulate the following necessary and sufficient conditions of the existence of weakly associated Hamiltonian forms.

Theorem 3.6.4. *A Hamiltonian form H weakly associated with an almost regular Lagrangian L exists if and only if the fibred manifold (2.1.32):*

$$\widehat{L} : J^1Q \to N_L, \tag{3.6.14}$$

admits a global section.

In particular, any point of V^*Q possesses an open neighborhood U such that there exists a complete set of local Hamiltonian forms on U which are weakly associated with an almost regular Lagrangian L. Moreover, one can construct a complete set of local L-associated Hamiltonian forms on U [138].

3.7 Quadratic Lagrangian and Hamiltonian systems

Let us study an important case of the almost regular quadratic Lagrangian L (2.3.1). We show that there exists a complete set of Hamiltonian forms associated with L.

Given the almost regular quadratic Lagrangian L (2.3.1), there is the splitting (2.3.10) of a phase space V^*Q. It takes the form

$$V^*Q = \mathcal{R}(V^*Q) \underset{Q}{\oplus} \mathcal{P}(V^*Q) = \operatorname{Ker}\sigma_0 \underset{Q}{\oplus} N_L, \tag{3.7.1}$$

$$p_i = \mathcal{R}_i + \mathcal{P}_i = [p_i - a_{ij}\sigma_0^{jk}p_k] + [a_{ij}\sigma_0^{jk}p_k], \tag{3.7.2}$$

where $\sigma = \sigma_0 + \sigma_1$ is the linear bundle map (2.3.7) whose summands σ_0 and σ_1 satisfy the relations (2.3.11). These relations lead to the equalities

$$\sigma_0^{jk}\mathcal{R}_k = 0, \qquad \sigma_1^{jk}\mathcal{P}_k = 0. \tag{3.7.3}$$

It is readily observed that, with respect to the coordinates $\mathcal{R}_i$ and $\mathcal{P}_i$ (3.7.2), the Lagrangian constraint space (2.3.2) is defined by the equations

$$\mathcal{R}_i = p_i - a_{ij}\sigma_0^{jk}p_k = 0. \tag{3.7.4}$$

Given the linear map σ (2.3.7) and the arbitrary connection Γ (2.3.3), let us consider the morphism

$$\Phi = \widehat{H}_\Gamma + \sigma : V^*Q \to J^1Q, \qquad \Phi = \partial_t + (\Gamma^i + \sigma^{ij}p_j)\partial_i, \tag{3.7.5}$$

and the Hamiltonian form

$$\begin{aligned} H(\sigma,\Gamma) &= -\Phi\rfloor\Theta + \Phi^*L \qquad (3.7.6)\\ &= p_i dq^i - \left[p_i\Gamma^i + \frac{1}{2}\sigma_0{}^{ij}p_ip_j + \sigma_1{}^{ij}p_ip_j - c'\right]dt \\ &= (\mathcal{R}_i + \mathcal{P}_i)dq^i - \left[(\mathcal{R}_i + \mathcal{P}_i)\Gamma^i + \frac{1}{2}\sigma_0^{ij}\mathcal{P}_i\mathcal{P}_j + \sigma_1^{ij}\mathcal{R}_i\mathcal{R}_j - c'\right]dt. \end{aligned}$$

Theorem 3.7.1. *The Hamiltonian form (3.7.6) is weakly associated with the Lagrangian (2.3.1) (and (2.3.5)), and it is L-associated if $\sigma_1 = 0$.*

Proof. A direct computation shows that the Hamiltonian form (3.7.6) satisfies the condition (3.6.3) and the condition (3.6.4) on the constraint space (3.7.4). This condition holds everywhere on V^*Q if $\sigma_1 = 0$. □

Theorem 3.7.2. *The Hamiltonian forms*

$$H(\sigma, \Gamma) = p_i dq^i - \left[p_i \Gamma^i + \frac{1}{2} \sigma_0{}^{ij} p_i p_j - c' \right] dt \tag{3.7.7}$$

parameterized by Lagrangian frame connections Γ *(2.3.3) constitute a complete set of L-associated Hamiltonian forms.*

Proof. Let s be an arbitrary section of $Q \to \mathbb{R}$, e.g., a solution of the Lagrange equation. There exists a connection Γ (2.3.3) such that the relation (3.6.13) holds. Namely, let us put $\Gamma = \mathcal{S} \circ \Gamma'$ where Γ' is a connection on $Q \to \mathbb{R}$ which has s as an integral section. □

3.8 Hamiltonian conservation laws

As was mentioned above, integrals of motion in Lagrangian mechanics usually come from variational symmetries of a Lagrangian (Theorem 2.5.3), though not all integrals of motion are of this type (Section 2.5). In Hamiltonian mechanics, all integrals of motion are conserved generalized symmetry currents (Theorem 3.8.12 below).

An *integral of motion* of a Hamiltonian system (V^*Q, H) is defined as a smooth real function F on V^*Q which is an integral of motion of the Hamilton equation (3.3.22) – (3.3.23) (Section 1.10). Its Lie derivative

$$\mathbf{L}_{\gamma_H} F = \partial_t F + \{\mathcal{H}, F\}_V \tag{3.8.1}$$

along the Hamilton vector field γ_H (3.3.21) vanishes in accordance with the equation (1.10.7). Given the Hamiltonian vector field ϑ_F of F with respect to the Poisson bracket (3.3.7), it is easily justified that

$$[\gamma_H, \vartheta_F] = \vartheta_{\mathbf{L}_{\gamma_H} F}. \tag{3.8.2}$$

Consequently, the Hamiltonian vector field of an integral of motion is a symmetry of the Hamilton equation (3.3.22) – (3.3.23).

One can think of the formula (3.8.1) as being the *evolution equation* of Hamiltonian non-relativistic mechanics. In contrast with the autonomous evolution equation (3.2.5), the right-hand side of the equation (3.8.1) is not reduced to the Poisson bracket $\{,\}_V$.

Given a Hamiltonian system (V^*Q, H), let $(T^*Q, \mathcal{H}^*)$ be an equivalent homogeneous Hamiltonian system. It follows from the equality (3.4.6) that

$$\zeta^*(\mathbf{L}_{\gamma_H} F) = \{\mathcal{H}^*, \zeta^* F\}_T = \zeta^*(\partial_t F + \{\mathcal{H}, F\}_V) \tag{3.8.3}$$

for any function $F \in C^\infty(V^*Q)$. This formula is equivalent to the evolution equation (3.8.1). It is called the *homogeneous evolution equation.*

Proposition 3.8.1. *A function $F \in C^\infty(V^*Q)$ is an integral of motion of a Hamiltonian system (V^*Q, H) if and only if its pull-back $\zeta^* F$ onto T^*Q is an integral of motion of a homogeneous Hamiltonian system $(T^*Q, \mathcal{H}^*)$.*

Proof. It follows from the equality (3.8.3) that

$$\{\mathcal{H}^*, \zeta^* F\}_T = \zeta^*(\mathbf{L}_{\gamma_H} F) = 0. \tag{3.8.4}$$

□

Proposition 3.8.2. *If F and F' are integrals of motion of a Hamiltonian system, their Poisson bracket $\{F, F'\}_V$ also is an integral of motion.*

Proof. This fact results from the equalities (3.3.8) and (3.8.4). □

Consequently, integrals of motion of a Hamiltonian system (V^*Q, H) constitute a real Lie subalgebra of the Poisson algebra $C^\infty(V^*Q)$.

Let us turn to Hamiltonian conservation laws. We are based on the fact that the Hamilton equation (3.3.22) – (3.3.23) also is the Lagrange equation of the Lagrangian L_H (3.5.1). Therefore, one can study conservation laws in Hamiltonian mechanics similarly to those in Lagrangian mechanics [110].

Since the Hamilton equation (3.3.22) – (3.3.23) is of first order, we restrict our consideration to classical symmetries, i.e., vector fields on V^*Q. In this case, all conserved generalized symmetry currents are integrals of motion.

Let

$$\upsilon = u^t \partial_t + \upsilon^i \partial_i + \upsilon_i \partial^i, \qquad u^t = 0, 1, \tag{3.8.5}$$

be a vector field on a phase space V^*Q. Its prolongation onto $V^*Q \times_Q J^1Q$ (Remark 3.5.1) reads

$$J^1\upsilon = u^t \partial_t + \upsilon^i \partial_i + \upsilon_i \partial^i + d_t \upsilon^i \partial_i^t.$$

Then the first variational formula (2.5.11) for the Lagrangian L_H (3.5.1) takes the form

$$\begin{aligned} -u^t \partial_t \mathcal{H} - \upsilon^i \partial_i \mathcal{H} + \upsilon_i (q_t^i - \partial^i \mathcal{H}) + p_i d_t \upsilon^i \\ = -(\upsilon^i - q_t^i u^t)(p_{ti} + \partial_i \mathcal{H}) + (\upsilon_i - p_{ti} u^t)(q_t^i - \partial^i \mathcal{H}) \\ + d_t(p_i \upsilon^i - u^t \mathcal{H}). \end{aligned} \tag{3.8.6}$$

If υ (3.8.5) is a variational symmetry, i.e.,

$$\mathbf{L}_{J^1\upsilon}L_H = d_H\sigma,$$

we obtain the weak conservation law, *called the Hamiltonian conservation law*,

$$0 \approx d_t\mathfrak{T}_\upsilon \tag{3.8.7}$$

of the generalized symmetry current (2.5.17) which reads.

$$\mathfrak{T}_\upsilon = p_i\upsilon^i - u^t\mathcal{H} - \sigma. \tag{3.8.8}$$

This current is an integral of motion of a Hamiltonian system.

The converse also is true. Let F be an integral of motion, i.e.,

$$\mathbf{L}_{\gamma_H}F = \partial_t F + \{\mathcal{H}, F\}_V = 0. \tag{3.8.9}$$

We aim to show that there is a variational symmetry υ of L_H such that $F = \mathfrak{T}_\upsilon$ is a conserved generalized symmetry current along υ.

In accordance with Proposition 2.5.1, the vector field υ (3.8.5) is a variational symmetry if and only if

$$\upsilon^i(p_{ti} + \partial_i\mathcal{H}) - \upsilon_i(q^i_t - \partial^i\mathcal{H}) + u^t\partial_t\mathcal{H} = d_t(\mathfrak{T}_u + u^t\mathcal{H}). \tag{3.8.10}$$

A glance at this equality shows the following.

Proposition 3.8.3. *The vector field υ (3.8.5) is a variational symmetry only if*

$$\partial^i\upsilon_i = -\partial_i\upsilon^i. \tag{3.8.11}$$

For instance, if the vector field υ (3.8.5) is projectable onto Q (i.e., its components υ^i are independent of momenta p_i), we obtain that $u_i = -p_j\partial_i u^j$. Consequently, υ is the canonical lift $\widetilde{u}$ (3.5.3) onto V^*Q of the vector field u (2.5.4) on Q. Moreover, let $\widetilde{u}$ be a variational symmetry of a Lagrangian L_H. It follows at once from the equality (3.8.10) that $\widetilde{u}$ is an exact symmetry of L_H. The corresponding conserved symmetry current reads

$$\mathfrak{T}_{\widetilde{u}p} = p_iu^i - u^t\mathcal{H}. \tag{3.8.12}$$

We agree to call the vector field u (2.5.4) the *Hamiltonian symmetry* if its canonical lift $\widetilde{u}$ (3.5.3) onto V^*Q is a variational (consequently, exact) symmetry of the Lagrangian L_H (3.5.1). If a Hamiltonian symmetry is vertical, the corresponding conserved symmetry current $\mathfrak{T}_{\widetilde{u}} = p_iu^i$ is called the *Noether current*.

Proposition 3.8.4. *The Hamilton vector field γ_H (3.3.21) is a unique variational symmetry of L_H whose conserved generalized symmetry current equals zero.*

It follows that, given a non-vertical variational symmetry υ, $u^t = 1$, of a Lagrangian L_H, there exists a vertical variational symmetry $\upsilon - \gamma_H$ possessing the same generalized conserved symmetry current $\mathfrak{T}_\upsilon = \mathfrak{T}_{\upsilon-\gamma_H}$ as υ.

Theorem 3.8.1. *Any integral of motion F of a Hamiltonian system (V^*Q, H) is a generalized conserved current $F = \mathfrak{T}_{\vartheta_F}$ of the Hamiltonian vector field ϑ_F (3.1.4) of F.*

Proof. If $\upsilon = \vartheta_F$ and $\mathfrak{T}_{\vartheta_F} = F$, the relation (3.8.10) is satisfied owing to the equality (3.8.9). □

It follows from Theorem 3.8.1 that the Lie algebra of integrals of motion of a Hamiltonian system in Proposition 3.8.2 coincides with the Lie algebra of conserved generalized symmetry currents with respect to the bracket

$$\{F, F'\}_V = \{\mathfrak{T}_{\vartheta_F}, \mathfrak{T}_{\vartheta_{F'}}\}_V = \mathfrak{T}_{[\vartheta_F, \vartheta_{F'}]}.$$

In accordance with Theorem 3.8.1, any integral of motion of a Hamiltonian system can be treated as a conserved generalized current along a vertical variational symmetry. However, this is not convenient for the study of energy conservation laws.

Let $\mathcal{E}_\Gamma$ (3.3.18) be the Hamiltonian function of a Hamiltonian system relative to a reference frame Γ. Given bundle coordinates adapted to Γ, its evolution equation (3.8.1) takes the form

$$\mathbf{L}_{\gamma_H}\mathcal{E}_\Gamma = \partial_t \mathcal{E}_\Gamma = \partial_t \mathcal{H}. \tag{3.8.13}$$

It follows that, a Hamiltonian function $\mathcal{E}_\Gamma$ relative to a reference frame Γ is an integral of motion if and only if a Hamiltonian, written with respect to the coordinates adapted to Γ, is time-independent. One can think of $\mathcal{E}_\Gamma$ as being the energy function relative to a reference frame Γ [36; 106; 110; 139]. Indeed, by virtue of Theorem 3.8.1, if $\mathcal{E}_\Gamma$ is an integral of motion, it is a conserved generalized symmetry current of the variational symmetry

$$\gamma_H + \vartheta_{\mathcal{E}_\Gamma} = -(\partial_t + \Gamma^i \partial_i - p_j \partial_i \Gamma^j \partial^i) = -\widetilde{\Gamma}.$$

This is the canonical lift (3.5.3) onto V^*Q of the vector field $-\Gamma$ (1.1.18) on Q. Consequently, $-\widetilde{\Gamma}$ is an exact symmetry, and $-\Gamma$ is a Hamiltonian symmetry.

Example 3.8.1. Let us consider the *Kepler system* on the configuration space Q (2.5.29) in Example 2.5.4. Its phase space is

$$V^*Q = \mathbb{R} \times \mathbb{R}^6$$

coordinated by (t, q^i, p_i). The Lagrangian (2.5.30) and (2.5.34) of the Kepler system is hyperregular. The associated Hamiltonian form reads

$$H = p_i dq^i - \left[\frac{1}{2}\left(\sum_i (p_i)^2\right) - \frac{1}{r}\right] dt. \tag{3.8.14}$$

The corresponding Lagrangian L_H (3.5.1) is

$$L_H = \left[p_i q_t^i - \frac{1}{2}\left(\sum_i (p_i)^2\right) + \frac{1}{r}\right] dt. \tag{3.8.15}$$

The Kepler system possesses the following integrals of motion [49]:

- an energy function $\mathcal{E} = \mathcal{H}$;
- orbital momenta

$$M_b^a = q^a p_b - q^b p_a; \tag{3.8.16}$$

- components of the Rung–Lenz vector

$$A^a = \sum_b (q^a p_b - q^b p_a) p_b - \frac{q^a}{r}. \tag{3.8.17}$$

These integrals of motions are the conserved currents of:

- the exact symmetry ∂_t,
- the exact vertical symmetries

$$\upsilon_b^a = q^a \partial_b - q^b \partial_a - p_b \partial^a + p_a \partial^b, \tag{3.8.18}$$

- the variational vertical symmetries

$$\upsilon^a = \sum_b [p_b \upsilon_b^a + (q^a p_b - q^b p_a)\partial_b] + \partial_b \left(\frac{q^a}{r}\right) \partial^b, \tag{3.8.19}$$

respectively. Note that the vector fields υ_b^a (3.8.18) are the canonical lift (3.5.3) onto V^*Q of the vector fields

$$u_b^a = q^a \partial_b - q^b \partial_a$$

on Q. Thus, these vector fields are vertical Hamiltonian symmetries, and integrals of motion M_b^a (3.8.16) are the Noether currents.

Let us remind that, in contrast with the Rung–Lenz vector (3.8.19) in Hamiltonian mechanics, the Rung–Lenz vector (2.5.35) in Lagrangian mechanics fails to come from variational symmetries of a Lagrangian. There is the following relation between Lagrangian and Hamiltonian symmetries if they are the same vector fields on a configuration space Q.

Theorem 3.8.2. *Let a Hamiltonian form H be associated with an almost regular Lagrangian L. Let r be a solution of the Hamilton equation (3.3.22) – (3.3.23) for H which lives in the Lagrangian constraint space N_L. Let $s = \pi_\Pi \circ r$ be the corresponding solution of the Lagrange equation for L so that the relation (3.6.13) holds. Then, for any vector field u (2.5.4) on a fibre bundle $Q \to \mathbb{R}$, we have*

$$\mathfrak{T}_{\widetilde{u}}(r) = \mathfrak{T}_u(\pi_\Pi \circ r), \qquad \mathfrak{T}_{\widetilde{u}}(\widehat{L} \circ J^1 s) = \mathfrak{T}_u(s), \tag{3.8.20}$$

where $\mathfrak{T}_u$ is the symmetry current (2.5.21) on J^1Y and $\mathfrak{T}_{\widetilde{u}}$ is the symmetry current (3.8.12) on V^*Q.

Proof. The proof is straightforward. □

By virtue of Theorems 3.6.2 – 3.6.3, it follows that:

• if $\mathfrak{T}_u$ in Theorem 3.8.2 is a conserved symmetry current, then the symmetry current $\mathfrak{T}_{\widetilde{u}}$ (3.8.20) is conserved on solutions of the Hamilton equation which live in the Lagrangian constraint space;

• if $\mathfrak{T}_{\widetilde{u}}$ in Theorem 3.8.2 is a conserved symmetry current, then the symmetry current $\mathfrak{T}_u$ (3.8.20) is conserved on solutions s of the Lagrange equation which obey the condition (3.6.13).

3.9 Time-reparametrized mechanics

We have assumed above that the base $\mathbb{R}$ of a configuration space of non-relativistic mechanics is parameterized by a coordinate t with the transition functions $t \to t' = t+$ const. Here, we consider an arbitrary reparametrization of time

$$t \to t' = f(t) \tag{3.9.1}$$

which is discussed in some models of quantum mechanics [83].

In the case of an arbitrary time reparametrization (3.9.1), a configuration space of non-relativistic mechanics is a fibre bundle $Q \to R$ over a one-dimensional base R, diffeomorphic to $\mathbb{R}$. Let R be coordinated by t with the transition functions (3.9.1). In contrast with $\mathbb{R}$, the base R admits neither the standard vector field ∂_t nor the standard one-form dt. We can not use the simplifications mentioned in Remark 1.1.1 and, therefore, should strictly follow the (polysymplectic) Hamiltonian formalism on fibre bundles [55; 68; 138]. Nevertheless, Hamiltonian formalism of time-reparametrized mechanics possesses some peculiarities because of a one-dimensional base R.

• There exists the canonical tangent-valued one-form

$$\theta_R = dt \otimes \partial_t$$

on the base R of a configuration space of time-reparametrized mechanics.

• The velocity space J^1Q of time-reparametrized mechanics is not an affine subbundle of the tangent bundle TQ, whereas a phase space is isomorphic to the vertical cotangent bundle $V^*Q \to Q$. It follows that a phase space of time-reparametrized mechanics is provided with the canonical Poisson structure (3.3.7). Moreover, this Poisson structure is invariant under time reparametrization (3.9.1) which, consequently, is a canonical transformation.

• A phase space V^*Q is endowed with the canonical polysymplectic form

$$\Lambda = dp_i \wedge dq^i \wedge \theta_R.$$

Then the notions of a Hamiltonian connection and a Hamiltonian form are the repetitions of those in Hamiltonian field theory [68]. At the same time, since the homogeneous Legendre bundle of time-reparametrized mechanics is the cotangent bundle T^*Q of Q, Hamiltonian forms and Hamilton equations of time-reparametrized mechanics are defined as those in Section 3.3. The difference is only that the Hamiltonian function $\mathcal{E}_\Gamma$ in the splitting (3.3.17) is a density, but not a function under the transformations (3.9.1).

• Since a Lagrangian and a Hamiltonian of time-reparametrized mechanics are densities under the transformations (3.9.1), one should introduce a volume element on the base R in order to construct them in an explicit form. A key problem of models with time reparametrization lies in the fact that the time axis R of time-reparametrized mechanics has no canonical volume element. Another problem is concerned with a mass tensor. Since a velocity space J^1Q of time-reparametrized mechanics is an affine bundle $J^1Q \to Q$ modelled over the vector bundle $T^*R \underset{Q}{\otimes} VQ$, a mass tensor fails to be invariant under time reparametrization (3.9.1).

Chapter 4

Algebraic quantization

Algebraic quantum theory follows the hypothesis that an autonomous quantum system can be characterized by a topological involutive algebra A and continuous positive forms f on A treated as mean values of quantum observables. In quantum mechanics, C^*-algebras are considered. In accordance with the Gelfand–Naimark–Segal (henceforth *GNS*) construction, any positive form on a C^*-algebra A determines its cyclic representation by bounded operators in a Hilbert space (Section 4.1.5).

Quantum non-relativistic mechanics is phrased in the geometric terms of Banach and Hilbert manifolds and locally trivial Hilbert and C^*-algebra bundles over smooth finite-dimensional manifolds, e.g., $\mathbb{R}$ [65; 148]. For instance, this is the case of time-dependent quantum systems (Section 4.4) and quantum models depending on classical parameters (Section 9.3).

4.1 GNS construction

We start with a brief exposition of the conventional GNS representation of C^*-algebras [33; 65].

4.1.1 *Involutive algebras*

A complex algebra A is called *involutive*, if it is provided with an *involution* $*$ such that

$$(a^*)^* = a, \quad (a + \lambda b)^* = a^* + \overline{\lambda} b^*, \quad (ab)^* = b^* a^*, \quad a, b \in A, \quad \lambda \in \mathbb{C}.$$

Let us recall the standard terminology. An element $a \in A$ is *normal* if $aa^* = a^* a$, and it is *Hermitian* or *self-adjoint* (Section 4.1.6) if $a^* = a$. If

A is a unital algebra, a normal element such that

$$aa^* = a^*a = \mathbf{1}$$

is called *unitary*.

An involutive algebra A is called a *normed* algebra (resp. a *Banach algebra*) if it is a normed (resp. complete normed) vector space whose norm $\|.\|$ obeys the multiplicative conditions

$$\|ab\| \leq \|a\|\|b\|, \qquad \|a^*\| = \|a\|, \qquad a, b \in A.$$

A Banach involutive algebra A is called a C^**-algebra* if

$$\|a\|^2 = \|a^*a\|$$

for all $a \in A$. If A is a unital C^*-algebra, then $\|\mathbf{1}\| = 1$. A C^*-algebra is provided with the normed topology, i.e., it is a topological involutive algebra.

Let $\mathcal{I}$ be a closed two-sided ideal of a C^*-algebra A. Then $\mathcal{I}$ is *self-adjoint*, i.e., $\mathcal{I}^* = \mathcal{I}$. Endowed with the quotient norm, the quotient $A/\mathcal{I}$ is a C^*-algebra.

Remark 4.1.1. It should be emphasized that by a morphism of normed involutive algebras is customarily meant a morphism of the underlying involutive algebras, without any condition on the norms and continuity. At the same time, an isomorphism of normed algebras means always an isometric morphism. Any morphism ϕ of C^*-algebras is automatically continuous due to the property

$$\|\phi(a)\| \leq \|a\|, \qquad a \in A. \tag{4.1.1}$$

Any involutive algebra A can be extended to a unital algebra $\widetilde{A} = \mathbb{C} \oplus A$ by the adjunction of the identity $\mathbf{1}$ to A (Remark 11.1.1). The unital extension of A also is an involutive algebra with respect to the operation

$$(\lambda\mathbf{1} + a)^* = (\overline{\lambda}\mathbf{1} + a^*), \qquad \lambda \in \mathbb{C}, \quad a \in A.$$

If A is a normed involutive algebra, a norm on A is extended to $\widetilde{A}$, but not uniquely. If A is a C^*-algebra, a norm on A is uniquely prolonged to the norm

$$\|\lambda\mathbf{1} + a\| = \sup_{\|a'\| \leq 1} \|\lambda a' + aa'\|$$

on $\widetilde{A}$ which makes $\widetilde{A}$ into a C^*-algebra.

Remark 4.1.2. Any C^*-algebra admits an *approximate identity*. This is a family $\{u_\iota\}_{\iota\in I}$ of elements of A, indexed by a directed set I, which possesses the following properties:

(i) $\|u_\iota\| < 1$ for all $\iota \in I$,

(ii) $\|u_\iota a - a\| \to 0$ and $\|a u_\iota - a\| \to 0$ for every $a \in A$.

It should be noted that the existence of an approximate identity is an essential condition for many results. In particular, the GNS representation is relevant to Banach involutive algebras with an approximate identity. However, there is no loss of generality if we restrict our study of the GNS representation to C^*-algebras because any Banach involutive algebra A with an approximate identity defines the so called enveloping C^*-algebra $A^\dagger$ such that there is one-to-one correspondence between the representations of A and those of $A^\dagger$ [33].

Remark 4.1.3. Unless otherwise stated, by a *tensor product* $A\otimes A'$ of C^*-algebras A and A' is meant their minimal (or spatial tensor) product. This is the C^*-algebra defined as the completion of the tensor product of involutive algebras A and A' with respect to the minimal norm which obeys the condition

$$||a\otimes a'|| = ||a||\,||a'||. \qquad a\in A, \qquad a'\in A.$$

For instance, if A and A' are operator algebras in Hilbert spaces E and E', this norm is exactly the operator norm (4.1.8) of operators in the tensor product $E\otimes E'$ of Hilbert spaces E and E'. In general, there are several ways of completing the algebraic tensor product of C^*-algebras in order to obtain a C^*-algebra [137].

4.1.2 *Hilbert spaces*

An important example of C^*-algebras is the algebra $B(E)$ of bounded (and, equivalently, continuous) operators in a Hilbert space E. Every closed involutive subalgebra of $B(E)$ is a C^*-algebra and, conversely, every C^*-algebra is isomorphic to a C^*-algebra of this type (see Theorem 4.1.1 below).

Let us recall the basic facts on pre-Hilbert and Hilbert spaces [17].

A *Hermitian form* on a complex vector space E is defined as a sesquilinear form $\langle .|.\rangle$ such that

$$\langle e|e'\rangle = \overline{\langle e'|e\rangle}, \qquad \langle \lambda e|e'\rangle = \langle e|\overline{\lambda}e'\rangle = \lambda\langle e|e'\rangle, \quad e,e'\in E, \quad \lambda\in\mathbb{C}.$$

Remark 4.1.4. There exists another convention where $\langle e|\lambda e'\rangle = \lambda\langle e|e'\rangle$.

A Hermitian form $\langle .|. \rangle$ is said to be *positive* if $\langle e|e \rangle \geq 0$ for all $e \in E$. Throughout the book, all Hermitian forms are assumed to be positive. A Hermitian form is called *non-degenerate* if the equality $\langle e|e \rangle = 0$ implies $e = 0$. A complex vector space endowed with a (positive) Hermitian form is called a *pre-Hilbert* space. Morphisms of pre-Hilbert spaces, by definition, are isometric.

A Hermitian form provides E with the topology determined by the seminorm

$$\|e\| = \langle e|e \rangle^{1/2}. \tag{4.1.2}$$

Hence, a pre-Hilbert space is Hausdorff if and only if the Hermitian form $\langle .|. \rangle$ is non-degenerate, i.e., the seminorm (4.1.2) is a norm. In this case, the Hermitian form $\langle .|. \rangle$ is called a *scalar product*.

A family $\{e_i\}_I$ of elements of a pre-Hilbert space E is called *orthonormal* if its members are mutually orthogonal and $\|e_i\| = 1$ for all $i \in I$. Given an element $e \in E$, there exists at most a countable set of elements e_i of an orthonormal family such that $\langle e|e_i \rangle \neq 0$ and

$$\sum_{i \in I} \langle e|e_i \rangle^2 \leq \|e\|^2.$$

A family $\{e_i\}_I$ is called *total* if it spans a dense subset of E or, equivalently, if the condition $\langle e|e_i \rangle = 0$ for all $i \in I$ implies $e = 0$. A total orthonormal family in a Hausdorff pre-Hilbert space is called a *basis* for E. Given a basis $\{e_i\}_I$, any element $e \in E$ admits the decomposition

$$e = \sum_{i \in I} \langle e|e_i \rangle e_i, \qquad \|e\|^2 = \sum_{i \in I} |\langle e|e_i \rangle|^2.$$

A basis for a pre-Hilbert space need not exist.

Proposition 4.1.1. *Every Hausdorff pre-Hilbert space, satisfying the first axiom of countability (e.g., if it is second-countable), has a countable orthonormal basis.*

Remark 4.1.5. The notion of a basis for a pre-Hilbert space differs from that of an algebraic basis for a vector space.

A *Hilbert space* is defined as a complete Hausdorff pre-Hilbert space. Any Hausdorff pre-Hilbert space can be completed to a Hilbert space so that its basis, if any, also is a basis for its completion. Every Hilbert space has a basis, and any orthonormal family in a Hilbert space can be extended to its basis. All bases for a Hilbert space have the same cardinal number, called the *Hilbert dimension*. Moreover, given two bases for a Hilbert space, there is its isomorphism sending these bases to each other. A Hilbert space has a countable basis if and only if it is separable. Then it is called a *separable Hilbert space*. A separable Hilbert space is second-countable.

Remark 4.1.6. Unless otherwise stated, by a Hilbert space is meant a complex Hilbert space. A complex Hilbert space $(E, \langle .|. \rangle)$ seen as a real vector space $E_{\mathbb{R}}$ is provided with a real scalar product

$$(e, e') = \frac{1}{2}(\langle e|e' \rangle + \langle e'|e \rangle) = \mathrm{Re}\,\langle e|e' \rangle, \tag{4.1.3}$$

which makes $E_{\mathbb{R}}$ into a *real Hilbert space*. It also is a Banach real space. Conversely, the complexification $E = \mathbb{C} \otimes V$ of a real Hilbert space $(V, (.,.))$ is a complex Hilbert space with respect to the Hermitian form

$$\langle e_1 + ie_2|e_1' + ie_2' \rangle = (e_1, e_1') + i((e_2, e_1') - (e_1, e_2')) + (e_2, e_2'). \tag{4.1.4}$$

The following are the standard constructions of new Hilbert spaces from old ones.

• Let $(E^\iota, \langle .|. \rangle_{E^\iota})$ be a set of Hilbert spaces and $\sum E^\iota$ denote the direct sum of vector spaces E^ι. For any two elements $e = (e^\iota)$ and $e' = (e'^\iota)$ of $\sum E^\iota$, the sum

$$\langle e|e' \rangle_\oplus = \sum_\iota \langle e^\iota | e'^\iota \rangle_{E^\iota} \tag{4.1.5}$$

is finite, and defines a non-degenerate Hermitian form on $\sum E^\iota$. The completion $\oplus E^\iota$ of $\sum E^\iota$ with respect to this form is a Hilbert space, called the *Hilbert sum* of $\{E^\iota\}$. This is a subspace of the Cartesian product $\prod E^\iota$ which consists of the elements $e = (e^\iota)$ such that

$$\sum_\iota \|e^\iota\|_{E^\iota} < \infty.$$

The union of bases for Hilbert spaces E^ι is a basis for their Hilbert sum $\oplus E^\iota$.

• Let $(E, \langle .|. \rangle_E)$ and $(H, \langle .|. \rangle_H)$ be Hilbert spaces. Their *tensor product* $E \otimes H$ is defined as the completion of the tensor product of vector spaces E and H with respect to the scalar product

$$\langle w_1|w_2 \rangle_\otimes = \sum_{\iota,\beta} \langle e_1^\iota | e_2^\beta \rangle_E \langle h_1^\iota | h_2^\beta \rangle_H,$$

$$w_1 = \sum_\iota e_1^\iota \otimes h_1^\iota, \quad w_2 = \sum_\beta e_2^\beta \otimes h_2^\beta, \quad e_1^\iota, e_2^\beta \in E, \quad h_1^\iota, h_2^\beta \in H.$$

Let $\{e_i\}$ and $\{h_j\}$ be bases for E and H, respectively. Then $\{e_i \otimes h_j\}$ is a basis for $E \otimes H$.

• Let E' be the topological dual of a Hilbert space E. Then the assignment

$$e \to \overline{e}(e') = \langle e'|e\rangle, \qquad e, e' \in E, \tag{4.1.6}$$

defines an *antilinear* bijection of E onto E', i.e., $\overline{\lambda e} = \overline{\lambda}\overline{e}$. The dual E' of a Hilbert space is a Hilbert space provided with the scalar product

$$\langle \overline{e}|\overline{e}'\rangle' = \langle e'|e\rangle \tag{4.1.7}$$

such that the morphism (4.1.6) is isometric. The E' is called the *dual Hilbert space*, and is usually denoted by $\overline{E}$. A Hilbert space E and its dual E' seen as real Hilbert and Banach spaces are isomorphic to each other.

4.1.3 *Operators in Hilbert spaces*

Unless otherwise stated (Section 4.1.6), we deal with bounded *operators* $a \in B(E)$ in a Hilbert space E. They are continuous, and *vice versa*. Bounded operators are provided with the *operator norm*

$$\|a\| = \sup_{\|e\|_E = 1} \|ae\|_E, \qquad a \in B(E). \tag{4.1.8}$$

This norm makes the involutive algebra $B(E)$ of bounded operators in a Hilbert space E into a C^*-algebra. The corresponding topology on $B(E)$ is called the *normed operator topology*.

One also provides $B(E)$ with the *strong* and *weak operator topologies*, determined by the families of seminorms

$$\{p_e(a) = \|ae\|, \quad e \in E\},$$
$$\{p_{e,e'}(a) = |\langle ae|e'\rangle|, \quad e, e' \in E\},$$

respectively. The normed operator topology is finer than the strong one which, in turn, is finer than the weak operator topology. The strong and weak operator topologies on the subgroup $U(E) \subset B(E)$ of unitary operators coincide with each other.

Remark 4.1.7. It should be emphasized that $B(E)$ fails to be a topological algebra with respect to strong and weak operator topologies. Nevertheless, the involution in $B(E)$ also is continuous with respect to the weak operator topology, while the operations

$$B(E) \ni a \to aa' \in B(E),$$
$$B(E) \ni a \to a'a \in B(E),$$

where a' is a fixed element of $B(E)$, are continuous with respect to all the above mentioned operator topologies.

Remark 4.1.8. Let N be a subset of $B(E)$. The *commutant* N' of N is the set of elements of $B(E)$ which commute with all elements of N. It is a subalgebra of $B(E)$. Let $N'' = (N')'$ denote the *bicommutant*. Clearly, $N \subset N''$. An involutive subalgebra B of $B(E)$ is called a *von Neumann algebra* if $B = B''$. This property holds if and only if B is strongly (or, equivalently, weakly) closed in $B(E)$ [33]. For instance, $B(E)$ is a von Neumann algebra. Since a strongly (weakly) closed subalgebra of $B(E)$ also is closed with respect to the normed operator topology on $B(E)$, any von Neumann algebra is a C^*-algebra.

Remark 4.1.9. An operator in a Hilbert space E is called *completely continuous* if it is *compact*, i.e., it sends any bounded set into a set whose closure is compact. An operator $a \in B(E)$ is completely continuous if and only if it can be represented by the series

$$a(e) = \sum_{k=1}^{\infty} \lambda_k \langle e|e_k\rangle e_k, \qquad (4.1.9)$$

where e_k are elements of a basis for E and λ_k are positive numbers which tend to zero as $k \to \infty$. For instance, every *degenerate operator* (i.e., an operator of finite rank which sends E onto its finite-dimensional subspace) is completely continuous. Moreover, the set $T(E)$ of completely continuous operators in E is the completion of the set of degenerate operators with respect to the operator norm (4.1.8). Every completely continuous operator can be written as $a = UT$, where U is a unitary operator and T is a *positive* completely continuous operator, i.e., $\langle Te|e\rangle \geq 0$ for all $e \in E$.

4.1.4 *Representations of involutive algebras*

In this Section, we consider a *representation* of and involutive algebra A by bounded operators in a Hilbert space [33; 128]. It is a morphism π of an involutive algebra A to the algebra $B(E)$ of bounded operators in a Hilbert space E, called the *carrier space* of π. Representations throughout are assumed to be *non-degenerate*, i.e., there is no element $e \neq 0$ of E such that $Ae = 0$ or, equivalently, AE is dense in E. A representation π of an involutive algebra A is uniquely prolonged to a representation $\widetilde{\pi}$ of the unital extension $\widetilde{A}$ of A.

Theorem 4.1.1. *If A is a C^*-algebra, there exists its isomorphic representation.*

Two representations π_1 and π_2 of an involutive algebra A in Hilbert spaces E_1 and E_2 are said to be *equivalent* if there is an isomorphism $\gamma : E_1 \to E_2$ such that

$$\pi_2(a) = \gamma \circ \pi_1(a) \circ \gamma^{-1}, \qquad a \in A.$$

Let $\{\pi^\iota\}$ be a family of representations of an involutive algebra A in Hilbert spaces E^ι. If the set of numbers $\|\pi^\iota(a)\|$ is bounded for each $a \in A$, one can construct the continuous linear operator $\pi(a)$ in the Hilbert sum $\oplus E^\iota$ which induces $\pi^\iota(a)$ in each E^ι. For instance, this is the case of a C^*-algebra A due to the property (4.1.1). Then π is a representation of A in $\oplus E^\iota$, called the *Hilbert sum* of representations π^λ.

Given a representation π of an involutive algebra A in a Hilbert space E, an element $\theta \in E$ is said to be a *cyclic vector* for π if the closure of $\pi(A)\theta$ is equal to E. Accordingly, π is called a *cyclic representation.*

Theorem 4.1.2. *Every representation of an involutive algebra A is a Hilbert sum of cyclic representations.*

A representation π of an involutive algebra A in a Hilbert space E is called *topologically irreducible* if the following equivalent conditions hold:

- the only closed subspaces of E invariant under $\pi(A)$ are 0 and E;
- the commutant of $\pi(A)$ in $B(E)$ is the set of scalar operators;
- every non-zero element of E is a cyclic vector for π.

Let us recall that irreducibility of π in the algebraic sense means that the only subspaces of E invariant under $\pi(A)$ are 0 and E. If A is a C^*-algebra, the notions of topologically and *algebraically* irreducible representations are equivalent. Therefore, we will further speak on *irreducible representations* of a C^*-algebra without the above mentioned qualification.

An algebraically irreducible representation π of an involutive algebra A is characterized by its kernel $\operatorname{Ker}\pi \subset A$. This is a two-sided ideal, called *primitive*. The assignment

$$\widehat{A} \ni \pi \to \operatorname{Ker}\pi \in \operatorname{Prim}(A) \qquad (4.1.10)$$

defines the canonical surjection of the set $\widehat{A}$ of the equivalence classes of algebraically irreducible representations of an involutive algebra A onto the set $\operatorname{Prim}(A)$ of primitive ideals of A. It follows that algebraically irreducible representations with different kernels are necessarily inequivalent.

The set $\mathrm{Prim}(A)$ is equipped with the so called *Jacobson topology* [33]. This topology is not Hausdorff, but it obeys the *Fréchet axiom*, i.e., for any two distinct points of $\mathrm{Prim}(A)$, there is a neighborhood of one of the points which does not contain the other. Then the set $\widehat{A}$ is endowed with the coarsest topology such that the surjection (4.1.10) is continuous. It is called the *spectrum* of an involutive algebra A.

Proposition 4.1.2. *If the spectrum $\widehat{A}$ satisfies the Fréchet axiom (e.g., $\widehat{A}$ is Hausdorff), the map $\widehat{A} \to \mathrm{Prim}(A)$ is a homeomorphism, i.e., algebraically irreducible representations with the same kernel are equivalent.*

Proposition 4.1.3. *If an involutive algebra A is unital, $Prim(A)$ and $\widehat{A}$ are quasi-compact, i.e., they satisfy the Borel–Lebesgue axiom, but need not be Hausdorff.*

Proposition 4.1.4. *The spectrum $\widehat{A}$ of a C^*-algebra A is a locally quasi-compact space.*

A C^*-algebra is said to be *elementary* if it is isomorphic to the algebra $T(E) \subset B(E)$ of compact operators in some Hilbert space E. Every non-trivial irreducible representation of an elementary C^* algebra $A = T(E)$ is equivalent to its isomorphic representation by compact operators in E [33]. Hence, the spectrum of an elementary algebra is a singleton set.

4.1.5 *GNS representation*

Let f be a complex form on an involutive algebra A. It is called *positive* if $f(a^*a) \geq 0$ for all $a \in A$. Given a positive form f, the Hermitian form

$$\langle a|b\rangle = f(b^*a), \qquad a,b \in A, \tag{4.1.11}$$

makes A into a pre-Hilbert space. In particular, the relation

$$|f(b^*a)|^2 \leq f(a^*a)f(b^*b), \qquad a,b \in A, \tag{4.1.12}$$

holds. If A is a normed involutive algebra, positive continuous forms on A are provided with the norm

$$\|f\| = \sup_{\|a\|=1} |f(a)|, \qquad a \in A.$$

One says that f is a *state* of A if $\|f\| = 1$. Positive forms on a C^*-algebra are continuous. Conversely, a continuous form f on an unital C^*-algebra is positive if and only if $f(\mathbf{1}) = \|f\|$. In particular, it is a state if and only if $f(\mathbf{1}) = 1$.

For instance, let A be an involutive algebra, π its representation in a Hilbert space E, and θ an element of E. Then the map

$$\omega_\theta : a \to \langle \pi(a)\theta|\theta\rangle \tag{4.1.13}$$

is a positive form on A. It is called the *vector form* determined by π and θ. This vector form is a state if the vector θ is normalized. Let ω_{θ_1} and ω_{θ_2} be two vector forms on A determined by representations π_1 in E_1 and π_2 in E_2. If $\omega_{\theta_1} = \omega_{\theta_2}$, there exists a unique isomorphism of E_1 to E_2 which sends π_1 to π_2 and $\theta_1 \in E_1$ to $\theta_2 \in E_2$.

The following theorem states that, conversely, any positive form on a C^*-algebra equals a vector form determined by some representation of A called the *GNS representation* [33].

Theorem 4.1.3. *Let f be a positive form on a C^*-algebra A. It is extended to a unique positive form $\widetilde f$ on the unital extension $\widetilde A$ of A such that $\widetilde f(\mathbf{1}) = \|f\|$. Let N_f be a left ideal of $\widetilde A$ consisting of those elements $a \in A$ such that $\widetilde f(a^*a) = 0$. The quotient $\widetilde A/N_f$ is a Hausdorff pre-Hilbert space with respect to the Hermitian form obtained from $\widetilde f(b^*a)$ (4.1.11) by passage to the quotient. We abbreviate with E_f the completion of $\widetilde A/N_f$ and with θ_f the canonical image of $\mathbf{1} \in \widetilde A$ in $\widetilde A/N_f \subset E_f$. For each $a \in \widetilde A$, let $\tau(a)$ be the operator in $\widetilde A/N_f$ obtained from the left multiplication by a in $\widetilde A$ by passage to the quotient. Then the following hold.*

(i) Each $\tau(a)$ has a unique extension to an operator $\pi_f(a)$ in the Hilbert space E_f.

(ii) The map $a \to \pi_f(a)$ is a representation of A in E_f.

(iii) The representation π_f admits a cyclic vector θ_f.

(iv) $f(a) = \langle \pi(a)\theta_f|\theta_f\rangle$ for each $a \in A$.

The representation π_f and the cyclic vector θ_f in Theorem 4.1.3 are said to be *determined by the form* f, and the form f equals the vector form determined by π_f and θ_f. Conversely, given a representation π of A in a Hilbert space E and a cyclic vector θ for π, let ω be the vector form on A determined by π and θ. Let π_ω and θ_ω be the representation in E_ω and the vector of E_ω determined by ω in accordance with Theorem 4.1.3. Then there is a unique isomorphism of E to E_ω which sends π to π_ω and θ to θ_ω.

Example 4.1.1. In particular, any cyclic representation of a C^*-algebra A is a summand of the *universal representation* $\oplus_f \pi_f$ of A, where f runs through all positive forms on A.

It may happen that different positive forms on a C^*-algebra determine the same representation as follows.

Proposition 4.1.5. *(i) Let A be a C^*-algebra and f a positive form on A which determines a representation π_f of A and its cyclic vector θ_f. Then for any $b \in A$, the positive form $a \to f(b^*ab)$ on A determines the same representation π_f.*

(ii) Conversely, any vector form f' on A determined by the representation π_f is the limit $a \to F(b_i^* a b_i)$, where $\{b_i\}$ is a convergent sequence with respect to the normed topology on A.

Now let us specify positive forms on a C^*-algebra A which determine its irreducible representations.

A positive form f' on an involutive algebra A is said to be *dominated* by a positive form f if $f - f'$ is a positive form. A non-zero positive form f on an involutive algebra A is called *pure* if every positive form f' on A which is dominated by f reads λf, $0 \leq \lambda \leq 1$.

Theorem 4.1.4. *The representation of π_f of a C^*-algebra A determined by a positive form f on A is irreducible if and only if f is a pure form [33].*

In particular, any vector form determined by a vector of a carrier space of an irreducible representation is a pure form. Therefore, it may happen that different pure forms determine the same irreducible representation.

Theorem 4.1.5. *(i) Pure states f_1 and f_2 of a C^*-algebra A yield equivalent representations of A if and only if there exists an unitary element U of the unital extension $\widetilde{A}$ of A such that*

$$f_2(a) = f_1(U^* a U), \qquad a \in A.$$

(ii) Conversely, let π be an irreducible representation of a C^*-algebra A in a Hilbert space E. Given two different elements θ_1 and θ_2 of E (they are cyclic for π), the vector forms on A determined by (π, θ_1) and (π, θ_2) are equal if and only if there exists $\lambda \in \mathbb{C}$, $|\lambda| = 1$, such that $\theta_1 = \lambda\theta_2$.

(iii) There is one-to-one correspondence between the pure states of a C^*-algebra A associated to the same irreducible representation π of A in a Hilbert space E and the one-dimensional complex subspaces of E. It follows that these states constitute the projective Hilbert space PE in Section 4.3.5.

Let $P(A)$ denote the set of pure states of a C^*-algebra A. Theorem 4.1.5 implies a surjection $P(A) \to \widehat{A}$, where $\widehat{A}$ is the spectrum of A. This surjection is a bijection if and only if any irreducible representation of A is one-dimensional, i.e., A is a commutative C^*-algebra. In this case, A is the C^*-algebra of continuous complex functions vanishing at infinity on $\widehat{A}$, while a pure state on A is a Dirac measure ε_x, $x \in \widehat{A}$, on $\widehat{A}$, i.e., $\varepsilon_x(a) = a(x)$ for all $a \in A$.

Being a subset of the topological dual A' of the Banach space A, the set $P(A)$ is provided with the normed topology. However, one usually refers to $P(A)$ equipped with the weak* topology. In this case, the canonical surjection $P(A) \to \widehat{A}$ is continuous and open [33].

4.1.6 *Unbounded operators*

There are algebras whose representations in Hilbert spaces need not be normed. Therefore, let us consider a generalization of the conventional GNS representation of C^*-algebras to unnormed topological involutive algebras.

By an *operator* in a Hilbert (or Banach) space E is meant a linear morphism a of a dense subspace $D(a)$ of E to E. The $D(a)$ is called a *domain* of an operator a. One says that an operator b on $D(b)$ is an *extension* of an operator a in $D(a)$ if $D(a) \subset D(b)$ and $b|_{D(a)} = a$. For the sake of brevity, let us write $a \subset b$. An operator a is said to be *bounded* in $D(a)$ if there exists a real number r such that

$$\|ae\| \leq r\|e\|, \qquad e \in D(a).$$

If otherwise, it is called *unbounded*. Any bounded operator in a domain $D(a)$ is uniquely extended to a bounded operator everywhere in E. Therefore, by bounded operators in E are usually meant bounded (continuous) operators defined everywhere in E.

An operator a in a domain $D(a)$ is called *closed* if the condition that a sequence $\{e_i\} \subset D(a)$ converges to $e \in E$ and that the sequence $\{ae_i\}$ does to $e' \in E$ implies that $e \in D(a)$ and $e' = ae$. Of course, any operator defined everywhere in E is closed. An operator a in a domain $D(a)$ is called *closable* if it can be extended to a closed operator. The *closure* of a closable operator a is defined as the minimal closed extension of a.

Operators a and b in E are called *adjoint* if

$$\langle ae|e'\rangle = \langle e|be'\rangle, \quad e \in D(a), \quad e' \in D(b).$$

Any operator a has a *maximal adjoint operator* a^*, which is closed. Of course, $a \subset a^{**}$ and $b^* \subset a^*$ if $a \subset b$. An operator a is called *symmetric* if it is adjoint to itself, i.e., $a \subset a^*$. Hence, a symmetric operator is closable.

One can obtain the following chain of extensions of a symmetric operator:

$$a \subset \overline{a} \subset a^{**} \subset a^* = \overline{a}^* = a^{***}.$$

In particular, if a is a symmetric operator, so are $\overline{a}$ and a^{**}. At the same time, the maximal adjoint operator a^* of a symmetric operator a need not be symmetric. A symmetric operator a is called *self-adjoint* if $a = a^*$, and it is called *essentially self-adjoint* if $\overline{a} = a^* = \overline{a}^*$. It should be emphasized that a symmetric operator a is sometimes called essentially self-adjoint if $a^{**} = a^*$. We here follow the terminology of [129; 130]. If a is a closed operator, the both notions coincide. For bounded operators, the notions of symmetric, self-adjoint and essentially self-adjoint operators coincide.

Let E be a Hilbert space. The pair (B, D) of a dense subspace D of E and a unital algebra B of (unbounded) operators in E is called the *Op^*-algebra* (*O^*-algebra* in the terminology of [146]) on the domain D if, whenever $b \in B$, we have:

(i) $D(b) = D$ and $bD \subset D$,

(ii) $D \subset D(b^*)$,

(iii) $b^*|_D \subset B$ [86; 129].

The algebra B is provided with the involution $b \to b^+ = b^*|_D$, and its elements are closable.

A representation $\pi(A)$ of a topological involutive algebra A in a Hilbert space E is an Op^*-algebra if there exists a dense subspace $D(\pi) \subset E$ such that

$$D(\pi) = D(\pi(a))$$

for all $a \in A$ and this representation is *Hermitian*, i.e., $\pi(a^*) \subset \pi(a)^*$ for all $a \in A$. In this case, one also considers the representations

$$\overline{\pi} : a \to \overline{\pi}(a) = \overline{\pi(a)}|_{D(\overline{\pi})}, \qquad D(\overline{\pi}) = \bigcap_{a \in A} D(\overline{\pi(a)}),$$

$$\pi^* : a \to \pi^*(a) = \pi(a^*)^*|_{D(\pi^*)}, \qquad D(\pi^*) = \bigcap_{a \in A} D(\pi(a)^*),$$

$$\pi^{**} : a \to \pi^{**}(a) = \pi^*(a^*)^*|_{D(\pi^{**})}, \qquad D(\pi^{**}) = \bigcap_{a \in A} D(\pi^*(a)^*),$$

called the *closure of a representation* π, an *adjoint representation* and a *second adjoint representation*, respectively. There are the representation extensions

$$\pi \subset \overline{\pi} \subset \pi^{**} \subset \pi^*,$$

where $\pi_1 \subset \pi_2$ means $D(\pi_1) \subset D(\pi_2)$. The representations $\overline{\pi}$ and π^{**} are Hermitian, while $\pi^* = \overline{\pi}^* = \pi^{***}$. A Hermitian representation $\pi(A)$ is said to be *closed* if $\pi = \overline{\pi}$, and it is *self-adjoint* if $\pi = \pi^*$. Herewith, a representation $\pi(A)$ is closed (resp. self-adjoint) if one of operators $\pi(A)$ is closed (resp. self-adjoint).

The representation domain $D(\pi)$ is endowed with the *graph-topology*. This is generated by the neighborhoods of the origin

$$U(M,\varepsilon) = \left\{ x \in D(\pi) \, : \, \sum_{a\in M} \|\pi(a)x\| < \varepsilon \right\},$$

where M is a finite subset of elements of A. All operators of $\pi(A)$ are continuous with respect to this topology. Let us note that the graph-topology is finer than the relative topology on $D(\pi) \subset E$, unless all operators $\pi(a)$, $a \in A$, are bounded [146].

Let $\overline{N}^g$ denote the closure of a subset $N \subset D(\pi)$ with respect to the graph-topology. An element $\theta \in D(\pi)$ is called *strongly cyclic* (cyclic in the terminology of [146]) if

$$D(\pi) \subset \overline{(\pi(A)\theta)}^g.$$

Then the GNS representation Theorem 4.1.3 can be generalized as follows [86; 146].

Theorem 4.1.6. *Let A be a unital topological involutive algebra and f a positive continuous form on A such that $f(\mathbf{1}) = 1$ (i.e., f is a state). There exists a strongly cyclic Hermitian representation (π_f, θ_f) of A such that*

$$\phi(a) = \langle \pi(a)\theta_\phi | \theta_\phi \rangle, \qquad a \in A.$$

4.2 Automorphisms of quantum systems

Let us consider uniformly and strongly continuous one-parameter groups of automorphisms of C^*-algebras. In particular, they characterize evolution of quantum systems. Forthcoming Remarks 4.2.1 and 4.2.2 explain why we restrict our consideration to these automorphism groups.

Remark 4.2.1. Let V be a Banach space and $B(V)$ the set of bounded endomorphisms of V. The normed, strong and weak operator topologies on $B(V)$ are defined in the same manner as in Section 4.1.3. Automorphisms of a C^*-algebra obviously are its isometries as a Banach space. Any weakly continuous one-parameter group of endomorphism of a Banach space also is strongly continuous and their weak and strong generators coincide with each other [19].

Remark 4.2.2. There is the following relation between morphisms of a C^*-algebra A and the set $E(A)$ of its states which is a convex subset of the topological dual A' of A. A linear morphism γ of a C^*-algebra A as a vector space is called the *Jordan morphism* if the relations

$$\gamma(ab+ba) = \gamma(a)\gamma(b) + \gamma(b)\gamma(a), \qquad \phi(a^*) = \gamma(a)^*, \qquad a, b \in A,$$

hold. One can show the following [39]. Let γ be a Jordan automorphism of a unital C^*-algebra A. It yields the dual weakly* continuous *affine* bijection γ' of $E(A)$ onto itself, i.e.,

$$\gamma'(\lambda f + (1-\lambda)f') = \lambda\gamma'(f) + (1-\lambda)\gamma'(f'),$$
$$f, f', \in E(A), \qquad \lambda \in [0,1].$$

Conversely, any such a map of $E(A)$ is the dual to some Jordan automorphism of A. However, we are not concerned with groups of Jordan automorphisms because of the following fact. If G is a connected group of weakly continuous Jordan automorphisms of a unital C^*-algebra A which is provided with a weak operator topology, then it is a weakly continuous group of automorphisms of A.

One says that a one-parameter group $G(\mathbb{R})$ is a *uniformly* (resp. *strongly*) *continuous group* of automorphisms of a C^*-algebra A if it is a range of a continuous map of $\mathbb{R}$ to the group $\mathrm{Aut}\,(A)$ of automorphisms of A which is provided with the normed (resp. strong) operator topology, and whose action on A is separately continuous. A problem is that, if a curve $G(\mathbb{R})$ in $\mathrm{Aut}\,(A)$ is continuous with respect to the normed operator topology, then the curve $G(\mathbb{R})(a)$ for any $a \in A$ is continuous in the C^*-algebra A, but the converse is not true. At the same time, a curve $G(\mathbb{R})$ is continuous in $\mathrm{Aut}\,(A)$ with respect to the strong operator topology if and only if the curve $G(\mathbb{R})(a)$ for any $a \in A$ is continuous in A. By this reason, strongly continuous one-parameter groups of automorphisms of C^*-algebras are most interesting. However, the infinitesimal generator of such a group fails to be bounded, unless this group is uniformly continuous.

Remark 4.2.3. If $G(\mathbb{R})$ is a strongly continuous one-parameter group of automorphisms of a C^*-algebra A, there are the following continuous maps [19]:

- $\mathbb{R} \ni t \to \langle G_t(a), f\rangle \in \mathbb{C}$ is continuous for all $a \in A$ and $f \in A'$;
- $A \ni a \to G_t(a) \in A$ is continuous for all $t \in \mathbb{R}$;
- $\mathbb{R} \ni t \to G_t(a) \in A$ is continuous for all $a \in A$.

Let A be a C^*-algebra. Without a loss of generality, we assume that A is a unital algebra. The space of *derivations* of A is provided with the involution $u \to u^*$ defined by the equality

$$\delta^*(a) = -\delta(a^*)^*, \qquad a \in \mathcal{A}. \tag{4.2.1}$$

Throughout this Section, by a derivation δ of A is meant an (unbounded) *symmetric* derivation of A (i.e., $\delta(a^*) = \delta(a)^*$, $a \in A$) which is defined on a dense involutive subalgebra $D(\delta)$ of A. If a derivation δ on $D(\delta)$ is bounded, it is extended to a bounded derivation everywhere on A. Conversely, every derivation defined everywhere on a C^*-algebra is bounded [33]. For instance, any *inner derivation*

$$\delta(a) = i[b, a],$$

where b is a Hermitian element of A, is bounded. There is the following relation between bounded derivations of a C^*-algebra A and one-parameter groups of automorphisms of A [19].

Theorem 4.2.1. *Let δ be a derivation of a C^*-algebra A. The following assertions are equivalent:*

- *δ is defined everywhere and, consequently, is bounded;*
- *δ is the infinitesimal generator of a uniformly continuous one-parameter group $[G_t]$ of automorphisms of the C^*-algebra A.*

Furthermore, for any representation π of A in a Hilbert space E, there exists a bounded self-adjoint operator $\mathcal{H} \in \pi(A)''$ in E and the uniformly continuous representation

$$\pi(G_t) = \exp(-it\mathcal{H}), \qquad t \in \mathbb{R}, \tag{4.2.2}$$

of the group $[G_t]$ in E such that

$$\pi(\delta(a)) = -i[\mathcal{H}, \pi(a)], \qquad a \in A, \tag{4.2.3}$$

$$\pi(G_t(a)) = e^{-it\mathcal{H}}\pi(a)e^{it\mathcal{H}}, \qquad t \in \mathbb{R}. \tag{4.2.4}$$

A C^*-algebra need not admit non-zero bounded derivations. For instance, no commutative C^*-algebra possesses bounded derivations. The following is the relation between (unbounded) derivations of a C^*-algebra A and strongly continuous one-parameter groups of automorphisms of A [18; 129; 130].

Theorem 4.2.2. *Let δ be a closable derivation of a C^*-algebra A. Its closure $\overline{\delta}$ is an infinitesimal generator of a strongly continuous one-parameter group of automorphisms of A if and only if*

(i) the set $(\mathbf{1} + \lambda\delta)(D(\delta)$ for any $\lambda \in \mathbb{R} \setminus \{0\}$ is dense in A,

(ii) $\|(\mathbf{1} + \lambda\delta)(a)\| \geq \|a\|$ for any $\lambda \in \mathbb{R}$ and any $a \in A$.

It should be noted that, if A is a unital algebra and δ is its closable derivation, then $\mathbf{1} \in D(\delta)$.

Let us mention a more convenient sufficient condition for a derivation of a C^*-algebra to be an infinitesimal generator of a strongly continuous one-parameter group of its automorphisms. A derivation δ of a C^*-algebra A is called *well-behaved* if, for each element $a \in D(\delta)$, there exists a state f of A such that

$$f(a) = \|a\|, \qquad f(\delta(a)) = 0.$$

If δ is a well-behaved derivation, it is closable [92], and it obeys the condition (ii) of Theorem 4.2.2 [18; 129; 130]. Then we come to the following.

Proposition 4.2.1. *If δ is a well-behaved derivation of a C^*-algebra A and it obeys condition (i) of Theorem 4.2.2, its closure $\overline{\delta}$ is an infinitesimal generator of a strongly continuous one-parameter group of automorphisms of A.*

For instance, a derivation δ is well-behaved if it is *approximately inner*, i.e., there exists a sequence of self-adjoint elements $\{b_n\}$ in A such that

$$\delta(a) = \lim_n i[b_n, a], \qquad a \in A.$$

In contrast with the case of a uniformly continuous one-parameter group of automorphisms of a C^*-algebra A, a representation of A does not imply necessarily a unitary representation (4.2.2) of a strongly continuous one-parameter group of automorphisms of A, unless the following.

Proposition 4.2.2. *Let G_t be a strongly continuous one-parameter group of automorphisms of a C^*-algebra A and δ its infinitesimal generator. Let A admit a state f such that*

$$|f(\delta(a))| \leq \lambda[f(a^*a) + f(aa^*)]^{1/2} \tag{4.2.5}$$

for all $a \in A$ and a positive number λ, and let (π_f, θ_f) be a cyclic representation of A in a Hilbert space E_f determined by f. Then there exist a self-adjoint operator $\mathcal{H}$ in a domain $D(\mathcal{H}) \subset A\theta_f$ in E_f and a strongly continuous unitary representation (4.2.2) of G_t in E_f which fulfils the relations (4.2.3) – (4.2.4) for $\pi = \pi_f$.

Let us note that the condition (4.2.5) of Theorem 4.2.2 is sufficient in order that the derivation δ is closable [92].

There is a general problem of a unitary representation of an automorphism group of a C^*-algebra.

For instance, let $B(E)$ be the C^*-algebra of bounded operators in a Hilbert space E. All its automorphisms are inner. Any (unitary) automorphism U of a Hilbert space E yields the inner automorphism

$$a \to UaU^{-1}, \qquad a \in B(E), \tag{4.2.6}$$

of $B(E)$. Herewith, the automorphism (4.2.6) is the identity if and only if $U = \lambda\mathbf{1}$, $|\lambda| = 1$, is a scalar operator in E. It follows that the group of automorphisms of $B(E)$ is the quotient

$$PU(E) = U(E)/U(1), \tag{4.2.7}$$

called the *projective unitary group* of the unitary group $U(E)$ with respect to the circle subgroup $U(1)$. Therefore, given a group G of automorphisms of the C^*-algebra $B(E)$, the representatives U_g in $U(E)$ of elements $g \in G$ constitute a group up to *phase multipliers*, i.e.,

$$U_g U_{g'} = \exp[i\alpha(g,g')]U_{gg'}, \qquad \alpha(g,g') \in \mathbb{R}.$$

Nevertheless, if G is a one-parameter weakly* continuous group of automorphisms of $B(E)$ whose infinitesimal generator is a bounded derivation of $B(E)$, one can choose the phase multipliers

$$\exp[i\alpha(g,g')] = 1.$$

Representations of groups by unitary operators up to phase multipliers are called *projective representations* [24; 159].

In a general setting, let G be a group and $\mathcal{A}$ a commutative algebra. An $\mathcal{A}$-*multiplier* of G is a map $\xi : G \times G \to \mathcal{A}$ such that

$$\xi(\mathbf{1}_G, g) = \xi(g, \mathbf{1}_G) = \mathbf{1}_{\mathcal{A}}, \qquad g \in G,$$

$$\xi(g_1, g_2g_3)\xi(g_2,g_3) = \xi(g_1,g_2)\xi(g_1g_2,g_3), \qquad g_i \in G.$$

For instance,

$$\xi : G \times G \to \mathbf{1}_{\mathcal{A}} \in \mathcal{A}$$

is a multiplier. Two A-multipliers ξ and ξ' are said to be *equivalent* if there exists a map $f : G \to \mathcal{A}$ such that

$$\xi(g_1,g_2) = \frac{f(g_1g_2)}{f(g_1)f(g_2)}\xi'(g_1,g_2), \qquad g_i \in G.$$

An $\mathcal{A}$-multiplier is called *exact* if it is equivalent to the multiplier $\xi = \mathbf{1}_{\mathcal{A}}$. The set of $\mathcal{A}$-multipliers is an Abelian group with respect to the pointwise multiplication, and the set of exact multipliers is its subgroup.

Proposition 4.2.3. *Let G be a simply connected locally compact Lie group. Each $U(1)$-multiplier ξ of G is brought into the form $\xi = \exp i\alpha$, where α is an $\mathbb{R}$-multiplier. Moreover, ξ is exact if and only if α is well. Any $\mathbb{R}$-multiplier of G is equivalent to a smooth one [24].*

Let G be a locally compact group of strongly continuous automorphisms of a C^*-algebra A. Let $M(A)$ denote the *multiplier algebra* of A, i.e., the largest C^*-algebra containing A as an *essential ideal*, i.e., if $a \in M(A)$ and $ab = 0$ for all $b \in A$, then $a = 0$. For instance, $M(A) = A$ if A is a unital algebra. Let ξ be a multiplier of G with values in the center of $M(A)$. A *G-covariant representation* π of A [34] is a representation π of A (and, consequently, $M(A)$) in a Hilbert space E together with a projective representation of G by unitary operators $U(g)$, $g \in G$, in E such that

$$\pi(g(a)) = U(g)\pi(a)U^*(g), \qquad U(g)U(g') = \pi(\xi(g,g'))U(gg').$$

4.3 Banach and Hilbert manifolds

We start with the notion of a real Banach manifold [100; 155]. Banach manifolds are defined similarly to finite-dimensional smooth manifolds, but they are modelled on Banach spaces, not necessarily finite-dimensional.

4.3.1 *Real Banach spaces*

Let us recall some particular properties of (infinite-dimensional) real Banach spaces (see Section 11.7 for topological vector spaces). Let us note that a finite-dimensional Banach space is always provided with an Euclidean norm.

• Given Banach spaces E and H, every continuous bijective linear map of E to H is an isomorphism of topological vector spaces.

• Given a Banach space E, let F be its closed subspace. One says that F *splits* in E if there exists a closed complement F' of F such that $E = F \oplus F'$. In particular, finite-dimensional and finite-codimensional subspaces split in E. As a consequence, any subspace of a finite-dimensional space splits.

• Let E and H be Banach spaces and $f : E \to H$ a continuous injection. One says that f splits if there exists an isomorphism

$$g : H \to H_1 \times H_2$$

such that $g \circ f$ yields an isomorphism of E onto $H_1 \times \{0\}$.

• Given Banach spaces $(E, ||.||_E)$ and $(H, ||.||_H)$, one can provide the set $\mathrm{Hom}^0(E,H)$ of continuous linear morphisms of E to H with the norm

$$||f|| = \sup_{||z||_E=1} ||f(z)||_H, \qquad f \in \mathrm{Hom}^0(E,H). \tag{4.3.1}$$

In particular, the norm (11.7.1) on the topological dual E' of E is of this type. If E, H and F are Banach spaces, the bilinear map

$$\mathrm{Hom}^0(E,F) \times \mathrm{Hom}^0(F,H) \to \mathrm{Hom}^0(E,H),$$

obtained by the composition $f \circ g$ of morphisms $\gamma \in \mathrm{Hom}^0(E,F)$ and $f \in \mathrm{Hom}^0(F,H)$, is continuous.

• Let $(E, \|.\|_E)$ and $(H, \|.\|_H)$ be real Banach spaces. One says that a continuous map $f : E \to H$ (not necessarily linear and isometric) is a *differentiable function* between E and H if, given a point $z \in E$, there exists an $\mathbb{R}$-linear continuous map

$$df(z) : E \to H$$

(not necessarily isometric) such that

$$f(z') = f(z) + df(z)(z' - z) + o(z' - z),$$
$$\lim_{\|z'-z\|_E \to 0} \frac{\|o(z'-z)\|_H}{\|z'-z\|_E} = 0,$$

for any z' in some open neighborhood U of z. For instance, any continuous linear morphism f of E to H is differentiable and $df(z)z = f(z)$. The linear map $df(z)$ is called a *differential* of f at a point $z \in U$. Given an element $v \in E$, we obtain the map

$$E \ni z \to \partial_v f(z) = df(z)v \in H, \tag{4.3.2}$$

called the *derivative* of a function f along a vector $v \in E$. One says that f is two-times differentiable if the map (4.3.2) is differentiable for any $v \in E$. Similarly, r-times differentiable and infinitely differentiable (smooth) functions on a Banach space are defined. The composition of smooth maps is a smooth map.

The following *inverse mapping theorem* enables one to consider smooth Banach manifolds and bundles similarly to the finite-dimensional ones.

Theorem 4.3.1. *Let $f : E \to H$ be a smooth map such that, given a point $z \in E$, the differential $df(z) : E \to H$ is an isomorphism of topological vector spaces. Then f is a local isomorphism at z.*

4.3.2 *Banach manifolds*

Let us turn to the notion of a Banach manifold, without repeating the statements true for both finite-dimensional and Banach manifolds.

Definition 4.3.1. A *Banach manifold* $\mathcal{B}$ modelled on a Banach space B is defined as a topological space which admits an atlas of charts $\Psi_{\mathcal{B}} = \{(U_\iota, \phi_\iota)\}$, where the maps ϕ_ι are homeomorphisms of U_ι onto open subsets of the Banach space B, while the transition functions $\phi_\zeta\phi_\iota^{-1}$ from $\phi_\iota(U_\iota \cap U_\zeta) \subset B$ to $\phi_\zeta(U_\iota \cap U_\zeta) \subset B$ are smooth. Two atlases of a Banach manifold are said to be equivalent if their union also is an atlas.

Unless otherwise stated, Banach manifolds are assumed to be connected paracompact Hausdorff topological spaces. A locally compact Banach manifold is necessarily finite-dimensional.

Remark 4.3.1. Let us note that a paracompact Banach manifold admits a smooth partition of unity if and only if its model Banach space does. For instance, this is the case of (real) separable Hilbert spaces. Therefore, we restrict our consideration to Hilbert manifolds modelled on separable Hilbert spaces.

Any open subset U of a Banach manifold $\mathcal{B}$ is a Banach manifold whose atlas is the restriction of an atlas of $\mathcal{B}$ to U.

Morphisms of Banach manifolds are defined similarly to those of smooth finite-dimensional manifolds. However, the notion of the immersion and submersion need a certain modification (see Definition 4.3.2 below).

Tangent vectors to a smooth Banach manifold $\mathcal{B}$ are introduced by analogy with tangent vectors to a finite-dimensional one. Given a point $z \in \mathcal{B}$, let us consider the pair $(v; (U_\iota, \phi_\iota))$ of a vector $v \in B$ and a chart $(U_\iota \ni z, \phi_\iota)$ on a Banach manifold $\mathcal{B}$. Two pairs $(v; (U_\iota, \phi_\iota))$ and $(v'; (U_\zeta, \phi_\zeta))$ are said to be equivalent if

$$v' = d(\phi_\zeta\phi_\iota^{-1})(\phi_\iota(z))v. \tag{4.3.3}$$

The equivalence classes of such pairs make up the *tangent space* $T_z\mathcal{B}$ to a Banach manifold $\mathcal{B}$ at a point $z \in \mathcal{B}$. This tangent space is isomorphic to the topological vector space B. Tangent spaces to a Banach manifold $\mathcal{B}$ are assembled into the *tangent bundle* $T\mathcal{B}$ of $\mathcal{B}$. It is a Banach manifold modelled over the Banach space $B \oplus B$ which possesses the transition functions

$$(\phi_\zeta\phi_\iota^{-1}, d(\phi_\zeta\phi_\iota^{-1})).$$

Any morphism $f : \mathcal{B} \to \mathcal{B}'$ of Banach manifolds yields the corresponding tangent morphism of the tangent bundles $Tf : T\mathcal{B} \to T\mathcal{B}'$.

Definition 4.3.2. Let $f : \mathcal{B} \to \mathcal{B}'$ be a morphism of Banach manifolds.

(i) It is called an immersion at a point $z \in \mathcal{B}$ if the tangent morphism Tf at z is injective and splits.

(ii) A morphism f is called a submersion at a point $z \in \mathcal{B}$ if Tf at z is surjective and its kernel splits.

The range of a surjective submersion f of a Banach manifold is a submanifold, though f need not be an isomorphism onto a submanifold, unless f is an imbedding.

4.3.3 *Banach vector bundles*

One can think of a surjective submersion $\pi : \mathcal{B} \to \mathcal{B}'$ of Banach manifolds as a *Banach fibred manifold*. For instance, the product $\mathcal{B} \times \mathcal{B}'$ of Banach manifolds is a Banach fibred manifold with respect to pr_1 and pr_2.

Let $\mathcal{B}$ be a Banach manifold and E a Banach space. The definition of a (locally trivial) vector bundle with the typical fibre E and the base $\mathcal{B}$ is a repetition of that of finite-dimensional smooth vector bundles. Such a vector bundle Y is a Banach manifold and $Y \to \mathcal{B}$ is a surjective submersion. It is called the *Banach vector bundle*. The above mentioned tangent bundle $T\mathcal{B}$ of a Banach manifold exemplifies a Banach vector bundle over $\mathcal{B}$.

The Whitney sum, the tensor product, and the exterior product of Banach vector bundles are defined as those of smooth vector bundles. In particular, since the topological dual E' of a Banach space E is a Banach space with respect to the norm (11.7.1), one can associate to each Banach vector bundle $Y_E \to \mathcal{B}$ the dual $Y_E^* = Y_{E'}$ with the typical fibre E'. For instance, the dual of the tangent bundle $T\mathcal{B}$ of a Banach manifold $\mathcal{B}$ is the *cotangent bundle* $T^*\mathcal{B}$.

Sections of the tangent bundle $T\mathcal{B} \to \mathcal{B}$ of a Banach manifold are called *vector fields* on a Banach manifold $\mathcal{B}$. They form a locally free module $\mathcal{T}_1(\mathcal{B})$ over the ring $C^\infty(\mathcal{B})$ of smooth real functions on $\mathcal{B}$. Every vector field ϑ on a Banach manifold $\mathcal{B}$ determines a derivation of the $\mathbb{R}$-ring $C^\infty(\mathcal{B})$ by the formula

$$f(z) \to \partial_\vartheta f(z) = df(z)\vartheta(z), \qquad z \in \mathcal{B}.$$

Different vector fields yield different derivations. It follows that $\mathcal{T}_1(\mathcal{B})$ possesses a structure of a real Lie algebra, and there is its monomorphism

$$\mathcal{T}_1(\mathcal{B}) \to \mathfrak{d}C^\infty(\mathcal{B}) \tag{4.3.4}$$

to the derivation module of the $\mathbb{R}$-ring $C^\infty(\mathcal{B})$.

Let us consider the Chevalley–Eilenberg complex of the real Lie algebra $\mathcal{T}_1(\mathcal{B})$ with coefficients in $C^\infty(\mathcal{B})$ and its subcomplex $\mathcal{O}^*[\mathcal{T}_1(\mathcal{B})]$ of $C^\infty(\mathcal{B})$-multilinear skew-symmetric maps by analogy with the complex $\mathcal{O}^*[\mathfrak{d}\mathcal{A}]$ in Section 11.6. This subcomplex is a differential calculus over an $\mathbb{R}$-ring $C^\infty(\mathcal{B})$ where the Chevalley–Eilenberg coboundary operator d (11.6.8) and the product (11.6.9) read

$$d\phi(\vartheta_0,\ldots,\vartheta_r) = \sum_{i=0}^{r}(-1)^i\partial_{\vartheta_i}(\phi(\vartheta_0,\ldots,\widehat{\vartheta}_i,\ldots,\vartheta_r)) \tag{4.3.5}$$
$$+\sum_{i<j}(-1)^{i+j}\phi([\vartheta_i,\vartheta_j],\vartheta_0,\ldots,\widehat{\vartheta}_i,\ldots,\widehat{\vartheta}_j,\ldots,\vartheta_k),$$

$$\phi\wedge\phi'(\vartheta_1,...,\vartheta_{r+s}) \tag{4.3.6}$$
$$=\sum_{i_1<\cdots<i_r;j_1<\cdots<j_s}\mathrm{sgn}^{i_1\cdots i_rj_1\cdots j_s}_{1\cdots r+s}\phi(\vartheta_{i_1},\ldots,\vartheta_{i_r})\phi'(\vartheta_{j_1},\ldots,\vartheta_{j_s}),$$
$$\phi\in\mathcal{O}^r[\mathcal{T}_1(\mathcal{B})],\qquad \phi'\in\mathcal{O}^s[\mathcal{T}_1(\mathcal{B})],\qquad \vartheta_i\in\mathcal{T}_1(\mathcal{B}).$$

There are the familiar relations

$$\vartheta\rfloor df=\partial_\vartheta f,\qquad f\in C^\infty(\mathcal{B}),\qquad \vartheta\in\mathcal{T}_1(\mathcal{B}),$$
$$d(\phi\wedge\phi')=d\phi\wedge\phi'+(-1)^{|\phi|}\phi\wedge d\phi',\qquad \phi,\phi'\in\mathcal{O}^*[\mathcal{T}_1(\mathcal{B})].$$

The differential calculus $\mathcal{O}^*[\mathcal{T}_1(\mathcal{B})]$ contains the following subcomplex. Let $\mathcal{O}^1(\mathcal{B})$ be the $C^\infty(\mathcal{B})$-module of global sections of the cotangent bundle $T^*\mathcal{B}$ of $\mathcal{B}$. Obviously, there is its monomorphism

$$\mathcal{O}^1(\mathcal{B})\to\mathfrak{d}C^\infty(\mathcal{B})^* \tag{4.3.7}$$

to the dual of the derivation module $\mathfrak{d}C^\infty(\mathcal{B})$. Furthermore, let $\overset{r}{\wedge}T^*\mathcal{B}$ be the r-degree exterior product of the cotangent bundle $T^*\mathcal{B}$ and $\mathcal{O}^r(\mathcal{B})$ the $C^\infty(\mathcal{B})$-module of its sections. Let $\mathcal{O}^*(\mathcal{B})$ be the direct sum of $C^\infty(\mathcal{B})$-modules $\mathcal{O}^r(\mathcal{B})$, $r\in\mathbb{N}$, where we put $\mathcal{O}^0(\mathcal{B})=C^\infty(\mathcal{B})$. Elements of $\mathcal{O}^*(\mathcal{B})$ obviously are $C^\infty(\mathcal{B})$-multilinear skew-symmetric maps of $\mathcal{T}_1(\mathcal{B})$ to $C^\infty(\mathcal{B})$. Therefore, the Chevalley–Eilenberg differential d (4.3.5) and the exterior product (4.3.6) of elements of $\mathcal{O}^*(\mathcal{B})$ are well defined. Moreover, one can show that $d\phi$ and $\phi\wedge\phi'$, $\phi,\phi'\in\mathcal{O}^*(\mathcal{B})$, also are elements of $\mathcal{O}^*(\mathcal{B})$. Thus, $\mathcal{O}^*(\mathcal{B})$ is a differential graded commutative algebra, called the algebra of *exterior forms* on a Banach manifold $\mathcal{B}$.

At the same time, one can consider the Chevalley–Eilenberg differential calculus $\mathcal{O}^*[\mathfrak{d}C^\infty(\mathcal{B})]$ over the $\mathbb{R}$-ring $C^\infty(\mathcal{B})$. Because of the monomorphism (4.3.4), we have a homomorphism of $C^\infty(\mathcal{B})$-modules

$$\mathcal{O}^1[\mathfrak{d}C^\infty(\mathcal{B})]=\mathfrak{d}C^\infty(\mathcal{B})^*\to\mathcal{T}_1(\mathcal{B})^*=\mathcal{O}^1[\mathcal{T}_1(\mathcal{B})]\leftarrow\mathcal{O}^1(\mathcal{B}). \tag{4.3.8}$$

It follows that the differential calculi $\mathcal{O}^*[\mathcal{T}_1(\mathcal{B})]$, $\mathcal{O}^*(\mathcal{B})$ and $\mathcal{O}^1[\mathfrak{d}C^\infty(\mathcal{B})]$ over the $\mathbb{R}$-ring $C^\infty(\mathcal{B})$ are not mutually isomorphic in general. However, it is readily observed that the minimal differential calculi in $\mathcal{O}^*[\mathcal{T}_1(\mathcal{B})]$ and $\mathcal{O}^*(\mathcal{B})$ coincide with the minimal Chevalley–Eilenberg differential calculus $\mathcal{O}^*C^\infty(\mathcal{B})$ over the $\mathbb{R}$-ring $C^\infty(\mathcal{B})$ because they are generated by the elements df, $f \in C^\infty(\mathcal{B})$, where d is the restriction (4.3.5) to $\mathcal{T}_1(\mathcal{B})$ of the Chevalley–Eilenberg coboundary operator (11.6.8).

A *connection* on a Banach manifold $\mathcal{B}$ is defined as a connection on the $C^\infty(\mathcal{B})$-module $\mathcal{T}_1(\mathcal{B})$ [155]. In accordance with Definition 11.5.4, it is an $\mathbb{R}$-module morphism

$$\nabla : \mathcal{T}_1(\mathcal{B}) \to \mathcal{O}^1C^\infty(\mathcal{B}) \otimes \mathcal{T}_1(\mathcal{B}),$$

which obeys the Leibniz rule

$$\nabla(f\vartheta) = df \otimes \vartheta + f\nabla(\vartheta), \qquad f \in C^\infty(\mathcal{B}), \qquad \vartheta \in \mathcal{T}_1(\mathcal{B}). \tag{4.3.9}$$

In view of the inclusions,

$$\mathcal{O}^1C^\infty(\mathcal{B}) \subset \mathcal{O}^1(\mathcal{B}) \subset \mathcal{T}_1(\mathcal{B})^*, \qquad \mathcal{T}_1(\mathcal{B}) \subset \mathcal{T}_1(\mathcal{B})^{**} \subset \mathcal{O}^1(\mathcal{B})^*,$$

it is however convenient to define a connection on a Banach manifold as an $\mathbb{R}$-module morphism

$$\nabla : \mathcal{T}_1(\mathcal{B}) \to \mathcal{O}^1(\mathcal{B}) \otimes \mathcal{T}_1(\mathcal{B}), \tag{4.3.10}$$

which obeys the Leibniz rule (4.3.9).

4.3.4 *Hilbert manifolds*

Hilbert manifolds are particular Banach manifolds modelled on complex Hilbert spaces, which are assumed to be separable (Remark 4.3.1).

Remark 4.3.2. We refer the reader to [100] for the theory of real Hilbert and (infinite-dimensional) Riemannian manifolds. A *real Hilbert manifold* is a Banach manifold $\mathcal{B}$ modelled on a real Hilbert space V (Remark 4.1.6). It is assumed to be connected Hausdorff and paracompact space admitting the partition of unity by smooth functions (this is the case of a separable V). A Riemannian metric on $\mathcal{B}$ is defined as a smooth section g of the tensor bundle $\overset{2}{\vee} T^*\mathcal{B}$ such that $g(z)$ is a positive non-degenerate continuous bilinear form on the tangent space $T_z\mathcal{B}$. This form yields the maps $T_z\mathcal{B} \to T^*_z\mathcal{B}$ and $T^*_z\mathcal{B} \to T_z\mathcal{B}$. It is said to be non-degenerate if these maps are continuous isomorphisms. In infinite-dimensional geometry, the most of local results follow from general arguments analogous to those in the finite-dimensional case. In particular, a Riemannian metric makes $\mathcal{B}$ into a metric space. Just as in the finite-dimensional case, $\mathcal{B}$ admits a unique Levi–Civita connection. The global theory of real Hilbert manifolds is more intricate. For instance, an infinite-dimensional real (and, consequently, complex) Hilbert space V is proved to be diffeomorphic to $V \setminus \{0\}$, and the unit sphere in V is a deformation retract of V [10].

A complex Hilbert space $(E, \langle.|.\rangle)$ can be seen as a real Hilbert space

$$E \ni v \to v_{\mathbb{R}} \in E_{\mathbb{R}}, \qquad (v_{\mathbb{R}}, v'_{\mathbb{R}}) = \mathrm{Re}\,\langle v|v'\rangle,$$

in Remark 4.1.6 equipped with the complex structure $Jv_{\mathbb{R}} = (iv)_{\mathbb{R}}$. We have

$$(Jv_{\mathbb{R}}, Jv'_{\mathbb{R}}) = (v_{\mathbb{R}}, v'_{\mathbb{R}}), \qquad (Jv_{\mathbb{R}}, v'_{\mathbb{R}}) = \mathrm{Im}\,(v'_{\mathbb{R}}, v_{\mathbb{R}}).$$

Let $E_{\mathbb{C}} = \mathbb{C} \otimes E_{\mathbb{R}}$ denote the complexification of $E_{\mathbb{R}}$ provided with the Hermitian form $\langle.|.\rangle_{\mathbb{C}}$ (4.1.4). The complex structure J on $E_{\mathbb{R}}$ is naturally extended to $E_{\mathbb{C}}$ by letting $J \circ i = i \circ J$. Then $E_{\mathbb{C}}$ is split into the two complex subspaces

$$\begin{aligned} E_{\mathbb{C}} &= E^{1,0} \oplus E^{0,1}, \\ E^{1,0} &= \{v_{\mathbb{R}} - iJv_{\mathbb{R}} \,:\, v_{\mathbb{R}} \in E_{\mathbb{R}}\}, \\ E^{0,1} &= \{v_{\mathbb{R}} + iJv_{\mathbb{R}} \,:\, v_{\mathbb{R}} \in E_{\mathbb{R}}\}, \end{aligned} \tag{4.3.11}$$

which are mutually orthogonal with respect to the Hermitian form $\langle.|.\rangle_{\mathbb{C}}$. Since

$$\begin{aligned} \langle v_{\mathbb{R}} - iJv_{\mathbb{R}})|v'_{\mathbb{R}} - iJv'_{\mathbb{R}}\rangle &= 2\langle v|v'\rangle, \\ \langle v_{\mathbb{R}} + iJv_{\mathbb{R}}|v'_{\mathbb{R}} + iJv'_{\mathbb{R}}\rangle &= 2\langle v'|v\rangle, \end{aligned}$$

there are the following linear and antilinear isometric bijections

$$\begin{aligned} E \ni v \to v_{\mathbb{R}} &\to \frac{1}{\sqrt{2}}(v_{\mathbb{R}} - iJv_{\mathbb{R}}) \in E^{1,0}, \\ E \ni v \to v_{\mathbb{R}} &\to \frac{1}{\sqrt{2}}(v_{\mathbb{R}} + iJv_{\mathbb{R}}) \in E^{0,1}. \end{aligned}$$

They make $E^{1,0}$ and $E^{0,1}$ isomorphic to the Hilbert space E and the dual Hilbert space $\overline{E}$, respectively. Hence, the decomposition (4.3.11) takes the form

$$E_{\mathbb{C}} = E \oplus \overline{E}. \tag{4.3.12}$$

The complex structure J on the direct sum (4.3.12) reads

$$J : E \oplus \overline{E} \ni v + \overline{u} \to iv - i\overline{u} \in E \oplus \overline{E}, \tag{4.3.13}$$

where E and $\overline{E}$ are the (holomorphic and antiholomorphic) eigenspaces of J characterized by the eigenvalues i and $-i$, respectively.

Let f be a function (not necessarily linear) from a Hilbert space E to a Hilbert space H. It is said to be *differentiable* if the corresponding function $f_\mathbb{R}$ between the real Banach spaces $E_\mathbb{R}$ and $H_\mathbb{R}$ is differentiable. Let $df_\mathbb{R}(z)$, $z \in E_\mathbb{R}$, be the differential (4.3.2) of $f_\mathbb{R}$ on $E_\mathbb{R}$ which is a continuous linear morphism

$$E_\mathbb{R} \ni v_\mathbb{R} \to df_\mathbb{R}(z)v_\mathbb{R} \in H_\mathbb{R}$$

between real topological vector spaces $E_\mathbb{R}$ and $H_\mathbb{R}$. This morphism is naturally extended to the $\mathbb{C}$-linear morphism

$$E_\mathbb{C} \ni v_\mathbb{C} \to df_\mathbb{R}(z)v_\mathbb{C} \in H_\mathbb{C} \tag{4.3.14}$$

between the complexifications of $E_\mathbb{R}$ and $H_\mathbb{R}$. In view of the decomposition (4.3.12), one can introduce the $\mathbb{C}$-linear maps

$$\partial f_\mathbb{R}(z)(v + \overline{u}) = df_\mathbb{R}(z)v, \qquad \overline{\partial} f(z)(v + \overline{u}) = df_\mathbb{R}(z)\overline{u}$$

from $E \oplus \overline{E}$ to $H_\mathbb{C}$ such that

$$df_\mathbb{R}(z)v_\mathbb{C} = df_\mathbb{R}(z)(v + \overline{u}) = \partial f_\mathbb{R}(z)v + \overline{\partial} f_\mathbb{R}(z)\overline{u}.$$

Let us split

$$f_\mathbb{R}(z) = f(z) + \overline{f}(z)$$

in accordance with the decomposition $H_\mathbb{C} = H \oplus \overline{H}$. Then the morphism (4.3.14) takes the form

$$df_\mathbb{R}(z)(v + \overline{u}) = \partial f(z)v + \overline{\partial} f(z)\overline{u} + \partial \overline{f}(z)v + \overline{\partial}\,\overline{f}(z)\overline{u}, \tag{4.3.15}$$

where $\partial \overline{f} = \overline{\overline{\partial} f}$, $\overline{\partial}\,\overline{f} = \overline{\partial f}$. A function $f : E \to H$ is said to be *holomorphic* (resp. *antiholomorphic*) if it is differentiable and $\overline{\partial} f(z) = 0$ (resp. $\partial f(z) = 0$) for all $z \in E$. A holomorphic function is smooth, and is given by the Taylor series. If f is a holomorphic function, then the morphism (4.3.15) is split into the sum

$$df_\mathbb{R}(z)(v + \overline{u}) = \partial f(z)v + \overline{\partial}\,\overline{f}(z)\overline{u}$$

of morphisms $E \to H$ and $\overline{E} \to \overline{H}$.

Example 4.3.1. Let f be a complex function on a Hilbert space E. Then

$$f_\mathbb{R} = (\operatorname{Re} f, \operatorname{Im} f)$$

is a map of E to $\mathbb{R}^2$. The differential $df_\mathbb{R}(z)$, $z \in E$, of $f_\mathbb{R}$ yields the complex linear morphism

$$E \oplus \overline{E} \ni v_\mathbb{C} \to (d\mathrm{Re}\, f(z)v_\mathbb{C}, d\mathrm{Im}\, f(z)v_\mathbb{C}) \to d(\mathrm{Re}\, f + i\mathrm{Im}\, f)(z)v_\mathbb{C} \in \mathbb{C},$$

which is regarded as a differential $df(z)$ of a complex function f on a Hilbert space E.

A *Hilbert manifold* $\mathcal{P}$ modelled on a Hilbert space E is defined as a real Banach manifold modelled on the Banach space $E_\mathbb{R}$ which admits an atlas $\{(U_\iota, \phi_\iota)\}$ with holomorphic transition functions $\phi_\zeta\phi_\iota^{-1}$. Let $CT\mathcal{P}$ denote the *complexified tangent bundle* of a Hilbert manifold $\mathcal{P}$. In view of the decomposition (4.3.12), each fibre $CT_z\mathcal{P}$, $z \in \mathcal{P}$, of $CT\mathcal{P}$ is split into the direct sum

$$CT_z\mathcal{P} = T_z\mathcal{P} \oplus \overline{T}_z\mathcal{P}$$

of subspaces $T_z\mathcal{P}$ and $\overline{T}_z\mathcal{P}$, which are topological complex vector spaces isomorphic to the Hilbert space E and the dual Hilbert space $\overline{E}$, respectively. The spaces $CT_z\mathcal{P}$, $T_z\mathcal{P}$ and $\overline{T}_z\mathcal{P}$ are respectively called the *complex*, *holomorphic* and *antiholomorphic tangent spaces* to a Hilbert manifold $\mathcal{P}$ at a point $z \in \mathcal{P}$. Since transition functions of a Hilbert manifold are holomorphic, the complex tangent bundle $CT\mathcal{P}$ is split into a sum

$$CT\mathcal{P} = T\mathcal{P} \oplus \overline{T}\mathcal{P}$$

of *holomorphic* and *antiholomorphic* subbundles, together with the antilinear bundle automorphism

$$T\mathcal{P} \oplus \overline{T}\mathcal{P} \ni v + \overline{u} \to \overline{v} + u \in T\mathcal{P} \oplus \overline{T}\mathcal{P}$$

and the complex structure

$$J : T\mathcal{P} \oplus \overline{T}\mathcal{P} \ni v + \overline{u} \to iv - i\overline{u} \in T\mathcal{P} \oplus \overline{T}\mathcal{P}. \qquad (4.3.16)$$

Sections of the complex tangent bundle $CT\mathcal{P} \to \mathcal{P}$ are called *complex vector fields* on a Hilbert manifold $\mathcal{P}$. They constitute the locally free module $C\mathcal{T}_1(\mathcal{P})$ over the ring $\mathbb{C}^\infty(\mathcal{P})$ of smooth complex functions on $\mathcal{P}$. Every complex vector field $\vartheta + \overline{v}$ on $\mathcal{P}$ yields a derivation

$$f(z) \to df(z)(\vartheta + v) = \partial f(z)\vartheta(z) + \overline{\partial} f(z)v(z), \qquad f \in \mathbb{C}^\infty(\mathcal{P}), \quad z \in \mathcal{P},$$

of the $\mathbb{C}$-ring $\mathbb{C}^\infty(\mathcal{P})$.

The (topological) dual of the complex tangent bundle $CT\mathcal{P}$ is the *complex cotangent bundle* $CT^*\mathcal{P}$ of $\mathcal{P}$. Its fibres $CT^*_z\mathcal{P}$, $z \in \mathcal{P}$, are topological complex vector spaces isomorphic to $E\oplus\overline{E}$. Since Hilbert spaces are reflexive, the complex tangent bundle $CT\mathcal{P}$ is the dual of $CT^*\mathcal{P}$. The complex cotangent bundle $CT^*\mathcal{P}$ is split into the sum

$$CT^*\mathcal{P} = T^*\mathcal{P}\oplus\overline{T}^*\mathcal{P} \tag{4.3.17}$$

of *holomorphic* and *antiholomorphic* subbundles, which are the annihilators of antiholomorphic and holomorphic tangent bundles $\overline{T}\mathcal{P}$ and $T\mathcal{P}$, respectively. Accordingly, $CT^*\mathcal{P}$ is provided with the complex structure J via the relation

$$\langle v, Jw\rangle = \langle Jv, w\rangle, \qquad v\in CT_z\mathcal{P}, \qquad w\in CT^*_z\mathcal{P}, \qquad z\in\mathcal{P}.$$

Sections of the complex cotangent bundle $CT^*\mathcal{P}\to\mathcal{P}$ constitute a locally free $\mathbb{C}^\infty(\mathcal{P})$-module $\mathcal{O}^1(\mathcal{P})$. It is the $\mathbb{C}^\infty(\mathcal{P})$-dual

$$\mathcal{O}^1(\mathcal{P}) = C\mathcal{T}_1(\mathcal{P})^* \tag{4.3.18}$$

of the module $C\mathcal{T}_1(\mathcal{P})$ of complex vector fields on $\mathcal{P}$, and *vice versa.*

Similarly to the case of a Banach manifold, let us consider the differential calculi $\mathcal{O}^*[\mathcal{T}_1(\mathcal{P})]$, $\mathcal{O}^*(\mathcal{P})$ (further denoted by $\mathcal{C}^*(\mathcal{P})$) and $\mathcal{O}^1[\mathfrak{d}C^\infty(\mathcal{P})]$ over the $\mathbb{C}$-ring $\mathbb{C}^\infty(\mathcal{P})$. Due to the isomorphism (4.3.18), $\mathcal{O}^*[\mathcal{T}_1(\mathcal{P})]$ is isomorphic to $\mathcal{C}^*(\mathcal{P})$, whose elements are called *complex exterior forms* on a Hilbert manifold $\mathcal{P}$. The exterior differential d on these forms is the Chevalley–Eilenberg coboundary operator

$$d\phi(\vartheta_0,\ldots,\vartheta_k) = \sum_{i=0}^{k}(-1)^i d\phi(\vartheta_0,\ldots,\widehat{\vartheta}_i,\ldots,\vartheta_k)\vartheta_i \tag{4.3.19}$$
$$+\sum_{i<j}(-1)^{i+j}\phi([\vartheta_i,\vartheta_j],\vartheta_0,\ldots,\widehat{\vartheta}_i,\ldots,\widehat{\vartheta}_j,\ldots,\vartheta_k), \quad \vartheta_i\in C\mathcal{T}_1(\mathcal{P}).$$

In view of the splitting (4.3.17), the differential graded algebra $\mathcal{C}^*(\mathcal{P})$ admits the decomposition

$$\mathcal{C}^*(\mathcal{P}) = \mathop{\oplus}_{p,q=0} \mathcal{C}^{p,q}(\mathcal{P})$$

into subspaces $\mathcal{C}^{p,q}(\mathcal{P})$ of *p-holomorphic* and *q-antiholomorphic* forms. Accordingly, the exterior differential d on $\mathcal{C}^*(\mathcal{P})$ is split into a sum $d=\partial+\overline{\partial}$ of holomorphic and antiholomorphic differentials

$$\partial: \mathcal{C}^{p,q}(\mathcal{P})\to\mathcal{C}^{p+1,q}(\mathcal{P}), \qquad \overline{\partial}: \mathcal{C}^{p,q}(\mathcal{P})\to\mathcal{C}^{p,q+1}(\mathcal{P}),$$
$$\partial\circ\partial=0, \qquad \overline{\partial}\circ\overline{\partial}=0, \qquad \partial\circ\overline{\partial}+\overline{\partial}\circ\partial=0.$$

A *Hermitian metric* on a Hilbert manifold $\mathcal{P}$ is defined as a complex bilinear form g on fibres of the complex tangent bundle $CT\mathcal{P}$ which obeys the following conditions:

• g is a smooth section of the tensor bundle $CT^*\mathcal{P} \otimes CT^*\mathcal{P} \to \mathcal{P}$;

• $g(\vartheta_z, \vartheta'_z) = 0$ if complex tangent vectors $\vartheta_z, \vartheta'_z \in CT_z\mathcal{P}$ are simultaneously holomorphic or antiholomorphic;

• $g(\vartheta_z, \overline{\vartheta}_z) > 0$ for any non-vanishing complex tangent vector $\vartheta_z \in CT_z\mathcal{P}$;

• the bilinear form $g(\vartheta_z, \vartheta'_z)$, $\vartheta_z, \vartheta'_z \in CT_z\mathcal{P}$, defines a norm topology on the complex tangent space $CT_z\mathcal{P}$ which is equivalent to its Hilbert space topology.

As an immediate consequence of this definition, we obtain

$$\overline{g(\vartheta_z, \vartheta'_z)} = g(\overline{\vartheta}_z, \overline{\vartheta}'_z), \qquad g(J\vartheta_z, J\vartheta'_z) = g(\vartheta_z, \vartheta'_z).$$

A Hermitian metric exists, e.g., on paracompact Hilbert manifolds modelled on separable Hilbert spaces.

The above mentioned properties of a Hermitian metric on a Hilbert manifold are similar to properties of a Hermitian metric on a finite-dimensional complex manifold [65]. Therefore, one can think of the pair $(\mathcal{P}, g)$ as being an infinite-dimensional *Hermitian manifold*.

A Hermitian manifold $(\mathcal{P}, g)$ is endowed with a non-degenerate exterior two-form

$$\Omega(\vartheta_z, \vartheta'_z) = g(J\vartheta_z, \vartheta'_z), \qquad \vartheta_z, \vartheta'_z \in CT_z\mathcal{P}, \qquad z \in \mathcal{P}, \tag{4.3.20}$$

called the *fundamental form* of the Hermitian metric g. This form satisfies the relations

$$\overline{\Omega(\vartheta_z, \vartheta'_z)} = \Omega(\overline{\vartheta}_z, \overline{\vartheta}'_z), \qquad \Omega(J_z\vartheta_z, J_z\vartheta'_z) = \Omega(\vartheta_z, \vartheta'_z).$$

If Ω (4.3.20) is a closed (i.e., symplectic) form, the Hermitian metric g is called a *Kähler metric* and Ω a *Kähler form*. Accordingly, $(\mathcal{P}, g, \Omega)$ is said to be an infinite-dimensional *Kähler manifold*.

A Kähler metric g and its Kähler form Ω on a Hilbert manifold $\mathcal{P}$ yield the bundle isomorphisms

$$g^\flat : CT\mathcal{P} \ni \nu \to \nu\rfloor g \in CT^*\mathcal{P},$$
$$\Omega^\flat : CT\mathcal{P} \ni \nu \to -\nu\rfloor\Omega \in CT^*\mathcal{P}.$$

Let $g^\sharp$ and $\Omega^\sharp$ denote the inverse bundle isomorphisms $CT^*\mathcal{P} \to CT\mathcal{P}$. They possess the properties

$$\Omega^\sharp = Jg^\sharp,$$
$$g^\sharp(w_z)\rfloor w'_z = g^\sharp(w'_z)\rfloor w_z,$$
$$\Omega^\sharp(w_z)\rfloor w'_z = -\Omega^\sharp(w'_z)\rfloor w_z, \quad w_z, w'_z \in CT^*\mathcal{P}.$$

In particular, every smooth complex function $f \in \mathbb{C}^\infty(\mathcal{P})$ on a Kähler manifold $(\mathcal{P}, g)$ determines:

- the complex vector field

$$g^\sharp(df) = g^\sharp(\partial f) + g^\sharp(\overline{\partial} f), \tag{4.3.21}$$

which is split into holomorphic and antiholomorphic parts $g^\sharp(\overline{\partial} f)$ and $g^\sharp(\partial f)$;

- the *complex Hamiltonian vector field*

$$\Omega^\sharp(df) = J(g^\sharp(df)) = -ig^\sharp(\partial f) + ig^\sharp(\overline{\partial} f); \tag{4.3.22}$$

- the Poisson bracket

$$\{f, f'\} = \Omega^\sharp(df)\rfloor df', \qquad f, f' \in \mathbb{C}^\infty(\mathcal{P}). \tag{4.3.23}$$

By analogy with the case of a Banach manifold, we modify Definition 11.5.4 and define a *connection* ∇ on a Hilbert manifold $\mathcal{P}$ as a $\mathbb{C}$-module morphism

$$\nabla : C\mathcal{T}_1(\mathcal{P}) \to \mathcal{C}^1(\mathcal{P}) \otimes C\mathcal{T}_1(\mathcal{P}),$$

which obeys the Leibniz rule

$$\nabla(f\vartheta) = df \otimes \vartheta + f\nabla(\vartheta), \qquad f \in \mathbb{C}^\infty(\mathcal{P}), \qquad \vartheta \in C\mathcal{T}_1(\mathcal{P}).$$

Similarly, a connection is introduced on any $\mathbb{C}^\infty(\mathcal{P})$-module, e.g., on sections of tensor bundles over a Hilbert manifold $\mathcal{P}$. Let D and $\overline{D}$ denote the holomorphic and antiholomorphic parts of ∇, and let $\nabla_\vartheta = \vartheta\rfloor\nabla$, D_ϑ and $\overline{D}_\vartheta$ be the corresponding covariant derivatives along a complex vector field ϑ on $\mathcal{P}$. For any complex vector field $\vartheta = \nu + \overline{\upsilon}$ on $\mathcal{P}$, we have the relations

$$D_\vartheta = \nabla_\nu, \qquad \overline{D}_\vartheta = \overline{\nabla}_{\overline{\upsilon}}, \qquad D_{J\vartheta} = iD_\vartheta, \qquad \overline{D}_{J\vartheta} = -i\overline{D}_\vartheta.$$

Proposition 4.3.1. *Given a Kähler manifold $(\mathcal{P}, g)$, there always exists a metric connection on $\mathcal{P}$ such that*

$$\nabla g = 0, \qquad \nabla\Omega = 0, \qquad \nabla J = 0,$$

where J is regarded as a section of the tensor bundle $CT^*\mathcal{P} \otimes CT\mathcal{P}$.

Example 4.3.2. If $\mathcal{P} = E$ is a Hilbert space, then

$$CT\mathcal{P} = E \times (E \oplus \overline{E}).$$

A Hermitian form $\langle.|.\rangle$ on E defines the constant Hermitian metric

$$g : (E \oplus \overline{E}) \times (E \oplus \overline{E}) \to \mathbb{C},$$

$$g(\vartheta, \vartheta') = \langle v|u'\rangle + \langle v'|u\rangle, \quad \vartheta = v + \overline{u}, \quad \vartheta' = v' + \overline{u}', \tag{4.3.24}$$

on $\mathcal{P} = E$. The associated fundamental form (4.3.20) reads

$$\Omega(\vartheta, \vartheta') = i\langle v|u'\rangle - i\langle v'|u\rangle. \tag{4.3.25}$$

It is constant on E. Therefore, $d\Omega = 0$ and g (4.3.24) is a Kähler metric. The metric connection on E is trivial, i.e., $\nabla = d$, $D = \partial$, $\overline{D} = \overline{\partial}$.

4.3.5 Projective Hilbert space

Given a Hilbert space E, a *projective Hilbert space* PE is made up by complex one-dimensional subspaces (i.e., *complex rays*) of E. This is a Hilbert manifold with the following standard atlas. For any non-zero element $x \in E$, let us denote by x a point of PE such that $x \in \mathrm{x}$. Then each normalized element $h \in E$, $||h|| = 1$, defines a chart (U_h, ϕ_h) of the projective Hilbert space PE such that

$$U_h = \{\mathrm{x} \in PE \; : \; \langle x|h\rangle \neq 0\}, \qquad \phi_h(\mathrm{x}) = \frac{x}{\langle x|h\rangle} - h. \tag{4.3.26}$$

The image of U_h in the Hilbert space E is the one-codimensional closed (Hilbert) subspace

$$E_h = \{z \in E \; : \; \langle z|h\rangle = 0\}, \tag{4.3.27}$$

where $z(\mathrm{x}) + h \in \mathrm{x}$. In particular, given a point $\mathrm{x} \in PE$, one can choose the centered chart E_h, $h \in \mathrm{x}$, such that $\phi_h(\mathrm{x}) = 0$. Hilbert spaces E_h and $E_{h'}$ associated to different charts U_h and $U_{h'}$ are isomorphic. The transition function between them is a holomorphic function

$$z'(\mathrm{x}) = \frac{z(\mathrm{x}) + h}{\langle z(\mathrm{x}) + h|h'\rangle} - h', \qquad \mathrm{x} \in U_h \cap U_{h'}, \tag{4.3.28}$$

from $\phi_h(U_h \cap U_{h'}) \subset E_h$ to $\phi_{h'}(U_h \cap U_{h'}) \subset E_{h'}$. The set of the charts $\{(U_h, \phi_h)\}$ with the transition functions (4.3.28) provides a holomorphic atlas of the projective Hilbert space PE. The corresponding coordinate transformations for the tangent vectors to PE at $\mathrm{x} \in PE$ reads

$$v' = \frac{1}{\langle x|h'\rangle}[\langle x|h\rangle v - x\langle v|h\rangle]. \tag{4.3.29}$$

The projective Hilbert space PE is homeomorphic to the quotient of the unitary group $U(E)$ equipped with the normed operator topology by the stabilizer of a ray of E. It is connected and simply connected [26].

The projective Hilbert space PE admits a unique Hermitian metric g such that the corresponding distance function on PE is

$$\rho(\mathrm{x}, \mathrm{x}') = \sqrt{2}\arccos(|\langle x|x'\rangle|), \tag{4.3.30}$$

where x, x' are normalized elements of E. It is a Kähler metric, called the *Fubini–Studi metric*. Given a coordinate chart (U_h, ϕ_h), this metric reads

$$\begin{aligned} g_{\mathrm{FS}}(\vartheta_1, \vartheta_2) = & \frac{\langle v_1|u_2\rangle + \langle v_2|u_1\rangle}{1 + \|z\|^2} \\ & - \frac{\langle z|u_2\rangle\langle v_1|z\rangle + \langle z|u_1\rangle\langle v_2|z\rangle}{(1 + \|z\|^2)^2}, \qquad z \in E_h, \end{aligned} \tag{4.3.31}$$

for any complex tangent vectors $\vartheta_1 = v_1 + \overline{u}_1$ and $\vartheta_2 = v_2 + \overline{u}_2$ in CT_zPE. The corresponding Kähler form is given by the expression

$$\Omega_{\rm FS}(\vartheta_1, \vartheta_2) = i\frac{\langle v_1|u_2\rangle - \langle v_2|u_1\rangle}{1+\|z\|^2} - i\frac{\langle z|u_2\rangle\langle v_1|z\rangle - \langle z|u_1\rangle\langle v_2|z\rangle}{(1+\|z\|^2)^2}. \qquad (4.3.32)$$

It is readily justified that the expressions (4.3.31) – (4.3.32) are preserved under the transition functions (4.3.28) – (4.3.29). Written in the coordinate chart centered at a point $z(\mathrm{x}) = 0$, these expressions come to the expressions (4.3.24) and (4.3.25), respectively.

4.4 Hilbert and C^*-algebra bundles

This Section addresses particular Banach vector bundles whose fibres are C^*-algebras (seen as Banach spaces) and Hilbert spaces, but a base is a finite-dimensional smooth manifold.

Note that sections of a Banach vector bundle $\mathcal{B} \to Q$ over a smooth finite-dimensional manifold Q constitute a locally free $C^\infty(Q)$-module $\mathcal{B}(Q)$. Following the proof of Serre–Swan Theorem 11.5.2 [65], one can show that it is a projective $C^\infty(Q)$-module. In a general setting, we therefore can consider projective locally free $C^\infty(Q)$-modules, locally generated by a Banach space. In contrast with the case of projective $C^\infty(X)$ modules of finite rank, such a module need not be a module of sections of some Banach vector bundle.

Let $\mathcal{C} \to Q$ be a locally trivial topological fibre bundle over a finite-dimensional smooth real manifold Q whose typical fibre is a C^*-algebra A, regarded as a real Banach space, and whose transition functions are smooth. Namely, given two trivialization charts (U_1, ψ_1) and (U_2, ψ_2) of $\mathcal{C}$, we have the smooth morphism of Banach manifolds

$$\psi_1 \circ \psi_2^{-1} : U_1 \cap U_2 \times A \to U_1 \cap U_2 \times A,$$

where

$$\psi_1 \circ \psi_2^{-1}|_{q\in U_1\cap U_2}$$

is an automorphism of A. We agree to call $\mathcal{C} \to Q$ a *bundle of C^*-algebras*. It is a Banach vector bundle. The $C^\infty(Q)$-module $\mathcal{C}(Q)$ of smooth sections of this fibre bundle is a unital involutive algebra with respect to fibrewise operations. Let us consider a subalgebra $A(Q) \subset \mathcal{C}(Q)$ which consists of sections of $\mathcal{C} \to Q$ vanishing at infinity on Q. It is provided with the norm

$$||\alpha|| = \sup_{q\in Q} ||\alpha(q)|| < \infty, \qquad \alpha \in A(Q), \qquad (4.4.1)$$

but fails to be complete. Nevertheless, one extends $A(Q)$ to a C^*-algebra of continuous sections of $\mathcal{C} \to Q$ vanishing at infinity on a locally compact space Q as follows.

Let $\mathcal{C} \to Q$ be a topological bundle of C^*-algebras over a locally compact topological space Q, and let $\mathcal{C}^0(Q)$ denote the involutive algebra of its continuous sections. This algebra exemplifies a locally trivial continuous *field of C^*-algebras* in [33]. Its subalgebra $A^0(Q)$ of sections vanishing at infinity on Q is a C^*-algebra with respect to the norm (4.4.1). It is called a *C^*-algebra defined by a continuous field of C^*-algebras.* There are several important examples of C^*-algebras of this type. For instance, any commutative C^*-algebra is isomorphic to the algebra of continuous complex functions vanishing at infinity on its spectrum.

Hilbert bundles over a smooth manifold are similarly defined. Let $\mathcal{E} \to Q$ be a locally trivial topological fibre bundle over a finite-dimensional smooth real manifold Q whose typical fibre is a Hilbert space E, regarded as a real Banach space, and whose transition functions are smooth functions taking their values in the unitary group $U(E)$ equipped with the normed operator topology. We agree to call $\mathcal{E} \to Q$ a *Hilbert bundle.* It is a Banach vector bundle. Smooth sections of $\mathcal{E} \to Q$ make up a $C^\infty(Q)$-module $\mathcal{E}(Q)$, called a *Hilbert module.* Continuous sections of $\mathcal{E} \to Q$ constitute a locally trivial continuous field of Hilbert spaces [33].

There are the following relations between bundles of C^*-algebras and Hilbert bundles.

Let $T(E) \subset B(E)$ be the C^*-algebra of compact (completely continuous) operators in a Hilbert space E (Remark 4.1.9). Every automorphism ϕ of E yields the corresponding automorphism

$$T(E) \to \phi T(E) \phi^{-1}$$

of the C^*-algebra $T(E)$. Therefore, given a Hilbert bundle $\mathcal{E} \to Q$ with transition functions

$$E \to \rho_{\iota\zeta}(q) E, \qquad q \in U_\iota \cap U_\zeta,$$

over a cover $\{U_\iota\}$ of Q, we have the associated locally trivial bundle of elementary C^*-algebras $T(E)$ with transition functions

$$T(E) \to \rho_{\alpha\beta}(q) T(E) (\rho_{\alpha\beta}(q))^{-1}, \qquad q \in U_\alpha \cap U_\beta, \tag{4.4.2}$$

which are proved to be continuous with respect to the normed operator topology on $T(E)$ [33]. The proof is based on the following facts.

• The set of *degenerate operators* (i.e., operators of finite rank) is dense in $T(E)$.

• Any operator of finite rank is a linear combination of operators

$$P_{\xi,\eta} : \zeta \to \langle\zeta|\eta\rangle\xi, \qquad \xi,\eta,\zeta \in E,$$

and even of the projectors P_ξ onto $\xi \in E$.

• Let $\xi_1,\ldots,\xi_{2n}$ be variable vectors of E. If ξ_i, $i=1,\ldots,2n$, converges to η_i (or, more generally, $\langle\xi_i|\xi_j\rangle$ converges to $\langle\eta_i|\eta_j\rangle$ for any i and j), then

$$P_{\xi_1,\xi_2} + \cdots + P_{\xi_{2n-1},\xi_{2n}}$$

uniformly converges to

$$P_{\eta_1,\eta_2} + \cdots + P_{\eta_{2n-1},\eta_{2n}}.$$

Note that, given a Hilbert bundle $\mathcal{E} \to Q$, the associated bundle of C^*-algebras $B(E)$ of bounded operators in E fails to be constructed in general because the transition functions (4.4.2) need not be continuous.

The opposite construction however meets a topological obstruction as follows [22; 23].

Let $\mathcal{C} \to Q$ be a bundle of C^*-algebras whose typical fibre is an elementary C^*-algebra $T(E)$ of compact operators in a Hilbert space E. One can think of $\mathcal{C} \to Q$ as being a topological fibre bundle with the structure group of automorphisms of $T(E)$. This is the projective unitary group $PU(E)$ (4.2.7). With respect to the normed operator topology, the groups $U(E)$ and $PU(E)$ are the Banach Lie groups [84]. Moreover, $U(E)$ is contractible if a Hilbert space E is infinite-dimensional [97]. Let $(U_\alpha, \rho_{\alpha\beta})$ be an atlas of the fibre bundle $\mathcal{C} \to Q$ with $PU(E)$-valued transition functions $\rho_{\alpha\beta}$. These transition functions give rise to the maps

$$\overline{\rho}_{\alpha\beta} : U_\alpha \cap U_\beta \to U(E),$$

which however fail to be transition functions of a fibre bundle with the structure group $U(E)$ because they need not satisfy the cocycle condition. Their failure to be so is measured by the $U(1)$-valued cocycle

$$e_{\alpha\beta\gamma} = \overline{g}_{\beta\gamma}\overline{g}^{-1}_{\alpha\gamma}\overline{g}_{\alpha\beta}.$$

This cocycle defines a class $[e]$ in the cohomology group $H^2(Q;U(1)_Q)$ of the manifold Q with coefficients in the sheaf $U(1)_Q$ of continuous maps of Q to $U(1)$. This cohomology class vanishes if and only if there exists a Hilbert bundle associated to $\mathcal{C}$. Let us consider the short exact sequence of sheaves

$$0 \to \mathbb{Z} \longrightarrow C^0_Q \xrightarrow{\gamma} U(1)_Q \to 0,$$

where C^0_Q is the sheaf of continuous real functions on Q and the morphism γ reads

$$\gamma : C^0_Q \ni f \to \exp(2\pi i f) \in U(1)_Q.$$

This exact sequence yields the long exact sequence of the sheaf cohomology groups [68; 85]:

$$0 \to \mathbb{Z} \longrightarrow C^0_Q \longrightarrow U(1)_Q \longrightarrow H^1(Q;\mathbb{Z}) \longrightarrow \cdots$$
$$H^p(Q;\mathbb{Z}) \longrightarrow H^p(Q;C^0_Q) \longrightarrow H^p(Q;U(1)_Q) \longrightarrow H^{p+1}(Q;\mathbb{Z}) \longrightarrow \cdots,$$

where $H^*(Q;\mathbb{Z})$ is cohomology of Q with coefficients in the constant sheaf $\mathbb{Z}$. Since the sheaf C^0_Q is fine and acyclic, we obtain at once from this exact sequence the isomorphism of cohomology groups

$$H^2(Q;U(1)_Q) = H^3(Q;\mathbb{Z}).$$

The image of $[e]$ in $H^3(Q;\mathbb{Z})$ is called the *Dixmier–Douady class* [33]. One can show that the negative $-[e]$ of the Dixmier–Douady class is the obstruction class of the lift of $PU(E)$-principal bundles to the $U(E)$-principal ones [23].

4.5 Connections on Hilbert and C^*-algebra bundles

There are different notions of a connection on Hilbert and C^*-algebra bundles whose equivalence is not so obvious as in the case of finite-dimensional bundles. These are connections on structure modules of sections, connections as a horizontal splitting and principal connections.

Given a bundle of C^*-algebras $\mathcal{C} \to Q$ with a typical fibre A over a smooth real manifold Q, the involutive algebra $\mathcal{C}(Q)$ of its smooth sections is a $C^\infty(Q)$-algebra. Therefore, one can introduce a connection on the fibre bundle $\mathcal{C} \to Q$ as a connection on the $C^\infty(Q)$-algebra $\mathcal{C}(Q)$. In accordance with Definition 11.5.3, such a *connection* assigns to each vector field τ on Q a symmetric derivation ∇_τ of the involutive algebra $\mathcal{C}(Q)$ which obeys the Leibniz rule

$$\nabla_\tau(f\alpha) = (\tau\rfloor df)\alpha + f\nabla_\tau\alpha, \qquad f \in \mathbb{C}^\infty(Q), \qquad \alpha \in \mathcal{C}(Q),$$

and the condition

$$\nabla_\tau\alpha^* = (\nabla_\tau\alpha)^*.$$

Let us recall that two such connections ∇_τ and ∇'_τ differ from each other in a derivation of the $C^\infty(Q)$-algebra $\mathcal{C}(Q)$. Then, given a trivialization chart

$$\mathcal{C}|_U = U \times A$$

of $\mathcal{C} \to Q$, a connection on $\mathcal{C}(Q)$ can be written in the form

$$\nabla_\tau = \tau^m(q)(\partial_m - \delta_m(q)), \qquad q \in U, \tag{4.5.1}$$

where (q^m) are local coordinates on Q and $\delta_m(q)$ for all $q \in U$ are symmetric bounded derivations of the C^*-algebra A.

Remark 4.5.1. Bearing in mind the discussion in Section 4.2, one should assume that, in physical models, the derivations $\delta_m(q)$ in the expression (4.5.1) are unbounded in general. This leads us to the notion of a *generalized connection* on bundles of C^*-algebras [6].

Let $\mathcal{E} \to Q$ be a Hilbert bundle with a typical fibre E and $\mathcal{E}(Q)$ the $C^\infty(Q)$-module of its smooth sections. Then a *connection on a Hilbert bundle* $\mathcal{E} \to Q$ is defined as a connection ∇ on the module $\mathcal{E}(Q)$. In accordance with Definition 11.5.2, such a connection assigns to each vector field τ on Q a first order differential operator ∇_τ on $\mathcal{E}(Q)$ which obeys both the Leibniz rule

$$\nabla_\tau(f\psi) = (\tau\rfloor df)\psi + f\nabla_\tau\psi, \qquad f \in \mathbb{C}^\infty(Q), \qquad \psi \in \mathcal{E}(Q),$$

and the additional condition

$$\langle(\nabla_\tau\psi)(q)|\psi(q)\rangle + \langle\psi(q)|(\nabla_\tau\psi)(q)\rangle = \tau(q)\rfloor d\langle\psi(q)|\psi(q)\rangle. \tag{4.5.2}$$

Given a trivialization chart $\mathcal{E}|_U = U \times E$ of $\mathcal{E} \to Q$, a connection on $\mathcal{E}(Q)$ reads

$$\nabla_\tau = \tau^m(q)(\partial_m + i\mathcal{H}_m(q)), \qquad q \in U, \tag{4.5.3}$$

where $\mathcal{H}_m(q)$ for all $q \in U$ are bounded self-adjoint operators in a Hilbert space E.

In a more general setting, let $\mathcal{B} \to Q$ be a Banach vector bundle over a finite-dimensional smooth manifold Q and $\mathcal{B}(X)$ the locally free $C^\infty(Q)$-module of its smooth sections $s(q)$. By virtue of Definition 11.5.2, a connection on $\mathcal{B}(Q)$ assigns to each vector field τ on Q a first order differential operator ∇_τ on $\mathcal{B}(Q)$ which obeys the Leibniz rule

$$\nabla_\tau(fs) = (\tau\rfloor df)s + f\nabla_\tau s, \qquad f \in \mathbb{C}^\infty(Q), \qquad s \in \mathcal{B}(Q). \tag{4.5.4}$$

One can show that such a connection exists ([65], Proposition 1.8.11). Connections (4.5.1) and (4.5.3) exemplify connections on Banach vector bundles $\mathcal{C} \to Q$ and $\mathcal{E} \to Q$, but they obey additional conditions because these bundles possess additional structures of a C^*-algebra bundle and a Hilbert bundle, respectively. In particular, the connection (4.5.3) is a principal connection whose second term is an element of the Lie algebra of the unitary group $U(E)$.

In a different way, a connection on a Banach vector bundle $\mathcal{B} \to Q$ can be defined as a splitting of the exact sequence

$$0 \to V\mathcal{B} \to T\mathcal{B} \to TQ \underset{Q}{\otimes} \mathcal{B} \to 0,$$

where $V\mathcal{B}$ denotes the vertical tangent bundle of $\mathcal{B} \to Q$. In the case of finite-dimensional vector bundles, both definitions are equivalent. This equivalence is extended to the case of Banach vector bundles over a finite-dimensional base. We leave the proof of this fact outside the scope of our exposition because it involves the notion of jets of Banach fibre bundles.

Turn now to principal connections. Given a Banach-Lie group G, a principal bundle over a finite-dimensional smooth manifold Q, a principal connection, its curvature form and that a holonomy group are defined similarly to those in the case of finite-dimensional Lie groups. The main difference lies in the facts that there are Banach-Lie algebras without Lie groups and the holonomy group of a principal connection need not be a Lie group. Referring the reader to [96] for theory of Lie groups and principal bundles modelled over so called convenient locally convex vector spaces (including Fréchet spaces), we here formulate some statements adapted to the case of Banach-Lie groups and Banach principal bundles over a finite-dimensional manifold.

• Any Banach-Lie group G admits an exponential mapping which is a diffeomorphism of a neighborhood of 0 in the Lie algebra $\mathfrak{g}$ of G onto a neighborhood of the unit in G. In a general setting, one can always associate to a Banach-Lie algebra a local Banach-Lie group which however fails to be extended to the global one in general [84].

• Let G be a Banach-Lie group and $\mathfrak{g}$ its Lie algebra. If $\mathfrak{h}$ is a closed Lie subalgebra of $\mathfrak{g}$, there exists a unique connected closed Banach-Lie subgroup H of G with the Lie algebra $\mathfrak{h}$ [134].

• Given a Banach-Lie group, the definition of a *G-principal bundle* $P \to Q$ over a finite-dimensional smooth manifold Q, a *principal connection with a structure Banach-Lie group* and its curvature form in [96] follows those in the case of a locally compact Lie group [93]. A principal connection Γ on P defines the global parallel transport and a holonomy group. In particular, the following generalizations of the reduction theorem ([93], Theorem 7.1) and the Ambrose–Singer theorem ([93], Theorem 8.1) to Banach principal bundles hold [160].

Theorem 4.5.1. *Let $P \to Q$ be a principal bundle with a Banach-Lie structure group G over a simply connected finite-dimensional manifold Q. Let H be a Banach-Lie subgroup of G. Let us assume that there exists a principal connection on P whose curvature form ω possesses the following property. For any smooth one-parameter family of horizontal paths Hc_s starting at a point $p \in P$ and arbitrary smooth vector fields u, u' on Q,*

$$[0,1]^2 \ni s,t \to \omega_{c_s(t)}(u,u') \tag{4.5.5}$$

is a smooth $\mathfrak{h}$-valued map. Then the structure group G of P is reduced to H.

Theorem 4.5.2. *Let us consider closed Lie subalgebras of the Lie algebra $\mathfrak{g}$ which contain the range of the map (4.5.5). Their overlap is the minimal closed Lie subalgebra $\mathfrak{g}_{\rm red}$ of $\mathfrak{g}$ possessing this property. The corresponding Banach-Lie group $G_{\rm red}$ is the minimal Banach-Lie group which contains the holonomy group of a connection Γ. By virtue of Theorem 4.5.1, the structure group Γ of P is reduced to $G_{\rm red}$.*

• Given a trivialization chart of a Banach principal bundle $P \to Q$ with a structure Banach-Lie group G, a principal connection on P is represented by a $\mathfrak{g}$-valued local connection one-form $\Gamma_m dq^m$ with the corresponding transition functions. Let

$$\mathcal{B} = (P \times V)/G$$

be a Banach vector bundle associated with P whose typical fibre V is a Banach space provided with a continuous effective left action of the structure group G. Then a principal connection Γ on P yields a connection on $\mathcal{B}$ given by the first order differential operators

$$\nabla_\tau = \tau^m(\partial_m - \Gamma_m) \tag{4.5.6}$$

on the $C^\infty(Q)$ module $\mathcal{B}(Q)$ of sections of $\mathcal{B} \to Q$ which obey the Leibniz rule (4.5.4).

For instance, let $G = U(E)$ be the unitary group of a Hilbert space E. Its Lie algebra consists of the operators $i\mathcal{H}$, where $\mathcal{H}$ are bounded self-adjoint operators in the Hilbert space E. It follows that a $U(E)$-principal connection takes the form (4.5.3).

In conclusion, let us mention the straightforward definition of a connection on a Hilbert bundle as a parallel displacement along paths lifted from a base [88]. Roughly speaking, such a connection corresponds to parallel displacement operators whose infinitesimal generators are (4.5.3). Due to the condition (4.5.2), these operators are unitary. If a path is closed, we come to the notion of a holonomy group of a connection on a Hilbert bundle.

4.6 Instantwise quantization

As it is shown in Section 5.3.3, geometric quantization of Hamiltonian non-relativistic mechanics takes a form of instantwise quantization, and results in a quantum system described by a Hilbert bundle over the time axis $\mathbb{R}$. This Section addresses the evolution of such quantum systems which can be viewed as a parallel displacement along time.

It should be emphasized that, in quantum mechanics based on the Schrödinger and Heisenberg equations, the physical time plays a role of a classical parameter. Indeed, all relations between operators in quantum mechanics are simultaneous, while computation of mean values of operators in a quantum state does not imply integration over time. It follows that, at each instant $t \in \mathbb{R}$, there is an instantaneous quantum system characterized by some C^*-algebra A_t. Thus, we come to *instantwise quantization*. Let us suppose that all instantaneous C^*-algebras A_t are isomorphic to some unital C^*-algebra A. Furthermore, let they constitute a locally trivial smooth bundle $\mathcal{C}$ of C^*-algebras over the time axis $\mathbb{R}$. Its typical fibre is A. This bundle of C^*-algebras is trivial, but need not admit a canonical trivialization in general. One can think of its different trivializations as being associated to different reference frames.

Let us describe evolution of quantum systems in the framework of instantwise quantization. Given a bundle of C^*-algebras $\mathcal{C} \to \mathbb{R}$, this evolution can be regarded as a parallel displacement with respect to some connection on $\mathcal{C} \to \mathbb{R}$ [6; 65; 140]. Following Section 4.5, we define ∇ as a connection on the involutive $C^\infty(\mathbb{R})$-algebra $\mathcal{C}(\mathbb{R})$ of smooth sections of $\mathcal{C} \to \mathbb{R}$. It assigns to the standard vector field ∂_t on $\mathbb{R}$ a symmetric derivation ∇_t of $\mathcal{C}(\mathbb{R})$ which obeys the Leibniz rule

$$\nabla_t(f\alpha) = \partial_t f\alpha + f\nabla_t\alpha, \qquad \alpha \in \mathcal{C}(\mathbb{R}), \qquad f \in C^\infty(\mathbb{R}),$$

and the condition

$$\nabla_t\alpha^* = (\nabla_t\alpha)^*.$$

Given a trivialization $\mathcal{C} = \mathbb{R} \times A$, a connection ∇_t reads

$$\nabla_t = \partial_t - \delta(t), \tag{4.6.1}$$

where $\delta(t)$, $t \in \mathbb{R}$, are symmetric derivations of a C^*-algebra A, i.e.,

$$\delta_t(ab) = \delta_t(a)b + a\delta_t(b), \qquad \delta_t(a^*) = \delta_t(a)^*, \qquad a, b \in A.$$

We say that a section α of the bundle of C^*-algebras $\mathcal{C} \to \mathbb{R}$ is an integral section of the connection ∇_t if

$$\nabla_t\alpha(t) = [\partial_t - \delta(t)]\alpha(t) = 0. \tag{4.6.2}$$

One can think of the equation (4.6.2) as being the *Heisenberg equation* describing quantum evolution.

In particular, let the derivations $\delta(t) = \delta$ in the Heisenberg equation (4.6.2) be the same for all $t \in \mathbb{R}$, and let δ be an infinitesimal generator of a strongly continuous one-parameter group $[G_t]$ of automorphisms of the C^*-algebra A (Theorem 4.2.2). A pair $(A, [G_t])$ is called the *C^*-dynamic system*. It describes evolution of an autonomous quantum system. Namely, for any $a \in A$, the curve $\alpha(t) = G_t(a)$, $t \in \mathbb{R}$, in A is a unique solution with the initial value $\alpha(0) = a$ of the Heisenberg equation (4.6.2).

It should be emphasized that, if a derivation δ is unbounded, the connection ∇_t (4.6.1) is not defined everywhere on the algebra $\mathcal{C}(\mathbb{R})$. In this case, we deal with a generalized connection. It is given by operators of a parallel displacement, whose generators however are ill defined [6]. Moreover, it may happen that a representation π of the C^*-algebra A does not carry out a representation of the automorphism group $[G_t]$ (Proposition 4.2.2). Therefore, quantum evolution described by the conservative Heisenberg equation, whose solution is a strongly (but not uniformly) continuous dynamic system $(A, [G_t])$, need not be described by the Schrödinger equation (see Remark 4.6.1 below).

If δ is a bounded derivation of a C^*-algebra A, the Heisenberg and Schrödinger pictures of evolution of an autonomous quantum system are equivalent. Namely, by virtue of Theorem 4.2.1, δ is an infinitesimal generator of a uniformly continuous one-parameter group $[G_t]$ of automorphisms of A, and *vice versa*. For any representation π of A in a Hilbert space E, there exists a bounded self-adjoint operator $\mathcal{H}$ in E (called the *Hamilton operator*) such that

$$\pi(\delta(a)) = -i[\mathcal{H}, \pi(a)], \qquad \pi(G_t) = \exp(-it\mathcal{H}), \qquad a \in A. \tag{4.6.3}$$

The corresponding *autonomous Schrödinger equation* reads

$$(\partial_t + i\mathcal{H})\psi = 0, \tag{4.6.4}$$

where ψ is a section of the trivial Hilbert bundle $\mathbb{R} \times E \to \mathbb{R}$. Its solution with an initial value $\psi(0) \in E$ is

$$\psi(t) = \exp[-it\mathcal{H}]\psi(0). \tag{4.6.5}$$

Remark 4.6.1. If the derivation δ is unbounded, but obeys the assumptions of Proposition 4.2.2, we also obtain the unitary representation (4.6.3) of the group $[G_t]$, but the curve $\psi(t)$ (4.6.5) need not be differentiable, and the Schrödinger equation (4.6.4) is ill defined.

Let us return to the general case of a quantum system characterized by a bundle of C^*-algebras $\mathcal{C} \to \mathbb{R}$ with a typical fibre A. Let us suppose that a phase Hilbert space of a quantum system is preserved under evolution, i.e., instantaneous C^*-algebras A_t are endowed with representations equivalent to some representation of the C^*-algebra A in a Hilbert space E. Then quantum evolution can be described by means of the Schrödinger equation as follows.

Let us consider a smooth Hilbert bundle $\mathcal{E} \to \mathbb{R}$ with the typical fibre E and a connection ∇ on the $C^\infty(\mathbb{R})$-module $\mathcal{E}(\mathbb{R})$ of smooth sections of $\mathcal{E} \to \mathbb{R}$ (Section 4.5). This connection assigns to the standard vector field ∂_t on $\mathbb{R}$ an $\mathbb{R}$-module endomorphism ∇_t of $\mathcal{E}(\mathbb{R})$ which obeys the Leibniz rule

$$\nabla_t(f\psi) = \partial_t f\psi + f\nabla_t\psi, \qquad \psi \in \mathcal{E}(\mathbb{R}), \qquad f \in C^\infty(\mathbb{R}),$$

and the condition

$$\langle(\nabla_t\psi)(t)|\psi(t)\rangle + \langle\psi(t)|(\nabla_t\psi)(t)\rangle = \partial_t\langle\psi(t)|\psi(t)\rangle.$$

Given a trivialization $\mathcal{E} = \mathbb{R} \times E$, the connection ∇_t reads

$$\nabla_t\psi = (\partial_t + i\mathcal{H}(t))\psi, \tag{4.6.6}$$

where $\mathcal{H}(t)$ are bounded self-adjoint operators in E for all $t \in \mathbb{R}$. It is a $U(E)$-principal connection.

We say that a section ψ of the Hilbert bundle $\mathcal{E} \to \mathbb{R}$ is an integral section of the connection ∇_t (4.6.6) if it fulfils the equation

$$\nabla_t\psi(t) = (\partial_t + i\mathcal{H}(t))\psi(t) = 0. \tag{4.6.7}$$

One can think of this equation as being the *Schrödinger equation* for the Hamilton operator $\mathcal{H}(t)$. Its solution with an initial value $\psi(0) \in E$ exists and reads

$$\psi(t) = U(t)\psi(0), \tag{4.6.8}$$

where $U(t)$ is an *operator of a parallel displacement* with respect to the connection (4.6.6). This operator is a differentiable section of the trivial bundle

$$\mathbb{R} \times U(E) \to \mathbb{R},$$

which obeys the equation

$$\partial_t U(t) = -i\mathcal{H}(t) \circ U(t), \qquad U(0) = \mathbf{1}. \tag{4.6.9}$$

The operator $U(t)$ plays a role of the *evolution operator*. It is given by the *time-ordered exponential*

$$U(t) = T \exp\left[-i \int_0^t \mathcal{H}(t')dt'\right], \qquad (4.6.10)$$

which uniformly converges in the operator norm [29]. Under certain conditions, $U(t)$ can be written as a true exponential

$$U(t) = \exp S(t)$$

of an anti-Hermitian operator $S(t)$ which is expressed as the *Magnus series*

$$S(t) = \sum_{k=1}^{\infty} S_k(t)$$

of multiple integrals of nested commutators [98; 126].

It should be emphasized that the evolution operator $U(t)$ has been defined with respect to a given trivialization of a Hilbert bundle $\mathcal{E} \to \mathbb{R}$.

Chapter 5

Geometric quantization

To quantize classical Hamiltonian systems, one usually follows canonical quantization which replaces the Poisson bracket $\{f, f'\}$ of smooth functions with the bracket $[\widehat{f}, \widehat{f'}]$ of Hermitian operators in a Hilbert space such that Dirac's condition (0.0.4) holds. Canonical quantization of Hamiltonian non-relativistic mechanics on a configuration space $Q \to \mathbb{R}$ is geometric quantization [57; 65]. It takes a form of instantwise quantization phrased in terms of Hilbert bundles over $\mathbb{R}$ (Section 5.4.3).

We start with the standard geometric quantization of symplectic manifolds (Section 5.1). This is the case of autonomous Hamiltonian systems. In particular, we refer to geometric quantization of the cotangent bundle (Section 5.2). Developed for symplectic manifolds [38; 148], the geometric quantization technique has been generalized to Poisson manifolds in terms of contravariant connections [156; 157]. Though there is one-to-one correspondence between the Poisson structures on a smooth manifold and its symplectic foliations, geometric quantization of a Poisson manifold need not imply quantization of its symplectic leaves [158].

Geometric quantization of symplectic foliations disposes of these problems (Section 5.3). A quantum algebra of a symplectic foliation also is a quantum algebra of the associated Poisson manifold such that its restriction to each symplectic leaf is a quantum algebra of this leaf. Thus, geometric quantization of a symplectic foliation provides leafwise quantization of a Poisson manifold. For instance, this is the case of Hamiltonian systems whose symplectic leaves are indexed by non-quantizable variables, e.g., instants of time (Section 5.4.3) and classical parameters (Section 9.3).

For the sake of simplicity, symplectic and Poisson manifolds throughout this Chapter are assumed to be simple connected (see Remark 5.1.2). Geometric quantization of toroidal cylinders possessing a non-trivial first homotopy group is considered in Section 7.8.

5.1 Geometric quantization of symplectic manifolds

Geometric quantization of a symplectic manifold falls into the following three steps: prequantization, polarization and metaplectic correction.

Let (Z,Ω) be a $2m$-dimensional simply connected symplectic manifold. Let $C\to Z$ be a *complex line bundle* whose typical fibre is $\mathbb{C}$. It is coordinated by (z^λ,c) where c is a complex coordinate.

Proposition 5.1.1. *By virtue of the well-known theorems [85; 109], the structure group of a complex line bundle $C\to Z$ is reducible to $U(1)$ such that:*

• *given a bundle atlas of $C\to Z$ with $U(1)$-valued transition functions and associated bundle coordinates (z^λ,c), there exists a Hermitian fibre metric*

$$g(c,c)=c\bar c \tag{5.1.1}$$

in C;

• *for any Hermitian fibre metric g in $C\to Z$, there exists a bundle atlas of $C\to Z$ with $U(1)$-valued transition functions such that g takes the form (5.1.1) with respect to the associated bundle coordinates.*

Let $\mathcal{K}$ be a linear connection on a fibre bundles $C\to Z$. It reads

$$\mathcal{K}=dz^\lambda\otimes(\partial_\lambda+\mathcal{K}_\lambda c\partial_c), \tag{5.1.2}$$

where $\mathcal{K}_\lambda$ are local complex functions on Z. The corresponding covariant differential $D^{\mathcal{K}}$ (11.4.8) takes the form

$$D^{\mathcal{K}}=(c_\lambda-\mathcal{K}_\lambda c)dz^\lambda\otimes\partial_c. \tag{5.1.3}$$

The curvature two-form (11.4.18) of the connection K (5.1.2) reads

$$R=\frac{1}{2}(\partial_\nu\mathcal{K}_\mu-\partial_\mu\mathcal{K}_\nu)cdz^\nu\wedge dz^\mu\otimes\partial_c. \tag{5.1.4}$$

Proposition 5.1.2. *A connection A on a complex line bundle $C\to Z$ is a $U(1)$-principal connection if and only if there exists an A-invariant Hermitian fibre metric g in C, i.e.,*

$$d_H(g(c,c))=g(D^Ac,c)+g(c,D^Ac).$$

With respect to the bundle coordinates (z^λ,c) in Proposition 5.1.1, this connection reads

$$A=dz^\lambda\otimes(\partial_\lambda+iA_\lambda c\partial_c), \tag{5.1.5}$$

where A_λ are local real functions on Z.

The curvature R (5.1.4) of the $U(1)$-principal connection A (5.1.5) defines the *first Chern characteristic form*

$$c_1(A) = -\frac{1}{4\pi}(\partial_\nu A_\mu - \partial_\mu A_\nu)cdz^\nu \wedge dz^\mu, \tag{5.1.6}$$

$$R = -2\pi i c_1 \otimes u_C, \tag{5.1.7}$$

where

$$u_C = c\partial_c \tag{5.1.8}$$

is the Liouville vector field (11.2.33) on C. The Chern form (5.1.6) is closed, but it need not be exact because $A_\mu dz^\mu$ is not a one-form on Z in general.

Definition 5.1.1. A complex line bundle $C \to Z$ over a symplectic manifold (Z, Ω) is called a *prequantization bundle* if a form $(2\pi)^{-1}\Omega$ on Z belongs to the first Chern characteristic class of C.

A prequantization bundle, by definition, admits a $U(1)$-principal connection A, called an *admissible connection*, whose curvature R (5.1.4) obeys the relation

$$R = -i\Omega \otimes u_C, \tag{5.1.9}$$

called the *admissible condition*.

Remark 5.1.1. A criterion of the existence of an admissible connection is based on the fact that the Chern form c_1 is a representative of an integral cohomology class in the de Rham cohomology group $H^2_{\rm DR}(Z)$. Consequently, a symplectic manifold (Z, Ω) admits a prequantization bundle $C \to Z$ and an admissible connection if and only if the symplectic form Ω belongs to an integral de Rham cohomology class.

Remark 5.1.2. Let A be the admissible connection (5.1.5) and $B = B_\mu dz^\mu$ a closed one-form on Z. Then

$$A' = A + icB \otimes \partial_c \tag{5.1.10}$$

also is an admissible connection. Since a manifold Z is assumed to be simply connected, a closed one-form B is exact. In this case, connections A and A' (5.1.10) are *gauge conjugate* . This means that there is a vertical principal automorphism Φ of a complex line bundle C and a C-associated $U(1)$-principal bundle P such that $A' = J^1\Phi \circ A$, where A and A' are treated as sections of the jet bundle $J^1P \to P$ [109].

Given an admissible connection A, one can assign to each function $f \in C^\infty(Z)$ the C-valued first order differential operator $\widehat{f}$ on a fibre bundle $C \to Z$ in accordance with *Kostant–Souriau formula*

$$\widehat{f} = -i\vartheta_f \rfloor D^A - fu_C = -[i\vartheta_f^\lambda(c_\lambda - iA_\lambda c) + fc]\partial_c, \tag{5.1.11}$$

where D^A is the covariant differential (5.1.3) and ϑ_f is the Hamiltonian vector field of f. It is easily justified that the operators (5.1.11) obey Dirac's condition (0.0.4) for all elements f of the Poisson algebra $C^\infty(Z)$.

Remark 5.1.3. In order to obtain Dirac's condition with the physical coefficients

$$[\widehat{f}, \widehat{f'}] = -i\hbar\widehat{\{f, f'\}}, \tag{5.1.12}$$

one should take the operators

$$\widehat{f} = -\left[i\hbar\vartheta_f \rfloor D^A + \frac{1}{\hbar}fc\right]\partial_c.$$

The Kostant–Souriau formula (5.1.11) is called *prequantization* because, in order to obtain Hermitian operators $\widehat{f}$ (5.1.11) acting on a Hilbert space, one should restrict both a class of functions $f \in C^\infty(Z)$ and a class of sections of $C \to Z$ in consideration as follows.

Given a symplectic manifold (Z, Ω), by its *polarization* is meant a maximal involutive distribution $\mathbf{T} \subset TZ$ such that

$$\Omega(\vartheta, \upsilon) = 0, \qquad \vartheta, \upsilon \in \mathbf{T}_z, \qquad z \in Z.$$

This term also stands for the algebra $\mathcal{T}_\Omega$ of sections of the distribution $\mathbf{T}$. We denote by $\mathcal{A}_\mathcal{T}$ the subalgebra of the Poisson algebra $C^\infty(Z)$ which consists of the functions f such that

$$[\vartheta_f, \mathcal{T}_\Omega] \subset \mathcal{T}_\Omega.$$

It is called the *quantum algebra* of a symplectic manifold (Z, Ω). Elements of this algebra only are quantized.

In order to obtain the carrier space of the algebra $\mathcal{A}_\mathcal{T}$, let us assume that Z is oriented and that its cohomology $H^2(Z; \mathbb{Z}_2)$ with coefficients in the constant sheaf $\mathbb{Z}_2$ vanishes. In this case, one can consider the *metalinear complex line bundle* $\mathcal{D}_{1/2}[Z] \to Z$ characterized by a bundle atlas $\{(U; z^\lambda, r)\}$ with the transition functions

$$r' = Jr, \qquad J\overline{J} = \left|\det\left(\frac{\partial z^\mu}{\partial z'^\nu}\right)\right|. \tag{5.1.13}$$

Global sections ρ of this bundle are called the *half-densities* on Z [38; 165]. Note that the metalinear bundle $\mathcal{D}_{1/2}[Z] \to Z$ admits the canonical lift of any vector field u on Z such that the corresponding Lie derivative of its sections reads

$$\mathbf{L}_u = u^\lambda \partial_\lambda + \frac{1}{2}\partial_\lambda u^\lambda. \tag{5.1.14}$$

Given an admissible connection A, the prequantization formula (5.1.11) is extended to sections $s \otimes \rho$ of the fibre bundle

$$C \underset{Z}{\otimes} \mathcal{D}_{1/2}[Z] \to Z \tag{5.1.15}$$

as follows:

$$\begin{aligned} \widehat{f}(s \otimes \rho) &= (-i\nabla_{\vartheta_f} - f)(s \otimes \rho) = (\widehat{f}s) \otimes \rho + s \otimes \mathbf{L}_{\vartheta_f}\rho, \\ \nabla_{\vartheta_f}(s \otimes \rho) &= (\nabla^A_{\vartheta_f} s) \otimes \rho + s \otimes \mathbf{L}_{\vartheta_f}\rho, \end{aligned} \tag{5.1.16}$$

where $\mathbf{L}_{\vartheta_f}\rho$ is the Lie derivative (5.1.14) acting on half-densities. This extension is said to be the *metaplectic correction*, and the tensor product (5.1.15) is called the *quantization bundle*. One can think of its sections ϱ as being C-valued *half-forms*. It is readily observed that the operators (5.1.16) obey Dirac's condition (0.0.4). Let us denote by $\mathfrak{E}_Z$ a complex vector space of sections ϱ of the fibre bundle $C \otimes \mathcal{D}_{1/2}[Z] \to Z$ of compact support such that

$$\begin{aligned} &\nabla_\upsilon \varrho = 0, \qquad \upsilon \in \mathcal{T}_\Omega, \\ &\nabla_\upsilon \varrho = \nabla_\upsilon(s \otimes \rho) = (\nabla^A_\upsilon s) \otimes \rho + s \otimes \mathbf{L}_\upsilon \rho. \end{aligned} \tag{5.1.17}$$

Lemma 5.1.1. *For any function $f \in \mathcal{A}_\mathcal{T}$ and an arbitrary section $\varrho \in \mathfrak{E}_Z$, the relation $\widehat{f}\varrho \in \mathfrak{E}_Z$ holds.*

Thus, we have a representation of the quantum algebra $\mathcal{A}_\mathcal{T}$ in the space $\mathfrak{E}_Z$. Therefore, by quantization of a function $f \in \mathcal{A}_\mathcal{T}$ is meant the restriction of the operator $\widehat{f}$ (5.1.16) to $\mathfrak{E}_Z$. It should be emphasized that a non-zero space $\mathfrak{E}_Z$ need not exist (see Section 5.2).

Let g be an A-invariant Hermitian fibre metric in $C \to Z$ in accordance with Proposition 5.1.2. If $\mathfrak{E}_Z \neq 0$, the Hermitian form

$$\langle s_1 \otimes \rho_1 | s_2 \otimes \rho_2 \rangle = \int_Z g(s_1, s_2)\rho_1\overline{\rho}_2 \tag{5.1.18}$$

brings $\mathfrak{E}_Z$ into a pre-Hilbert space. Its completion $\overline{\mathfrak{E}}_Z$ is called a *quantum Hilbert space*, and the operators $\widehat{f}$ (5.1.16) in this Hilbert space are Hermitian.

5.2 Geometric quantization of a cotangent bundle

Let us consider the standard geometric quantization of a cotangent bundle [38; 148; 165].

Let M be an m-dimensional simply connected smooth manifold coordinated by (q^i). Its cotangent bundle T^*M is simply connected. It is provided with the canonical symplectic form Ω_T (3.1.3) written with respect to holonomic coordinates $(q^i, p_i = \dot{q}_i)$ on T^*M. Let us consider the trivial complex line bundle

$$C = T^*M \times \mathbb{C} \to T^*M. \tag{5.2.1}$$

The canonical symplectic form (3.1.3) on T^*M is exact, i.e., it has the same zero de Rham cohomology class as the first Chern class of the trivial $U(1)$-bundle C (5.2.1). Therefore, C is a prequantization bundle in accordance with Definition 5.1.1.

Coordinated by (q^i, p_i, c), this bundle is provided with the admissible connection (5.1.5):

$$A = dp_j \otimes \partial^j + dq^j \otimes (\partial_j - ip_j c\partial_c) \tag{5.2.2}$$

such that the condition (5.1.9) is satisfied. The corresponding A-invariant fibre metric in C is given by the expression (5.1.1). The covariant derivative of sections s of the prequantization bundle C (5.2.1) relative to the connection A (5.2.2) along the vector field $u = u^j\partial_j + u_j\partial^j$ on T^*M reads

$$\nabla_u s = u^j(\partial_j + ip_j)s + u_j\partial^j s. \tag{5.2.3}$$

Given a function $f \in C^\infty(T^*M)$ and its Hamiltonian vector field

$$\vartheta_f = \partial^i f\partial_i - \partial_i f\partial^i,$$

the covariant derivative (5.2.3) along ϑ_f is

$$\nabla_{\vartheta_f} s = \partial^i f(\partial_i + ip_i)s - \partial_i f\partial^i s.$$

With the connection A (5.2.2), the prequantization (5.1.11) of elements f of the Poisson algebra $C^\infty(T^*M)$ takes the form

$$\widehat{f} = -i\partial^j f(\partial_j + ip_j) + i\partial_j f\partial^j - f. \tag{5.2.4}$$

Let us note that, since the complex line bundle (5.2.1) is trivial, its sections are simply smooth complex functions on T^*M. Then the prequantum operators (5.2.4) can be written in the form

$$\widehat{f} = -i\mathbf{L}_{\vartheta_f} + (\mathbf{L}_\upsilon f - f), \tag{5.2.5}$$

where $\upsilon = p_j\partial^j$ is the Liouville vector field (11.2.33) on $T^*M \to M$.

It is readily observed that the vertical tangent bundle VT^*M of the cotangent bundle $T^*M \to M$ provides a polarization of T^*M. Certainly, it is not a unique polarization of T^*M (see Section 6.5). We call VT^*M the *vertical polarization.* The corresponding quantum algebra $\mathcal{A}_T \subset C^\infty(T^*M)$ consists of affine functions of momenta

$$f = a^i(q^j)p_i + b(q^j) \tag{5.2.6}$$

on T^*M. Their Hamiltonian vector fields read

$$\vartheta_f = a^i\partial_i - (p_j\partial_i a^j + \partial_i b)\partial^i. \tag{5.2.7}$$

We call $\mathcal{A}_T$ the *quantum algebra of a cotangent bundle.*

Since the Jacobain of holonomic coordinate transformations of the cotangent bundle T^*M equals 1, the geometric quantization of T^*M need no metaplectic correction. Consequently, the quantum algebra $\mathcal{A}_T$ of the affine functions (5.2.6) acts on the subspace $\mathfrak{E}_{T^*M} \subset C^\infty(T^*M)$ of complex functions of compact support on T^*M which obey the condition (5.1.17):

$$\nabla_\upsilon s = \upsilon_i\partial^i s = 0, \qquad \mathcal{T}_\Omega \ni \upsilon = \upsilon_i\partial^i.$$

A glance at this equality shows that elements of $\mathfrak{E}_{T^*M}$ are independent of momenta p_i, i.e., they are the pull-back of complex functions on M with respect to the fibration $T^*M \to M$. These functions fail to be of compact support, unless $s = 0$. Consequently, the carrier space $\mathfrak{E}_{T^*M}$ of the quantum algebra $\mathcal{A}_T$ is reduced to zero. One can overcome this difficulty as follows.

Given the canonical zero section $\widehat{0}(M)$ of the cotangent bundle $T^*M \to M$, let

$$C_M = \widehat{0}(M)^*C \tag{5.2.8}$$

be the pull-back of the complex line bundle C (5.2.1) over M. It is a trivial complex line bundle $C_M = M \times \mathbb{C}$ provided with the pull-back Hermitian fibre metric $g(c, c') = c\overline{c}'$ and the pull-back (11.4.7):

$$A_M = \widehat{0}(M)^*A = dq^j \otimes (\partial_j - ip_jc\partial_c)$$

of the connection A (5.2.2) on C. Sections of C_M are smooth complex functions on M. One can consider a representation of the quantum algebra $\mathcal{A}_T$ of the affine functions (5.2.6) in the space of complex functions on M by the prequantum operators (5.2.4):

$$\widehat{f} = -ia^j\partial_j - b.$$

However, this representation need a metaplectic correction.

Let us assume that M is oriented and that its cohomology $H^2(M;\mathbb{Z}_2)$ with coefficients in the constant sheaf $\mathbb{Z}_2$ vanishes. Let $\mathcal{D}_{1/2}[M]$ be the metalinear complex line over M. Since the complex line bundle C_M (5.2.8) is trivial, the quantization bundle (5.1.15):

$$C_M \underset{M}{\otimes} \mathcal{D}_{1/2}[M] \to M \tag{5.2.9}$$

is isomorphic to $\mathcal{D}_{1/2}[M]$.

Because the Hamiltonian vector fields (5.2.7) of functions f (5.2.6) project onto vector fields $a^j\partial_j$ on M and $\mathbf{L}_\upsilon f - f = -b$ in the formula (5.2.5) is a function on M, one can assign to each element f of the quantum algebra $\mathcal{A}_T$ the following first order differential operator in the space $\mathcal{D}_{1/2}(M)$ of complex half-densities ρ on M:

$$\widehat{f}\rho = (-i\mathbf{L}_{a^j\partial_j} - b)\rho = (-ia^j\partial_j - \frac{i}{2}\partial_j a^j - b)\rho, \tag{5.2.10}$$

where $\mathbf{L}_{a^j\partial_j}$ is the Lie derivative (5.1.14) of half-densities. A glance at the expression (5.2.10) shows that it is the *Schrödinger representation* of the quantum algebra A_T of the affine functions (5.2.6). We call $\widehat{f}$ (5.2.10) the *Schrödinger operators*.

Let $\mathfrak{E}_M \subset \mathcal{D}_{1/2}(M)$ be a space of complex half-densities ρ of compact support on M and $\overline{\mathfrak{E}}_M$ the completion of $\mathfrak{E}_M$ with respect to the non-degenerate Hermitian form

$$\langle\rho|\rho'\rangle = \int_Q \rho\overline{\rho}'. \tag{5.2.11}$$

The (unbounded) Schrödinger operators (5.2.10) in the domain $\mathfrak{E}_M$ of the Hilbert space $\overline{\mathfrak{E}}_M$ are Hermitian.

5.3 Leafwise geometric quantization

As was mentioned above, the geometric quantization technique has been generalized to Poisson manifolds in terms of contravariant connections [156; 157], but geometric quantization of a Poisson manifold need not imply quantization of its symplectic leaves [158].

• Firstly, contravariant connections fail to admit the pull-back operation. Therefore, prequantization of a Poisson manifold does not determine straightforwardly prequantization of its symplectic leaves.

• Secondly, polarization of a Poisson manifold is defined in terms of sheaves of functions, and it need not be associated to any distribution. As a consequence, its pull-back onto a leaf is not polarization of a symplectic manifold in general.

• Thirdly, a quantum algebra of a Poisson manifold contains the center of a Poisson algebra. However, there are models where quantization of this center has no physical meaning. For instance, a center of the Poisson algebra of a mechanical system with classical parameters consists of functions of these parameters [58].

Geometric quantization of symplectic foliations disposes of these problems. A quantum algebra $\mathcal{A}_\mathcal{F}$ of a symplectic foliation $\mathcal{F}$ also is a quantum algebra of the associated Poisson manifold such that its restriction to each symplectic leaf F is a quantum algebra of F. Thus, geometric quantization of a symplectic foliation provides leafwise quantization of a Poisson manifold [58; 65].

Geometric quantization of a symplectic foliation is phrased in terms of leafwise connections on a foliated manifold (see Definition 5.3.1 below). Firstly, we have seen that homomorphisms of the de Rham cohomology of a Poisson manifold both to the de Rham cohomology of its symplectic leaf and the LP cohomology factorize through the leafwise de Rham cohomology (Propositions 3.1.3 and 3.1.4). Secondly, any leafwise connection on a complex line bundle over a Poisson manifold is proved to come from a connection on this bundle (Theorem 5.3.1). Using these facts, we state the equivalence of prequantization of a Poisson manifold to prequantization of its symplectic foliation (Remark 5.3.2), which also yields prequantization of each symplectic leaf (Proposition 5.3.2). We show that polarization of a symplectic foliation is associated to particular polarization of a Poisson manifold (Proposition 5.3.3), and its restriction to any symplectic leaf is polarization of this leaf (Proposition 5.3.4). Therefore, a quantum algebra of a symplectic foliation is both a quantum algebra of a Poisson manifold and, restricted to each symplectic leaf, a quantum algebra of this leaf.

We define metaplectic correction of a symplectic foliation so that its quantum algebra is represented by Hermitian operators in the pre-Hilbert module of leafwise half-forms, integrable over the leaves of this foliation.

5.3.1 *Prequantization*

Let $(Z, \{,\})$ be a Poisson manifold and $(\mathcal{F}, \Omega_\mathcal{F})$ its symplectic foliation such that $\{,\} = \{,\}_\mathcal{F}$ (Section 3.1.5). Let leaves of $\mathcal{F}$ be simply connected.

Prequantization of a symplectic foliation $(\mathcal{F}, \Omega_{\mathcal{F}})$ provides a representation

$$f \to \widehat{f}, \qquad [\widehat{f}, \widehat{f'}] = -i\widehat{\{f, f'\}}_{\mathcal{F}}, \tag{5.3.1}$$

of the Poisson algebra $(C^\infty(Z), \{,\}_{\mathcal{F}})$ by first order differential operators on sections s of some complex line bundle $C \to Z$, called the *prequantization bundle*. These operators are given by the Kostant–Souriau *prequantization* formula

$$\widehat{f} = -i\nabla^{\mathcal{F}}_{\vartheta_f} s + \varepsilon f s, \qquad \vartheta_f = \Omega^\sharp_{\mathcal{F}}(\widetilde{d}f), \qquad \varepsilon \neq 0, \tag{5.3.2}$$

where $\nabla^{\mathcal{F}}$ is an admissible leafwise connection on $C \to Z$ such that its curvature form $\widetilde{R}$ (5.3.11) obeys the *admissible condition*

$$\widetilde{R} = i\varepsilon\Omega_{\mathcal{F}} \otimes u_C, \tag{5.3.3}$$

where u_C is the Liouville vector field (5.1.8) on C.

Using the above mentioned fact that any leafwise connection comes from a connection, we provide the cohomology analysis of this condition, and show that prequantization of a symplectic foliation yields prequantization of its symplectic leaves.

Remark 5.3.1. If Z is a symplectic manifold whose symplectic foliation is reduced to Z itself, the formulas (5.3.2) – (5.3.3), $\varepsilon = -1$, of leafwise prequantization restart the formulas (5.1.11) and (5.1.9) of geometric quantization of a symplectic manifold Z.

Let $S_{\mathcal{F}}(Z) \subset C^\infty(Z)$ be a subring of functions constant on leaves of a foliation $\mathcal{F}$, and let $\mathcal{T}_1(\mathcal{F})$ be the real Lie algebra of global sections of the tangent bundle $T\mathcal{F} \to Z$ to $\mathcal{F}$. It is the Lie $S_{\mathcal{F}}(Z)$-algebra of derivations of $C^\infty(Z)$, regarded as a $S_{\mathcal{F}}(Z)$-ring.

Definition 5.3.1. In the framework of the leafwise differential calculus $\mathfrak{F}^*(Z)$ (3.1.21), a (linear) *leafwise connection* on a complex line bundle $C \to Z$ is defined as a connection $\nabla^{\mathcal{F}}$ on the $C^\infty(Z)$-module $C(Z)$ of global sections of this bundle, where $C^\infty(Z)$ is regarded as an $S_{\mathcal{F}}(Z)$-ring (see Definition 11.5.2). It associates to each element $\tau \in \mathcal{T}_1(\mathcal{F})$ an $S_{\mathcal{F}}(Z)$-linear endomorphism $\nabla^{\mathcal{F}}_\tau$ of $C(Z)$ which obeys the Leibniz rule

$$\nabla^{\mathcal{F}}_\tau(fs) = (\tau\rfloor\widetilde{d}f)s + f\nabla^{\mathcal{F}}_\tau(s), \qquad f \in C^\infty(Z), \qquad s \in C(Z). \tag{5.3.4}$$

A linear connection on $C \to Z$ can be equivalently defined as a connection on the module $C(Z)$ which assigns to each vector field $\tau \in \mathcal{T}_1(Z)$ on Z an $\mathbb{R}$-linear endomorphism of $C(Z)$ obeying the Leibniz rule (5.3.4).

Restricted to $\mathcal{T}_1(\mathcal{F})$, it obviously yields a leafwise connection. In order to show that any leafwise connection is of this form, we appeal to an alternative definition of a leafwise connection in terms of leafwise forms.

The inverse images $\pi^{-1}(F)$ of leaves F of the foliation $\mathcal{F}$ of Z provide a (regular) foliation $C_\mathcal{F}$ of the line bundle C. Given the (holomorphic) tangent bundle $TC_\mathcal{F}$ of this foliation, we have the exact sequence of vector bundles

$$0 \to VC \underset{C}{\longrightarrow} TC_\mathcal{F} \underset{C}{\longrightarrow} C \underset{Z}{\times} T\mathcal{F} \to 0, \tag{5.3.5}$$

where VC is the (holomorphic) vertical tangent bundle of $C \to Z$.

Definition 5.3.2. A (linear) leafwise connection on the complex line bundle $C \to Z$ is a splitting of the exact sequence (5.3.5) which is linear over C.

One can choose an adapted coordinate atlas $\{(U_\xi; z^\lambda, z^i)\}$ (11.2.65) of a foliated manifold $(Z, \mathcal{F})$ such that U_ξ are trivialization domains of the complex line bundle $C \to Z$. Let (z^λ, z^i, c), $c \in \mathbb{C}$, be the corresponding bundle coordinates on $C \to Z$. They also are adapted coordinates on the foliated manifold $(C, C_\mathcal{F})$. With respect to these coordinates, a (linear) leafwise connection is represented by a $TC_\mathcal{F}$-valued leafwise one-form

$$A_\mathcal{F} = \widetilde{d}z^i \otimes (\partial_i + A_i c\partial_c), \tag{5.3.6}$$

where A_i are local complex functions on C.

The exact sequence (5.3.5) is obviously a subsequence of the exact sequence

$$0 \to VC \underset{C}{\longrightarrow} TC \underset{C}{\longrightarrow} C \underset{Z}{\times} TZ \to 0,$$

where TC is the holomorphic tangent bundle of C. Consequently, any connection

$$\mathcal{K} = dz^\lambda \otimes (\partial_\lambda + \mathcal{K}_\lambda c\partial_c) + dz^i \otimes (\partial_i + \mathcal{K}_i c\partial_c) \tag{5.3.7}$$

on the complex line bundle $C \to Z$ yields a leafwise connection

$$\mathcal{K}_\mathcal{F} = \widetilde{d}z^i \otimes (\partial_i + \mathcal{K}_i c\partial_c). \tag{5.3.8}$$

Theorem 5.3.1. *Any leafwise connection on the complex line bundle $C \to Z$ comes from a connection on it.*

Proof. Let $A_{\mathcal{F}}$ (5.3.6) be a leafwise connection on $C \to Z$ and $\mathcal{K}_{\mathcal{F}}$ (5.3.8) a leafwise connection which comes from some connection $\mathcal{K}$ (5.3.7) on $C \to Z$. Their affine difference over C is a section

$$S = A_{\mathcal{F}} - \mathcal{K}_{\mathcal{F}} = \widetilde{d}z^i \otimes (A_i - \mathcal{K}_i)c\partial_c$$

of the vector bundle

$$T\mathcal{F}^* \underset{C}{\otimes} VC \to C.$$

Given some splitting

$$B : \widetilde{d}z^i \to dz^i - B^i_\lambda dz^\lambda \tag{5.3.9}$$

of the exact sequence (11.2.67), the composition

$$(B \otimes \mathrm{Id}\, _{VC}) \circ S = (dz^i - B^i_\lambda dz^\lambda) \otimes (A_i - \mathcal{K}_i)c\partial_c : C \to T^*Z \underset{C}{\otimes} VC$$

is a soldering form on the complex line bundle $C \to Z$. Then

$$\begin{aligned}\mathcal{K} + (B \otimes \mathrm{Id}\, _{VC}) \circ S = &\\ dz^\lambda \otimes (\partial_\lambda + [\mathcal{K}_\lambda - B^i_\lambda(A_i - \mathcal{K}_i)]c\partial_c) + dz^i \otimes (\partial_i + A_i c\partial_c)&\end{aligned}$$

is a desired connection on $C \to Z$ which yields the leafwise connection $A_{\mathcal{F}}$ (5.3.6). □

In particular, it follows that Definitions 5.3.1 and 5.3.2 of a leafwise connection are equivalent, namely,

$$\nabla^{\mathcal{F}} s = \widetilde{d}s - A_i s\widetilde{d}z^i, \qquad s \in C(Z).$$

The *curvature of a leafwise connection* $\nabla^{\mathcal{F}}$ is defined as a $C^\infty(Z)$-linear endomorphism

$$\widetilde{R}(\tau, \tau') = \nabla^{\mathcal{F}}_{[\tau,\tau']} - [\nabla^{\mathcal{F}}_\tau, \nabla^{\mathcal{F}}_{\tau'}] = \tau^i \tau'^j R_{ij}, \quad R_{ij} = \partial_i A_j - \partial_j A_i, \tag{5.3.10}$$

of $C(Z)$ for any vector fields $\tau, \tau' \in \mathcal{T}_1(\mathcal{F})$. It is represented by the vertical-valued leafwise two-form

$$\widetilde{R} = \frac{1}{2} R_{ij} \widetilde{d}z^i \wedge \widetilde{d}z^j \otimes u_C. \tag{5.3.11}$$

If a leafwise connection $\nabla^{\mathcal{F}}$ comes from a connection ∇, its *curvature leafwise form* $\widetilde{R}$ (5.3.11) is an image $\widetilde{R} = i^*_{\mathcal{F}} R$ of the curvature form R (11.4.13) of the connection ∇ with respect to the morphism $i^*_{\mathcal{F}}$ (3.1.22).

Now let us turn to the admissible condition (5.3.3).

Lemma 5.3.1. *Let us assume that there exists a leafwise connection* $\mathcal{K}_{\mathcal{F}}$ *on the complex line bundle* $C \to Z$ *which fulfils the admissible condition*

(5.3.3). Then, for any Hermitian fibre metric g in $C \to Z$, there exists a leafwise connection $A^g_{\mathcal{F}}$ on $C \to Z$ which:

(i) satisfies the admissible condition (5.3.3),

(ii) preserves g,

(iii) comes from a $U(1)$-principal connection on $C \to Z$.

This leafwise connection $A^g_{\mathcal{F}}$ is called admissible.

Proof. Given a Hermitian fibre metric g in $C \to Z$, let $\Psi^g = \{(z^\lambda, z^i, c)\}$ an associated bundle atlas of C with $U(1)$-valued transition functions such that $g(c, c') = c\bar{c}'$ (Proposition 5.1.1). Let the above mentioned leafwise connection $\mathcal{K}_{\mathcal{F}}$ come from a linear connection $\mathcal{K}$ (5.3.7) on $C \to Z$ written with respect to the atlas Ψ^g. The connection $\mathcal{K}$ is split into the sum $A^g + \gamma$ where

$$A^g = dz^\lambda \otimes (\partial_\lambda + \mathrm{Im}(\mathcal{K}_\lambda)c\partial_c) + dz^i \otimes (\partial_i + \mathrm{Im}(\mathcal{K}_i)c\partial_c) \qquad (5.3.12)$$

is a $U(1)$-principal connection, preserving the Hermitian fibre metric g. The curvature forms R of $\mathcal{K}$ and R^g of A^g obey the relation $R^g = \mathrm{Im}(R)$. The connection A^g (5.3.12) defines the leafwise connection

$$A^g_{\mathcal{F}} = i^*_{\mathcal{F}}A = \widetilde{d}z^i \otimes (\partial_i + iA^g_i c\partial_c), \qquad iA^g_i = \mathrm{Im}(\mathcal{K}_i), \qquad (5.3.13)$$

preserving the Hermitian fibre metric g. Its curvature fulfils a desired relation

$$\widetilde{R}^g = i^*_{\mathcal{F}}R^g = \mathrm{Im}(i^*_{\mathcal{F}}R) = i\varepsilon\Omega_{\mathcal{F}} \otimes u_C. \qquad (5.3.14)$$

□

Since A^g (5.3.12) is a $U(1)$-principal connection, its curvature form R^g is related to the first Chern form of integral de Rham cohomology class by the formula (5.1.7). If the admissible condition (5.3.3) holds, the relation (5.3.14) shows that the leafwise cohomology class of the leafwise form $-(2\pi)^{-1}\varepsilon\Omega_{\mathcal{F}}$ is an image of an integral de Rham cohomology class with respect to the cohomology morphism $[i^*_{\mathcal{F}}]$ (3.1.23). Conversely, if a leafwise symplectic form $\Omega_{\mathcal{F}}$ on a foliated manifold $(Z, \mathcal{F})$ is of this type, there exist a prequantization bundle $C \to Z$ and a $U(1)$-principal connection A on $C \to Z$ such that the leafwise connection $i^*_{\mathcal{F}}A$ fulfils the relation (5.3.3). Thus, we have stated the following.

Proposition 5.3.1. *A symplectic foliation $(\mathcal{F}, \Omega_{\mathcal{F}})$ of a manifold Z admits the prequantization (5.3.2) if and only if the leafwise cohomology class of $-(2\pi)^{-1}\varepsilon\Omega_{\mathcal{F}}$ is an image of an integral de Rham cohomology class of Z.*

Remark 5.3.2. In particular, let (Z, w) be a Poisson manifold and $(\mathcal{F}, \Omega_{\mathcal{F}})$ its characteristic symplectic foliation. As is well-known, a Poisson manifold admits *prequantization* if and only if the LP cohomology class of the bivector field $(2\pi)^{-1}\varepsilon w$, $\varepsilon > 0$, is an image of an integral de Rham cohomology class with respect to the cohomology morphism $[w^\sharp]$ (3.1.20) [156; 157]. By virtue of Proposition 3.1.4, this morphism factorizes through the cohomology morphism $[i^*_{\mathcal{F}}]$ (3.1.23). Therefore, in accordance with Proposition 5.3.1, prequantization of a Poisson manifold takes place if and only if prequantization of its symplectic foliation does well, and both these prequantizations utilize the same *prequantization bundle* $C \to Z$. Herewith, each leafwise connection $\nabla^{\mathcal{F}}$ obeying the admissible condition (5.3.3) yields the admissible *contravariant connection*

$$\nabla^w_\phi = \nabla^{\mathcal{F}}_{w^\sharp(\phi)}, \qquad \phi \in \mathcal{O}^1(Z),$$

on $C \to Z$ whose curvature bivector equals $i\varepsilon w$. Clearly, $\nabla^{\mathcal{F}}$ and ∇^w lead to the same prequantization formula (5.3.2).

Let F be a leaf of a symplectic foliation $(\mathcal{F}, \Omega_{\mathcal{F}})$ provided with the symplectic form

$$\Omega_F = i^*_F \Omega_{\mathcal{F}}.$$

In accordance with Proposition 3.1.3 and the commutative diagram

$$\begin{array}{ccc} H^*(Z;\mathbb{Z}) & \longrightarrow & H^*_{\rm DR}(Z) \\ \big\downarrow & & \big\downarrow \\ H^*(F;\mathbb{Z}) & \longrightarrow & H^*_{\rm DR}(F) \end{array}$$

of groups of the de Rham cohomology $H^*_{\rm DR}(*)$ and the cohomology $H^*(*;\mathbb{Z})$ with coefficients in the constant sheaf $\mathbb{Z}$, the symplectic form $-(2\pi)^{-1}\varepsilon\Omega_F$ belongs to an integral de Rham cohomology class if a leafwise symplectic form $\Omega_{\mathcal{F}}$ fulfils the condition of Proposition 5.3.1. This states the following.

Proposition 5.3.2. *If a symplectic foliation admits prequantization, each its symplectic leaf does prequantization too.*

The corresponding *prequantization bundle* for F is the pull-back complex line bundle $i^*_F C$, coordinated by (z^i, c). Furthermore, let $A^g_{\mathcal{F}}$ (5.3.13) be a leafwise connection on the prequantization bundle $C \to Z$ which obeys Lemma 5.3.1, i.e., comes from a $U(1)$-principal connection A^g on $C \to Z$. Then the pull-back

$$A_F = i^*_F A^g = dz^i \otimes (\partial_i + i i^*_F(A^g_i) c\partial_c) \qquad (5.3.15)$$

of the connection A^g onto $i^*_F C \to F$ satisfies the *admissible condition*

$$R_F = i^*_F R = i\varepsilon\Omega_F,$$

and preserves the pull-back Hermitian fibre metric $i^*_F g$ in $i^*_{\mathcal{F}} C \to F$.

5.3.2 *Polarization*

Let us define *polarization* of a symplectic foliation $(\mathcal{F},\Omega_{\mathcal{F}})$ of a manifold Z as a maximal (regular) involutive distribution $\mathbf{T}\subset T\mathcal{F}$ on Z such that

$$\Omega_{\mathcal{F}}(u,v)=0, \qquad u,v\in\mathbf{T}_z, \qquad z\in Z. \tag{5.3.16}$$

Given the Lie algebra $\mathbf{T}(Z)$ of $\mathbf{T}$-subordinate vector fields on Z, let $\mathcal{A}_{\mathcal{F}}\subset C^\infty(Z)$ be the complexified subalgebra of functions f whose leafwise Hamiltonian vector fields ϑ_f (3.1.28) fulfil the condition

$$[\vartheta_f,\mathbf{T}(Z)]\subset\mathbf{T}(Z).$$

It is called the *quantum algebra* of a symplectic foliation $(\mathcal{F},\Omega_{\mathcal{F}})$ with respect to the polarization $\mathbf{T}$. This algebra obviously contains the center $S_{\mathcal{F}}(Z)$ of the Poisson algebra $(C^\infty(Z),\{,\}_{\mathcal{F}})$, and it is a Lie $S_{\mathcal{F}}(Z)$-algebra.

Proposition 5.3.3. *Every polarization* $\mathbf{T}$ *of a symplectic foliation* $(\mathcal{F},\Omega_{\mathcal{F}})$ *yields polarization of the associated Poisson manifold* (Z,w_Ω).

Proof. Let us consider the presheaf of local smooth functions f on Z whose leafwise Hamiltonian vector fields ϑ_f (3.1.28) are subordinate to $\mathbf{T}$. The sheaf Φ of germs of these functions is polarization of the Poisson manifold (Z,w_Ω) (see Remark 5.3.3 below). Equivalently, Φ is the sheaf of germs of functions on Z whose leafwise differentials are subordinate to the codistribution $\Omega^\flat_{\mathcal{F}}\mathbf{T}$. □

Remark 5.3.3. Let us recall that *polarization of a Poisson manifold* $(Z,\{,\})$ is defined as a sheaf $\mathbf{T}^*$ of germs of complex functions on Z whose stalks $\mathbf{T}^*_z$, $z\in Z$, are Abelian algebras with respect to the Poisson bracket $\{,\}$ [158]. Let $\mathbf{T}^*(Z)$ be the structure algebra of global sections of the sheaf $\mathbf{T}^*$; it also is called the Poisson polarization [156; 157]. A *quantum algebra* $\mathcal{A}$ associated to the Poisson polarization $\mathbf{T}^*$ is defined as a subalgebra of the Poisson algebra $C^\infty(Z)$ which consists of functions f such that

$$\{f,\mathbf{T}^*(Z)\}\subset\mathbf{T}^*(Z).$$

Polarization of a symplectic manifold yields its Poisson one.

Let us note that the polarization Φ in the proof of Proposition 5.3.3) need not be maximal, unless $\mathbf{T}$ is of maximal dimension $\dim\mathcal{F}/2$. It belongs to the following particular type of polarizations of a Poisson manifold. Since the cochain morphism $i^*_{\mathcal{F}}$ (3.1.22) is an epimorphism, the leafwise differential calculus $\mathfrak{F}^*$ is universal, i.e., the leafwise differentials $\widetilde{d}f$ of functions

$f \in C^\infty(Z)$ on Z make up a basis for the $C^\infty(Z)$-module $\mathfrak{F}^1(Z)$. Let $\Phi(Z)$ denote the structure $\mathbb{R}$-module of global sections of the sheaf Φ. Then the leafwise differentials of elements of $\Phi(Z)$ make up a basis for the $C^\infty(Z)$-module of global sections of the codistribution $\Omega^\flat_{\mathcal{F}}\mathbf{T}$. Equivalently, the leafwise Hamiltonian vector fields of elements of $\Phi(Z)$ constitute a basis for the $C^\infty(Z)$-module $\mathbf{T}(Z)$. Then one can easily show that polarization $\mathbf{T}$ of a symplectic foliation $(\mathcal{F}, \Omega_{\mathcal{F}})$ and the corresponding polarization Φ of the Poisson manifold (Z, w_Ω) in Proposition 5.3.3 define the same quantum algebra $\mathcal{A}_{\mathcal{F}}$.

Let (F, Ω_F) be a symplectic leaf of a symplectic foliation $(\mathcal{F}, \Omega_{\mathcal{F}})$. Given polarization $\mathbf{T} \to Z$ of $(\mathcal{F}, \Omega_{\mathcal{F}})$, its restriction

$$\mathbf{T}_F = i_F^*\mathbf{T} \subset i_F^* T\mathcal{F} = TF$$

to F is an involutive distribution on F. It obeys the condition

$$i_F^*\Omega_{\mathcal{F}}(u, v) = 0, \qquad u, v \in \mathbf{T}_{Fz}, \qquad z \in F,$$

i.e., it is *polarization* of the symplectic manifold (F, Ω_F). Thus, we have stated the following.

Proposition 5.3.4. *Polarization of a symplectic foliation defines polarization of each symplectic leaf.*

Clearly, the *quantum algebra* $\mathcal{A}_F$ of a symplectic leaf F with respect to the polarization $\mathbf{T}_F$ contains all elements $i_F^* f$ of the quantum algebra $\mathcal{A}_{\mathcal{F}}$ restricted to F.

5.3.3 *Quantization*

Since $\mathcal{A}_{\mathcal{F}}$ is the quantum algebra both of a symplectic foliation $(\mathcal{F}, \Omega_{\mathcal{F}})$ and the associated Poisson manifold (Z, w_Ω), let us follow the standard metaplectic correction technique [38; 165].

Assuming that Z is oriented and that $H^2(Z; \mathbb{Z}_2) = 0$, let us consider the metalinear complex line bundle $\mathcal{D}_{1/2}[Z] \to Z$ characterized by an atlas

$$\Psi_Z = \{(U; z^\lambda, z^i, r)\}$$

with the transition functions (5.1.13). Global sections ρ of this bundle are half-densities on Z. Their Lie derivative (5.1.14) along a vector field u on Z reads

$$\mathbf{L}_u\rho = u^\lambda\partial_\lambda\rho + u^i\partial_i\rho + \frac{1}{2}(\partial_\lambda u^\lambda + \partial_i u^i)\rho. \tag{5.3.17}$$

Given an admissible connection $A^g_{\mathcal{F}}$, the prequantization formula (5.3.2) is extended to sections $\varrho = s \otimes \rho$ of the fibre bundle

$$C \underset{Z}{\otimes} \mathcal{D}_{1/2}[Z] \tag{5.3.18}$$

as follows

$$\begin{aligned} \widehat{f} &= -i[(\nabla^{\mathcal{F}}_{\vartheta_f} + i\varepsilon f) \otimes \mathrm{Id} + \mathrm{Id} \otimes \mathbf{L}_{\vartheta_f}] \\ &= -i[\nabla^{\mathcal{F}}_{\vartheta_f} + i\varepsilon f + \frac{1}{2}\partial_i \vartheta^i_f], \qquad f \in \mathcal{A}_{\mathcal{F}}. \end{aligned} \tag{5.3.19}$$

This extension is the *metaplectic correction* of leafwise quantization. It is readily observed that the operators (5.3.19) obey Dirac's condition (5.3.1). Let us denote by $\mathfrak{E}_Z$ the complex space of sections ϱ of the fibre bundle (5.3.18) of compact support such that

$$(\nabla^{\mathcal{F}}_{\vartheta} \otimes \mathrm{Id} + \mathrm{Id} \otimes \mathbf{L}_{\vartheta})\varrho = (\nabla^{\mathcal{F}}_{\vartheta} + \frac{1}{2}\partial_i \vartheta^i)\varrho = 0$$

for all $\mathbf{T}$-subordinate leafwise Hamiltonian vector fields ϑ.

Lemma 5.3.2. *For any function $f \in \mathcal{A}_T$ and an arbitrary section $\varrho \in \mathfrak{E}_Z$, the relation $\widehat{f}\varrho \in \mathfrak{E}_Z$ holds.*

Thus, we have a representation of the quantum algebra $\mathcal{A}_{\mathcal{F}}$ in the space $\mathfrak{E}_Z$. Therefore, by quantization of a function $f \in \mathcal{A}_{\mathcal{F}}$ is meant the restriction of the operator $\widehat{f}$ (5.3.19) to $\mathfrak{E}_Z$.

The space $\mathfrak{E}_Z$ is provided with the non-degenerate Hermitian form

$$\langle \rho | \rho' \rangle = \int_Z \rho \rho', \tag{5.3.20}$$

which brings $\mathfrak{E}_Z$ into a pre-Hilbert space. Its completion carries a representation of the quantum algebra $\mathcal{A}_{\mathcal{F}}$ by (unbounded) Hermitian operators.

However, it may happen that the above quantization has no physical meaning because the Hermitian form (5.3.20) on the carrier space $\mathfrak{E}_Z$ and, consequently, the mean values of operators (5.3.19) are defined by integration over the whole manifold Z. For instance, it implies integration over time and classical parameters. Therefore, we suggest a different scheme of quantization of symplectic foliations.

Let us consider the exterior bundle $\overset{2m}{\wedge} T\mathcal{F}^*$, $2m = \dim \mathcal{F}$. Its structure group $GL(2m, \mathbb{R})$ is reducible to the group $GL^+(2m, \mathbb{R})$ since a symplectic foliation is oriented. One can regard this fibre bundle as being associated to a $GL(2m, \mathbb{C})$-principal bundle $P \to Z$. As earlier, let us assume that

$H^2(Z;\mathbb{Z}_2)=0$. Then the principal bundle P admits a two-fold covering principal bundle with the structure *metalinear group* $ML(2m,\mathbb{C})$ [38]. As a consequence, there exists a complex line bundle $\mathcal{D}_\mathcal{F}\to Z$ characterized by an atlas

$$\Psi_\mathcal{F}=\{(U_\xi;z^\lambda,z^i,r)\}$$

with the transition functions $r'=J_\mathcal{F}r$ such that

$$J_\mathcal{F}\overline{J}_\mathcal{F}=\det\left(\frac{\partial z'^i}{\partial z^j}\right). \tag{5.3.21}$$

One can think of its sections as being complex *leafwise half-densities* on Z. The *metalinear bundle* $\mathcal{D}_{1/2}[\mathcal{F}]\to Z$ admits the canonical lift of any $\mathbf{T}$-subordinate vector field u on Z. The corresponding Lie derivative of its sections reads

$$\mathbf{L}_u^\mathcal{F}=u^i\partial_i+\frac{1}{2}\partial_iu^i. \tag{5.3.22}$$

We define the *quantization bundle* as the tensor product

$$Y_\mathcal{F}=C\underset{Z}{\otimes}\mathcal{D}_{1/2}[\mathcal{F}]\to Z. \tag{5.3.23}$$

Its sections are C-valued *leafwise half-forms*. Given an admissible leafwise connection $A_\mathcal{F}^g$ and the Lie derivative $\mathbf{L}_u^\mathcal{F}$ (5.3.22), let us associate the first order differential operator

$$\begin{aligned}\widehat{f}&=-i[(\nabla_{\vartheta_f}^\mathcal{F}+i\varepsilon f)\otimes\mathrm{Id}+\mathrm{Id}\otimes\mathbf{L}_{\vartheta_f}^\mathcal{F}] \\ &=-i[\nabla_{\vartheta_f}^\mathcal{F}+i\varepsilon f+\frac{1}{2}\partial_i\vartheta_f^i], \qquad f\in\mathcal{A}_\mathcal{F},\end{aligned} \tag{5.3.24}$$

on sections $\varrho_\mathcal{F}$ of $Y_\mathcal{F}$ to each element of the quantum algebra $\mathcal{A}_\mathcal{F}$. A direct computation with respect to the local Darboux coordinates on Z proves the following.

Lemma 5.3.3. *The operators (5.3.24) obey Dirac's condition (5.3.1).*

Lemma 5.3.4. *If a section $\varrho_\mathcal{F}$ fulfils the condition*

$$(\nabla_\vartheta^\mathcal{F}\otimes\mathrm{Id}+\mathrm{Id}\otimes\mathbf{L}_\vartheta^\mathcal{F})\varrho_\mathcal{F}=(\nabla_\vartheta^\mathcal{F}+\frac{1}{2}\partial_i\vartheta^i)\varrho_\mathcal{F}=0 \tag{5.3.25}$$

for all $\mathbf{T}$-subordinate leafwise Hamiltonian vector field ϑ, then $\widehat{f}\varrho_\mathcal{F}$ for any $f\in\mathcal{A}_\mathcal{F}$ possesses the same property.

Let us restrict the representation of the quantum algebra $\mathcal{A}_\mathcal{F}$ by the operators (5.3.24) to the subspace $\mathfrak{E}_\mathcal{F}$ of sections $\varrho_\mathcal{F}$ of the quantization bundle (5.3.23) which obey the condition (5.3.25) and whose restriction to any leaf of $\mathcal{F}$ is of compact support. The last condition is motivated by the following.

Since $i_F^*T\mathcal{F}^* = T^*F$, the pull-back $i_F^*\mathcal{D}_{1/2}[\mathcal{F}]$ of $\mathcal{D}_{1/2}[\mathcal{F}]$ onto a leaf F is a metalinear bundle of half-densities on F. By virtue of Propositions 5.3.2 and 5.3.4, the pull-back $i_F^*Y_\mathcal{F}$ of the quantization bundle $Y_\mathcal{F} \to Z$ onto F is a quantization bundle for the symplectic manifold $(F, i_F^*\Omega_\mathcal{F})$. Given the pull-back connection A_F (5.3.15) and the polarization $\mathbf{T}_F = i_F^*\mathbf{T}$, this symplectic manifold is subject to the standard geometric quantization by the first order differential operators

$$\widehat{f} = -i(i_F^*\nabla^\mathcal{F}_{\vartheta_f} + i\varepsilon f + \frac{1}{2}\partial_i\vartheta^i_f), \qquad f \in \mathcal{A}_F, \tag{5.3.26}$$

on sections ϱ_F of $i_F^*Y_\mathcal{F} \to F$ of compact support which obey the condition

$$(i_F^*\nabla^\mathcal{F}_\vartheta + \frac{1}{2}\partial_i\vartheta^i)\varrho_F = 0 \tag{5.3.27}$$

for all $\mathbf{T}_F$-subordinate Hamiltonian vector fields ϑ on F. These sections constitute a pre-Hilbert space $\mathfrak{E}_F$ with respect to the Hermitian form

$$\langle \rho_F | \rho'_F \rangle = \int_F \varrho_F \varrho'_F.$$

The key point is the following.

Proposition 5.3.5. *We have $i_F^*\mathfrak{E}_\mathcal{F} \subset \mathfrak{E}_F$, and the relation*

$$i_F^*(\widehat{f}\varrho_\mathcal{F}) = \widehat{(i_F^*f)}(i_F^*\varrho_\mathcal{F}) \tag{5.3.28}$$

holds for all elements $f \in \mathcal{A}_\mathcal{F}$ and $\varrho_\mathcal{F} \in \mathfrak{E}_\mathcal{F}$.

Proof. One can use the fact that the expressions (5.3.26) and (5.3.27) have the same coordinate form as the expressions (5.3.24) and (5.3.25) where $z^\lambda =$const. □

The relation (5.3.28) enables one to think of the operators $\widehat{f}$ (5.3.24) as being the leafwise quantization of the $S_\mathcal{F}(Z)$-algebra $\mathcal{A}_\mathcal{F}$ in the *pre-Hilbert $S_\mathcal{F}(Z)$-module* $\mathfrak{E}_\mathcal{F}$ of leafwise half-forms.

5.4 Quantization of non-relativistic mechanics

Let us develop geometric quantization of Hamiltonian non-relativistic mechanics on a configuration space $Q \to \mathbb{R}$ which is assumed to be simply connected. In contrast with the existent geometric quantizations of non-relativistic mechanics [148; 165], we do not fix a trivialization

$$Q = \mathbb{R} \times M, \qquad V^*Q = \mathbb{R} \times T^*M. \tag{5.4.1}$$

The key point is that, in this case, the evolution equation is not reduced to the Poisson bracket on a phase space V^*Q, but can be expressed in the Poisson bracket on the homogeneous phase space T^*Q. Therefore, geometric quantization of Hamiltonian non-relativistic mechanics on a configuration space $Q \to \mathbb{R}$ requires compatible geometric quantization both of the symplectic cotangent bundle T^*Q and the Poisson vertical cotangent bundle V^*Q of Q.

The relation (3.3.8) defines the monomorphism of Poisson algebras

$$\zeta^* : (C^\infty(V^*Q), \{,\}_V) \to (C^\infty(T^*Q), \{,\}_T). \tag{5.4.2}$$

Therefore, a compatibility of geometric quantizations of T^*Q and V^*Q implies that this monomorphism is prolonged to a monomorphism of quantum algebras of V^*Q and T^*Q.

Of course, it seems natural to quantize $C^\infty(V^*Q)$ as a subalgebra (5.4.2) of the Poisson algebra $C^\infty(T^*Q)$. However, geometric quantization of the Poisson algebra $(C^\infty(T^*Q), \{,\}_T)$ need not imply that of its Poisson subalgebra $\zeta^*C^\infty(V^*Q)$.

We show that the standard prequantization of the cotangent bundle T^*Q (Section 5.2) yields the compatible prequantization of the Poisson manifold V^*Q such that the monomorphism ζ^* (5.4.2) is prolonged to a monomorphism of prequantum algebras. However, polarization of T^*Q need not induce any polarization of V^*Q, unless it contains the vertical cotangent bundle $V_\zeta T^*Q$ of the fibre bundle ζ (3.3.3) spanned by vectors ∂^0. A unique canonical real polarization of T^*Q, satisfying the above condition

$$V_\zeta T^*Q \subset \mathbf{T}, \tag{5.4.3}$$

is the vertical tangent bundle VT^*Q of $T^*Q \to Q$. The associated *quantum algebra* $\mathcal{A}_T$ consists of functions on T^*Q which are affine in momenta (p_0, p_i). We show that this vertical polarization of T^*Q yields polarization of a Poisson manifold V^*Q such that the corresponding *quantum algebra* $\mathcal{A}_V$ consists of functions on V^*Q which are affine in momenta p_i. It follows that $\mathcal{A}_V$ is a subalgebra of $\mathcal{A}_T$ under the monomorphism (5.4.2). After metaplectic correction, the compatible Schrödinger representations (5.4.15) of $\mathcal{A}_T$ and (5.4.17) of $\mathcal{A}_V$ by operators on complex half-densities on Q is obtained.

The physical relevance of the Schrödinger quantization of T^*Q however is open to question. The scalar product of half-densities on Q implies integration over time, though the time plays a role of a classical evolution parameter in quantum mechanics, based on Schrödinger and Heisenberg equations. At the same time, the Schrödinger quantization of V^*Q provides instantwise quantization of non-relativistic mechanics. Indeed, a glance at the Poisson bracket (3.3.7) shows that the Poisson algebra $C^\infty(V^*Q)$ is a Lie algebra over the ring $C^\infty(\mathbb{R})$ of functions of time alone, where algebraic operations in fact are instantwise operations depending on time as a parameter. We show that the Schrödinger quantization of the Poisson manifold V^*Q induces geometric quantization of its symplectic fibres V_t^*Q, $t \in \mathbb{R}$, such that the *quantum algebra* $\mathcal{A}_t$ of V_t^*Q consists of elements $f \in \mathcal{A}_V$ restricted to V_t^*Q. This agrees with the instantwise quantization of symplectic fibres $\{t\} \times T^*M$ of the direct product (5.4.1) in [148]. Moreover, the induced geometric quantization of fibres V_t^*Q, by construction, is determined by their injection to V^*Q, but not projection of V^*Q. Therefore, it is independent of the trivialization (5.4.1).

Let us turn now to quantization of the evolution equation (3.8.1) in non-relativistic mechanics. Since this equation is not reduced to the Poisson bracket, quantization of the Poisson manifold V^*Q fails to provide quantization of this evolution equation. Therefore, we quantize the equivalent homogeneous evolution equation (3.8.3) expressed in the Poisson bracket on the symplectic manifold T^*Z. A problem however is that the homogeneous Hamiltonian $\mathcal{H}^*$ (3.4.1) in the formula (3.8.3) does not belong to the algebra $\mathcal{A}_T$, unless it is affine in momenta. Let us assume that $\mathcal{H}^*$ is a polynomial of momenta. This is the case of all physical models. Then we show below that $\mathcal{H}^*$ can be represented by a finite sum of products of elements of $\mathcal{A}_T$, though this representation by no means is unique. Thereby, it can be quantized as an element of the *enveloping algebra* $\overline{\mathcal{A}}_T$ of the Lie algebra $\mathcal{A}_T$.

Remark 5.4.1. An ambiguity of an operator representation of a classical Hamiltonian is a well-known technical problem of Schrödinger quantization as like as any geometric quantization scheme, where a Hamiltonian does not preserve polarization (see [148] for a general, but rather sophisticated analysis of such Hamiltonians). One can include the homogeneous Hamiltonian $\mathcal{H}^*$ (3.4.1) in a quantum algebra by choosing polarization of T^*Q which contains the Hamiltonian vector field of $\mathcal{H}^*$. This polarization always exists, but does not satisfy the condition (5.4.3) and, therefore, does not define any polarization of the Poisson manifold V^*Q. Let us note that, given a trivialization (5.4.1), symplectic fibres V_t^*Q, $t \in \mathbb{R}$, of the Poisson bundle $V^*Q \to \mathbb{R}$ can be provided with the instantwise polarization spanned by vectors

$$(\partial_1\mathcal{H}\partial^1 - \partial^1\mathcal{H}\partial_1, \cdots, \partial_m\mathcal{H}\partial^m - \partial^m\mathcal{H}\partial_m).$$

However, this polarization need not be regular. It is a standard polarization in autonomous Hamiltonian mechanics of one-dimensional systems, but it requires an exclusive analysis of each physical model.

Given a homogeneous Hamiltonian $\mathcal{H}^*$ (3.4.1) and its representative $\overline{\mathcal{H}}^*$ in $\overline{\mathcal{A}}_T$, the map

$$\nabla : f \to \{\overline{\mathcal{H}}^*, f\}_T$$

is a derivation of the enveloping algebra $\overline{\mathcal{A}}_V \subset \overline{\mathcal{A}}_T$ of the Lie algebra $\mathcal{A}_V$. Moreover, this derivation obeys the Leibniz rule

$$\nabla(rf) = \partial_t r f + r\nabla f, \qquad r \in C^\infty(\mathbb{R}),$$

and, consequently, is a connection on the instantwise algebra $\overline{\mathcal{A}}_V$. Since this property is preserved under quantization, geometric quantization of non-relativistic mechanics leads to its instantwise quantization (Section 4.6).

5.4.1 *Prequantization of T^*Q and V^*Q*

We start with the standard prequantization of the cotangent bundle T^*Q coordinated by

$$(q^\lambda, p_\lambda) = (q^0 = t, q^i, p_0, p_i)$$

(Section 5.2). Since the symplectic form Ω_T on T^*Q is exact and, consequently, belongs to the zero de Rham cohomology class, a prequantization bundle is the trivial complex line bundle

$$C = T^*Q \times \mathbb{C} \to T^*Q \tag{5.4.4}$$

of zero Chern class. Coordinated by $(q^\lambda, p_\lambda, c)$, this bundle is provided with the admissible linear connection (5.2.2):

$$A = dp_\lambda \otimes \partial^\lambda + dq^\lambda \otimes (\partial_\lambda - ip_\lambda c\partial_c), \tag{5.4.5}$$

whose curvature form equals $-i\Omega_T \otimes u_C$. The A-invariant Hermitian fibre metric in $C \to Q$ is given by the expression (5.1.1). The covariant derivative of sections s of the prequantization bundle C (5.4.4) relative to the connection A (5.4.5) along the vector field u on T^*Q reads

$$\nabla_u s = u^\lambda(\partial_\lambda + ip_\lambda)s + u_\lambda \partial^\lambda s. \tag{5.4.6}$$

Given a function $f \in C^\infty(T^*Q)$ and its Hamiltonian vector field

$$\vartheta_f = \partial^\lambda f \partial_\lambda - \partial_\lambda f \partial^\lambda$$

the covariant derivative (5.4.6) along ϑ_f is

$$\nabla_{\vartheta_f} s = \partial^\lambda f(\partial_\lambda + ip_\lambda)s - \partial_\lambda f \partial^\lambda s.$$

By virtue of the Kostant–Souriau formula (5.1.11), one assigns to each function $f \in C^\infty(T^*Q)$ the first order differential operator (5.2.4):

$$\widehat{f}(s) = -i(\nabla_{\vartheta_f} + if)s = [-i(\partial^\lambda f \partial_\lambda - \partial_\lambda f \partial^\lambda) + (p_\lambda \partial^\lambda f) - f]s, \tag{5.4.7}$$

on sections s of the prequantization bundle C (5.4.4). These operators satisfy Dirac's condition (0.0.4). The prequantum operators (5.4.7) for elements f of the Poisson subalgebra

$$\zeta^* C^\infty(V^*Q) \subset C^\infty(T^*Q)$$

read

$$\widehat{f}(s) = [-i(\partial^k f \partial_k - \partial_\lambda f \partial^\lambda) + (p_k \partial^k f - f)]s. \tag{5.4.8}$$

Let us turn now to prequantization of the Poisson manifold $(V^*Q, \{,\}_V)$. The Poisson bivector w of the Poisson structure (3.3.7) on V^*Q is

$$w = \partial^k \wedge \partial_k = -[w, u_V]_{\rm SN}, \tag{5.4.9}$$

where $[,]_{\rm SN}$ is the Schouten–Nijenhuis bracket and $u_V = p_i \partial^i$ is the Liouville vector field on the vertical cotangent bundle $V^*Q \to Q$. The relation (5.4.9) shows that the Poisson bivector w is $\widehat{w}$-exact (see the formula (3.1.17)) and, consequently, possesses the zero LP cohomology class. Therefore, let us consider the trivial complex line bundle

$$C_V = V^*Q \times \mathbb{C} \to V^*Q \tag{5.4.10}$$

of zero Chern class. Since the line bundles C (5.4.4) and C_V (5.4.10) are trivial, C can be seen as the pull-back $\zeta^* C_V$ of C_V, while C_V is isomorphic to the pull-back h^*C of C with respect to a section h (3.3.13) of the affine bundle (3.3.3). Since $C_V = h^*C$ and since the covariant derivative of the connection A (5.4.5) along the fibres of ζ (3.3.3) is trivial, let us consider the pull-back

$$h^*A = dp_k \otimes \partial^k + dq^k \otimes (\partial_k - ip_k c\partial_c) + dt \otimes (\partial_t - i\mathcal{H}c\partial_c) \qquad (5.4.11)$$

of the connection A (5.4.5) onto $C_V \to V^*Q$. This connection defines the *contravariant derivative*

$$\nabla_\phi s_V = \nabla_{w^\sharp\phi} s_V \qquad (5.4.12)$$

of sections s_V of $C_V \to V^*Q$ along one-forms ϕ on V^*Q. This contravariant derivative corresponds to a *contravariant connection* A_V on the line bundle $C_V \to V^*Q$ [157]. Since the vector fields $w^\sharp\phi = \phi^k\partial_k - \phi_k\partial^k$ are vertical on $V^*Q \to \mathbb{R}$, this contravariant connection does not depend on the choice of a section h. By virtue of the relation (5.4.12), the curvature bivector of A_V is equal to $-iw$ [158], i.e., A_V is an admissible connection for the Poisson structure on V^*Q. Then the Kostant–Souriau formula

$$\widehat{f}(s_V) = (-i\nabla_{\vartheta_f} - f)s_V = [-i(\partial^k f\partial_k - \partial_k f\partial^k) + (p_k\partial^k f - f)]s_V \quad (5.4.13)$$

defines prequantization of the Poisson manifold V^*Q. In particular, the prequantum operators of functions $f \in C^\infty(\mathbb{R})$ of time alone are reduced to the multiplication $\widehat{f}s_V = fs_V$. Consequently, the prequantum algebra of V^*Q inherits the structure of a $C^\infty(\mathbb{R})$-algebra.

It is immediately observed that the prequantum operator $\widehat{f}$ (5.4.13) coincides with the prequantum operator $\widehat{\zeta^*f}$ (5.4.8) restricted to the pull-back sections $s = \zeta^*s_V$. Thus, the above mentioned prequantization of the Poisson algebra $C^\infty(V^*Q)$ is equivalent to its prequantization as a subalgebra of the Poisson algebra $C^\infty(T^*Q)$.

Let us note that, since the complex line bundles C (5.4.4) and C_V (5.4.10) are trivial, their sections are simply smooth complex functions on T^*Q and V^*Q, respectively. Then the prequantum operators (5.4.7) and (5.4.13) can be written in the form

$$\widehat{f} = -i\mathbf{L}_{\vartheta_f} + (f - \mathbf{L}_\upsilon f), \qquad (5.4.14)$$

where υ is the Liouville vector field $\upsilon = p_\lambda\partial^\lambda$ on $T^*Q \to Q$ or $\upsilon = p_k\partial^k$ on $V^*Q \to Q$.

5.4.2 *Quantization of T^*Q and V^*Q*

Given compatible prequantizations of the cotangent bundle T^*Q and the vertical cotangent bundle V^*Q, let us now construct their compatible polarizations and quantizations. We assume that Q is an oriented manifold and that the cohomology $H^2(Q;\mathbb{Z}_2)$ is trivial.

Let $\mathbf{T}^*$ be polarization of the Poisson manifold $(T^*Q, \{,\}_T)$ (Remark 5.3.3). Its direct image in V^*Q with respect to the fibration ζ (3.3.3) is polarization of the Poisson manifold $(V^*Q, \{,\}_V)$ if the germs of $\mathbf{T}^*$ are constant along the fibres of ζ [158], i.e., are germs of functions independent of the momentum coordinate p_0. It follows that the corresponding symplectic polarization $\mathbf{T}$ of T^*Q is vertical with respect to the fibration $T^*Q \to V^*Q$.

The vertical polarization $\mathbf{T} = VT^*Q$ of T^*Q obeys this condition. The associated quantum algebra $\mathcal{A}_T \subset C^\infty(T^*Q)$ consists of functions which are affine in momenta p_λ. The algebra $\mathcal{A}_T$ acts by operators (5.4.14) on the space of smooth complex functions s on T^*Q which fulfill the relation $\nabla_u s = 0$ for any $\mathbf{T}$-valued (i.e., vertical) vector field $u = u_\lambda \partial^\lambda$ on the cotangent bundle $T^*Q \to Q$. Clearly, these functions are the pull-back of complex functions on Q with respect to the fibration $T^*Q \to Q$.

Following the general metaplectic technique, we come to complex half-densities on Q which are sections of the metalinear bundle $\mathcal{D}_{1/2}[Q] \to Q$ over Q. Then the formula (5.4.14), where $\mathbf{L}_{\vartheta_f}$ is the Lie derivative of half-densities, defines the Schrödinger representation

$$\begin{aligned} &\widehat{f}\rho = (-i\mathbf{L}_{a^\lambda\partial_\lambda} - b)\rho = \left(-ia^\lambda\partial_\lambda - \frac{i}{2}\partial_\lambda a^\lambda - b\right)\rho, \qquad (5.4.15)\\ &f = a^\lambda(q^\mu)p_\lambda + b(q^\mu) \in \mathcal{A}_T, \end{aligned}$$

of the quantum algebra $\mathcal{A}_T$ by operators in the space $\mathcal{D}_{1/2}(Q)$ of complex half-densities ρ on Q.

From now on, we assume that a coordinate atlas of Q and a bundle atlas of $\mathcal{D}_{1/2}[Q] \to Q$ are defined on the same covering of Q, e.g., by contractible open sets.

Let $\mathfrak{E}_Q \subset \mathcal{D}_{1/2}(Q)$ consist of half-densities of compact support, and let $\overline{\mathfrak{E}}_Q$ be its completion with respect to the non-degenerate Hermitian form

$$\langle\rho|\rho'\rangle = \int_Q \rho\overline{\rho}'. \qquad (5.4.16)$$

The (unbounded) Schrödinger operators (5.4.15) in the domain $\mathfrak{E}_Q$ in the Hilbert space $\overline{\mathfrak{E}}_Q$ are Hermitian.

The vertical polarization of T^*Q defines the polarization $\mathbf{T}^*_V$ of the Poisson manifold V^*Q which contains the germs of functions, constant on the fibres of $V^*Q \to Q$. The associated quantum algebra $\mathcal{A}_V$ consists of functions on V^*Q which are affine in momenta. It is a $C^\infty(\mathbb{R})$-algebra. This algebra acts by operators (5.4.14) on the space of smooth complex functions s_V on V^*Q which fulfill the relation $\nabla_u s_V = 0$ for any vertical vector field $u = u_i\partial^i$ on $V^*Q \to Q$. These functions also are the pull-back of complex functions on Q with respect to the fibration $V^*Q \to Q$. Similarly to the case of $\mathcal{A}_T$, we obtain the Schrödinger representation of the quantum algebra $\mathcal{A}_V$ by the operators

$$\widehat{f}\rho = (-i\mathbf{L}_{a^k\partial_k} + b)\rho = \left(-ia^k\partial_k - \frac{i}{2}\partial_k a^k - b\right)\rho, \qquad (5.4.17)$$

$$f = a^k(q^\mu)p_k + b(q^\mu) \in \mathcal{A}_V,$$

on half-densities on Q and in the above mentioned Hilbert space $\overline{E}$. Moreover, a glance at the expressions (5.4.15) and (5.4.17) shows that (5.4.17) is the representation of $\mathcal{A}_V$ as a subalgebra of the quantum algebra $\mathcal{A}_T$.

5.4.3 *Instantwise quantization of* V^*Q

As was mentioned above, the physical relevance of the space of half-densities on Q with the scalar product (5.4.16) is open to question. At the same time, the representation (5.4.17) preserves the structure of $\mathcal{A}_V$ as a $C^\infty(\mathbb{R})$-algebra. Therefore, let us show that this representation defines the leafwise quantization of the symplectic foliation $V^*Q \to \mathbb{R}$ which takes the form of instantwise quantization of $\mathcal{A}_V$.

(i) The prequantization (5.4.13) of a Poisson manifold V^*Q yields prequantization of its symplectic leaves V_t^*Q, $t \in \mathbb{R}$, as follows. The symplectic structure on V_t^*Q is

$$\Omega_t = (h \circ i_t)^*\Omega_T = dp_k \wedge dq^k, \qquad (5.4.18)$$

where h is an arbitrary section of the fibre bundle ζ (3.3.3) and $i_t : V_t^*Q \to V^*Q$ is the natural imbedding. Since $w^\sharp\phi$ is a vertical vector field on $V^*Q \to \mathbb{R}$ for any one-form ϕ on V^*Q, the contravariant derivative (5.4.12) defines a connection along each fibre V_t^*Q, $t \in \mathbb{R}$, of the Poisson bundle $V^*Q \to \mathbb{R}$. It is the pull-back

$$A_t = i_t^*h^*A = dp_k \otimes \partial^k + dq^k \otimes (\partial_k - ip_k c\partial_c)$$

of the connection h^*A (5.4.11) onto the trivial pull-back line bundle

$$i_t^*C_V = V_t^*Q \times \mathbb{C} \to V_t^*Q.$$

It is readily observed that this connection is admissible for the symplectic structure (5.4.18) on V_t^*Q, and provides prequantization of the symplectic manifold (V_t^*Q, Ω_t) by the formula

$$\widehat{f}_t = -i\mathbf{L}_{\vartheta_{f_t}} + (\mathbf{L}_{\vartheta_t} - f_t) = -i(\partial^k f_t\partial_k - \partial_k f_t\partial^k) + (p_k\partial^k f_t - f_t), \quad (5.4.19)$$

where

$$\vartheta_{f_t} = \partial^k f_t \partial_k - \partial_k f_t \partial^k$$

is the Hamiltonian vector field of a function f_t on V_t^*Q with respect to the symplectic form Ω_t (5.4.18). The operators (5.4.19) act on smooth complex functions s_t on V_t^*Q. In particular, let f_t, s_t and $(\widehat{f}s)_t$ be the restriction to V_t^*Q of a real function f and complex functions s and $\widehat{f}(s)$ on V^*Q, respectively. We obtain from the formulas (5.4.13) and (5.4.19) that $(\widehat{f}s)_t = \widehat{f}_t s_t$. This equality shows that the prequantization (5.4.13) of the Poisson manifold V^*Q is leafwise prequantization.

(ii) Let $\mathbf{T}_V^*$ be the above mentioned polarization of the Poisson manifold V^*Q. It yields the pull-back polarization $\mathbf{T}_t^* = i_t^*\mathbf{T}_V^*$ of a fibre V_t^*Q with respect to the Poisson morphism

$$i_t : V_t^*Q \to V^*Q.$$

The corresponding distribution $\mathbf{T}_t$ coincides with the vertical tangent bundle of the fibre bundle $V_t^*Q \to Q_t$. The associated quantum algebra $\mathcal{A}_t$ consists of functions on V_t^*Q which are affine in momenta. In particular, the restriction to V_t^*Q of any element of the quantum algebra $\mathcal{A}_V$ of V^*Q obeys this condition and, consequently, belongs to $\mathcal{A}_t$. Conversely, any element of $\mathcal{A}_t$ is of this type. For instance, using a trivialization (5.4.1) and the corresponding surjection $\pi_t : V^*Q \to V_t^*Q$, one can define the pull-back $\pi_t^* f_t$ of a function $f_t \in \mathcal{A}_t$ which belongs to the quantum algebra $\mathcal{A}_V$ and $f_t = i_t^*(\pi_t^* f_t)$. Thus, $\mathcal{A}_t = i_t^*\mathcal{A}_V$ and, therefore, the polarization $\mathbf{T}_V^*$ of the Poisson bundle $V^*Q \to \mathbb{R}$ is fibrewise polarization.

(iii) The Jacobian S of transition function between coordinate charts $(U; t, q^k)$ and $(U'; t, q'^k)$ possesses the property

$$S = \det \begin{pmatrix} 1 & \partial_t q'^k \\ 0 & (\partial_i q'^k) \end{pmatrix} = \det(\partial_i q'^k). \tag{5.4.20}$$

It follows that the metalinear complex line bundle $\mathcal{D}_{1/2}[Q] \to Q$ with transition functions J such that $J\overline{J} = S$ on $U \cap U'$ also is the metalinear bundle of fibrewise half-densities on a fibre bundle $Q \to \mathbb{R}$.

(iv) Any atlas $\{(U; t, q^k)\}$ of bundle coordinates on a fibre bundle $Q \to \mathbb{R}$ induces a coordinate atlas $\{(Q_t \cap U; q^k)\}$ of its fibre Q_t, $t \in \mathbb{R}$. Due to the equality (5.4.20), the Jacobian S on Q coincides with the Jacobian S_t of the transition function between coordinate charts $(Q_t \cap U; q^k)$ and $(Q_t \cap U'; q'^k)$ on Q_t at points of $Q_t \cap U \cap U'$. It follows that, for any fibre Q_t of Q, the pull-back $i_t^*\mathcal{D} \to Q_t$ of the complex line bundle $\mathcal{D} \to Q$ of complex densities on Q with transition functions $v' = Sv$ is the complex line bundle of complex densities on Q_t with transition functions $S_t = S|_{Q_t}$. Accordingly, any density L on Q yields the pull-back section $L_t = L \circ i_t$ of the line bundle $i^*\mathcal{D} \to Q_t$, i.e., L_t is a density on Q_t. The pull-back $L \to L_t$ takes the coordinate form

$$L = \mathcal{L}(t, q^k) d^m q \wedge dt \to L_t = \mathcal{L}(t, q^k) d^m q|_{t=\text{const}} = \mathcal{L}_t(q^k) \overline{d}^m q,$$

where $\{\overline{d}q^k\}$ are holonomic fibre bases for V^*Q. It is maintained under transformations of bundle coordinates on Q.

Let $\mathcal{D}_{1/2}[Q] \to Q$ be the metalinear complex line bundle over Q in item (iii). Its pull-back $i_t^*\mathcal{D}_{1/2}[Q]$ is a complex line bundle over a fibre Q_t, $t \in \mathbb{R}$, with transition functions $J_t = J|_{Q_t}$. These transition functions obey the relation

$$J_t \overline{J}_t = S|_{Q_t} = S_t,$$

i.e., $i_t^*\mathcal{D}_{1/2}[Q] \to Q_t$ is the metalinear complex line bundle over Q_t. Then the formula (5.4.19) defines the Schrödinger representation of the quantum algebra $\mathcal{A}_t$ of the symplectic fibre Q_t by (unbounded) Hermitian operators

$$\widehat{f}_t \rho_t = (-i\mathbf{L}_{a^k \partial_k} - b)\rho_t = \left(-ia^k \partial_k - \frac{i}{2}\partial_k a^k - b\right)\rho_t, \qquad (5.4.21)$$
$$f_t = a^k(q^i)p_k + b(q^i) \in \mathcal{A}_t,$$

in the Hilbert space $\overline{\mathfrak{E}}_t$ which is the completion of the pre-Hilbert space $\mathfrak{E}_t$ of half-densities on Q_t of compact support with respect to the scalar product

$$\langle \rho_t | \rho'_t \rangle = \int_{Q_t} \rho_t \overline{\rho}'_t.$$

If Q_t is compact, the operators (5.4.21) in $\overline{\mathfrak{E}}_t$ are self-adjoint. Pre-Hilbert spaces $\mathfrak{E}_t$ constitute a trivial bundle over $\mathbb{R}$.

As in the case of densities in item (iv), any half-density ρ on Q yields the section $\rho \circ i_t$ of the pull-back bundle $i_t^*\mathcal{D}_{1/2}[Q] \to Q_t$, i.e, a half-density on Q_t. Given an element $f \in \mathcal{A}_V$ and its pull-back $f_t = i_t^* f \in \mathcal{A}_t$, we obtain from the formulas (5.4.17) and (5.4.21) that

$$\widehat{f}\rho \circ i_t = \widehat{f}_t(\rho \circ i_t).$$

This equality shows that the Schrödinger quantization of the Poisson manifold V^*Q can be seen as the instantwise quantization.

Following this interpretation and bearing in mind that $\rho \in \mathcal{D}_{1/2}[Q]$ are fibrewise half-densities on $Q \to \mathbb{R}$, let us choose the carrier space $\mathfrak{E}_R$ of the Schrödinger representation (5.4.17) of $\mathcal{A}_V$ which consists of complex half-densities ρ on Q such that, for any $t \in \mathbb{R}$, the half-density $\rho \circ i_t$ on Q_t is of compact support. It is a pre-Hilbert $C^\infty(\mathbb{R})$-module with respect to the fibrewise Hermitian form

$$\langle \rho | \rho' \rangle_t = \int\limits_{Q_t} \rho \overline{\rho}'. \qquad (5.4.22)$$

The pre-Hilbert module $\mathfrak{E}_R$ also is the carrier space for the quantum algebra $\mathcal{A}_T$, but its action in $\mathfrak{E}_R$ is not instantwise.

5.4.4 *Quantization of the evolution equation*

Let us turn now to quantization of the evolution equation (3.8.3). As was mentioned above, the problem is that, in the framework of the Schrödinger quantization, the homogeneous Hamiltonian $\mathcal{H}^*$ (3.4.1) does not belong to the quantum algebra $\mathcal{A}_T$, unless it is affine in momenta. Let us restrict our consideration to the physically relevant case of $\mathcal{H}^*$, polynomial in momenta. We aim to show that such $\mathcal{H}^*$ is decomposed in a finite sum of products of elements of the algebra $\mathcal{A}_T$.

Let f be a smooth function on T^*Q which is a polynomial of momenta p_λ. A glance at the transformation laws (2.2.4) shows that it is a sum of homogeneous polynomials of fixed degree in momenta. Therefore, it suffices to justify a desired decomposition of an arbitrary homogeneous polynomial F of degree $k > 1$ on T^*Q. We use the fact that the cotangent bundle T^*Q admits a finite bundle atlas (Theorem 11.2.7). Let $\{U_\xi\}$, $\xi = 1, \ldots, r$, be the corresponding open cover of Q and $\{f_\xi\}$ a smooth partition of unity subordinate to this cover. Let us put

$$l_\xi = f_\xi (f_1^k + \cdots + f_r^k)^{-1/k}.$$

It is readily observed that $\{l_\xi^k\}$ also is a partition of unity subordinate to $\{U_i\}$. Let us consider the local polynomials

$$F_\xi = F|_{U_\xi} = \sum_{(\alpha_1 \ldots \alpha_k)} a_\xi^{\alpha_1 \ldots \alpha_k}(q) p_{\alpha_1} \cdots p_{\alpha_k}, \qquad q \in U_\xi.$$

Then we obtain a desired decomposition

$$F = \sum_\xi l_\xi^k F_\xi = \sum_\xi \sum_{(\alpha_1 \ldots \alpha_k)} [l_\xi a_\xi^{\alpha_1 \ldots \alpha_k} p_{\alpha_1}][l_\xi p_{\alpha_2}] \cdots [l_\xi p_{\alpha_k}], \qquad (5.4.23)$$

where all terms $l_\xi a_\xi^{\alpha_1 \ldots \alpha_k} p_{\alpha_1}$ and $l_\xi p_\alpha$ are smooth functions on T^*Q. Clearly, the decomposition (5.4.23) by no means is unique.

The decomposition (5.4.23) shows that one can associate to a polynomial homogeneous Hamiltonian $\mathcal{H}^*$ an element $\overline{\mathcal{H}}^*$ of the enveloping algebra $\overline{\mathcal{A}}_T$ of the Lie algebra $\mathcal{A}_T$. Let us recall that $\overline{\mathcal{A}}$ consists of finite sums of tensor products of elements of $\mathcal{A}_T$ modulo the relations

$$f \otimes f' - f' \otimes f - \{f, f'\}_T = 0.$$

To be more precise, a representative $\overline{\mathcal{H}}^*$ belongs to $\mathcal{A}_T + \overline{\mathcal{A}}_V$, where $\overline{\mathcal{A}}_V$ is the enveloping algebra of the Lie algebra $\mathcal{A}_V \subset \mathcal{A}_T$. The enveloping algebra $\overline{\mathcal{A}}_V$ is provided with the anti-automorphism

$$* : f_1 \otimes \cdots \otimes f_k \to (-1)^k f_k \otimes \cdots \otimes f_1,$$

and one can always make a representative $\overline{\mathcal{H}}^*$ Hermitian.

Since Dirac's condition (0.0.4) holds, the Schrödinger representation of the Lie algebras $\mathcal{A}_T$ and $\mathcal{A}_V$ in the pre-Hilbert module $\mathfrak{E}_R$ is naturally extended to their enveloping algebras $\overline{\mathcal{A}}_T$ and $\overline{\mathcal{A}}_V$, and provides the quantization $\widehat{\mathcal{H}}^*$ of a homogeneous Hamiltonian $\mathcal{H}^*$.

Moreover, since $\widehat{p}_0 = -i\partial_t$, the operator $i\widehat{\mathcal{H}}^*$ obeys the Leibniz rule

$$i\widehat{\mathcal{H}}^*(r\rho) = \partial_t r\rho + r(i\widehat{\mathcal{H}}^*\rho), \qquad r \in C^\infty(\mathbb{R}), \qquad \rho \in \mathfrak{E}_R. \tag{5.4.24}$$

Therefore, it is a connection on the $C^\infty(\mathbb{R})$-module $\mathfrak{E}_R$. Then the quantum constraint

$$i\widehat{\mathcal{H}}^*\rho = 0, \qquad \rho \in \mathfrak{E}_R, \tag{5.4.25}$$

plays a role of the *Schrödinger equation* (4.6.7) in quantum non-relativistic mechanics.

Given an operator $\widehat{\mathcal{H}}^*$, the bracket

$$\nabla\widehat{f} = i[\widehat{\mathcal{H}}^*, \widehat{f}] \tag{5.4.26}$$

defines a derivation of the quantum algebra $\overline{\mathcal{A}}_V$. Since $\widehat{p}_0 = -i\partial_t$, the derivation (5.4.26) obeys the Leibniz rule

$$\nabla(r\widehat{f}) = \partial_t r\widehat{f} + r\nabla\widehat{f}, \qquad r \in C^\infty(\mathbb{R}).$$

Therefore, it is a connection on the instantwise algebra $\overline{\mathcal{A}}_V$. In particular, $\widehat{f}$ is parallel with respect to the connection (5.4.26) if

$$[\widehat{\mathcal{H}}^*, \widehat{f}] = 0. \tag{5.4.27}$$

By analogy with the equation (4.6.2), one can think of this equality as being the *Heisenberg equation* in quantum non-relativistic mechanics. It is readily observed that an operator $\widehat{f}$ is a solution of the Heisenberg equation (5.4.27) if and only if it preserves the subspaces of solutions of the Schrödinger equation (5.4.25). We call $\widehat{\mathcal{H}}^*$ the *Heisenberg operator*.

5.5 Quantization with respect to different reference frames

In accordance with the Schrödinger representation (5.2.10), the homogeneous Hamiltonian (3.4.1):

$$\mathcal{H}^* = p_0 + \mathcal{H}, \tag{5.5.1}$$

is quantized as the operator

$$\widehat{\mathcal{H}}^* = \widehat{p}_0 + \widehat{\mathcal{H}} = -i\partial_t + \widehat{\mathcal{H}}. \tag{5.5.2}$$

A problem is that the decomposition (5.5.1) and the corresponding splitting (5.5.2) of the Heisenberg operator $\widehat{\mathcal{H}}^*$ are ill defined.

At the same time, any reference frame Γ yields the decomposition

$$\mathcal{H}^* = (p_0 + \mathcal{H}_\Gamma) + (\mathcal{H} - \mathcal{H}_\Gamma) = \mathcal{H}^*_\Gamma + \mathcal{E}_\Gamma,$$

where $\mathcal{H}_\Gamma$ is the Hamiltonian (3.3.16) and $\mathcal{E}_\Gamma$ (3.3.18) is the energy function relative to a reference frame Γ (Remark 3.4.1). Accordingly, we obtain the splitting of the Heisenberg operator

$$\widehat{\mathcal{H}}^* = \widehat{\mathcal{H}}^*_\Gamma + \widehat{\mathcal{E}}_\Gamma,$$

where

$$\widehat{\mathcal{H}}^*_\Gamma = -i\partial_t - i\Gamma^k\partial_k - \frac{i}{2}\partial_k\Gamma^k \tag{5.5.3}$$

and $\widehat{\mathcal{E}}_\Gamma$ is the *operator of energy* relative to a reference frame Γ [110].

Note that the homogeneous Hamiltonian $\mathcal{H}^*_\Gamma$ (3.3.16) is affine in momenta and, therefore, it belongs to the quantum algebra $\mathcal{A}_T$ of T^*Q. Its Schrödinger representation (5.5.3) is well defined. Written with respect to Γ-adapted coordinates, it takes the form $\widehat{\mathcal{H}}^*_\Gamma = -i\partial_t$.

Remark 5.5.1. Any connection Γ (1.1.18) on a configuration bundle $Q \to \mathbb{R}$ induces the connection (5.4.26):

$$\widehat{\nabla}_\Gamma \widehat{f} = i[\widehat{\mathcal{H}}_\Gamma, \widehat{f}] \tag{5.5.4}$$

on the algebra $\overline{\mathcal{A}}_V$ which also is a connection on the quantum algebra $\mathcal{A}_V \subset \overline{\mathcal{A}}_V$. The corresponding Schrödinger equation (5.4.25) reads

$$-i\left(\partial_t + \Gamma^k\partial_k + \frac{1}{2}\partial_k\Gamma^k\right)\rho = 0.$$

Its solutions are half-densities $\rho \in \mathfrak{C}_R$ which, written relative to Γ-adapted coordinates $(t, \overline{q}^j)$, are time-independent, i.e., $\rho = \rho(\overline{q}^j)$.

Given a reference frame Γ, the energy function $\mathcal{E}_\Gamma$ is quantized as $\widehat{\mathcal{E}}_\Gamma = \widehat{\mathcal{H}}^* - \widehat{\mathcal{H}}^*_\Gamma$. As a consequence, the Schrödinger equation (5.4.25) reads

$$(\widehat{\mathcal{H}}_\Gamma + \widehat{\mathcal{E}}_\Gamma)\rho = -i\left(\partial_t + \Gamma^k\partial_k + \frac{1}{2}\partial_k\Gamma^k\right)\rho + \widehat{\mathcal{E}}_\Gamma\rho = 0. \tag{5.5.5}$$

For instance, let a classical Hamiltonian system be autonomous, and let Γ be a reference frame such that the energy function $\mathcal{E}_\Gamma$ is time-independent relative to Γ-adapted coordinates. In this case, the Schrödinger equation (5.5.5) takes the familiar form

$$(-i\partial_t + \widehat{\mathcal{E}}_\Gamma)\rho = 0. \tag{5.5.6}$$

It follows from the Heisenberg equation (5.4.27) that a a quantum Hamiltonian system is autonomous if and only if there exists a reference frame Γ such that

$$[\widehat{\mathcal{H}}^*, \widehat{\mathcal{E}}_\Gamma] = 0.$$

Given different reference frames Γ and Γ', the operators of energy $\widehat{\mathcal{E}}_\Gamma$ and $\widehat{\mathcal{E}}_{\Gamma'}$ obey the relation

$$\widehat{\mathcal{H}}^*_\Gamma + \widehat{\mathcal{E}}_\Gamma = \widehat{\mathcal{H}}^*_{\Gamma'} + \widehat{\mathcal{E}}_{\Gamma'} \tag{5.5.7}$$

taking the form

$$\widehat{\mathcal{E}}_{\Gamma'} = \widehat{\mathcal{E}}_\Gamma - i(\Gamma^k - \Gamma'^k)\partial_k - \frac{i}{2}\partial_k(\Gamma^k - \Gamma'^k). \tag{5.5.8}$$

In particular, let $\widehat{\mathcal{E}}_\Gamma$ be a time-independent energy operator of an autonomous Hamiltonian system, and let ρ_E be its eigenstate of eigenvalue E, i.e., $\widehat{\mathcal{E}}_\Gamma\rho_E = E\rho_E$. Then the energy of this state relative to a reference frame Γ' at an instant t is

$$\begin{aligned}\langle\rho_E|\widehat{\mathcal{E}}_{\Gamma'}\rho_E\rangle &= E + i\langle\rho_E|\left(\Gamma'^k\partial_k + \frac{1}{2}\partial_k\Gamma'^k\right)\rho_E\rangle_t \\ &= E + i\int_{Q_t}\overline{\rho}_E\left(\Gamma'^k(q^j,t)\partial_k + \frac{1}{2}\partial_k\Gamma'^k(q^j,t)\right)\rho_E.\end{aligned}$$

Example 5.5.1. Let us consider a Hamiltonian system on $Q = \mathbb{R}\times U$, where $U \subset \mathbb{R}^m$ is an open domain equipped with coordinates (q^i). These coordinates yield a reference frame on Q given by the connection Γ such that $\Gamma^i = 0$ with respect to these coordinates. Let it be an autonomous Hamiltonian system whose energy function $\mathcal{E}_\Gamma$, written relative to coordinates (t, q^i), is time-independent. Let us consider a different reference frame on Q given by the connection

$$\Gamma' = dt\otimes(\partial_t + G^i\partial_i), \qquad G^i = \text{const}, \tag{5.5.9}$$

on Q. The Γ'-adapted coordinates (t, q'^j) obey the equations (1.6.1) – (1.6.2) which read

$$G^i = \frac{\partial q^i(t, q'^j)}{\partial t}, \qquad \frac{\partial q'^j(t, q^i)}{\partial q^k} G^k + \frac{\partial q'^j(t, q^i)}{\partial t} = 0. \qquad (5.5.10)$$

We obtain $q'^i = q^i - G^i t$. For instance, this is the case of inertial frames. Given by the relation (3.3.19), the energy function relative to the reference frame Γ' (5.5.9) reads

$$\mathcal{E}_{\Gamma'} = \mathcal{E}_\Gamma - G^k p_k.$$

Accordingly, the relation (5.5.8) between operators of energy $\widehat{\mathcal{E}}_{\Gamma'}$ and $\widehat{\mathcal{E}}_\Gamma$ takes the form

$$\widehat{\mathcal{E}}_{\Gamma'} = \widehat{\mathcal{E}}_\Gamma + iG^k \partial_k. \qquad (5.5.11)$$

Let ρ_E be an eigenstate of the energy operator $\widehat{\mathcal{E}}_\Gamma$. Then its energy with respect to the reference frame Γ' (5.5.9) is $E - G^k P_k$, where

$$P_k = \langle \rho_E | \widehat{p}_k \rho_E \rangle_t$$

are momenta of this state. This energy is time-independent.

In particular, the following condition holds in many physical models. Given an eigenstate ρ_E of the energy operator $\widehat{\mathcal{E}}_\Gamma$ and a reference frame Γ' (5.5.9), there is the equality

$$\begin{aligned}\widehat{\mathcal{E}}_{\Gamma'}(\widehat{p}^j, q^j)\rho_E &= (\widehat{\mathcal{E}}_\Gamma(\widehat{p}^j, q^j) - G^k(q^j)\widehat{p}_k)\rho_E \\ &= (\widehat{\mathcal{E}}_\Gamma(\widehat{p}^j + A_j, q^j) + B)\rho_E, \qquad A_j, B = \text{const.}\end{aligned}$$

Then $\exp(-iA_j q^j)\rho_E$ is an eigenstate of the energy operator $\widehat{\mathcal{E}}_{\Gamma'}$ possessing the eigenvalue $E + B$.

For instance, any Hamiltonian

$$\widehat{\mathcal{H}} = \widehat{\mathcal{E}}_\Gamma = \frac{1}{2}(m^{-1})^{ij}\widehat{p}_i \circ \widehat{p}_j + V(q^j)$$

quadratic in momenta $\widehat{p}_i$ with a non-degenerate constant mass tensor m^{ij} obeys this condition. Namely, we have

$$A_i = -m_{ij}G^j, \qquad B = -\frac{1}{2}m_{ij}G^i G^j.$$

Let us consider a massive point particle in an Euclidean space $\mathbb{R}^3$ in the presence of a central potential $V(r)$. Let $\mathbb{R}^3$ be equipped with the spherical coordinates (r, ϕ, θ). These coordinates define an inertial reference frame

Γ such that $\Gamma^r = \Gamma^\phi = \Gamma^\theta = 0$. The Hamiltonian of the above mentioned particle with respect to this reference frame reads

$$\widehat{\mathcal{H}} = \widehat{\mathcal{E}}_\Gamma = \frac{1}{m}\left(-\frac{1}{r}\partial_r - \frac{1}{2}\partial_r^2 + \frac{\widehat{I}^2}{r^2}\right) + V(r), \qquad (5.5.12)$$

where $\widehat{I}$ is the square of the angular momentum operator. Let us consider a rotatory reference frame $\Gamma'^\phi = \omega =$const, given by the adapted coordinates $(r, \phi' = \phi - \omega t, \theta)$. The operator of energy relative to this reference frame is

$$\widehat{\mathcal{E}}_{\Gamma'} = \widehat{\mathcal{E}}_\Gamma + i\omega\partial_\phi. \qquad (5.5.13)$$

Let $\rho_{E,n,l}$ be an eigenstate of the energy operator $\widehat{\mathcal{E}}_\Gamma$ (5.5.12) possessing its eigenvalue E, the eigenvalue n of the angular momentum operator $\widehat{I}_3 = \widehat{p}_\phi$, and the eigenvalue $l(l+1)$ of the operator $\widehat{I}^2$. Then $\rho_{E,n,l}$ also is an eigenstate of the energy operator $\widehat{\mathcal{E}}_{\Gamma'}$ with the eigenvalue $E' = E - n\omega$.

Chapter 6

Constraint Hamiltonian systems

In Section 3.6, we have observed that Hamiltonian systems associated with non-regular Lagrangian systems are necessarily characterized by constraints. This Chapter is devoted to Hamiltonian systems with time-dependent constraints and their geometric quantization. Let us note that, in Chapter 10, Hamiltonian relativistic mechanics is treated as such a constraint system.

6.1 Autonomous Hamiltonian systems with constraints

We start with constraints in autonomous Hamiltonian mechanics.

Let (Z, Ω) be a $2m$-dimensional symplectic manifold and $\mathcal{H}$ a Hamiltonian on Z. Let N be a $(2m-n)$-dimensional closed imbedded submanifold of Z called a *primary constraint space* or, simply, a *constraint space*. We consider the following two types of autonomous constraint systems:

• a *constraint Hamiltonian system*

$$S_{\mathcal{H}}|_N = \bigcup_{z\in N} \{v \in T_z N : v\rfloor\Omega + d\mathcal{H}(z) = 0\}, \qquad (6.1.1)$$

whose solutions are solutions of the Hamiltonian system $(\Omega, \mathcal{H})$ (3.2.6) on a manifold N which live in the tangent bundle TN of N;

• a *Dirac constraint system*

$$S_{i_N^*\mathcal{H}} = \bigcup_{z\in N} \{v \in T_z N : v\rfloor i_N^*(\Omega + d\mathcal{H}(z)) = 0\}, \qquad (6.1.2)$$

which is the restriction of a Hamiltonian system $(\Omega, \mathcal{H})$ on Z to a constraint space N, i.e., it is the presymplectic Hamiltonian system $(i_N^*\Omega, i_N^*\mathcal{H})$ (3.2.10) on N.

Remark 6.1.1. If a non-zero presymplectic form $i_N^*\mathcal{H}$ of a Dirac constraint system (6.1.2) is of constant rank, N is necessarily coisotropic.

This Section addresses constraint Hamiltonian systems (6.1.1).

Given a closed imbedded submanifold N of a symplectic manifold (Z,Ω), let us consider the set

$$I_N = \operatorname{Ker} i_N^* \subset C^\infty(Z) \tag{6.1.3}$$

of functions f on Z which vanish on N, i.e., $i_N^* f = 0$. It is an ideal of the $\mathbb{R}$-ring $C^\infty(Z)$. Then, since N is a closed imbedded submanifold of Z, we have the ring isomorphism

$$C^\infty(Z)/I_N = C^\infty(N). \tag{6.1.4}$$

Let us consider a space of all vector fields u on Z restrictable to vector fields on N, i.e., $u|_N \subset TN$. It is

$$\mathcal{T}_N = \{u \in \mathcal{T}(Z) : \ u\rfloor df \in I_N, \ f \in I_N\}. \tag{6.1.5}$$

Then we obtain at once that the Hamiltonian vector field ϑ_f of a function f on Z belongs to $\mathcal{T}_N$ if and only if

$$\vartheta_f\rfloor dg = \{f,g\} \in I_N, \qquad g \in I_N.$$

Hence, the functions whose Hamiltonian vector fields are restrictable to vector fields on N constitute the set

$$I(N) = \{f \in C^\infty(Z) : \ \{f,g\} \in I_N, \ g \in I_N\}, \tag{6.1.6}$$

called the *normalizer* of I_N. Owing to the Jacobi identity, the normalizer (6.1.6) is a Poisson subalgebra of $C^\infty(Z)$. Let us put

$$I'(N) = I(N) \cap I_N. \tag{6.1.7}$$

This is a Poisson subalgebra of $I(N)$ which is non-zero since $I^2 \subset I'(N)$ by virtue of the Leibniz rule.

Let us assume that the sets $w^\sharp(\operatorname{Ann} TN)$ and

$$\mathcal{C}(N) = w^\sharp(\operatorname{Ann} TN) \cap TN \tag{6.1.8}$$

of the tangent bundle TN of a constraint space N are distributions. All sections of $w^\sharp(\operatorname{Ann} TN) \to N$ are the restriction to N of Hamiltonian vector fields of elements of I_N, while all sections of $\mathcal{C}(N) \to N$ are the restriction to N of Hamiltonian vector fields of elements of $I'(N)$. In particular, if N is coisotropic, then $I_N \subset I(N)$, i.e., $I_N = I'(N)$ is a Poisson subalgebra of $C^\infty(Z)$.

Lemma 6.1.1. *The distribution $\mathcal{C}(N)$ is involutive [157].*

Theorem 6.1.1. *Let the foliation determined by $\mathcal{C}(N)$ be simple, i.e., a fibred manifold $N \to P$. Then there is the ring isomorphism*

$$C^\infty(P) = I(N)/I'(N). \qquad (6.1.9)$$

Since the quotient in the right-hand side of this isomorphism is a Poisson algebra, a base P is provided with a Poisson structure [90].

Theorem 6.1.1 describes a particular case of *Poisson reduction* which, in a general setting, is formulated in the following algebraic terms [65; 90].

Definition 6.1.1. Given a Poisson manifold Z, let J be an ideal of the Poisson algebra $C^\infty(Z)$ as an associative algebra, J'' its normalizer (6.1.6), and $J' = J'' \cap J'$. One says that the Poisson algebra J''/J' is the reduction of the Poisson algebra $C^\infty(Z)$ via the ideal J.

In accordance with this definition, an ideal J of a Poisson algebra $C^\infty(Z)$ is said to be *coisotropic* if J is a Poisson subalgebra of $\mathcal{P}$.

Remark 6.1.2. The following local relations are useful in the sequel. Let a constraint space N be locally given by the equations

$$f_a(z) = 0, \qquad a = 1, \ldots, n, \qquad (6.1.10)$$

where $f_a(z)$ are local functions on Z called the *primary constraints*. Let us consider the ideal $I_N \subset C^\infty(Z)$ (6.1.3) of functions vanishing on N. It is locally generated by the constraints f_a, and its elements are locally written in the form

$$f = \sum_{a=1}^{n} g^a f_a, \qquad (6.1.11)$$

where g^a are functions on Z. We agree to call $\{f_a\}$ a *local basis* for the ideal I_N. Let dI_N be the submodule of the $C^\infty(Z)$-module $\mathcal{O}^1(Z)$ of one-forms on Z which is locally generated by the exterior differentials df of functions $f \in I_N$. Its elements are finite sums

$$\sigma = \sum_i g^i df_i, \qquad f_i \in I_N, \qquad g^i \in C^\infty(Z).$$

In view of the formula (6.1.11), they are given by local expressions

$$\sigma = \sum_{a=1}^{n} (g^a df_a + f_a \phi^a), \qquad (6.1.12)$$

where g^a are functions and ϕ^a are one-forms on Z.

Turn now to the constraint Hamiltonian system (6.1.1). Its solution obviously exists if a Hamiltonian vector field $\vartheta_{\mathcal{H}}$, restricted to a constraint space N, is tangent to N. Then integral curves of the Hamiltonian vector field $\vartheta_{\mathcal{H}}$ do not leave N. This condition is fulfilled if and only if

$$\{\mathcal{H}, I_N\} \subset I_N, \qquad (6.1.13)$$

i.e., if and only if the Hamiltonian $\mathcal{H}$ belongs to the normalizer $I(N)$ (6.1.6) of the ideal I_N. With respect to a local basis $\{f_a\}$ of the ideal I_N, the relation (6.1.13) reads

$$\vartheta_{\mathcal{H}} \rfloor df_a = \{\mathcal{H}, f_a\} = \sum_{c=1}^{n} g_a^c f_c, \qquad (6.1.14)$$

where g_a^c are functions on Z. If the relation (6.1.13) (and, consequently, (6.1.14)) fails to hold, one introduces *secondary constraints*

$$f_a^{(2)} = \{\mathcal{H}, f_a\} = 0.$$

If a collection of primary and secondary constraints is not closed (i.e., $\{\mathcal{H}, f_a^{(2)}\}$ is not expressed in f_a and $f_a^{(2)}$) let us add the *tertiary constraints*

$$f_a^{(3)} = \{\mathcal{H}, \{\mathcal{H}, f_a\}\} = 0,$$

and so on. If a solution of the constraint Hamiltonian system exists anywhere on N, the procedure is stopped after a finite number of steps by constructing a *complete system of constraints.* This complete system of constraints defines the *final constraint space*, where the Hamiltonian vector field $\vartheta_{\mathcal{H}}$ is not transversal to the primary constraint space N.

From the algebraic viewpoint, we have obtained the minimal extension $I_{\rm fin}$ of the ideal I_N such that $\{\mathcal{H}, I_{\rm fin}\} \subset I_{\rm fin}$.

In algebraic terms, a solution of a constraint Hamiltonian system can be reformulated as follows. Let N be a closed imbedded submanifold of a symplectic manifold (Z, Ω) and I_N the ideal of functions vanishing everywhere on N (though any ideal of the ring $C^\infty(Z)$ can be utilized). All elements of I_N are said to be *constraints.* One aims to find a Hamiltonian, called *admissible*, on Z such that a symplectic Hamiltonian system $(\Omega, \mathcal{H})$ has a solution everywhere on N, i.e., N is a final constraint space for $(\Omega, \mathcal{H})$. In accordance with the condition (6.1.13), only an element of the normalizer $I(N)$ (6.1.6) is an admissible Hamiltonian. However, the normalizer $I(N)$ also contains constraints $I(N) \cap I_N$. In order to separate Hamiltonians and constraints, let us consider the overlap $I'(N)$ (6.1.7). Its elements are called the *first-class constraints*, while the remaining elements of I_N

are the *second-class constraints*. As was mentioned above, the set $I'(N)$ of first-class constraints is a Poisson subalgebra of the normalizer $I(N)$ and, consequently, of the Poisson algebra $C^\infty(Z)$ on a symplectic manifold (Z, Ω). Let us also recall that $I_N^2 \subset I'(N)$, i.e., products of second-class constraints are the first-class ones. Admissible Hamiltonians which is not reduced to first-class constraints are representatives of non-zero elements of the quotient $I(N)/I'(N)$, which is the reduction of the Poisson algebra $C^\infty(Z)$ via the ideal I_N in accordance with Definition 6.1.1.

If $\mathcal{H}$ is an admissible Hamiltonian, the constraint Hamiltonian systems $(Z, \Omega, \mathcal{H}, N)$ is equivalent to the presymplectic Hamiltonian system $(N, i_N^*\Omega, i_N^*\mathcal{H})$, i.e., their solutions coincide.

Example 6.1.1. If N is a coisotropic submanifold of Z, then $I_N \subset I(N)$ and $I'(N) = I_N$. Therefore, all constraints are of first-class. The presymplectic form $i_N^*\Omega$ on N is of constant rank. Let its characteristic foliation be simple, i.e., it defines a fibration $\pi : N \to P$ over a symplectic manifold (P, Ω_P). In view of the isomorphism (6.1.9), one can think of elements of the quotient $I(N)/I'(N)$ as being the Hamiltonians on a base P. It follows that the restriction of an admissible Hamiltonian $\mathcal{H}$ to the constraint space N coincides with the pull-back onto N of some Hamiltonian $\overline{\mathcal{H}}$ on P, i.e., $i_N^*\mathcal{H} = \pi^*\overline{\mathcal{H}}$. Thus, $(N, i_N^*\Omega, i_N^*\mathcal{H})$ is a gauge-invariant Hamiltonian system which is equivalent to the reduced Hamiltonian system $(P, \Omega_P, \overline{\mathcal{H}})$, and the original constraint Hamiltonian system $(Z, \Omega, \mathcal{H}, N)$ is so if $\mathcal{H}$ is an admissible Hamiltonian.

6.2 Dirac constraints

As was mentioned above, the Dirac constraint system $S_{N^*\mathcal{H}}$ (6.1.2) really is the pull-back presymplectic Hamiltonian system

$$(\Omega_N = i_N^*\Omega, \mathcal{H}_N = i_N^*\mathcal{H})$$

on the primary constraint space $N \subset Z$. By virtue of Proposition 3.2.1, it has a solution only at the points of the subset

$$N_2 = \{z \in N \,:\, u\rfloor d\mathcal{H}_N(z) = 0,\, u \in \operatorname{Ker}{}_z\Omega_N\},$$

which is assumed to be a manifold. Such a solution however need not be tangent to N_2. Then the above mentioned constraint algorithm for presymplectic Hamiltonian systems can be called into play. Nevertheless, one can say something more since the presymplectic system $S_{N^*\mathcal{H}}$ (6.1.2) on N is the pull-back of the symplectic one on Z.

Let us assume that a $(2m-n)$-dimensional closed imbedded submanifold N of Z is already a desired *final constraint space* of the Dirac constraint system (6.1.2), i.e., the equation

$$v\rfloor\Omega_N + d\mathcal{H}_N(z) = 0, \qquad v \in T_z N, \tag{6.2.1}$$

has a solution at each point $z \in N$. As was mentioned above, this is equivalent to the injection

$$\operatorname{Ker}\Omega_N = TN \cap \operatorname{Orth}_\Omega TN \subset \operatorname{Ker} d\mathcal{H}_N. \tag{6.2.2}$$

Let us reformulate this condition in algebraic terms of the ideal of constraints I_N (6.1.3), its normalizer $I(N)$ (6.1.6) and the Poisson algebra of first-class constraints $I'(N)$ (6.1.7). It is readily observed that, restricted to N, Hamiltonian vector fields ϑ_f of elements f of $I'(N)$ with respect to the symplectic form Ω on Z take their values into $TN \cap \operatorname{Orth}_\Omega TN$ [90]. Then the condition (6.2.2) can be written in the form

$$\{\mathcal{H}, I'(N)\} \subset I_N. \tag{6.2.3}$$

At the same time, $\{\mathcal{H}, I_N\} \not\subset I_N$ in general. This relation shows that, though the Dirac constraint system $(\Omega_N, \mathcal{H}_N)$ on N has a solution, the Hamiltonian vector field $\vartheta_\mathcal{H}$ of a Hamiltonian $\mathcal{H}$ on Z is not necessarily tangent to N, and its restriction to N need not be such a solution. The goal is to find a constraint $f \in I_N$ such that the modified Hamiltonian $\mathcal{H} + f$ would satisfy the condition

$$\{\mathcal{H} + f, I_N\} \subset I_N \tag{6.2.4}$$

and, consequently, the condition

$$\{\mathcal{H} + f, I'(N)\} \subset I_N. \tag{6.2.5}$$

It is called a *generalized Hamiltonian system.*

The condition (6.2.5) is fulfilled for all $f \in I_N$, while (6.2.4) is an equation for a second-class constraint f. Therefore, its solution implies separating first- and second-class constraints. A general difficulty lies in the fact that the set of elements generating $I_N^2 \subset I'(N)$ is necessarily infinitely reducible [90]. At the same time, the Hamiltonian vector fields of elements of I_N^2 vanish on the constraint space N. Therefore, one can employ the following procedure [120].

Since N is a $(2m-n)$-dimensional closed imbedded submanifold of Z, the ideal I_N is locally generated by a finite basis $\{f_a\}$, $a = 1, \ldots, n$, whose elements determine N by the local equations (6.1.10). Let the presymplectic form Ω_N be of constant rank $2m-n-k$. It defines a k-dimensional characteristic foliation of N. Since $N \subset Z$ is closed, there locally exist k linearly independent vector fields u_b on Z which, restricted to N, are tangent to the leaves of this foliation. They read

$$u_b = \sum_{a=1}^{n} g_b^a \vartheta_{f_a}, \qquad b = 1, \ldots, k,$$

where g_b^a are local functions on Z and ϑ_{f_a} are Hamiltonian vector fields of the constraints f_a. Then one can choose a new local basis $\{\phi_b\}$, $b = 1, \ldots, n$, for I_N where the first k functions take the form

$$\phi_b = \sum_{a=1}^{n} g_b^a f_a, \qquad b = 1, \ldots, k.$$

Let ϑ_{ϕ_b} be their Hamiltonian vector fields. One can easily justify that

$$\vartheta_{\phi_b}|_N = u_b|_N, \qquad b = 1, \ldots, k.$$

It follows that the constraints ϕ_b, $b = 1, \ldots, k$, belong to $I'(N) \setminus I_N^2$, i.e., they are first-class constraints, while the remaining ones $\phi_{k+1}, \ldots, \phi_n$ are of second-class. We have the relations

$$\{\phi_b, \phi_c\} = \sum_{a=1}^{n} C_{bc}^a \phi_a, \qquad b = 1, \ldots, k, \qquad c = 1, \ldots, n,$$

where C_{bc}^a are local functions on Z. It should be emphasized that the first-class constraints $\phi_1, \cdots, \phi_k$ do not constitute any local basis for $I'(N)$.

Now let us consider a local Hamiltonian on Z

$$\mathcal{H}' = \mathcal{H} + \sum_{a=1}^{n} \lambda^a \phi_a, \tag{6.2.6}$$

where λ^a are functions on Z. Since $\mathcal{H}$ obeys the condition (6.2.5), we find

$$\{\mathcal{H}, \phi_b\} = \sum_{a=1}^{n} B_b^a \phi_a, \qquad b = 1 \ldots, k,$$

where B_b^a are functions on Z. Then the equation (6.2.4) takes the form

$$\{\mathcal{H}, \phi_c\} + \sum_{a=k+1}^{n} \lambda^a \{\phi_a, \phi_c\} = \sum_{b=1}^{n} D_c^b \phi_b, \qquad c = k+1, \ldots n, \tag{6.2.7}$$

where D_c^b are functions on Z. It is a system of linear algebraic equations for the coefficients λ^a, $a = k+1, \ldots n$, of second-class constraints. These coefficients are defined uniquely by the equations (6.2.7), while the coefficients λ^a, $a = 1, \ldots, k$, of first-class constraints in the Hamiltonian $\mathcal{H}'$ (6.2.6) remain arbitrary.

Then, restricted to a constraint space N, the Hamiltonian vector field of the Hamiltonian $\mathcal{H}'$ (6.2.6) on Z provides a local solution of the Dirac constraint system on N.

We refer the reader to [120] for a global variant of the above procedure.

A generalized Hamiltonian system $(Z, \Omega, \mathcal{H} + f, N)$ is a constraint Hamiltonian system with an admissible Hamiltonian $\mathcal{H} + f$. It is equivalent to the original Dirac constraint system.

Example 6.2.1. Let a final constraint space N be a coisotropic submanifold of the symplectic manifold (Z, Ω). Then $I_N = I'(N)$, i.e., there are only first-class constraints. In this case, the Hamiltonian vector fields both of the Hamiltonian $\mathcal{H}$ and all the Hamiltonians $\mathcal{H} + f$, $f \in I_N$, provide solutions of the Dirac constraint system on N. If the characteristic foliation of the presymplectic form $i_N^*\mathcal{H}$ on N is simple, we have the reduced Hamiltonian system equivalent to the original Dirac constraint one (see Example 6.1.1).

Example 6.2.2. If N is a symplectic submanifold of Z, then $I'(N) = I_N^2$. Therefore, all constraints are of second-class, and the Hamiltonian (6.2.6) of a generalized Hamiltonian system is defined uniquely.

6.3 Time-dependent constraints

Given a non-relativistic mechanical system on a configuration bundle $Q \to \mathbb{R}$, time-dependent constraints on a phase space V^*Q can be described similarly to those in autonomous Hamiltonian mechanics.

Let N be a closed imbedded subbundle

$$i_N : N \to V^*Q$$

of a fibre bundle $V^*Q \to \mathbb{R}$, treated as a constraint space. It is neither Lagrangian nor symplectic submanifold with respect to the Poisson structure on V^*Q in general. Let us consider the ideal I_N of real functions f on V^*Q which vanish on N, i.e., $i_N^* f = 0$. Its elements are constraints. There is the isomorphism

$$C^\infty(V^*Q)/I_N = C^\infty(N) \tag{6.3.1}$$

of associative commutative algebras. Let $I(N)$ be the normalize (6.1.6) and $I'(N)$ the set (6.1.7) of first-class constraints, while the remaining elements of I_N are the second-class constraints.

Remark 6.3.1. Let N be a coisotropic submanifold of V^*Q, i.e., $w^\sharp(\mathrm{Ann}\, TN) \subset TN$. Then $I_N = I'(N)$, i.e., all constraints are of first class.

The relation (3.3.8) enables us to extend the constraint algorithms of autonomous mechanics and time-dependent mechanics on a product $\mathbb{R}\times M$ (see [25; 103]) to mechanical systems subject to time-dependent transformations.

Let H be a Hamiltonian form on a phase space V^*Q. In accordance with the relation (3.4.6), solutions of the Hamilton equation (3.3.22) – (3.3.23) does not leave the constraint space N if

$$\{\mathcal{H}^*, \zeta^* I_N\}_T \subset \zeta^* I_N. \tag{6.3.2}$$

If the relation (6.3.2) fails to hold, let us introduce secondary constraints $\{\mathcal{H}^*, \zeta^* f\}_T$, $f \in I_N$, which belong to $\zeta^* C^\infty(V^*Q)$. If the collection of primary and secondary constraints is not closed with respect to the relation (6.3.2), let us add the tertiary constraints $\{\mathcal{H}^*, \{\mathcal{H}^*, \zeta^* f_a\}_T\}_T$, and so on.

Let us assume that N is a final constraint space for a Hamiltonian form H. If a Hamiltonian form H satisfies the relation (6.3.2), so is a Hamiltonian form

$$H_f = H - f dt \tag{6.3.3}$$

where $f \in I'(N)$ is a first-class constraint. Though Hamiltonian forms H and H_f coincide with each other on the constraint space N, the corresponding Hamilton equations have different solutions on the constraint space N because

$$dH|_N \neq dH_f|_N.$$

At the same time, we have

$$d(i_N^* H) = d(i_N^* H_f).$$

Therefore, let us introduce the *constrained Hamiltonian form*

$$H_N = i_N^* H_f \tag{6.3.4}$$

which is the same for all $f \in I'(N)$. Let us note that H_N (6.3.4) is not a true Hamiltonian form on $N \to \mathbb{R}$ in general. On sections r of the fibre bundle $N \to \mathbb{R}$, we can write the equation

$$r^*(u_N \rfloor dH_N) = 0, \tag{6.3.5}$$

where u_N is an arbitrary vertical vector field on $N \to \mathbb{R}$. It is called the *constrained Hamilton equation*.

Proposition 6.3.1. *For any Hamiltonian form H_f (6.3.3), every solution of the Hamilton equation which lives in the constraint space N is a solution of the constrained Hamilton equation (6.3.5).*

Proof. The constrained Hamilton equation can be written as

$$r^*(u_N\rfloor di_N^* H_f) = r^*(u_N\rfloor dH_f|_N) = 0. \tag{6.3.6}$$

It differs from the Hamilton equation (3.3.22) – (3.3.23) for H_f restricted to N which reads

$$r^*(u\rfloor dH_f|_N) = 0, \tag{6.3.7}$$

where r is a section of $N \to \mathbb{R}$ and u is an arbitrary vertical vector field on $V^*Q \to \mathbb{R}$. A solution r of the equation (6.3.7) satisfies obviously the weaker condition (6.3.6). □

Remark 6.3.2. One also can consider the problem of constructing a generalized Hamiltonian system, similar to that for Dirac constraint system in autonomous mechanics [106]. Let H satisfy the condition $\{\mathcal{H}^*, \zeta^* I'(N)\}_T \subset I_N$, whereas $\{\mathcal{H}^*, \zeta^* I_N\}_T \not\subset I_N$. The goal is to find a constraint $f \in I_N$ such that the modified Hamiltonian form $H - f dt$ would satisfy the condition

$$\{\mathcal{H}^* + \zeta^* f, \zeta^* I_N\}_T \subset \zeta^* I_N.$$

This is an equation for a second-class constraint f.

The construction above, except the isomorphism (6.3.1), can be applied to any ideal J of $C^\infty(V^*Q)$. Then one says that the Poisson algebra J''/J' (see Definition 6.1.1) is the reduction of the Poisson algebra $C^\infty(V^*Q)$ via the ideal J. In particular, if an ideal J is coisotropic (i.e., a Poisson algebra), it is a Poisson subalgebra of the normalize J'' (6.1.6), and it coincides with J'.

Example 6.3.1. Let $\mathcal{A}$ be a Lie algebra of integrals of motion of a Hamiltonian system (V^*Q, H) (see Proposition 3.8.2). Let $I_\mathcal{A}$ denote the ideal of $C^\infty(V^*Q)$ generated by these integrals of motion. It is readily observed that this ideal is coisotropic. Then one can think of $I_\mathcal{A}$ as being an ideal of first-class constraints which form a complete system.

6.4 Lagrangian constraints

As was mentioned above, Hamiltonian systems associated with non-regular Lagrangian systems are constraint systems in general.

Let L be an almost regular Lagrangian on a velocity space J^1Q and

$$N_L = \widehat{L}(J^1Q) \subset V^*Q$$

the corresponding Lagrangian constraint space. In view of Theorem 3.6.4, let us assume that the fibred manifold

$$\widehat{L} : J^1Q \to N_L \tag{6.4.1}$$

admits a global section. Then there exist Hamiltonian forms weakly associated with a Lagrangian L. Theorems 3.6.2 – 3.6.3 establish the relation between the solutions of the Lagrange equation (2.1.25) for L and the solutions of the Hamilton equation (3.3.22) – (3.3.23) for H which live in the Lagrangian constraint space N_L. Therefore, let us consider the constrained Hamilton equation on N_L and compare its solutions with the solutions of the Lagrange (2.1.25) for L.

Given a global section Ψ of the fibred manifold (6.4.1), let us consider the pull-back constrained form

$$H_N = \Psi^* H_L = i_N^* H \tag{6.4.2}$$

on N_L. By virtue of Lemma 3.6.1, this form does not depend on the choice of a section of the fibred manifold (3.6.14) and, consequently, $H_L = \widehat{L}^* H_N$. For sections r of the fibre bundle $N_L \to \mathbb{R}$, one can write the constrained Hamilton equation (6.3.5):

$$r^*(u_N \rfloor dH_N) = 0, \tag{6.4.3}$$

where u_N is an arbitrary vertical vector field on $N_L \to \mathbb{R}$. This equation possesses the following important property.

Theorem 6.4.1. *For any Hamiltonian form H weakly associated with an almost regular Lagrangian L, every solution r of the Hamilton equation which lives in the Lagrangian constraint space N_L, is a solution of the constrained Hamilton equation (6.4.3).*

Proof. Such a Hamiltonian form H defines the global section $\Psi = \widehat{H} \circ i_N$ of the fibred manifold (6.4.1). Since $H_N = i_N^* H$ due to the relation (3.6.11), the constrained Hamilton equation can be written as

$$r^*(u_N \rfloor di_N^* H) = r^*(u_N \rfloor dH|_{N_L}) = 0. \tag{6.4.4}$$

Note that this equation differs from the Hamilton equation (3.3.25) restricted to N_L. This reads

$$r^*(u\rfloor dH|_{N_L}) = 0, \tag{6.4.5}$$

where r is a section of $N_L \to \mathbb{R}$ and u is an arbitrary vertical vector field on $V^*Q \to \mathbb{R}$. A solution r of the equations (6.4.5) obviously satisfies the weaker condition (6.4.4). □

Theorem 6.4.2. *The constrained Hamilton equation (6.4.3) is equivalent to the Hamilton–De Donder equation (2.2.14).*

Proof. Since $\widehat{L} = \zeta \circ \widehat{H}_L$ (2.2.6), the fibration ζ (2.2.5) yields a surjection of Z_L (2.2.3) onto N_L. Given a section Ψ of the fibred manifold (6.4.1), we have the morphism

$$\widehat{H}_L \circ \Psi : N_L \to Z_L.$$

By virtue of Lemma (3.6.1), this is a surjection such that

$$\zeta \circ \widehat{H}_L \circ \Psi = \mathrm{Id}\, N_L.$$

Hence, $\widehat{H}_L \circ \Psi$ is a bundle isomorphism over Q which is independent of the choice of a global section Ψ. Combination of (2.2.13) and (6.4.2) results in

$$H_N = (\widehat{H}_L \circ \Psi)^* \Xi_L$$

that leads to a desired equivalence. □

This proof gives something more. Namely, since Z_L and N_L are isomorphic, the homogeneous Legendre map $\widehat{H}_L$ (2.2.2) fulfils the conditions of Theorem 2.2.1. Then combining Theorem 2.2.1 and Theorem 6.4.2, we obtain the following.

Theorem 6.4.3. *Let L be an almost regular Lagrangian such that the fibred manifold (3.6.14) has a global section. A section $\overline{s}$ of the jet bundle $J^1Q \to \mathbb{R}$ is a solution of the Cartan equation (2.2.11) if and only if $\widehat{L} \circ \overline{s}$ is a solution of the constrained Hamilton equation (6.4.3).*

Theorem 6.4.3 also is a corollary of Lemma 6.4.1 below. The constrained Hamiltonian form H_N (6.4.2) defines the *constrained Lagrangian*

$$L_N = h_0(H_N) = (J^1 i_N)^* L_H \tag{6.4.6}$$

on the jet manifold $J^1 N_L$ of the fibre bundle $N_L \to \mathbb{R}$.

Lemma 6.4.1. *There are the relations*

$$\widetilde{L} = (J^1\widehat{L})^* L_N, \qquad L_N = (J^1\Psi)^* \widetilde{L}, \tag{6.4.7}$$

where $\widetilde{L}$ is the Lagrangian (2.2.7).

The Lagrange equation for the constrained Lagrangian L_N (6.4.6) is equivalent to the constrained Hamilton equation (6.4.3) and, by virtue of Lemma 6.4.1, is quasi-equivalent to the Cartan equation (2.2.9) – (2.2.10).

Example 6.4.1. Let us consider the almost regular quadratic Lagrangian L (2.3.1). The corresponding Lagrangian constraint space N_L is defined by the equations (3.7.4). There is a complete set of Hamiltonian forms $H(\sigma,\Gamma)$ (3.7.6) weakly associated with L. All of them define the same constrained Hamiltonian form

$$H_N = \mathcal{P}_i dq^i - \left[\frac{1}{2}\sigma_0^{ij}\mathcal{P}_i\mathcal{P}_j - c'\right] dt$$

and the constrained Lagrangian

$$L_N = \left[\mathcal{P}_i q_t^i - \frac{1}{2}\sigma_0^{ij}\mathcal{P}_i\mathcal{P}_j + c'\right] dt.$$

6.5 Geometric quantization of constraint systems

We start with autonomous constraint systems. Let (Z,Ω) be a symplectic manifold and $i_N : N \to Z$ its closed imbedded submanifold such that the presymplectic form $i_N^*\Omega$ on N is non-zero. We assume that N is a final constraint space and $\mathcal{H}$ is an admissible Hamiltonian on Z. In this case, the constraint Hamiltonian system $(Z,\Omega,\mathcal{H},N)$ is equivalent to the Dirac constraint system $(N, i_N^*\Omega, i_N^*\mathcal{H})$. Therefore, it seems natural to quantize a symplectic manifold (Z,Ω) and, afterwards, replace classical constraints with the quantum ones.

In algebraic quantum theory, quantum constraints are described as follows [77; 78]. Let E be a Hilbert space and $\mathcal{H} \in B(E)$ a Hermitian operator in E. By a *quantum constraint* is meant the condition

$$\mathcal{H}e = 0, \qquad e \in E. \tag{6.5.1}$$

A Hermitian operator $\mathcal{H}$ defines the unitary operator $\exp(i\mathcal{H})$. Then the quantum constraint (6.5.1) is equivalent to the condition

$$\exp(i\mathcal{H})e = e.$$

In a general setting, let A be a unital C^*-algebra and $\mathcal{I}$ some subset of its unitary elements called *state conditions*. Let $S_\mathcal{I}$ denote a set of states f of A such that $f(a) = 1$ for all $a \in \mathcal{I}$. They are called *Dirac states*. One has proved that $f \in S_\mathcal{I}$ if and only if

$$f(ba) = f(ab) = f(b)$$

for any $a \in \mathcal{I}$ and $b \in A$ [77]. In particular, if $f \in S_{\mathcal{I}}$, it follows at once from the relation (4.1.12) that

$$|f(b(a-\mathbf{1}))|^2 \leq f(bb^*)f((a-\mathbf{1})(a^*-\mathbf{1}))=0.$$

One can similarly show that, if $a, a' \in \mathcal{I}$ and $f \in S_{\mathcal{I}}$, then

$$f((a-\mathbf{1})(a'-\mathbf{1}))=0.$$

Thereby, elements $a-\mathbf{1}$, $a \in \mathcal{I}$, generate an algebra which belongs to $\operatorname{Ker} f$ for any $f \in S_{\mathcal{I}}$. The completion of this algebra in A is a C^*-algebra $A_{\mathcal{I}}$ such that $f(a)=0$ for all $a \in A_{\mathcal{I}}$ and $f \in \mathcal{I}$. The following theorem provides the important criterion of the existence of Dirac states [77].

Theorem 6.5.1. *The set of Dirac states $S_{\mathcal{I}}$ is not empty if and only if $\mathbf{1} \notin A_{\mathcal{I}}$.*

Let us return to quantization of constraint systems. In a general setting, one studies geometric quantization of a presymplectic manifold via its symplectic realization. There are the following two variants of this quantization [5; 12; 71].

(i) Let (N, ω) be a presymplectic manifold. There exists its imbedding

$$i_N : N \to Z \tag{6.5.2}$$

into a symplectic manifold (Z, Ω) such that $\omega = i_N^*\Omega$. This imbedding is not unique and different symplectic realizations (Z, Ω) of (N, ω) fail to be isomorphic. They lead to non-equivalent quantizations of a presymplectic manifold (N, ω). For instance, if a presymplectic form ω is of constant rank, one can quantize a presymplectic manifold (N, ω) via its canonical coisotropic imbedding in Proposition 3.1.1 [71]. Geometric quantization of (N, ω) via its imbedding into T^*N has been studied in [12].

Given an imbedding i_N (6.5.2), we have a constraint system where classical constraints are smooth functions on Z vanishing on N. They constitute an ideal I_N of the associative ring $C^\infty(Z)$. Then one usually attempts to provide geometric quantization of a symplectic manifold (Z, Ω) in the presence of quantum constraints, but meets the problem how to associate quantum constraints to the classical ones.

• Firstly, prequantization procedure $f \to \widehat{f}$ does not preserve the associative multiplication of functions. Consequently, prequantization $\widehat{I}_N$ of the ideal I_N of classical constraints fails to be an ideal in a prequantum algebra, i.e., if $f \in I_N$ then $f'f \in I_N$ for any $f' \in C^\infty(Z)$, but $\widehat{f'}\widehat{f} \notin \widehat{I}_N$ in general. Therefore, one has to choose some set of constraints $\phi_1, \ldots \phi_n \in I_N$,

$n = \dim N$, which (locally) defines N by the equations $\phi_i = 0$ and associate to them the quantum constraint conditions $\widehat{\phi}_i\psi = 0$ on *admissible states.* Though another set of constraints $\{\phi'_i\}$ characterizes the same constraint space N as $\{\phi_i\}$, the quantum constraints $\{\widehat{\phi}_i\}$ and $\{\widehat{\phi}'_i\}$ define different subspaces of a prequantum space in general. By the same reason, the direct adaptation of the notion of first and second class constraints to the quantum framework fails [77].

• Secondly, given a set $\{\phi_i\}$ of classical constraints, one should choose a compatible polarization of the symplectic manifold (Z, Ω) such that prequantum operators $\widehat{\phi}_i$ belong to the quantum algebra. Different sets of constraints imply different compatible polarizations in general. Moreover, a compatible polarization need not exist.

• If a presymplectic form ω is of constant rank and its characteristic foliation is simple, there is a different symplectic realization (P, Ω) of (N, ω) via a fibration $N \to P$ (see Proposition 3.1.2 and Example 6.1.1). Then the reduced symplectic manifold (P, Ω) is quantized [5].

Let us apply the above mentioned quantization procedures to the Poisson manifold $(V^*Q, \{,\}_V)$ in Section 3.3. A glance at the equation (3.3.20) shows that one can think of the vector field γ_H as being the Hamiltonian vector field of a zero Hamiltonian with respect to the presymplectic form dH on V^*Q. Therefore, one can examine quantization of the presymplectic manifold (V^*Q, dH). Given a trivialization (5.4.1), this quantization has been studied in [165].

(i) We use the fact that the range

$$N_h = h(V^*Q)$$

of any section h (3.3.13) is a one-codimensional imbedded submanifold and, consequently, is coisotropic. It is given by the constraint

$$\mathcal{H}^* = p_0 + \mathcal{H}(t, q^k, p_k) = 0.$$

Then the geometric quantization of the presymplectic manifold (V^*Q, dH) consists in geometric quantization of the cotangent bundle T^*Q and setting the quantum constraint condition

$$\widehat{\mathcal{H}}^*\psi = 0 \tag{6.5.3}$$

on admissible states. It serves as the Shrödinger equation. The condition (6.5.3) implies that, in contrast with geometric quantization in Section 5.4, the Hamiltonian $\widehat{\mathcal{H}}^*$ always belongs to the quantum algebra of T^*Q. This takes place if one uses polarization of T^*Q which contains the Hamiltonian vector field $\vartheta_{\mathcal{H}^*}$ (3.4.4). Such a polarization of T^*Q always exists. Indeed, any section h (3.3.13) of the affine bundle ζ (3.3.3) defines a splitting

$$a_\lambda \partial^\lambda = a_k(\partial^k - \partial^k \mathcal{H} \partial^0) + (a_0 + a_k \partial^k \mathcal{H})\partial^0$$

of the vertical tangent bundle VT^*Q of $T^*Q \to Q$. Then elements $(\partial^k - \partial^k \mathcal{H} \partial^0)$ together with the Hamiltonian vector field $\vartheta_{\mathcal{H}^*}$ (3.4.4) span a polarization of T^*Q. Clearly, this polarization does not satisfy the condition (5.4.3), and does not define any polarization of the Poisson manifold V^*Q.

(ii) In application to (V^*Q, dH), the reduction procedure leads to quantization along classical solutions as follows. The kernel of dH is spanned by the vector field γ_H and, consequently, the presymplectic form dH is of constant rank. Its characteristic foliation is made up by integral curves of this vector field, i.e., solutions of Hamilton equations. If the vector field γ_H is complete, this foliation is simple, i.e., is a fibration of V^*Q over a symplectic manifold N of initial values. In this case, we come to the instantwise quantization when functions on V^*Q at a given instant $t \in \mathbb{R}$ are quantized as functions on N.

Chapter 7

Integrable Hamiltonian systems

Let us recall that the Liouville–Arnold (or Liouville–Mineur–Arnold) theorem for completely integrable systems [4; 101], the Poincaré – Lyapounov – Nekhoroshev theorem for partially integrable systems [51; 122] and the Mishchenko–Fomenko theorem for the superintegrable ones [16; 41; 115] state the existence of action-angle coordinates around a compact invariant submanifold of a Hamiltonian integrable system. However, their global extension meets a well-known topological obstruction [7; 30; 35],

In this Chapter, completely integrable, partially integrable and superintegrable Hamiltonian systems are described in a general setting of invariant submanifolds which need not be compact [44; 46; 47; 48; 62; 65; 143; 161]. In particular, this is the case of non-autonomous completely integrable and superintegrable systems [45; 59; 65].

Geometric quantization of completely integrable and superintegrable Hamiltonian systems with respect to action-angle variables is considered [43; 60; 65; 66]. Using this quantization, the non-adiabatic holonomy operator is constructed in Section 9.6.

Let us note that throughout all functions and maps are smooth. We are not concerned with the real-analytic case because a paracompact real-analytic manifold admits the partition of unity by smooth functions. As a consequence, sheaves of modules over real-analytic functions need not be acyclic that is essential for our consideration.

7.1 Partially integrable systems with non-compact invariant submanifolds

We start with partially integrable systems because: (i) completely integrable systems can be regarded both as particular partially integrable and superintegrable systems, (ii) invariant submanifolds of any superintegrable system are maximal integral manifolds of a certain partially integrable system (Proposition 7.3.2).

7.1.1 *Partially integrable systems on a Poisson manifold*

Completely integrable and superintegrable systems are considered with respect to a symplectic structure on a manifold which holds fixed from the beginning. A partially integrable system admits different compatible Poisson structures (see Theorem 7.1.2 below). Treating partially integrable systems, we therefore are based on a wider notion of the dynamical algebra [62; 65].

Let we have m mutually commutative vector fields $\{\vartheta_\lambda\}$ on a connected smooth real manifold Z which are independent almost everywhere on Z, i.e., the set of points, where the multivector field $\overset{m}{\wedge}\vartheta_\lambda$ vanishes, is nowhere dense. We denote by $\mathcal{S}\subset C^\infty(Z)$ the $\mathbb{R}$-subring of smooth real functions f on Z whose derivations $\vartheta_\lambda\rfloor df$ vanish for all ϑ_λ. Let $\mathcal{A}$ be an m-dimensional Lie $\mathcal{S}$-algebra generated by the vector fields $\{\vartheta_\lambda\}$. One can think of one of its elements as being an autonomous first order dynamic equation on Z and of the other as being its integrals of motion in accordance with Definition 1.10.1. By virtue of this definition, elements of $\mathcal{S}$ also are regarded as integrals of motion. Therefore, we agree to call $\mathcal{A}$ a *dynamical algebra*.

Given a commutative dynamical algebra $\mathcal{A}$ on a manifold Z, let G be the group of local diffeomorphisms of Z generated by the flows of these vector fields. The orbits of G are maximal *invariant submanifolds* of $\mathcal{A}$ (we follow the terminology of [153]). Tangent spaces to these submanifolds form a (non-regular) *distribution* $\mathcal{V}\subset TZ$ whose maximal *integral manifolds* coincide with orbits of G. Let $z\in Z$ be a *regular* point of the distribution $\mathcal{V}$, i.e., $\overset{m}{\wedge}\vartheta_\lambda(z)\neq 0$. Since the group G preserves $\overset{m}{\wedge}\vartheta_\lambda$, a maximal integral manifold M of $\mathcal{V}$ through z also is *regular* (i.e., its points are regular). Furthermore, there exists an open neighborhood U of M such that, restricted to U, the distribution $\mathcal{V}$ is an m-dimensional regular distribution on U. Being involutive, it yields a foliation $\mathfrak{F}$ of U. A regular open neighborhood U of an invariant submanifold of M is called *saturated* if any invariant submanifold through a point of U belongs to U. For instance, any compact invariant submanifold has such an open neighborhood.

Definition 7.1.1. Let $\mathcal{A}$ be an m-dimensional dynamical algebra on a regular Poisson manifold (Z, w). It is said to be a *partially integrable system* if:

(a) its generators ϑ_λ are Hamiltonian vector fields of some functions $S_\lambda \in \mathcal{S}$ which are independent almost everywhere on Z, i.e., the set of points where the m-form $\overset{m}{\wedge} dS_\lambda$ vanishes is nowhere dense;

(b) all elements of $\mathcal{S} \subset C^\infty(Z)$ are mutually *in involution*, i.e., their Poisson brackets equal zero.

It follows at once from this definition that the Poisson structure w is at least of rank $2m$, and that $\mathcal{S}$ is a commutative Poisson algebra. We call the functions S_λ in item (a) of Definition 7.1.1 the *generating functions of a partially integrable system*, which is uniquely defined by a family $(S_1, \ldots, S_m)$ of these functions.

If $2m = \dim Z$ in Definition 7.1.1, we have a *completely integrable system* on a symplectic manifold Z (see Definition 7.3.2 below).

If $2m < \dim Z$, there exist different Poisson structures on Z which bring a dynamical algebra $\mathcal{A}$ into a partially integrable system. Forthcoming Theorems 7.1.1 and 7.1.2 describe all these Poisson structures around a regular invariant submanifold $M \subset Z$ of $\mathcal{A}$ [62].

Theorem 7.1.1. *Let $\mathcal{A}$ be a dynamical algebra, M its regular invariant submanifold, and U a saturated regular open neighborhood of M. Let us suppose that:*

(i) the vector fields ϑ_λ on U are complete,

(ii) the foliation $\mathfrak{F}$ of U admits a transversal manifold Σ and its holonomy pseudogroup on Σ is trivial,

(iii) the leaves of this foliation are mutually diffeomorphic.

Then the following hold.

(I) The leaves of $\mathcal{F}$ are diffeomorphic to a toroidal cylinder

$$\mathbb{R}^{m-r} \times T^r, \qquad 0 \leq r \leq m. \tag{7.1.1}$$

(II) There exists an open saturated neighborhood of M, say U again, which is the trivial principal bundle

$$U = N \times (\mathbb{R}^{m-r} \times T^r) \xrightarrow{\pi} N \tag{7.1.2}$$

over a domain $N \subset \mathbb{R}^{\dim Z - m}$ with the structure group (7.1.1).

(III) If $2m \leq \dim Z$, there exists a Poisson structure of rank $2m$ on U such that $\mathcal{A}$ is a partially integrable system in accordance with Definition 7.1.1.

Proof. We follow the proof in [28; 101] generalized to the case of non-compact invariant submanifolds [62; 65].

(I). Since m-dimensional leaves of the foliation $\mathcal{F}$ admit m complete independent vector fields, they are locally affine manifolds diffeomorphic to a toroidal cylinder (7.1.1).

(II). By virtue of the condition (ii), the foliation $\mathfrak{F}$ of U is a fibred manifold [116]. Then one can always choose an open fibred neighborhood of its fibre M, say U again, over a domain N such that this fibred manifold

$$\pi : U \to N \tag{7.1.3}$$

admits a section σ. In accordance with the well-known theorem [125; 127] complete Hamiltonian vector fields ϑ_λ define an action of a simply connected Lie group G on Z. Because vector fields ϑ_λ are mutually commutative, it is the additive group $\mathbb{R}^m$ whose group space is coordinated by parameters s^λ of the flows with respect to the basis $\{e_\lambda = \vartheta_\lambda\}$ for its Lie algebra. The orbits of the group $\mathbb{R}^m$ in $U \subset Z$ coincide with the fibres of the fibred manifold (7.1.3). Since vector fields ϑ_λ are independent everywhere on U, the action of $\mathbb{R}^m$ on U is locally free, i.e., isotropy groups of points of U are discrete subgroups of the group $\mathbb{R}^m$. Given a point $x \in N$, the action of $\mathbb{R}^m$ on the fibre $M_x = \pi^{-1}(x)$ factorizes as

$$\mathbb{R}^m \times M_x \to G_x \times M_x \to M_x \tag{7.1.4}$$

through the free transitive action on M_x of the factor group $G_x = \mathbb{R}^m/K_x$, where K_x is the isotropy group of an arbitrary point of M_x. It is the same group for all points of M_x because $\mathbb{R}^m$ is a commutative group. Clearly, M_x is diffeomorphic to the group space of G_x. Since the fibres M_x are mutually diffeomorphic, all isotropy groups K_x are isomorphic to the group $\mathbb{Z}_r$ for some fixed $0 \leq r \leq m$. Accordingly, the groups G_x are isomorphic to the additive group (7.1.1). Let us bring the fibred manifold (7.1.3) into a principal bundle with the structure group G_0, where we denote $\{0\} = \pi(M)$. For this purpose, let us determine isomorphisms $\rho_x : G_0 \to G_x$ of the group G_0 to the groups G_x, $x \in N$. Then a desired fibrewise action of G_0 on U is defined by the law

$$G_0 \times M_x \to \rho_x(G_0) \times M_x \to M_x. \tag{7.1.5}$$

Generators of each isotropy subgroup K_x of $\mathbb{R}^m$ are given by r linearly independent vectors of the group space $\mathbb{R}^m$. One can show that there exist ordered collections of generators $(v_1(x), \ldots, v_r(x))$ of the groups K_x such that $x \to v_i(x)$ are smooth $\mathbb{R}^m$-valued fields on N. Indeed, given a vector $v_i(0)$ and a section σ of the fibred manifold (7.1.3), each field $v_i(x) = (s_i^\alpha(x))$ is a unique smooth solution of the equation

$$g(s_i^\alpha)\sigma(x) = \sigma(x), \qquad (s_i^\alpha(0)) = v_i(0),$$

on an open neighborhood of $\{0\}$. Let us consider the decomposition

$$v_i(0) = B_i^a(0)e_a + C_i^j(0)e_j, \qquad a = 1, \ldots, m-r, \qquad j = 1, \ldots, r,$$

where $C_i^j(0)$ is a non-degenerate matrix. Since the fields $v_i(x)$ are smooth, there exists an open neighborhood of $\{0\}$, say N again, where the matrices $C_i^j(x)$ are non-degenerate. Then

$$A(x) = \begin{pmatrix} \mathrm{Id} & (B(x) - B(0))C^{-1}(0) \\ 0 & C(x)C^{-1}(0) \end{pmatrix} \tag{7.1.6}$$

is a unique linear endomorphism

$$(e_a, e_i) \to (e_a, e_j)A(x)$$

of the vector space $\mathbb{R}^m$ which transforms the frame $\{v_\lambda(0)\} = \{e_a, v_i(0)\}$ into the frame $\{v_\lambda(x)\} = \{e_a, \vartheta_i(x)\}$, i.e.,

$$v_i(x) = B_i^a(x)e_a + C_i^j(x)e_j = B_i^a(0)e_a + C_i^j(0)[A_j^b(x)e_b + A_j^k(x)e_k].$$

Since $A(x)$ (7.1.6) also is an automorphism of the group $\mathbb{R}^m$ sending K_0 onto K_x, we obtain a desired isomorphism ρ_x of the group G_0 to the group G_x. Let an element g of the group G_0 be the coset of an element $g(s^\lambda)$ of the group $\mathbb{R}^m$. Then it acts on M_x by the rule (7.1.5) just as the element $g((A_x^{-1})^\lambda_\beta s^\beta)$ of the group $\mathbb{R}^m$ does. Since entries of the matrix A (7.1.6) are smooth functions on N, this action of the group G_0 on U is smooth. It is free, and $U/G_0 = N$. Then the fibred manifold (7.1.3) is a trivial principal bundle with the structure group G_0. Given a section σ of this principal bundle, its trivialization $U = N \times G_0$ is defined by assigning the points $\rho^{-1}(g_x)$ of the group space G_0 to the points $g_x\sigma(x)$, $g_x \in G_x$, of a fibre M_x. Let us endow G_0 with the standard coordinate atlas $(r^\lambda) = (t^a, \varphi^i)$ of the group (7.1.1). Then U admits the trivialization (7.1.2) with respect to the bundle coordinates (x^A, t^a, φ^i) where x^A, $A = 1, \ldots, \dim Z - m$, are coordinates on a base N. The vector fields ϑ_λ on U relative to these coordinates read

$$\vartheta_a = \partial_a, \qquad \vartheta_i = -(BC^{-1})_i^a(x)\partial_a + (C^{-1})_i^k(x)\partial_k. \tag{7.1.7}$$

Accordingly, the subring $\mathcal{S}$ restricted to U is the pull-back $\pi^*C^\infty(N)$ onto U of the ring of smooth functions on N.

(III). Let us split the coordinates (x^A) on N into some m coordinates (J_λ) and the rest $\dim Z - 2m$ coordinates (z^A). Then we can provide the toroidal domain U (7.1.2) with the Poisson bivector field

$$w = \partial^\lambda \wedge \partial_\lambda \tag{7.1.8}$$

of rank $2m$. The independent complete vector fields ∂_a and ∂_i are Hamiltonian vector fields of the functions $S_a = J_a$ and $S_i = J_i$ on U which are in involution with respect to the Poisson bracket

$$\{f, f'\} = \partial^\lambda f \partial_\lambda f' - \partial_\lambda f \partial^\lambda f' \tag{7.1.9}$$

defined by the bivector field w (7.1.8). By virtue of the expression (7.1.7), the Hamiltonian vector fields $\{\partial_\lambda\}$ generate the $\mathcal{S}$-algebra $\mathcal{A}$. Therefore, $(w, \mathcal{A})$ is a partially integrable system. □

Remark 7.1.1. Condition (ii) of Theorem 7.1.1 is equivalent to that $U \to U/G$ is a fibred manifold [116]. It should be emphasized that a fibration in invariant submanifolds is a standard property of integrable systems [4; 13; 20; 51; 59; 122]. If fibres of such a fibred manifold are assumed to be compact then this fibred manifold is a fibre bundle (Theorem 11.2.4) and vertical vector fields on it (e.g., in condition (i) of Theorem 7.1.1) are complete (Theorem 11.2.12).

7.1.2 *Bi-Hamiltonian partially integrable systems*

A Poisson structure in Theorem 7.1.1 is by no means unique. Given the toroidal domain U (7.1.2) provided with bundle coordinates (x^A, r^λ), it is readily observed that, if a Poisson bivector field on U satisfies Definition 7.1.1, it takes the form

$$w = w_1 + w_2 = w^{A\lambda}(x^B)\partial_A \wedge \partial_\lambda + w^{\mu\nu}(x^B, r^\lambda)\partial_\mu \wedge \partial_\nu. \tag{7.1.10}$$

The converse also holds as follows.

Theorem 7.1.2. *For any Poisson bivector field w (7.1.10) of rank $2m$ on the toroidal domain U (7.1.2), there exists a toroidal domain $U' \subset U$ such that a dynamical algebra $\mathcal{A}$ in Theorem 7.1.1 is a partially integrable system on U'.*

Remark 7.1.2. It is readily observed that any Poisson bivector field w (7.1.10) fulfills condition (b) in Definition 7.1.1, but condition (a) imposes a restriction on the toroidal domain U. The key point is that the characteristic foliation $\mathcal{F}$ of U yielded by the Poisson bivector fields w (7.1.10) is the pull-back of an m-dimensional foliation $\mathcal{F}_N$ of the base N, which is defined by the first summand w_1 (7.1.10) of w. With respect to the adapted coordinates (J_λ, z^A), $\lambda = 1, \ldots, m$, on the foliated manifold $(N, \mathcal{F}_N)$, the Poisson bivector field w reads

$$w = w^\mu_\nu(J_\lambda, z^A)\partial^\nu \wedge \partial_\mu + w^{\mu\nu}(J_\lambda, z^A, r^\lambda)\partial_\mu \wedge \partial_\nu. \tag{7.1.11}$$

Then condition (a) in Definition 7.1.1 is satisfied if $N' \subset N$ is a domain of a coordinate chart (J_λ, z^A) of the foliation $\mathcal{F}_N$. In this case, the dynamical algebra $\mathcal{A}$ on the toroidal domain $U' = \pi^{-1}(N')$ is generated by the Hamiltonian vector fields

$$\vartheta_\lambda = -w\lfloor dJ_\lambda = w^\mu_\lambda \partial_\mu \tag{7.1.12}$$

of the m independent functions $S_\lambda = J_\lambda$.

Proof. The characteristic distribution of the Poisson bivector field w (7.1.10) is spanned by the Hamiltonian vector fields

$$v^A = -w\lfloor dx^A = w^{A\mu}\partial_\mu \tag{7.1.13}$$

and the vector fields

$$w\lfloor dr^\lambda = w^{A\lambda}\partial_A + 2w^{\mu\lambda}\partial_\mu.$$

Since w is of rank $2m$, the vector fields ∂_μ can be expressed in the vector fields v^A (7.1.13). Hence, the characteristic distribution of w is spanned by the Hamiltonian vector fields v^A (7.1.13) and the vector fields

$$v^\lambda = w^{A\lambda}\partial_A. \tag{7.1.14}$$

The vector fields (7.1.14) are projected onto N. Moreover, one can derive from the relation $[w, w] = 0$ that they generate a Lie algebra and, consequently, span an involutive distribution $\mathcal{V}_N$ of rank m on N. Let $\mathcal{F}_N$ denote the corresponding foliation of N. We consider the pull-back $\mathcal{F} = \pi^*\mathcal{F}_N$ of this foliation onto U by the trivial fibration π [116]. Its leaves are the inverse images $\pi^{-1}(F_N)$ of leaves F_N of the foliation $\mathcal{F}_N$, and so is its characteristic distribution

$$T\mathcal{F} = (T\pi)^{-1}(\mathcal{V}_N).$$

This distribution is spanned by the vector fields v^λ (7.1.14) on U and the vertical vector fields on $U \to N$, namely, the vector fields v^A (7.1.13) generating the algebra $\mathcal{A}$. Hence, $T\mathcal{F}$ is the characteristic distribution of the Poisson bivector field w. Furthermore, since $U \to N$ is a trivial bundle, each leaf $\pi^{-1}(F_N)$ of the pull-back foliation $\mathcal{F}$ is the manifold product of a leaf F_N of N and the toroidal cylinder $\mathbb{R}^{k-m} \times T^m$. It follows that the foliated manifold $(U, \mathcal{F})$ can be provided with an adapted coordinate atlas

$$\{(U_\iota, J_\lambda, z^A, r^\lambda)\}, \qquad \lambda = 1, \ldots, k, \qquad A = 1, \ldots, \dim Z - 2m,$$

such that (J_λ, z^A) are adapted coordinates on the foliated manifold $(N, \mathcal{F}_N)$. Relative to these coordinates, the Poisson bivector field (7.1.10) takes the form (7.1.11). Let N' be the domain of this coordinate chart. Then the dynamical algebra $\mathcal{A}$ on the toroidal domain $U' = \pi^{-1}(N')$ is generated by the Hamiltonian vector fields ϑ_λ (7.1.12) of functions $S_\lambda = J_\lambda$□

Remark 7.1.3. Let us note that the coefficients $w^{\mu\nu}$ in the expressions (7.1.10) and (7.1.11) are affine in coordinates r^λ because of the relation $[w, w] = 0$ and, consequently, they are constant on tori.

Now, let w and w' be two different Poisson structures (7.1.10) on the toroidal domain (7.1.2) which make a commutative dynamical algebra $\mathcal{A}$ into different partially integrable systems $(w, \mathcal{A})$ and $(w', \mathcal{A})$.

Definition 7.1.2. We agree to call the triple $(w, w', \mathcal{A})$ a *bi-Hamiltonian partially integrable system* if any Hamiltonian vector field $\vartheta \in \mathcal{A}$ with respect to w possesses the same Hamiltonian representation

$$\vartheta = -w\lfloor df = -w'\lfloor df, \qquad f \in \mathcal{S}, \tag{7.1.15}$$

relative to w', and *vice versa*.

Definition 7.1.2 establishes a *sui generis* equivalence between the partially integrable systems $(w, \mathcal{A})$ and $(w', \mathcal{A})$. Theorem 7.1.3 below states that the triple $(w, w', \mathcal{A})$ is a bi-Hamiltonian partially integrable system in accordance with Definition 7.1.2 if and only if the Poisson bivector fields w and w' (7.1.10) differ only in the second terms w_2 and w'_2. Moreover, these Poisson bivector fields admit a recursion operator as follows.

Theorem 7.1.3. *(I) The triple $(w, w', \mathcal{A})$ is a bi-Hamiltonian partially integrable system in accordance with Definition 7.1.2 if and only if the Poisson bivector fields w and w' (7.1.10) differ in the second terms w_2 and w'_2. (II) These Poisson bivector fields admit a recursion operator.*

Proof. (I). It is easily justified that, if Poisson bivector fields w (7.1.10) fulfil Definition 7.1.2, they are distinguished only by the second summand w_2. Conversely, as follows from the proof of Theorem 7.1.2, the characteristic distribution of a Poisson bivector field w (7.1.10) is spanned by the vector fields (7.1.13) and (7.1.14). Hence, all Poisson bivector fields w (7.1.10) distinguished only by the second summand w_2 have the same characteristic distribution, and they bring $\mathcal{A}$ into a partially integrable system on the same toroidal domain U'. Then the condition in Definition 7.1.2 is easily justified. (II). The result follows from forthcoming Lemma 7.1.1. □

Given a smooth real manifold X, let w and w' be Poisson bivector fields of rank $2m$ on X, and let $w^\sharp$ and $w'^\sharp$ be the corresponding bundle homomorphisms (3.1.8). A tangent-valued one-form R on X yields bundle endomorphisms

$$R : TX \to TX, \qquad R^* : T^*X \to T^*X. \tag{7.1.16}$$

It is called a *recursion operator* if

$$w'^\sharp = R \circ w^\sharp = w^\sharp \circ R^*. \tag{7.1.17}$$

Given a Poisson bivector field w and a tangent valued one-form R such that $R \circ w^\sharp = w^\sharp \circ R^*$, the well-known sufficient condition for $R \circ w^\sharp$ to be a Poisson bivector field is that the Nijenhuis torsion (11.2.60) of R, seen as a tangent-valued one-form, and the Magri–Morosi concomitant of R and w vanish [27; 123]. However, as we will see later, recursion operators between Poisson bivector fields in Theorem 7.1.3 need not satisfy these conditions.

Lemma 7.1.1. *A recursion operator between Poisson structures of the same rank exists if and only if their characteristic distributions coincide.*

Proof. It follows from the equalities (7.1.17) that a recursion operator R sends the characteristic distribution of w to that of w', and these distributions coincide if w and w' are of the same rank. Conversely, let regular Poisson structures w and w' possess the same characteristic distribution $T\mathcal{F} \to TX$ tangent to a foliation $\mathcal{F}$ of X. We have the exact sequences (11.2.66) – (11.2.67). The bundle homomorphisms $w^\sharp$ and $w'^\sharp$ (3.1.8) factorize in a unique fashion (3.1.29) through the bundle isomorphisms $w^\sharp_\mathcal{F}$ and $w'^\sharp_\mathcal{F}$ (3.1.29). Let us consider the inverse isomorphisms

$$w^\flat_\mathcal{F} : T\mathcal{F} \to T\mathcal{F}^*, \qquad w'^\flat_\mathcal{F} : T\mathcal{F} \to T\mathcal{F}^* \tag{7.1.18}$$

and the compositions

$$R_\mathcal{F} = w'^\sharp_\mathcal{F} \circ w^\flat_\mathcal{F} : T\mathcal{F} \to T\mathcal{F}, \qquad R^*_\mathcal{F} = w^\flat_\mathcal{F} \circ w'^\sharp_\mathcal{F} : T\mathcal{F}^* \to T\mathcal{F}^*. \tag{7.1.19}$$

There is the obvious relation

$$w'^\sharp_\mathcal{F} = R_\mathcal{F} \circ w^\sharp_\mathcal{F} = w^\sharp_\mathcal{F} \circ R^*_\mathcal{F}.$$

In order to obtain a recursion operator (7.1.17), it suffices to extend the morphisms $R_{\mathcal{F}}$ and $R^*_{\mathcal{F}}$ (7.1.19) onto TX and T^*X, respectively. For this purpose, let us consider a splitting

$$\zeta : TX \to T\mathcal{F},$$
$$TX = T\mathcal{F} \oplus (\mathrm{Id} - i_{\mathcal{F}} \circ \zeta)TX = T\mathcal{F} \oplus E,$$

of the exact sequence (11.2.66) and the dual splitting

$$\zeta^* : T\mathcal{F}^* \to T^*X,$$
$$T^*X = \zeta^*(T\mathcal{F}^*) \oplus (\mathrm{Id} - \zeta^* \circ i^*_{\mathcal{F}})T^*X = \zeta^*(T\mathcal{F}^*) \oplus E',$$

of the exact sequence (11.2.67). Then the desired extensions are

$$R = R_{\mathcal{F}} \times \mathrm{Id}\, E, \qquad R^* = (\zeta^* \circ R^*_{\mathcal{F}}) \times \mathrm{Id}\, E'.$$

This recursion operator is invertible, i.e., the morphisms (7.1.16) are bundle isomorphisms. □

For instance, the Poisson bivector field w (7.1.10) and the Poisson bivector field

$$w_0 = w^{A\lambda}\partial_A \wedge \partial_\lambda$$

admit a recursion operator $w_0^\sharp = R \circ w^\sharp$ whose entries are given by the equalities

$$R^A_B = \delta^A_B, \qquad R^\mu_\nu = \delta^\mu_\nu, \qquad R^A_\lambda = 0, \qquad w^{\mu\lambda} = R^\lambda_B w^{B\mu}. \tag{7.1.20}$$

Its Nijenhuis torsion (11.2.60) fails to vanish, unless coefficients $w^{\mu\lambda}$ are independent of coordinates r^λ.

7.1.3 *Partial action-angle coordinates*

Given a partially integrable system $(w, \mathcal{A})$ in Theorem 7.1.2, the bivector field w (7.1.11) can be brought into the canonical form (7.1.8) with respect to partial action-angle coordinates in forthcoming Theorem 7.1.4. This theorem extends the Liouville–Arnold theorem to the case of a Poisson structure and a non-compact invariant submanifold [62; 65].

Theorem 7.1.4. *Given a partially integrable system $(w, \mathcal{A})$ on a Poisson manifold (U, w), there exists a toroidal domain $U' \subset U$ equipped with partial action-angle coordinates $(I_a, I_i, z^A, \tau^a, \phi^i)$ such that, restricted to U', a Poisson bivector field takes the canonical form*

$$w = \partial^a \wedge \partial_a + \partial^i \wedge \partial_i, \tag{7.1.21}$$

while the dynamical algebra $\mathcal{A}$ is generated by Hamiltonian vector fields of the action coordinate functions $S_a = I_a$, $S_i = I_i$.

Proof. First, let us employ Theorem 7.1.2 and restrict U to the toroidal domain, say U again, equipped with coordinates $(J_\lambda, z^A, r^\lambda)$ such that the Poisson bivector field w takes the form (7.1.11) and the algebra $\mathcal{A}$ is generated by the Hamiltonian vector fields ϑ_λ (7.1.12) of m independent functions $S_\lambda = J_\lambda$ in involution. Let us choose these vector fields as new generators of the group G and return to Theorem 7.1.1. In accordance with this theorem, there exists a toroidal domain $U' \subset U$ provided with another trivialization $U' \to N' \subset N$ in toroidal cylinders $\mathbb{R}^{m-r} \times T^r$ and endowed with bundle coordinates $(J_\lambda, z^A, r^\lambda)$ such that the vector fields ϑ_λ (7.1.12) take the form (7.1.7). For the sake of simplicity, let U', N' and y^λ be denoted U, N and $r^\lambda = (t^a, \varphi^i)$ again. Herewith, the Poisson bivector field w is given by the expression (7.1.11) with new coefficients. Let $w^\sharp : T^*U \to TU$ be the corresponding bundle homomorphism. It factorizes in a unique fashion (3.1.29):

$$w^\sharp : T^*U \xrightarrow{i^*_\mathcal{F}} T\mathcal{F}^* \xrightarrow{w^\sharp_\mathcal{F}} T\mathcal{F} \xrightarrow{i_\mathcal{F}} TU$$

through the bundle isomorphism

$$w^\sharp_\mathcal{F} : T\mathcal{F}^* \to T\mathcal{F}, \qquad w^\sharp_\mathcal{F} : \alpha \to -w(x)\lfloor\alpha.$$

Then the inverse isomorphisms $w^\flat_\mathcal{F} : T\mathcal{F} \to T\mathcal{F}^*$ provides the foliated manifold $(U, \mathcal{F})$ with the leafwise symplectic form

$$\Omega_\mathcal{F} = \Omega^{\mu\nu}(J_\lambda, z^A, t^a)\widetilde{d}J_\mu \wedge \widetilde{d}J_\nu + \Omega^\nu_\mu(J_\lambda, z^A)\widetilde{d}J_\nu \wedge \widetilde{d}r^\mu, \tag{7.1.22}$$

$$\Omega^\alpha_\mu w^\mu_\beta = \delta^\alpha_\beta, \qquad \Omega^{\alpha\beta} = -\Omega^\alpha_\mu \Omega^\beta_\nu w^{\mu\nu}. \tag{7.1.23}$$

Let us show that it is $\widetilde{d}$-exact. Let F be a leaf of the foliation $\mathcal{F}$ of U. There is a homomorphism of the de Rham cohomology $H^*_{\rm DR}(U)$ of U to the de Rham cohomology $H^*_{\rm DR}(F)$ of F, and it factorizes through the leafwise cohomology $H^*_\mathcal{F}(U)$ due to (3.1.33). Since N is a domain of an adapted coordinate chart of the foliation $\mathcal{F}_N$, the foliation $\mathcal{F}_N$ of N is a trivial fibre bundle

$$N = V \times W \to W.$$

Since $\mathcal{F}$ is the pull-back onto U of the foliation $\mathcal{F}_N$ of N, it also is a trivial fibre bundle

$$U = V \times W \times (\mathbb{R}^{k-m} \times T^m) \to W \tag{7.1.24}$$

over a domain $W \subset \mathbb{R}^{\dim Z - 2m}$. It follows that

$$H^*_{\rm DR}(U) = H^*_{\rm DR}(T^r) = H^*_\mathcal{F}(U).$$

Then the closed leafwise two-form $\Omega_{\mathcal{F}}$ (7.1.22) is exact due to the absence of the term $\Omega_{\mu\nu}dr^\mu \wedge dr^\nu$. Moreover, $\Omega_{\mathcal{F}} = \widetilde{d}\Xi$ where Ξ reads

$$\Xi = \Xi^\alpha(J_\lambda, z^A, r^\lambda)\widetilde{d}J_\alpha + \Xi_i(J_\lambda, z^A)\widetilde{d}\varphi^i$$

up to a $\widetilde{d}$-exact leafwise form. The Hamiltonian vector fields $\vartheta_\lambda = \vartheta^\mu_\lambda \partial_\mu$ (7.1.7) obey the relation

$$\vartheta_\lambda \rfloor \Omega_{\mathcal{F}} = -\widetilde{d}J_\lambda, \qquad \Omega^\alpha_\beta \vartheta^\beta_\lambda = \delta^\alpha_\lambda, \tag{7.1.25}$$

which falls into the following conditions

$$\Omega^\lambda_i = \partial^\lambda \Xi_i - \partial_i \Xi^\lambda, \tag{7.1.26}$$

$$\Omega^\lambda_a = -\partial_a \Xi^\lambda = \delta^\lambda_a. \tag{7.1.27}$$

The first of the relations (7.1.23) shows that Ω^α_β is a non-degenerate matrix independent of coordinates r^λ. Then the condition (7.1.26) implies that $\partial_i \Xi^\lambda$ are independent of φ^i, and so are Ξ^λ since φ^i are cyclic coordinates. Hence,

$$\Omega^\lambda_i = \partial^\lambda \Xi_i, \tag{7.1.28}$$

$$\partial_i \rfloor \Omega_{\mathcal{F}} = -\widetilde{d}\Xi_i. \tag{7.1.29}$$

Let us introduce new coordinates $I_a = J_a$, $I_i = \Xi_i(J_\lambda)$. By virtue of the equalities (7.1.27) and (7.1.28), the Jacobian of this coordinate transformation is regular. The relation (7.1.29) shows that ∂_i are Hamiltonian vector fields of the functions $S_i = I_i$. Consequently, we can choose vector fields ∂_λ as generators of the algebra $\mathcal{A}$. One obtains from the equality (7.1.27) that

$$\Xi^a = -t^a + E^a(J_\lambda, z^A)$$

and Ξ^i are independent of t^a. Then the leafwise Liouville form Ξ reads

$$\Xi = (-t^a + E^a(I_\lambda, z^A))\widetilde{d}I_a + E^i(I_\lambda, z^A)\widetilde{d}I_i + I_i\widetilde{d}\varphi^i.$$

The coordinate shifts

$$\tau^a = -t^a + E^a(I_\lambda, z^A), \qquad \phi^i = \varphi^i - E^i(I_\lambda, z^A)$$

bring the leafwise form $\Omega_{\mathcal{F}}$ (7.1.22) into the canonical form

$$\Omega_{\mathcal{F}} = \widetilde{d}I_a \wedge \widetilde{d}\tau^a + \widetilde{d}I_i \wedge \widetilde{d}\phi^i$$

which ensures the canonical form (7.1.21) of a Poisson bivector field w. □

7.1.4 *Partially integrable system on a symplectic manifold*

Let $\mathcal{A}$ be a commutative dynamical algebra on a $2n$-dimensional connected symplectic manifold (Z, Ω). Let it obey condition (a) in Definition 7.1.1. However, condition (b) is not necessarily satisfied, unless $m = n$, i.e., a system is completely integrable. Therefore, we modify a definition of partially integrable systems on a symplectic manifold.

Definition 7.1.3. A collection $\{S_1, \ldots, S_m\}$ of $m \leq n$ independent smooth real functions in involution on a symplectic manifold (Z, Ω) is called a *partially integrable system.*

Remark 7.1.4. By analogy with Definition 7.1.1, one can require that functions S_λ in Definition 7.1.3 are independent almost everywhere on Z. However, all theorems that we have proved above are concerned with partially integrable systems restricted to some open submanifold $Z' \subset Z$ of regular points of Z. Therefore, let us restrict functions S_λ to an open submanifold $Z' \subset Z$ where they are independent, and we obtain a partially integrable system on a symplectic manifold (Z', Ω) which obeys Definition 7.1.3. However, it may happen that Z' is not connected. In this case, we have different partially integrable systems on different components of Z'.

Given a partially integrable system (S_λ) in Definition 7.1.3, let us consider the map

$$S : Z \to W \subset \mathbb{R}^m. \tag{7.1.30}$$

Since functions S_λ are everywhere independent, this map is a submersion onto a domain $W \subset \mathbb{R}^m$, i.e., S (7.1.30) is a fibred manifold of fibre dimension $2n - m$. Hamiltonian vector fields ϑ_λ of functions S_λ are mutually commutative and independent. Consequently, they span an m-dimensional involutive distribution on Z whose maximal integral manifolds constitute an isotropic foliation $\mathcal{F}$ of Z. Because functions S_λ are constant on leaves of this foliation, each fibre of a fibred manifold $Z \to W$ (7.1.30) is foliated by the leaves of the foliation $\mathcal{F}$.

If $m = n$, we are in the case of a completely integrable system, and leaves of $\mathcal{F}$ are connected components of fibres of the fibred manifold (7.1.30).

The Poincaré – Lyapounov – Nekhoroshev theorem [51; 122] generalizes the Liouville – Arnold one to a partially integrable system if leaves of the foliation $\mathcal{F}$ are compact. It imposes a sufficient condition which Hamiltonian vector fields v_λ must satisfy in order that the foliation $\mathcal{F}$ is a fibred manifold [51; 52]. Extending the Poincaré – Lyapounov – Nekhoroshev theorem to the case of non-compact invariant submanifolds, we in fact assume from the beginning that these submanifolds form a fibred manifold [62; 65].

Theorem 7.1.5. *Let a partially integrable system $\{S_1, \ldots, S_m\}$ on a symplectic manifold (Z, Ω) satisfy the following conditions.*

(i) The Hamiltonian vector fields ϑ_λ of S_λ are complete.

(ii) The foliation $\mathcal{F}$ is a fibred manifold

$$\pi : Z \to N \tag{7.1.31}$$

whose fibres are mutually diffeomorphic.

Then the following hold.

(I) The fibres of $\mathcal{F}$ are diffeomorphic to the toroidal cylinder (7.1.1).

(II) Given a fibre M of $\mathcal{F}$, there exists its open saturated neighborhood U whose fibration (7.1.31) is a trivial principal bundle with the structure group (7.1.1).

(III) The neighborhood U is provided with the bundle (partial action-angle) coordinates

$$(I_\lambda, p_s, q^s, y^\lambda) \to (I_\lambda, p_s, q^s), \qquad \lambda = 1, \ldots, m, \quad s = 1, \ldots n - m,$$

such that: (i) the action coordinates (I_λ) (7.1.42) are expressed in the values of the functions (S_λ), (ii) the angle coordinates (y^λ) (7.1.45) are coordinates on a toroidal cylinder, and (iii) the symplectic form Ω on U reads

$$\Omega = dI_\lambda \wedge dy^\lambda + dp_s \wedge dq^s. \tag{7.1.32}$$

Proof. (I) The proof of parts (I) and (II) repeats exactly that of parts (I) and (II) of Theorem 7.1.1. As a result, let

$$\pi : U \to \pi(U) \subset N \tag{7.1.33}$$

be a trivial principal bundle with the structure group $\mathbb{R}^{m-r} \times T^r$, endowed with the standard coordinate atlas $(r^\lambda) = (t^a, \varphi^i)$. Then U (7.1.33) admits a trivialization

$$U = \pi(U) \times (\mathbb{R}^{m-r} \times T^r) \to \pi(U) \tag{7.1.34}$$

with respect to the fibre coordinates (t^a, φ^i). The Hamiltonian vector fields ϑ_λ on U relative to these coordinates read (7.1.7):

$$\vartheta_a = \partial_a, \qquad \vartheta_i = -(BC^{-1})_i^a(x)\partial_a + (C^{-1})_i^k(x)\partial_k. \tag{7.1.35}$$

In order to specify coordinates on the base $\pi(U)$ of the trivial bundle (7.1.34), let us consider the fibred manifold S (7.1.30). It factorizes as

$$S : U \xrightarrow{\pi} \pi(U) \xrightarrow{\pi'} S(U)$$

through the fibre bundle π. The map π' also is a fibred manifold. One can always restrict the domain $\pi(U)$ to a chart of the fibred manifold π', say $\pi(U)$ again. Then $\pi(U) \to S(U)$ is a trivial bundle $\pi(U) = S(U) \times V$, and so is $U \to S(U)$. Thus, we have the composite bundle

$$U = S(U) \times V \times (\mathbb{R}^{m-r} \times T^r) \to S(U) \times V \to S(U). \tag{7.1.36}$$

Let us provide its base $S(U)$ with the coordinates (J_λ) such that

$$J_\lambda \circ S = S_\lambda. \tag{7.1.37}$$

Then $\pi(U)$ can be equipped with the bundle coordinates (J_λ, x^A), $A = 1, \ldots, 2(n-m)$, and $(J_\lambda, x^A, t^a, \varphi^i)$ are coordinates on U (7.1.2). Since fibres of $U \to \pi(U)$ are isotropic, a symplectic form Ω on U relative to the coordinates $(J_\lambda, x^A, r^\lambda)$ reads

$$\begin{aligned}\Omega = {} & \Omega^{\alpha\beta} dJ_\alpha \wedge dJ_\beta + \Omega^\alpha_\beta dJ_\alpha \wedge dr^\beta \\ & + \Omega_{AB} dx^A \wedge dx^B + \Omega^\lambda_A dJ_\lambda \wedge dx^A + \Omega_{A\beta} dx^A \wedge dr^\beta.\end{aligned} \tag{7.1.38}$$

The Hamiltonian vector fields $\vartheta_\lambda = \vartheta^\mu_\lambda \partial_\mu$ (7.1.35) obey the relations $\vartheta_\lambda \rfloor \Omega = -dJ_\lambda$ which result in the coordinate conditions

$$\Omega^\alpha_\beta \vartheta^\beta_\lambda = \delta^\alpha_\lambda, \qquad \Omega_{A\beta} \vartheta^\beta_\lambda = 0. \tag{7.1.39}$$

The first of them shows that Ω^α_β is a non-degenerate matrix independent of coordinates r^λ. Then the second one implies that $\Omega_{A\beta} = 0$. By virtue of the well-known Künneth formula for the de Rham cohomology of manifold products, the closed form Ω (7.1.38) is exact, i.e., $\Omega = d\Xi$ where the Liouville form Ξ is

$$\Xi = \Xi^\alpha(J_\lambda, x^B, r^\lambda) dJ_\alpha + \Xi_i(J_\lambda, x^B) d\varphi^i + \Xi_A(J_\lambda, x^B, r^\lambda) dx^A.$$

Since $\Xi_a = 0$ and Ξ_i are independent of φ^i, it follows from the relations

$$\Omega_{A\beta} = \partial_A \Xi_\beta - \partial_\beta \Xi_A = 0$$

that Ξ_A are independent of coordinates t^a and are at most affine in φ^i. Since φ^i are cyclic coordinates, Ξ_A are independent of φ^i. Hence, Ξ_i are independent of coordinates x^A, and the Liouville form reads

$$\Xi = \Xi^\alpha(J_\lambda, x^B, r^\lambda) dJ_\alpha + \Xi_i(J_\lambda) d\varphi^i + \Xi_A(J_\lambda, x^B) dx^A. \tag{7.1.40}$$

Because entries Ω^α_β of $d\Xi = \Omega$ are independent of r^λ, we obtain the following.

(i) $\Omega^\lambda_i = \partial^\lambda \Xi_i - \partial_i \Xi^\lambda$. Consequently, $\partial_i \Xi^\lambda$ are independent of φ^i, and so are Ξ^λ since φ^i are cyclic coordinates. Hence, $\Omega^\lambda_i = \partial^\lambda \Xi_i$ and $\partial_i \rfloor \Omega = -d\Xi_i$. A glance at the last equality shows that ∂_i are Hamiltonian vector fields. It follows that, from the beginning, one can separate m generating functions on U, say S_i again, whose Hamiltonian vector fields are tangent to invariant tori. In this case, the matrix B in the expressions (7.1.6) and (7.1.35) vanishes, and the Hamiltonian vector fields ϑ_λ (7.1.35) read

$$\vartheta_a = \partial_a, \qquad \vartheta_i = (C^{-1})^k_i \partial_k. \tag{7.1.41}$$

Moreover, the coordinates t^a are exactly the flow parameters s^a. Substituting the expressions (7.1.41) into the first condition (7.1.39), we obtain

$$\begin{aligned}\Omega = \Omega^{\alpha\beta} dJ_\alpha \wedge dJ_\beta + dJ_a \wedge ds^a + C^i_k dJ_i \wedge d\varphi^k \\ + \Omega_{AB} dx^A \wedge dx^B + \Omega^\lambda_A dJ_\lambda \wedge dx^A.\end{aligned}$$

It follows that Ξ_i are independent of J_a, and so are $C^k_i = \partial^k \Xi_i$.

(ii) $\Omega^\lambda_a = -\partial_a \Xi^\lambda = \delta^\lambda_a$. Hence, $\Xi^a = -s^a + E^a(J_\lambda)$ and $\Xi^i = E^i(J_\lambda, x^B)$ are independent of s^a.

In view of items (i) – (ii), the Liouville form Ξ (7.1.40) reads

$$\begin{aligned}\Xi = (-s^a + E^a(J_\lambda, x^B))dJ_a + E^i(J_\lambda, x^B)dJ_i \\ + \Xi_i(J_j)d\varphi^i + \Xi_A(J_\lambda, x^B)dx^A.\end{aligned}$$

Since the matrix $\partial^k \Xi_i$ is non-degenerate, we can perform the coordinate transformations

$$\begin{aligned} & I_a = J_a, \qquad I_i = \Xi_i(J_j), \\ & r'^a = -s^a + E^a(J_\lambda, x^B), \quad r'^i = \varphi^i - E^j(J_\lambda, x^B)\frac{\partial J_j}{\partial I_i}. \end{aligned} \tag{7.1.42}$$

These transformations bring Ω into the form

$$\Omega = dI_\lambda \wedge dr'^\lambda + \Omega_{AB}(I_\mu, x^C)dx^A \wedge dx^B + \Omega^\lambda_A(I_\mu, x^C)dI_\lambda \wedge dx^A. \tag{7.1.43}$$

Since functions I_λ are in involution and their Hamiltonian vector fields ∂_λ mutually commute, a point $z \in M$ has an open neighborhood

$$U_z = \pi(U_z) \times O_z, \qquad O_z \subset \mathbb{R}^{m-r} \times T^r,$$

endowed with local Darboux coordinates $(I_\lambda, p_s, q^s, y^\lambda)$, $s = 1, \ldots, n-m$, such that the symplectic form Ω (7.1.43) is given by the expression

$$\Omega = dI_\lambda \wedge dy^\lambda + dp_s \wedge dq^s. \tag{7.1.44}$$

Here, $y^\lambda(I_\lambda, x^A, r'^\alpha)$ are local functions

$$y^\lambda = r'^\lambda + f^\lambda(I_\lambda, x^A) \tag{7.1.45}$$

on U_z. With the above-mentioned group G of flows of Hamiltonian vector fields ϑ_λ, one can extend these functions to an open neighborhood

$$\pi(U_z) \times \mathbb{R}^{k-m} \times T^m$$

of M, say U again, by the law

$$y^\lambda(I_\lambda, x^A, G(z)^\alpha) = G(z)^\lambda + f^\lambda(I_\lambda, x^A).$$

Substituting the functions (7.1.45) on U into the expression (7.1.43), one brings the symplectic form Ω into the canonical form (7.1.32) on U. □

Remark 7.1.5. If one supposes from the beginning that leaves of the foliation $\mathcal{F}$ are compact, the conditions of Theorem 7.1.5 can be replaced with that $\mathcal{F}$ is a fibred manifold (see Theorems 11.2.4 and 11.2.12).

7.1.5 *Global partially integrable systems*

As was mentioned above, there is a topological obstruction to the existence of global action-angle coordinates. Forthcoming Theorem 7.1.6 is a global generalization of Theorem 7.1.5 [110; 143].

Theorem 7.1.6. *Let a partially integrable system $\{S_1, \ldots, S_m\}$ on a symplectic manifold (Z, Ω) satisfy the following conditions.*

(i) The Hamiltonian vector fields ϑ_λ of S_λ are complete.

(ii) The foliation $\mathcal{F}$ is a fibre bundle

$$\pi : Z \to N. \tag{7.1.46}$$

(iii) Its base N is simply connected and the cohomology $H^2(N;\mathbb{Z})$ of N with coefficients in the constant sheaf $\mathbb{Z}$ is trivial.

Then the following hold.

(I) The fibre bundle (7.1.46) is a trivial principal bundle with the structure group (7.1.1), and we have a composite fibred manifold

$$S = \zeta \circ \pi : Z \longrightarrow N \longrightarrow W, \tag{7.1.47}$$

where $N \to W$ however need not be a fibre bundle.

(II) The fibred manifold (7.1.47) is provided with the global fibred action-angle coordinates

$$(I_\lambda, x^A, y^\lambda) \to (I_\lambda, x^A) \to (I_\lambda), \qquad \lambda = 1, \ldots, m, \quad A = 1, \ldots 2(n-m),$$

such that: (i) the action coordinates (I_λ) *(7.1.56) are expressed in the values of the functions* (S_λ) *and they possess identity transition functions, (ii) the angle coordinates* (y^λ) *(7.1.56) are coordinates on a toroidal cylinder, (iii) the symplectic form* Ω *on* U *reads*

$$\Omega = dI_\lambda \wedge dy^\lambda + \Omega^\lambda_A dI_\lambda \wedge dx^A + \Omega_{AB} dx^A \wedge dx^B. \tag{7.1.48}$$

Proof. Following part (I) of the proof of Theorems 7.1.1 and 7.1.5, one can show that a typical fibre of the fibre bundle (7.1.46) is the toroidal cylinder (7.1.1). Let us bring this fibre bundle into a principal bundle with the structure group (7.1.1). Generators of each isotropy subgroup K_x of $\mathbb{R}^m$ are given by r linearly independent vectors $u_i(x)$ of a group space $\mathbb{R}^m$. These vectors are assembled into an r-fold covering $K \to N$. This is a subbundle of the trivial bundle

$$N \times \mathbb{R}^m \to N \tag{7.1.49}$$

whose local sections are local smooth sections of the fibre bundle (7.1.49). Such a section over an open neighborhood of a point $x \in N$ is given by a unique local solution $s^\lambda(x')e_\lambda$, $e_\lambda = \vartheta_\lambda$, of the equation

$$g(s^\lambda)\sigma(x') = \exp(s^\lambda e_\lambda)\sigma(x') = \sigma(x'), \qquad s^\lambda(x)e_\lambda = u_i(x),$$

where σ is an arbitrary local section of the fibre bundle $Z \to N$ over an open neighborhood of x. Since N is simply connected, the covering $K \to N$ admits r everywhere different global sections u_i which are global smooth sections $u_i(x) = u_i^\lambda(x)e_\lambda$ of the fibre bundle (7.1.49). Let us fix a point of N further denoted by $\{0\}$. One can determine linear combinations of the functions S_λ, say again S_λ, such that $u_i(0) = e_i$, $i = m-r, \ldots, m$, and the group G_0 is identified to the group $\mathbb{R}^{m-r} \times T^r$. Let E_x denote an r-dimensional subspace of $\mathbb{R}^m$ passing through the points $u_1(x), \ldots, u_r(x)$. The spaces E_x, $x \in N$, constitute an r-dimensional subbundle $E \to N$ of the trivial bundle (7.1.49). Moreover, the latter is split into the Whitney sum of vector bundles $E \oplus E'$, where $E'_x = \mathbb{R}^m/E_x$ [85]. Then there is a global smooth section γ of the trivial principal bundle $N \times GL(m, \mathbb{R}) \to N$ such that $\gamma(x)$ is a morphism of E_0 onto E_x, where

$$u_i(x) = \gamma(x)(e_i) = \gamma_i^\lambda e_\lambda.$$

This morphism also is an automorphism of the group $\mathbb{R}^m$ sending K_0 onto K_x. Therefore, it provides a group isomorphism $\rho_x : G_0 \to G_x$. With these isomorphisms, one can define the fibrewise action of the group G_0 on Z given by the law

$$G_0 \times M_x \to \rho_x(G_0) \times M_x \to M_x. \tag{7.1.50}$$

Namely, let an element of the group G_0 be the coset $g(s^\lambda)/K_0$ of an element $g(s^\lambda)$ of the group $\mathbb{R}^m$. Then it acts on M_x by the rule (7.1.50) just as the coset $g((\gamma(x)^{-1})^\lambda_\beta s^\beta)/K_x$ of an element $g((\gamma(x)^{-1})^\lambda_\beta s^\beta)$ of $\mathbb{R}^m$ does. Since entries of the matrix γ are smooth functions on N, the action (7.1.50) of the group G_0 on Z is smooth. It is free, and $Z/G_0 = N$. Thus, $Z \to N$ (7.1.46) is a principal bundle with the structure group $G_0 = \mathbb{R}^{m-r} \times T^r$.

Furthermore, this principal bundle over a paracompact smooth manifold N is trivial as follows. In accordance with the well-known theorem [85], its structure group G_0 (7.1.1) is reducible to the maximal compact subgroup T^r, which also is the maximal compact subgroup of the group product $\overset{r}{\times} GL(1,\mathbb{C})$. Therefore, the equivalence classes of T^r-principal bundles ξ are defined as

$$c(\xi) = c(\xi_1 \oplus \cdots \oplus \xi_r) = (1 + c_1(\xi_1)) \cdots (1 + c_1(\xi_r))$$

by the Chern classes $c_1(\xi_i) \in H^2(N;\mathbb{Z})$ of $U(1)$-principal bundles ξ_i over N [85]. Since the cohomology group $H^2(N;\mathbb{Z})$ of N is trivial, all Chern classes c_1 are trivial, and the principal bundle $Z \to N$ over a contractible base also is trivial. This principal bundle can be provided with the following coordinate atlas.

Let us consider the fibred manifold $S : Z \to W$ (7.1.30). Because functions S_λ are constant on fibres of the fibre bundle $Z \to N$ (7.1.46), the fibred manifold (7.1.30) factorizes through the fibre bundle (7.1.46), and we have the composite fibred manifold (7.1.47). Let us provide the principal bundle $Z \to N$ with a trivialization

$$Z = N \times \mathbb{R}^{m-r} \times T^r \to N, \tag{7.1.51}$$

whose fibres are endowed with the standard coordinates $(r^\lambda) = (t^a, \varphi^i)$ on the toroidal cylinder (7.1.1). Then the composite fibred manifold (7.1.47) is provided with the fibred coordinates

$$\begin{aligned} &(J_\lambda, x^A, t^a, \varphi^i), && (7.1.52)\\ &\lambda = 1,\ldots,m, \quad A = 1,\ldots,2(n-m), \quad a = 1,\ldots,m-r, \quad i = 1,\ldots,r, \end{aligned}$$

where J_λ (7.1.37) are coordinates on the base W induced by Cartesian coordinates on $\mathbb{R}^m$, and (J_λ, x^A) are fibred coordinates on the fibred manifold $\zeta : N \to W$. The coordinates J_λ on $W \subset \mathbb{R}^m$ and the coordinates (t^a, φ^i) on the trivial bundle (7.1.51) possess the identity transition functions, while the transition function of coordinates (x^A) depends on the coordinates (J_λ) in general.

The Hamiltonian vector fields ϑ_λ on Z relative to the coordinates (7.1.52) take the form

$$\vartheta_\lambda = \vartheta^a_\lambda(x)\partial_a + \vartheta^i_\lambda(x)\partial_i. \tag{7.1.53}$$

Since these vector fields commute (i.e., fibres of $Z \to N$ are isotropic), the symplectic form Ω on Z reads

$$\Omega = \Omega^\alpha_\beta dJ_\alpha \wedge dr^\beta + \Omega_{\alpha A} dr^\alpha \wedge dx^A + \Omega^{\alpha\beta} dJ_\alpha \wedge dJ_\beta \tag{7.1.54}$$
$$+ \Omega^\alpha_A dJ_\alpha \wedge dx^A + \Omega_{AB} dx^A \wedge dx^B.$$

This form is exact (see Lemma 7.1.2 below). Thus, we can write

$$\Omega = d\Xi, \qquad \Xi = \Xi^\lambda(J_\alpha, x^B, r^\alpha) dJ_\lambda + \Xi_\lambda(J_\alpha, x^B) dr^\lambda \tag{7.1.55}$$
$$+ \Xi_A(J_\alpha, x^B, r^\alpha) dx^A.$$

Up to an exact summand, the Liouville form Ξ (7.1.55) is brought into the form

$$\Xi = \Xi^\lambda(J_\alpha, x^B, r^\alpha) dJ_\lambda + \Xi_i(J_\alpha, x^B) d\varphi^i + \Xi_A(J_\alpha, x^B, r^\alpha) dx^A,$$

i.e., it does not contain the term $\Xi_a dt^a$.

The Hamiltonian vector fields ϑ_λ (7.1.53) obey the relations $\vartheta_\lambda \rfloor \Omega = -dJ_\lambda$, which result in the coordinate conditions (7.1.39). Then following the proof of Theorem 7.1.5, we can show that a symplectic form Ω on Z is given by the expression (7.1.48) with respect to the coordinates

$$I_a = J_a, \qquad I_i = \Xi_i(J_j), \tag{7.1.56}$$
$$y^a = -\Xi^a = t^a - E^a(J_\lambda, x^B), \quad y^i = \varphi^i - \Xi^j(J_\lambda, x^B)\frac{\partial J_j}{\partial I_i}.$$

□

Lemma 7.1.2. *The symplectic form Ω (7.1.54) is exact.*

Proof. In accordance with the well-known Künneth formula, the de Rham cohomology group of the product (7.1.51) reads

$$H^2_{\rm DR}(Z) = H^2_{\rm DR}(N) \oplus H^1_{\rm DR}(N) \otimes H^1_{\rm DR}(T^r) \oplus H^2_{\rm DR}(T^r).$$

By the de Rham theorem [85], the de Rham cohomology $H^2_{\rm DR}(N)$ is isomorphic to the cohomology $H^2(N;\mathbb{R})$ of N with coefficients in the constant sheaf $\mathbb{R}$. It is trivial since

$$H^2(N;\mathbb{R}) = H^2(N;\mathbb{Z}) \otimes \mathbb{R}$$

where $H^2(N;\mathbb{Z})$ is trivial. The first cohomology group $H^1_{\rm DR}(N)$ of N is trivial because N is simply connected. Consequently, $H^2_{\rm DR}(Z) = H^2_{\rm DR}(T^r)$. Then the closed form Ω (7.1.54) is exact since it does not contain the term $\Omega_{ij} d\varphi^i \wedge d\varphi^j$. □

7.2 KAM theorem for partially integrable systems

Introducing an appropriate Poisson structure for a partially integrable system and using the methods in [20], one can extend the well-known KAM theorem to partially integrable systems [62].

Let $\{S_\lambda\}$, $\lambda = 1,\ldots,m$, be a partially integrable system on a $2n$-dimensional symplectic manifold (Z,Ω). Let M be its connected compact invariant submanifold which admits an open neighborhood satisfying Theorem 7.1.5. In this case, Theorem 7.1.5 comes to the above mentioned Nekhoroshev theorem. By virtue of this theorem, there exists an open neighborhood of M which is a trivial composite bundle

$$\pi : U = V \times W \times T^m \to V \times W \to V \tag{7.2.1}$$

(cf. (7.1.36)) over domains $W \subset \mathbb{R}^{2(n-m)}$ and $V \subset \mathbb{R}^m$. It is provided with the partial action-angle coordinates $(I_\lambda, x^A, \phi^\lambda)$, $\lambda = 1,\ldots,m$, $A = 1,\ldots,2(n-m)$, such that the symplectic form Ω on U reads

$$\Omega = dI_\lambda \wedge d\phi^\lambda + \Omega_{AB}(I_\mu, x^C)dx^A \wedge dx^B + \Omega^\lambda_A(I_\mu, x^C)dI_\lambda \wedge dx^A \tag{7.2.2}$$

(cf. (7.1.43)), while the generating functions S_λ depend only on the action coordinates I_μ.

Note that, in accordance with part (III) of Theorem 7.1.5, one can always restrict U to a Darboux coordinate chart provided with coordinates $(I_i, p_s, q^s; \varphi^i)$ such that the symplectic form Ω (7.2.2) takes the canonical form

$$\Omega = dI_\lambda \wedge d\varphi^\lambda + dp_s \wedge dq^s.$$

Then the partially integrable system $\{S_\lambda\}$ on this chart can be extended to a completely integrable system, e.g., $\{S_\lambda, p_s\}$, but its invariant submanifolds fail to be compact. Therefore, this is not the case of the KAM theorem.

Let $\mathcal{H}$ be a Hamiltonian of a partially integrable system on U (7.2.1) such that the generating functions S_λ are integrals of motion of $\mathcal{H}$. Therefore, $\mathcal{H}$ is independent of the angle variables. Let us assume that it depends on only the action ones. Then its Hamiltonian vector field

$$\xi = \partial^\mu \mathcal{H}(I_\lambda)\partial_\mu \tag{7.2.3}$$

with respect to the symplectic form Ω (7.2.2) yields the Hamilton equation

$$\dot{I}_\lambda = 0, \qquad \dot{x}^A = 0, \qquad \dot{\phi}^\mu = \partial^\mu \mathcal{H}(I_\lambda) \tag{7.2.4}$$

on U. Let us consider perturbations

$$\mathcal{H}' = \mathcal{H} + \mathcal{H}_1(I_\mu, x^A, \phi^\mu). \tag{7.2.5}$$

We assume the following. (i) The Hamiltonian $\mathcal{H}$ and its perturbations (7.2.5) are real analytic, although generalizations to the case of infinite and finite order of differentiability are possible [20]. (ii) The Hamiltonian $\mathcal{H}$ is non-degenerate, i.e., the frequency map

$$\omega : V \times W \ni (I_\mu, x^A) \to (\xi^\lambda(I_\mu)) \in \mathbb{R}^m$$

is of rank m.

Note that $\omega(V \times W) \subset \mathbb{R}^m$ is open and bounded. As usual, given $\gamma > 0$, let

$$\Omega_\gamma = \left\{ \omega \in \mathbb{R}^m \, : \, |\omega^\mu a_\mu| \geq \gamma \left(\sum_{\lambda=1}^{m} |a_\lambda| \right)^{-m-1}, \quad a \in \mathbb{Z}^m \setminus \{0\} \right\}$$

denote the Cantor set of non-resonant frequencies. The complement of $\Omega_\gamma \cap \omega(V \times W)$ in $\omega(V \times W)$ is dense and open, but its relative Lebesgue measure tends to zero with γ. Let us denote $\Gamma_\gamma = \omega^{-1}(\Omega_\gamma)$, which also is called the Cantor set.

A problem is that the Hamiltonian vector field of the perturbed Hamiltonian (7.2.5) with respect to the symplectic form Ω (7.2.2) leads to the Hamilton equation $\dot{x}^A \neq 0$ and, therefore, no torus (7.2.4) persists.

To overcome this difficulty, let us provide the toroidal domain U (7.2.1) with the degenerate Poisson structure given by the Poisson bivector field

$$w = \partial^\lambda \wedge \partial_\lambda \tag{7.2.6}$$

of rank $2m$. It is readily observed that, relative to w, all integrals of motion of the original partially integrable system $(\Omega, \{S_\lambda\})$ remain in involution and, moreover, they possess the same Hamiltonian vector fields ϑ_λ. In particular, a Hamiltonian $\mathcal{H}$ with respect to the Poisson structure (7.2.6) leads to the same Hamilton equation (7.2.4). Thus, we can think of the pair $(w, \{S_\lambda\})$ as being a partially integrable system on the Poisson manifold (U, w). The key point is that, with respect to the Poisson bivector field w (7.2.6), the Hamiltonian vector field of the perturbed Hamiltonian $\mathcal{H}'$ (7.2.5) is

$$\xi' = \partial^\lambda \mathcal{H}' \partial_\lambda - \partial_\lambda \mathcal{H}' \partial^\lambda, \tag{7.2.7}$$

and the corresponding autonomous first order dynamic equation on U reads

$$\dot{I}_\lambda = -\partial_\lambda \mathcal{H}'(I_\mu, x^B, \phi^\mu), \quad \dot{x}^A = 0, \quad \dot{\phi}^\lambda = \partial^\lambda \mathcal{H}'(I_\mu, s^B, \phi^j \mu). \tag{7.2.8}$$

This is a Hamilton equation with respect to the Poisson structure w (7.2.6), but it is not so relative to the original symplectic form Ω. Since $\dot{x}^A = 0$

and the toroidal domain U (7.2.1) is a trivial bundle over W, one can think of the dynamic equation (7.2.8) as being a perturbation of the dynamic equation (7.2.4) depending on parameters x^A. Furthermore, the Poisson manifold (U, w) is the product of a symplectic manifold $(V \times T^m, \Omega')$ with the symplectic form

$$\Omega' = dI_\lambda \wedge d\phi^\lambda \tag{7.2.9}$$

and a Poisson manifold $(W, w = 0)$ with the zero Poisson structure. Therefore, the equation (7.2.8) can be seen as a Hamilton equation on the symplectic manifold $(V \times T^m, \Omega')$ depending on parameters. Then one can apply the conditions of quasi-periodic stability of symplectic Hamiltonian systems depending on parameters [20] with respect to the perturbation (7.2.8).

In a more general setting, these conditions can be formulated as follows. Let $(w, \{S_\lambda\})$, $\lambda = 1, \ldots, m$, be a partially integrable system on a regular Poisson manifold (Z, w) of rank $2m$. Let M be its regular connected compact invariant submanifold, and let U be its toroidal neighborhood U (7.2.1) in Theorem 7.1.4 provided with the partial action-angle coordinates $(I_\lambda, x^A, \phi^\lambda)$ such that the Poisson bivector w on U takes the canonical form (7.2.6). The following result is a reformulation of that in ([20], Section 5c), where $P = W$ is a parameter space and σ is the symplectic form (7.2.9) on $V \times T^m$.

Theorem 7.2.1. *Given a torus* $\{0\} \times T^m$, *let*

$$\xi = \xi^\lambda(I_\mu, x^A)\partial_\lambda \tag{7.2.10}$$

(cf. (7.2.3)) be a real analytic Hamiltonian vector field whose frequency map

$$\omega : V \times W \ni (I_\mu, x^A) \to \xi^\lambda(I_\mu, x^A) \in \mathbb{R}^m$$

is of maximal rank at $\{0\}$. *Then there exists a neighborhood* $N_0 \subset V \times W$ *of* $\{0\}$ *such that, for any real analytic Hamiltonian vector field*

$$\widetilde{\xi} = \widetilde{\xi}_\lambda(I_\mu, x^A, \phi^\mu)\partial^\lambda + \widetilde{\xi}^\lambda(I_\mu, x^A, \phi^\mu)\partial_\lambda$$

(cf. (7.2.7)) sufficiently near ξ *(7.2.10) in the real analytic topology, the following holds. Given the Cantor set* $\Gamma_\gamma \subset N_0$, *there exists the* $\widetilde{\xi}$*-invariant Cantor set* $\widetilde{\Gamma} \subset N_0 \times T^m$ *which is a* C^∞*-near-identity diffeomorphic image of* $\Gamma_\gamma \times T^m$.

Theorem 7.2.1 is an extension of the KAM theorem [101] to partially integrable systems on Poisson manifolds (Z, w). Given a partially integrable system $(\Omega, \{S_\lambda\})$ on a symplectic manifold (Z, Ω), Theorem 7.2.1 enables one to obtain its perturbations (7.2.7) possessing a large number of invariant tori, though these perturbations are not Hamiltonian.

7.3 Superintegrable systems with non-compact invariant submanifolds

In comparison with partially integrable and completely integrable systems integrals of motion of a superintegrable system need not be in involution. We consider superintegrable systems on a symplectic manifold. Completely integrable systems are particular superintegrable systems.

Definition 7.3.1. Let (Z, Ω) be a $2n$-dimensional connected symplectic manifold, and let $(C^\infty(Z), \{,\})$ be the Poisson algebra of smooth real functions on Z. A subset

$$F = (F_1, \ldots, F_k), \qquad n \leq k < 2n, \tag{7.3.1}$$

of the Poisson algebra $C^\infty(Z)$ is called a *superintegrable system* if the following conditions hold.

(i) All the functions F_i (called the *generating functions of a superintegrable system*) are independent, i.e., the k-form $\overset{k}{\wedge} dF_i$ nowhere vanishes on Z. It follows that the map $F : Z \to \mathbb{R}^k$ is a submersion, i.e.,

$$F : Z \to N = F(Z) \tag{7.3.2}$$

is a fibred manifold over a domain (i.e., contractible open subset) $N \subset \mathbb{R}^k$ endowed with the coordinates (x_i) such that $x_i \circ F = F_i$.

(ii) There exist smooth real functions s_{ij} on N such that

$$\{F_i, F_j\} = s_{ij} \circ F, \qquad i, j = 1, \ldots, k. \tag{7.3.3}$$

(iii) The matrix function $\mathbf{s}$ with the entries s_{ij} (7.3.3) is of constant corank $m = 2n - k$ at all points of N.

Remark 7.3.1. We restrict our consideration to the case of generating functions which are independent everywhere on a symplectic manifold Z (see Remarks 7.1.4 and 7.3.2).

If $k = n$, then $\mathbf{s} = 0$, and we are in the case of completely integrable systems as follows.

Definition 7.3.2. The subset F, $k = n$, (7.3.1) of the Poisson algebra $C^\infty(Z)$ on a symplectic manifold (Z, Ω) is called a completely integrable system if F_i are independent functions in involution.

If $k > n$, the matrix $\mathbf{s}$ is necessarily non-zero. Therefore, superintegrable systems also are called *non-commutative completely integrable systems*. If $k = 2n - 1$, a superintegrable system is called *maximally superintegrable*.

The following two assertions clarify the structure of superintegrable systems [41; 46].

Proposition 7.3.1. *Given a symplectic manifold* (Z, Ω)*, let* $F : Z \to N$ *be a fibred manifold such that, for any two functions* f*,* f' *constant on fibres of* F*, their Poisson bracket* $\{f, f'\}$ *is so. By virtue of Theorem 3.1.3,* N *is provided with an unique coinduced Poisson structure* $\{,\}_N$ *such that* F *is a Poisson morphism.*

Since any function constant on fibres of F is a pull-back of some function on N, the superintegrable system (7.3.1) satisfies the condition of Proposition 7.3.1 due to item (ii) of Definition 7.3.1. Thus, the base N of the fibration (7.3.2) is endowed with a coinduced Poisson structure of corank m. With respect to coordinates x_i in item (i) of Definition 7.3.1 its bivector field reads

$$w = s_{ij}(x_k)\partial^i \wedge \partial^j. \tag{7.3.4}$$

Proposition 7.3.2. *Given a fibred manifold* $F : Z \to N$ *in Proposition 7.3.1, the following conditions are equivalent [41; 104]:*

(i) the rank of the coinduced Poisson structure $\{,\}_N$ *on* N *equals* $2\dim N - \dim Z$,

(ii) the fibres of F *are isotropic,*

(iii) the fibres of F *are maximal integral manifolds of the involutive distribution spanned by the Hamiltonian vector fields of the pull-back* F^*C *of Casimir functions* C *of the coinduced Poisson structure (7.3.4) on* N.

It is readily observed that the fibred manifold F (7.3.2) obeys condition (i) of Proposition 7.3.2 due to item (iii) of Definition 7.3.1, namely, $k - m = 2(k - n)$.

Fibres of the fibred manifold F (7.3.2) are called the *invariant submanifolds.*

Remark 7.3.2. In many physical models, condition (i) of Definition 7.3.1 fails to hold. Just as in the case of partially integrable systems, it can be replaced with that a subset $Z_R \subset Z$ of regular points (where $\overset{k}{\wedge} dF_i \neq 0$) is open and dense. Let M be an invariant submanifold through a regular point $z \in Z_R \subset Z$. Then it is regular, i.e., $M \subset Z_R$. Let M admit a regular open saturated neighborhood U_M (i.e., a fibre of F through a point of U_M belongs to U_M). For instance, any compact invariant submanifold M has such a neighborhood U_M. The restriction of functions F_i to U_M defines a superintegrable system on U_M which obeys Definition 7.3.1. In this case, one says that a superintegrable system is considered around its invariant submanifold M.

Let (Z,Ω) be a $2n$-dimensional connected symplectic manifold. Given the superintegrable system (F_i) (7.3.1) on (Z,Ω), the well known Mishchenko – Fomenko theorem (Theorem 7.3.2) states the existence of (semi-local) generalized action-angle coordinates around its connected compact invariant submanifold [16; 41; 115]. The Mishchenko – Fomenko theorem is extended to superintegrable systems with non-compact invariant submanifolds (Theorem 7.3.1) [46; 48; 143]. These submanifolds are diffeomorphic to a toroidal cylinder

$$\mathbb{R}^{m-r}\times T^r, \qquad m=2n-k, \qquad 0\leq r\leq m. \tag{7.3.5}$$

Note that the Mishchenko – Fomenko theorem is mainly applied to superintegrable systems whose integrals of motion form a compact Lie algebra. The group generated by flows of their Hamiltonian vector fields is compact. Since a fibration of a compact manifold possesses compact fibres, invariant submanifolds of such a superintegrable system are compact. With Theorem 7.3.1, one can describe superintegrable Hamiltonian system with an arbitrary Lie algebra of integrals of motion (see Section 7.6).

Given a superintegrable system in accordance with Definition 7.3.1, the above mentioned generalization of the Mishchenko – Fomenko theorem to non-compact invariant submanifolds states the following.

Theorem 7.3.1. *Let the Hamiltonian vector fields ϑ_i of the functions F_i be complete, and let the fibres of the fibred manifold F (7.3.2) be connected and mutually diffeomorphic. Then the following hold.*

(I) The fibres of F (7.3.2) are diffeomorphic to the toroidal cylinder (7.3.5).

(II) Given a fibre M of F (7.3.2), there exists its open saturated neighborhood U_M which is a trivial principal bundle

$$U_M=N_M\times\mathbb{R}^{m-r}\times T^r \xrightarrow{F} N_M \tag{7.3.6}$$

with the structure group (7.3.5).

(III) The neighborhood U_M is provided with the bundle (generalized action-angle) coordinates $(I_\lambda,p_s,q^s,y^\lambda)$, $\lambda=1,\ldots,m$, $s=1,\ldots,n-m$, such that: (i) the generalized angle coordinates (y^λ) are coordinates on a toroidal cylinder, i.e., fibre coordinates on the fibre bundle (7.3.6), (ii) (I_λ,p_s,q^s) are coordinates on its base N_M where the action coordinates (I_λ)

are values of Casimir functions of the coinduced Poisson structure $\{,\}_N$ *on* N_M*, and (iii) the symplectic form* Ω *on* U_M *reads*

$$\Omega = dI_\lambda \wedge dy^\lambda + dp_s \wedge dq^s. \tag{7.3.7}$$

Proof. It follows from item (iii) of Proposition 7.3.2 that every fibre M of the fibred manifold (7.3.2) is a maximal integral manifolds of the involutive distribution spanned by the Hamiltonian vector fields υ_λ of the pull-back F^*C_λ of m independent Casimir functions $\{C_1, \ldots, C_m\}$ of the Poisson structure $\{,\}_N$ (7.3.4) on an open neighborhood N_M of a point $F(M) \in N$. Let us put $U_M = F^{-1}(N_M)$. It is an open saturated neighborhood of M. Consequently, invariant submanifolds of a superintegrable system (7.3.1) on U_M are maximal integral manifolds of the partially integrable system

$$C^* = (F^*C_1, \ldots, F^*C_m), \qquad 0 < m \leq n, \tag{7.3.8}$$

on a symplectic manifold (U_M, Ω). Therefore, statements (I) – (III) of Theorem 7.3.1 are the corollaries of Theorem 7.1.5. Its condition (i) is satisfied as follows. Let M' be an arbitrary fibre of the fibred manifold $F: U_M \to N_M$ (7.3.2). Since

$$F^*C_\lambda(z) = (C_\lambda \circ F)(z) = C_\lambda(F_i(z)), \qquad z \in M',$$

the Hamiltonian vector fields υ_λ on M' are $\mathbb{R}$-linear combinations of Hamiltonian vector fields ϑ_i of the functions F_i It follows that υ_λ are elements of a finite-dimensional real Lie algebra of vector fields on M' generated by the vector fields ϑ_i. Since vector fields ϑ_i are complete, the vector fields υ_λ on M' also are complete (see forthcoming Remark 7.3.3). Consequently, these vector fields are complete on U_M because they are vertical vector fields on $U_M \to N$. The proof of Theorem 7.1.5 shows that the action coordinates (I_λ) are values of Casimir functions expressed in the original ones C_λ. □

Remark 7.3.3. If complete vector fields on a smooth manifold constitute a basis for a finite-dimensional real Lie algebra, any element of this Lie algebra is complete [127].

Remark 7.3.4. Since an open neighborhood U_M (7.3.6) in item (II) of Theorem 7.3.1 is not contractible, unless $r = 0$, the generalized action-angle coordinates on U sometimes are called *semi-local.*

Remark 7.3.5. The condition of the completeness of Hamiltonian vector fields of the generating functions F_i in Theorem 7.3.1 is rather restrictive (see the Kepler system in Section 7.6). One can replace this condition with that the Hamiltonian vector fields of the pull-back onto Z of Casimir functions on N are complete.

If the conditions of Theorem 7.3.1 are replaced with that the fibres of the fibred manifold F (7.3.2) are compact and connected, this theorem restarts the Mishchenko – Fomenko one as follows.

Theorem 7.3.2. *Let the fibres of the fibred manifold F (7.3.2) be connected and compact. Then they are diffeomorphic to a torus T^m, and statements (II) – (III) of Theorem 7.3.1 hold.*

Remark 7.3.6. In Theorem 7.3.2, the Hamiltonian vector fields υ_λ are complete because fibres of the fibred manifold F (7.3.2) are compact. As well known, any vector field on a compact manifold is complete.

If F (7.3.1) is a completely integrable system, the coinduced Poisson structure on N equals zero, and the generating functions F_i are the pull-back of n independent functions on N. Then Theorems 7.3.2 and 7.3.1 come to the Liouville – Arnold theorem [4; 101] and its generalization (Theorem 7.3.3) to the case of non-compact invariant submanifolds [44; 65], respectively. In this case, the partially integrable system C^* (7.3.8) is exactly the original completely integrable system F.

Theorem 7.3.3. *Given a completely integrable system, F in accordance with Definition 7.3.2, let the Hamiltonian vector fields ϑ_i of the functions F_i be complete, and let the fibres of the fibred manifold F (7.3.2) be connected and mutually diffeomorphic. Then items (I) and (II) of Theorem 7.3.1 hold, and its item (III) is replaced with the following one.*

(III') The neighborhood U_M (7.3.6) where $m = n$ is provided with the bundle (generalized action-angle) coordinates (I_λ, y^λ), $\lambda = 1, \ldots, n$, such that the angle coordinates (y^λ) are coordinates on a toroidal cylinder, and the symplectic form Ω on U_M reads

$$\Omega = dI_\lambda \wedge dy^\lambda. \tag{7.3.9}$$

7.4 Globally superintegrable systems

To study a superintegrable system, one conventionally considers it with respect to generalized action-angle coordinates. A problem is that, restricted to an action-angle coordinate chart on an open subbundle U of the fibred manifold $Z \to N$ (7.3.2), a superintegrable system becomes different from the original one since there is no morphism of the Poisson algebra $C^\infty(U)$ on (U, Ω) to that $C^\infty(Z)$ on (Z, Ω). Moreover, a superintegrable system on U need not satisfy the conditions of Theorem 7.3.1 because it may happen that the Hamiltonian vector fields of the generating functions on U are not complete. To describe superintegrable systems in terms of generalized action-angle coordinates, we therefore follow the notion of a globally superintegrable system [143].

Definition 7.4.1. A superintegrable system F (7.3.1) on a symplectic manifold (Z, Ω) in Definition 7.3.1 is called *globally superintegrable* if there exist *global generalized action-angle coordinates*

$$(I_\lambda, x^A, y^\lambda), \qquad \lambda = 1, \ldots, m, \qquad A = 1, \ldots, 2(n-m), \tag{7.4.1}$$

such that: (i) the action coordinates (I_λ) are expressed in the values of some Casimir functions C_λ on the Poisson manifold $(N, \{,\}_N)$, (ii) the angle coordinates (y^λ) are coordinates on the toroidal cylinder (7.1.1), and (iii) the symplectic form Ω on Z reads

$$\Omega = dI_\lambda \wedge dy^\lambda + \Omega_{AB}(I_\mu, x^C) dx^A \wedge dx^B. \tag{7.4.2}$$

It is readily observed that the semi-local generalized action-angle coordinates on U in Theorem 7.3.1 are global on U in accordance with Definition 7.4.1.

Forthcoming Theorem 7.4.1 provides the sufficient conditions of the existence of global generalized action-angle coordinates of a superintegrable system on a symplectic manifold (Z, Ω) [110; 143]. It generalizes the well-known result for the case of compact invariant submanifolds [30; 41].

Theorem 7.4.1. *A superintegrable system F on a symplectic manifold (Z, Ω) is globally superintegrable if the following conditions hold.*

(i) Hamiltonian vector fields ϑ_i of the generating functions F_i are complete.

(ii) The fibred manifold F (7.3.2) is a fibre bundle with connected fibres.

(iii) Its base N is simply connected and the cohomology $H^2(V; \mathbb{Z})$ is trivial

(iv) The coinduced Poisson structure $\{,\}_N$ on a base N admits m independent Casimir functions C_λ.

Proof. Theorem 7.4.1 is a corollary of Theorem 7.1.6. In accordance with Theorem 7.1.6, we have a composite fibred manifold

$$Z \xrightarrow{F} N \xrightarrow{C} W, \tag{7.4.3}$$

where $C : N \to W$ is a fibred manifold of level surfaces of the Casimir functions C_λ (which coincides with the symplectic foliation of a Poisson manifold N). The composite fibred manifold (7.4.3) is provided with the adapted fibred coordinates $(J_\lambda, x^A, r^\lambda)$ (7.1.52), where J_λ are values of independent Casimir functions and $(r^\lambda) = (t^a, \varphi^i)$ are coordinates on a toroidal cylinder. Since $C_\lambda = J_\lambda$ are Casimir functions on N, the symplectic form Ω (7.1.54) on Z reads

$$\Omega = \Omega^\alpha_\beta dJ_\alpha \wedge r^\beta + \Omega_{\alpha A} dy^\alpha \wedge dx^A + \Omega_{AB} dx^A \wedge dx^B. \tag{7.4.4}$$

In particular, it follows that transition functions of coordinates x^A on N are independent of coordinates J_λ, i.e., $C : V \to W$ is a trivial bundle. By virtue of Lemma 7.1.2, the symplectic form (7.4.4) is exact, i.e., $\Omega = d\Xi$, where the Liouville form Ξ (7.1.55) is

$$\Xi = \Xi^\lambda(J_\alpha, y^\mu) dJ_\lambda + \Xi_i(J_\alpha) d\varphi^i + \Xi_A(x^B) dx^A.$$

Then the coordinate transformations (7.1.56):

$$\begin{aligned} &I_a = J_a, \qquad I_i = \Xi_i(J_j), \\ &y^a = -\Xi^a = t^a - E^a(J_\lambda), \qquad y^i = \varphi^i - \Xi^j(J_\lambda) \frac{\partial J_j}{\partial I_i}, \end{aligned} \tag{7.4.5}$$

bring Ω (7.4.4) into the form (7.4.2). In comparison with the general case (7.1.56), the coordinate transformations (7.4.5) are independent of coordinates x^A. Therefore, the angle coordinates y^i possess identity transition functions on N. □

Theorem 7.4.1 restarts Theorem 7.3.1 if one considers an open subset V of N admitting the Darboux coordinates x^A on the symplectic leaves of U.

Note that, if invariant submanifolds of a superintegrable system are assumed to be connected and compact, condition (i) of Theorem 7.4.1 is unnecessary since vector fields ϑ_λ on compact fibres of F are complete. Condition (ii) also holds by virtue of Theorem 11.2.4. In this case, Theorem 7.4.1 reproduces the well known result in [30].

If F in Theorem 7.4.1 is a completely integrable system, the coinduced Poisson structure on N equals zero, the generating functions F_i are the pull-back of n independent functions on N, and Theorem 7.4.1 takes the following form [110].

Theorem 7.4.2. *Let a completely integrable system $\{F_1, \ldots, F_n\}$ on a symplectic manifold (Z, Ω) satisfy the following conditions.*

(i) The Hamiltonian vector fields ϑ_i of F_i are complete.

(ii) The fibred manifold F (7.3.2) is a fibre bundle with connected fibres over a simply connected base N whose cohomology $H^2(N,\mathbb{Z})$ is trivial.

Then the following hold.

(I) The fibre bundle F (7.3.2) is a trivial principal bundle with the structure group $\mathbb{R}^{2n-r}\times T^r$.

(II) The symplectic manifold Z is provided with the global Darboux coordinates (I_λ, y^λ) such that $\Omega = dI_\lambda \wedge dy^\lambda$.

It follows from the proof of Theorem 7.1.6 that its condition (iii) and, accordingly, condition (iii) of Theorem 7.4.1 guarantee that fibre bundles F in conditions (ii) of these theorems are trivial. Therefore, Theorem 7.4.1 can be reformulated as follows.

Theorem 7.4.3. *A superintegrable system F on a symplectic manifold (Z,Ω) is globally superintegrable if and only if the following conditions hold.*

(i) The fibred manifold F (7.3.2) is a trivial fibre bundle.

(ii) The coinduced Poisson structure $\{,\}_N$ on a base N admits m independent Casimir functions C_λ such that Hamiltonian vector fields of their pull-back F^*C_λ are complete.

Remark 7.4.1. It follows from Remark 7.3.3 and condition (ii) of Theorem 7.4.3 that a Hamiltonian vector field of the the pull-back F^*C of any Casimir function C on a Poisson manifold N is complete.

7.5 Superintegrable Hamiltonian systems

In autonomous Hamiltonian mechanics, one considers superintegrable systems whose generating functions are integrals of motion, i.e., they are in involution with a Hamiltonian $\mathcal{H}$, and the functions $(\mathcal{H}, F_1, \ldots, F_k)$ are nowhere independent, i.e.,

$$\{\mathcal{H}, F_i\} = 0, \tag{7.5.1}$$

$$d\mathcal{H} \wedge (\overset{k}{\wedge} dF_i) = 0. \tag{7.5.2}$$

In order that an evolution of a Hamiltonian system can be defined at any instant $t \in \mathbb{R}$, one supposes that the Hamiltonian vector field of its Hamiltonian is complete. By virtue of Remark 7.4.1 and forthcoming Proposition 7.5.1, a Hamiltonian of a superintegrable system always satisfies this condition.

Proposition 7.5.1. *It follows from the equality (7.5.2) that a Hamiltonian $\mathcal{H}$ is constant on the invariant submanifolds. Therefore, it is the pull-back of a function on N which is a Casimir function of the Poisson structure (7.3.4) because of the conditions (7.5.1).*

Proposition 7.5.1 leads to the following.

Proposition 7.5.2. *Let $\mathcal{H}$ be a Hamiltonian of a globally superintegrable system provided with the generalized action-angle coordinates $(I_\lambda, x^A, y^\lambda)$ (2.3.15). Then a Hamiltonian $\mathcal{H}$ depends only on the action coordinates I_λ. Consequently, the Hamilton equation of a globally superintegrable system take the form*

$$\dot{y}^\lambda = \frac{\partial \mathcal{H}}{\partial I_\lambda}, \qquad I_\lambda = \text{const.}, \qquad x^A = \text{const.}$$

Following the original Mishchenko–Fomenko theorem, let us mention superintegrable systems whose generating functions $\{F_1, \ldots, F_k\}$ form a k-dimensional real Lie algebra $\mathfrak{g}$ of corank m with the commutation relations

$$\{F_i, F_j\} = c_{ij}^h F_h, \qquad c_{ij}^h = \text{const.} \tag{7.5.3}$$

Then F (7.3.2) is a momentum mapping of Z to the Lie coalgebra $\mathfrak{g}^*$ provided with the coordinates x_i in item (i) of Definition 7.3.1 [65; 79]. In this case, the coinduced Poisson structure $\{,\}_N$ coincides with the canonical Lie–Poisson structure on $\mathfrak{g}^*$ given by the Poisson bivector field

$$w = \frac{1}{2} c_{ij}^h x_h \partial^i \wedge \partial^j.$$

Let V be an open subset of $\mathfrak{g}^*$ such that conditions (i) and (ii) of Theorem 7.4.3 are satisfied. Then an open subset $F^{-1}(V) \subset Z$ is provided with the generalized action-angle coordinates.

Remark 7.5.1. Let Hamiltonian vector fields ϑ_i of the generating functions F_i which form a Lie algebra $\mathfrak{g}$ be complete. Then they define a locally free Hamiltonian action on Z of some simply connected Lie group G whose Lie algebra is isomorphic to $\mathfrak{g}$ [125; 127]. Orbits of G coincide with k-dimensional maximal integral manifolds of the regular distribution $\mathcal{V}$ on Z spanned by Hamiltonian vector fields ϑ_i [153]. Furthermore, Casimir functions of the Lie–Poisson structure on $\mathfrak{g}^*$ are exactly the coadjoint invariant functions on $\mathfrak{g}^*$. They are constant on orbits of the coadjoint action of G on $\mathfrak{g}^*$ which coincide with leaves of the symplectic foliation of $\mathfrak{g}^*$.

Theorem 7.5.1. *Let a globally superintegrable Hamiltonian system on a symplectic manifold Z obey the following conditions.*

(i) It is maximally superintegrable.

(ii) Its Hamiltonian $\mathcal{H}$ is regular, i.e, $d\mathcal{H}$ nowhere vanishes.

(iii) Its generating functions F_i constitute a finite dimensional real Lie algebra and their Hamiltonian vector fields are complete.

Then any integral of motion of this Hamiltonian system is the pull-back of a function on a base N of the fibration F (7.3.2). In other words, it is expressed in the integrals of motion F_i.

Proof. The proof is based on the following. A Hamiltonian vector field of a function f on Z lives in the one-codimensional regular distribution $\mathcal{V}$ on Z spanned by Hamiltonian vector fields ϑ_i if and only if f is the pull-back of a function on a base N of the fibration F (7.3.2). A Hamiltonian $\mathcal{H}$ brings Z into a fibred manifold of its level surfaces whose vertical tangent bundle coincide with $\mathcal{V}$. Therefore, a Hamiltonian vector field of any integral of motion of $\mathcal{H}$ lives in $\mathcal{V}$. □

It may happen that, given a Hamiltonian $\mathcal{H}$ of a Hamiltonian system on a symplectic manifold Z, we have different superintegrable Hamiltonian systems on different open subsets of Z. For instance, this is the case of the Kepler system.

7.6 Example. Global Kepler system

We consider the Kepler system on a plane $\mathbb{R}^2$ (see Example 3.8.1). Its phase space is $T^*\mathbb{R}^2 = \mathbb{R}^4$ provided with the Cartesian coordinates (q_i, p_i), $i = 1, 2$, and the canonical symplectic form

$$\Omega_T = \sum_i dp_i \wedge dq_i. \tag{7.6.1}$$

Let us denote

$$p = \left(\sum_i (p_i)^2\right)^{1/2}, \qquad r = \left(\sum_i (q^i)^2\right)^{1/2}, \qquad (p, q) = \sum_i p_i q_i.$$

An autonomous Hamiltonian of the Kepler system reads

$$\mathcal{H} = \frac{1}{2}p^2 - \frac{1}{r} \tag{7.6.2}$$

(cf. (3.8.14)). The Kepler system is a Hamiltonian system on a symplectic manifold

$$Z = \mathbb{R}^4 \setminus \{0\} \tag{7.6.3}$$

endowed with the symplectic form Ω_T (7.6.1).

Let us consider the functions

$$M_{12} = -M_{21} = q_1 p_2 - q_2 p_1, \tag{7.6.4}$$

$$A_i = \sum_j M_{ij} p_j - \frac{q_i}{r} = q_i p^2 - p_i(p,q) - \frac{q_i}{r}, \quad i = 1, 2, \tag{7.6.5}$$

on the symplectic manifold Z (7.6.3). As was mentioned in Example 3.8.1, they are integrals of motion of the Hamiltonian $\mathcal{H}$ (7.6.2) where M_{12} is an angular momentum and (A_i) is a Rung–Lenz vector. Let us denote

$$M^2 = (M_{12})^2, \qquad A^2 = (A_1)^2 + (A_a)^2 = 2M^2\mathcal{H} + 1. \tag{7.6.6}$$

Let $Z_0 \subset Z$ be a closed subset of points where $M_{12} = 0$. A direct computation shows that the functions (M_{12}, A_i) (7.6.4) – (7.6.5) are independent of an open submanifold

$$U = Z \setminus Z_0 \tag{7.6.7}$$

of Z. At the same time, the functions $(\mathcal{H}, M_{12}, A_i)$ are independent nowhere on U because it follows from the expression (7.6.6) that

$$\mathcal{H} = \frac{A^2 - 1}{2M^2} \tag{7.6.8}$$

on U (7.6.7). The well known dynamics of the Kepler system shows that the Hamiltonian vector field of its Hamiltonian is complete on U (but not on Z).

The Poisson bracket of integrals of motion M_{12} (7.6.4) and A_i (7.6.5) obeys the relations

$$\{M_{12}, A_i\} = \eta_{2i} A_1 - \eta_{1i} A_2, \tag{7.6.9}$$

$$\{A_1, A_2\} = 2\mathcal{H} M_{12} = \frac{A^2 - 1}{M_{12}}, \tag{7.6.10}$$

where η_{ij} is an Euclidean metric on $\mathbb{R}^2$. It is readily observed that these relations take the form (7.3.3). However, the matrix function s of the relations (7.6.9) – (7.6.10) fails to be of constant rank at points where $\mathcal{H} = 0$. Therefore, let us consider the open submanifolds $U_- \subset U$ where $\mathcal{H} < 0$ and U_+ where $\mathcal{H} > 0$. Then we observe that the Kepler system with the Hamiltonian $\mathcal{H}$ (7.6.2) and the integrals of motion (M_{ij}, A_i) (7.6.4) – (7.6.5) on U_- and the Kepler system with the Hamiltonian $\mathcal{H}$ (7.6.2) and the integrals of motion (M_{ij}, A_i) (7.6.4) – (7.6.5) on U_+ are superintegrable Hamiltonian systems. Moreover, these superintegrable systems can be brought into the form (7.5.3) as follows.

Let us replace the integrals of motions A_i with the integrals of motion

$$L_i = \frac{A_i}{\sqrt{-2\mathcal{H}}} \tag{7.6.11}$$

on U_-, and with the integrals of motion

$$K_i = \frac{A_i}{\sqrt{2\mathcal{H}}} \tag{7.6.12}$$

on U_+.

The superintegrable system (M_{12}, L_i) on U_- obeys the relations

$$\{M_{12}, L_i\} = \eta_{2i}L_1 - \eta_{1i}L_2, \tag{7.6.13}$$

$$\{L_1, L_2\} = -M_{12}. \tag{7.6.14}$$

Let us denote $M_{i3} = -L_i$ and put the indexes $\mu, \nu, \alpha, \beta = 1, 2, 3$. Then the relations (7.6.13) – (7.6.14) are brought into the form

$$\{M_{\mu\nu}, M_{\alpha\beta}\} = \eta_{\mu\beta}M_{\nu\alpha} + \eta_{\nu\alpha}M_{\mu\beta} - \eta_{\mu\alpha}M_{\nu\beta} - \eta_{\nu\beta}M_{\mu\alpha} \tag{7.6.15}$$

where $\eta_{\mu\nu}$ is an Euclidean metric on $\mathbb{R}^3$. A glance at the expression (7.6.15) shows that the integrals of motion M_{12} (7.6.4) and L_i (7.6.11) constitute the Lie algebra $\mathfrak{g} = so(3)$. Its corank equals 1. Therefore the superintegrable system (M_{12}, L_i) on U_- is maximally superintegrable. The equality (7.6.8) takes the form

$$M^2 + L^2 = -\frac{1}{2\mathcal{H}}. \tag{7.6.16}$$

The superintegrable system (M_{12}, K_i) on U_+ obeys the relations

$$\{M_{12}, K_i\} = \eta_{2i}K_1 - \eta_{1i}K_2, \tag{7.6.17}$$

$$\{K_1, K_2\} = M_{12}. \tag{7.6.18}$$

Let us denote $M_{i3} = -K_i$ and put the indexes $\mu, \nu, \alpha, \beta = 1, 2, 3$. Then the relations (7.6.17) – (7.6.18) are brought into the form

$$\{M_{\mu\nu}, M_{\alpha\beta}\} = \rho_{\mu\beta}M_{\nu\alpha} + \rho_{\nu\alpha}M_{\mu\beta} - \rho_{\mu\alpha}M_{\nu\beta} - \rho_{\nu\beta}M_{\mu\alpha} \tag{7.6.19}$$

where $\rho_{\mu\nu}$ is a pseudo-Euclidean metric of signature $(+, +, -)$ on $\mathbb{R}^3$. A glance at the expression (7.6.19) shows that the integrals of motion M_{12} (7.6.4) and K_i (7.6.12) constitute the Lie algebra $so(2, 1)$. Its corank equals 1. Therefore the superintegrable system (M_{12}, K_i) on U_+ is maximally superintegrable. The equality (7.6.8) takes the form

$$K^2 - M^2 = \frac{1}{2\mathcal{H}}. \tag{7.6.20}$$

Thus, the Kepler system on a phase space $\mathbb{R}^4$ falls into two different maximally superintegrable systems on open submanifolds U_- and U_+ of $\mathbb{R}^4$. We agree to call them the Kepler superintegrable systems on U_- and U_+, respectively.

Let us study the first one and put

$$F_1 = -L_1, \qquad F_2 = -L_2, \qquad F_3 = -M_{12}, \tag{7.6.21}$$
$$\{F_1, F_2\} = F_3, \qquad \{F_2, F_3\} = F_1, \qquad \{F_3, F_1\} = F_2.$$

We have a fibred manifold

$$F : U_- \to N \subset \mathfrak{g}^*, \tag{7.6.22}$$

which is the momentum mapping to the Lie coalgebra $\mathfrak{g}^* = so(3)^*$, endowed with the coordinates (x_i) such that integrals of motion F_i on $\mathfrak{g}^*$ read $F_i = x_i$. A base N of the fibred manifold (7.6.22) is an open submanifold of $\mathfrak{g}^*$ given by the coordinate condition $x_3 \neq 0$. It is a union of two contractible components defined by the conditions $x_3 > 0$ and $x_3 < 0$. The coinduced Lie–Poisson structure on N takes the form

$$w = x_2\partial^3 \wedge \partial^1 + x_3\partial^1 \wedge \partial^2 + x_1\partial^2 \wedge \partial^3. \tag{7.6.23}$$

The coadjoint action of $so(3)$ on N reads

$$\varepsilon_1 = x_3\partial^2 - x_2\partial^3, \quad \varepsilon_2 = x_1\partial^3 - x_3\partial^1, \quad \varepsilon_3 = x_2\partial^1 - x_1\partial^2. \tag{7.6.24}$$

The orbits of this coadjoint action are given by the equation

$$x_1^2 + x_2^2 + x_3^2 = \text{const}. \tag{7.6.25}$$

They are the level surfaces of the Casimir function

$$C = x_1^2 + x_2^2 + x_3^2$$

and, consequently, the Casimir function

$$h = -\frac{1}{2}(x_1^2 + x_2^2 + x_3^2)^{-1}. \tag{7.6.26}$$

A glance at the expression (7.6.16) shows that the pull-back F^*h of this Casimir function (7.6.26) onto U_- is the Hamiltonian $\mathcal{H}$ (7.6.2) of the Kepler system on U_-.

As was mentioned above, the Hamiltonian vector field of F^*h is complete. Furthermore, it is known that invariant submanifolds of the superintegrable Kepler system on U_- are compact. Therefore, the fibred manifold F (7.6.22) is a fibre bundle in accordance with Theorem 11.2.4. Moreover, this fibre bundle is trivial because N is a disjoint union of two contractible manifolds. Consequently, it follows from Theorem 7.4.3 that the Kepler superintegrable system on U_- is globally superintegrable, i.e., it admits global generalized action-angle coordinates as follows.

The Poisson manifold N (7.6.22) can be endowed with the coordinates

$$(I, x_1, \gamma), \qquad I < 0, \qquad \gamma \neq \frac{\pi}{2}, \frac{3\pi}{2}, \tag{7.6.27}$$

defined by the equalities

$$I = -\frac{1}{2}(x_1^2 + x_2^2 + x_3^2)^{-1}, \tag{7.6.28}$$

$$x_2 = \left(-\frac{1}{2I} - x_1^2\right)^{1/2} \sin\gamma, \qquad x_3 = \left(-\frac{1}{2I} - x_1^2\right)^{1/2} \cos\gamma.$$

It is readily observed that the coordinates (7.6.27) are Darboux coordinates of the Lie–Poisson structure (7.6.23) on U_-, namely,

$$w = \frac{\partial}{\partial x_1} \wedge \frac{\partial}{\partial \gamma}. \tag{7.6.29}$$

Let ϑ_I be the Hamiltonian vector field of the Casimir function I (7.6.28). By virtue of Proposition 7.3.2, its flows are invariant submanifolds of the Kepler superintegrable system on U_-. Let α be a parameter along the flow of this vector field, i.e.,

$$\vartheta_I = \frac{\partial}{\partial \alpha}. \tag{7.6.30}$$

Then U_- is provided with the generalized action-angle coordinates (I, x_1, γ, α) such that the Poisson bivector associated to the symplectic form Ω_T on U_- reads

$$W = \frac{\partial}{\partial I} \wedge \frac{\partial}{\partial \alpha} + \frac{\partial}{\partial x_1} \wedge \frac{\partial}{\partial \gamma}. \tag{7.6.31}$$

Accordingly, Hamiltonian vector fields of integrals of motion F_i (7.6.21) take the form

$$\vartheta_1 = \frac{\partial}{\partial \gamma},$$

$$\vartheta_2 = \frac{1}{4I^2}\left(-\frac{1}{2I} - x_1^2\right)^{-1/2} \sin\gamma \frac{\partial}{\partial \alpha} - x_1 \left(-\frac{1}{2I} - x_1^2\right)^{-1/2} \sin\gamma \frac{\partial}{\partial \gamma}$$
$$- \left(-\frac{1}{2I} - x_1^2\right)^{1/2} \cos\gamma \frac{\partial}{\partial x_1},$$

$$\vartheta_3 = \frac{1}{4I^2}\left(-\frac{1}{2I} - x_1^2\right)^{-1/2} \cos\gamma \frac{\partial}{\partial \alpha} - x_1 \left(-\frac{1}{2I} - x_1^2\right)^{-1/2} \cos\gamma \frac{\partial}{\partial \gamma}$$
$$+ \left(-\frac{1}{2I} - x_1^2\right)^{1/2} \sin\gamma \frac{\partial}{\partial x_1}.$$

A glance at these expressions shows that the vector fields ϑ_1 and ϑ_2 fail to be complete on U_- (see Remark 7.3.5).

One can say something more about the angle coordinate α. The vector field ϑ_I (7.6.30) reads

$$\frac{\partial}{\partial\alpha} = \sum_i \left(\frac{\partial\mathcal{H}}{\partial p_i}\frac{\partial}{\partial q_i} - \frac{\partial\mathcal{H}}{\partial q_i}\frac{\partial}{\partial p_i}\right).$$

This equality leads to the relations

$$\frac{\partial q_i}{\partial\alpha} = \frac{\partial\mathcal{H}}{\partial p_i}, \qquad \frac{\partial p_i}{\partial\alpha} = -\frac{\partial\mathcal{H}}{\partial q_i},$$

which take the form of the Hamilton equation. Therefore, the coordinate α is a cyclic time $\alpha = t\,\mathrm{mod}2\pi$ given by the well-known expression

$$\alpha = \phi - a^{3/2}e\sin(a^{-3/2}\phi), \qquad r = a(1 - e\cos(a^{-3/2}\phi)),$$
$$a = -\frac{1}{2I}, \qquad e = (1 + 2IM^2)^{1/2}.$$

Now let us turn to the Kepler superintegrable system on U_+. It is a globally superintegrable system with non-compact invariant submanifolds as follows.

Let us put

$$S_1 = -K_1, \qquad S_2 = -K_2, \qquad S_3 = -M_{12}, \tag{7.6.32}$$
$$\{S_1, S_2\} = -S_3, \qquad \{S_2, S_3\} = S_1, \qquad \{S_3, S_1\} = S_2.$$

We have a fibred manifold

$$S : U_+ \to N \subset \mathfrak{g}^*, \tag{7.6.33}$$

which is the momentum mapping to the Lie coalgebra $\mathfrak{g}^* = so(2,1)^*$, endowed with the coordinates (x_i) such that integrals of motion S_i on $\mathfrak{g}^*$ read $S_i = x_i$. A base N of the fibred manifold (7.6.33) is an open submanifold of $\mathfrak{g}^*$ given by the coordinate condition $x_3 \neq 0$. It is a union of two contractible components defined by the conditions $x_3 > 0$ and $x_3 < 0$. The coinduced Lie–Poisson structure on N takes the form

$$w = x_2\partial^3 \wedge \partial^1 - x_3\partial^1 \wedge \partial^2 + x_1\partial^2 \wedge \partial^3. \tag{7.6.34}$$

The coadjoint action of $so(2,1)$ on N reads

$$\varepsilon_1 = -x_3\partial^2 - x_2\partial^3, \qquad \varepsilon_2 = x_1\partial^3 + x_3\partial^1, \qquad \varepsilon_3 = x_2\partial^1 - x_1\partial^2.$$

The orbits of this coadjoint action are given by the equation

$$x_1^2 + x_2^2 - x_3^2 = \text{const}.$$

They are the level surfaces of the Casimir function

$$C = x_1^2 + x_2^2 - x_3^2$$

and, consequently, the Casimir function

$$h = \frac{1}{2}(x_1^2 + x_2^2 - x_3^2)^{-1}. \tag{7.6.35}$$

A glance at the expression (7.6.20) shows that the pull-back S^*h of this Casimir function (7.6.35) onto U_+ is the Hamiltonian $\mathcal{H}$ (7.6.2) of the Kepler system on U_+.

As was mentioned above, the Hamiltonian vector field of S^*h is complete. Furthermore, it is known that invariant submanifolds of the superintegrable Kepler system on U_+ are diffeomorphic to $\mathbb{R}$. Therefore, the fibred manifold S (7.6.33) is a fibre bundle in accordance with Theorem 11.2.4. Moreover, this fibre bundle is trivial because N is a disjoint union of two contractible manifolds. Consequently, it follows from Theorem 7.4.3 that the Kepler superintegrable system on U_+ is globally superintegrable, i.e., it admits global generalized action-angle coordinates as follows.

The Poisson manifold N (7.6.33) can be endowed with the coordinates

$$(I, x_1, \lambda), \qquad I > 0, \qquad \lambda \neq 0,$$

defined by the equalities

$$I = \frac{1}{2}(x_1^2 + x_2^2 - x_3^2)^{-1},$$

$$x_2 = \left(\frac{1}{2I} - x_1^2\right)^{1/2} \cosh\lambda, \qquad x_3 = \left(\frac{1}{2I} - x_1^2\right)^{1/2} \sinh\lambda.$$

These coordinates are Darboux coordinates of the Lie–Poisson structure (7.6.34) on N, namely,

$$w = \frac{\partial}{\partial\lambda} \wedge \frac{\partial}{\partial x_1}. \tag{7.6.36}$$

Let ϑ_I be the Hamiltonian vector field of the Casimir function I (7.6.28). By virtue of Proposition 7.3.2, its flows are invariant submanifolds of the Kepler superintegrable system on U_+. Let τ be a parameter along the flows of this vector field, i.e.,

$$\vartheta_I = \frac{\partial}{\partial\tau}. \tag{7.6.37}$$

Then U_+ (7.6.33) is provided with the generalized action-angle coordinates (I, x_1, λ, τ) such that the Poisson bivector associated to the symplectic form Ω_T on U_+ reads

$$W = \frac{\partial}{\partial I} \wedge \frac{\partial}{\partial\tau} + \frac{\partial}{\partial\lambda} \wedge \frac{\partial}{\partial x_1}. \tag{7.6.38}$$

Accordingly, Hamiltonian vector fields of integrals of motion S_i (7.6.32) take the form

$$\begin{aligned}
\vartheta_1 &= -\frac{\partial}{\partial\lambda},\\
\vartheta_2 &= \frac{1}{4I^2}\left(\frac{1}{2I}-x_1^2\right)^{-1/2}\cosh\lambda\frac{\partial}{\partial\tau} + x_1\left(\frac{1}{2I}-x_1^2\right)^{-1/2}\cosh\lambda\frac{\partial}{\partial\lambda}\\
&+\left(\frac{1}{2I}-x_1^2\right)^{1/2}\sinh\lambda\frac{\partial}{\partial x_1},\\
\vartheta_3 &= \frac{1}{4I^2}\left(\frac{1}{2I}-x_1^2\right)^{-1/2}\sinh\lambda\frac{\partial}{\partial\tau} + x_1\left(\frac{1}{2I}-x_1^2\right)^{-1/2}\sinh\lambda\frac{\partial}{\partial\lambda}\\
&+\left(\frac{1}{2I}-x_1^2\right)^{1/2}\cosh\lambda\frac{\partial}{\partial x_1}.
\end{aligned}$$

Similarly to the angle coordinate α (7.6.30), the generalized angle coordinate τ (7.6.37) obeys the Hamilton equation

$$\frac{\partial q_i}{\partial\tau} = \frac{\partial\mathcal{H}}{\partial p_i}, \qquad \frac{\partial p_i}{\partial\tau} = -\frac{\partial\mathcal{H}}{\partial q_i}.$$

Therefore, it is the time $\tau = t$ given by the well-known expression

$$\begin{aligned}
&\tau = s - a^{3/2}e\sinh(a^{-3/2}s), \qquad r = a(e\cosh(a^{-3/2}s)-1),\\
&a = \frac{1}{2I}, \qquad e = (1+2IM^2)^{1/2}.
\end{aligned}$$

7.7 Non-autonomous integrable systems

The generalization of Liouville – Arnold and Mishchenko – Fomenko theorems to the case of non-compact invariant submanifolds (Theorems 7.3.1 and 7.3.3) enables one to analyze completely integrable and superintegrable non-autonomous Hamiltonian systems whose invariant submanifolds are necessarily non-compact [59; 65].

Let us consider a non-autonomous mechanical system on a configuration space $Q \to \mathbb{R}$ in Section 3.3. Its phase space is the vertical cotangent bundle $V^*Q \to Q$ of $Q \to \mathbb{R}$ endowed with the Poisson structure $\{,\}_V$ (3.3.7).A Hamiltonian of a non-autonomous mechanical system is a section h (3.3.13) of the one-dimensional fibre bundle (3.3.3) – (3.3.6):

$$\zeta : T^*Q \to V^*Q, \tag{7.7.1}$$

where T^*Q is the cotangent bundle of Q endowed with the canonical symplectic form Ω_T (3.3.1). The Hamiltonian h (3.3.13) yields the pull-back

Hamiltonian form H (3.3.14) on V^*Q and defines the Hamilton vector field γ_H (3.3.21) on V^*Q. A smooth real function F on V^*Q is an integral of motion of a Hamiltonian system (V^*Q, H) if its Lie derivative $\mathbf{L}_{\gamma_H} F$ (3.8.1) vanishes.

Definition 7.7.1. A non-autonomous Hamiltonian system (V^*Q, H) of $n = \dim Q - 1$ degrees of freedom is called *superintegrable* if it admits $n \leq k < 2n$ integrals of motion $\Phi_1, \ldots, \Phi_k$, obeying the following conditions.

(i) All the functions Φ_α are independent, i.e., the k-form $d\Phi_1 \wedge \cdots \wedge d\Phi_k$ nowhere vanishes on V^*Q. It follows that the map

$$\Phi : V^*Q \to N = (\Phi_1(V^*Q), \ldots, \Phi_k(V^*Q)) \subset \mathbb{R}^k \qquad (7.7.2)$$

is a fibred manifold over a connected open subset $N \subset \mathbb{R}^k$.

(ii) There exist smooth real functions $s_{\alpha\beta}$ on N such that

$$\{\Phi_\alpha, \Phi_\beta\}_V = s_{\alpha\beta} \circ \Phi, \qquad \alpha, \beta = 1, \ldots, k. \qquad (7.7.3)$$

(iii) The matrix function with the entries $s_{\alpha\beta}$ (7.7.3) is of constant corank $m = 2n - k$ at all points of N.

In order to describe this non-autonomous superintegrable Hamiltonian system, we use the fact that there exists an equivalent autonomous Hamiltonian system $(T^*Q, \mathcal{H}^*)$ of $n + 1$ degrees of freedom on a symplectic manifold (T^*Q, Ω_T) whose Hamiltonian is the function $\mathcal{H}^*$ (3.4.1) (Theorem 3.4.1), and that this Hamiltonian system is superintegrable (Theorem 7.7.4). Our goal is the following.

Theorem 7.7.1. *Let Hamiltonian vector fields of the functions Φ_α be complete, and let fibres of the fibred manifold Φ (7.7.2) be connected and mutually diffeomorphic. Then there exists an open neighborhood U_M of a fibre M of Φ (7.7.2) which is a trivial principal bundle with the structure group*

$$\mathbb{R}^{1+m-r} \times T^r \qquad (7.7.4)$$

whose bundle coordinates are the generalized action-angle coordinates

$$(p_A, q^A, I_\lambda, t, y^\lambda), \qquad A = 1, \ldots, k - n, \qquad \lambda = 1, \ldots, m, \qquad (7.7.5)$$

such that:

(i) (t, y^λ) are coordinates on the toroidal cylinder (7.7.4),

(ii) the Poisson bracket $\{,\}_V$ on U_M reads

$$\{f, g\}_V = \partial^A f \partial_A g - \partial^A g \partial_A f + \partial^\lambda f \partial_\lambda g - \partial^\lambda g \partial_\lambda f,$$

(iii) a Hamiltonian $\mathcal{H}$ depends only on the action coordinates I_λ,

(iv) the integrals of motion $\Phi_1, \ldots \Phi_k$ are independent of coordinates (t, y^λ).

Let us start with the case $k = n$ of a completely integrable non-autonomous Hamiltonian system (Theorem 7.7.3).

Definition 7.7.2. A non-autonomous Hamiltonian system (V^*Q, H) of n degrees of freedom is said to be completely integrable if it admits n independent integrals of motion $F_1, \ldots, F_n$ which are in involution with respect to the Poisson bracket $\{,\}_V$ (3.3.7).

By virtue of the relations (3.3.10) and (3.8.2), the vector fields

$$(\gamma_H, \vartheta_{F_1}, \ldots, \vartheta_{F_n}), \qquad \vartheta_{F_\alpha} = \partial^i F_\alpha \partial_i - \partial_i F_\alpha \partial^i, \tag{7.7.6}$$

mutually commute and, therefore, they span an $(n+1)$-dimensional involutive distribution $\mathcal{V}$ on V^*Q. Let G be the group of local diffeomorphisms of V^*Q generated by the flows of vector fields (7.7.6). Maximal integral manifolds of $\mathcal{V}$ are the orbits of G and invariant submanifolds of vector fields (7.7.6). They yield a foliation $\mathcal{F}$ of V^*Q.

Let (V^*Q, H) be a non-autonomous Hamiltonian system and $(T^*Q, \mathcal{H}^*)$ an equivalent autonomous Hamiltonian system on T^*Q. An immediate consequence of the relations (3.3.8) and (3.4.6) is the following.

Theorem 7.7.2. *Given a non-autonomous completely integrable Hamiltonian system*

$$(\gamma_H, F_1, \ldots, F_n) \tag{7.7.7}$$

of n degrees of freedom on V^*Q, the associated autonomous Hamiltonian system

$$(\mathcal{H}^*, \zeta^* F_1, \ldots, \zeta^* F_n) \tag{7.7.8}$$

of $n+1$ degrees of freedom on T^*Q is completely integrable.

The Hamiltonian vector fields

$$(u_{\mathcal{H}^*}, u_{\zeta^* F_1}, \ldots, u_{\zeta^* F_m}), \qquad u_{\zeta^* F_\alpha} = \partial^i F_\alpha \partial_i - \partial_i F_\alpha \partial^i, \tag{7.7.9}$$

of the autonomous integrals of motion (7.7.8) span an $(n+1)$-dimensional involutive distribution $\mathcal{V}_T$ on T^*Q such that

$$T\zeta(\mathcal{V}_T) = \mathcal{V}, \qquad Th(\mathcal{V}) = \mathcal{V}_T|_{h(V^*Q) = I_0 = 0}, \tag{7.7.10}$$

where

$$Th : TV^*Q \ni (t, q^i, p_i, \dot t, \dot q^i, \dot p_i)$$
$$\to (t, q^i, p_i, I_0 = 0, \dot t, \dot q^i, \dot p_i, \dot I_0 = 0) \in TT^*Q.$$

It follows that, if M is an invariant submanifold of the non-autonomous completely integrable Hamiltonian system (7.7.7), then $h(M)$ is an invariant submanifold of the autonomous completely integrable Hamiltonian system (7.7.8).

In order do introduce generalized action-angle coordinates around an invariant submanifold M of the non-autonomous completely integrable Hamiltonian system (7.7.7), let us suppose that the vector fields (7.7.6) on M are complete. It follows that M is a locally affine manifold diffeomorphic to a toroidal cylinder

$$\mathbb{R}^{1+n-r} \times T^r. \tag{7.7.11}$$

Moreover, let assume that there exists an open neighborhood U_M of M such that the foliation $\mathcal{F}$ of U_M is a fibred manifold $\phi : U_M \to N$ over a domain $N \subset \mathbb{R}^n$ whose fibres are mutually diffeomorphic.

Because the morphism Th (7.7.10) is a bundle isomorphism, the Hamiltonian vector fields (7.7.9) on the invariant submanifold $h(M)$ of the autonomous completely integrable Hamiltonian system are complete. Since the affine bundle ζ (7.7.1) is trivial, the open neighborhood $\zeta^{-1}(U_M)$ of the invariant submanifold $h(M)$ is a fibred manifold

$$\widetilde{\phi} : \zeta^{-1}(U_M) = \mathbb{R} \times U_M \overset{(\mathrm{Id}\,\mathbb{R},\phi)}{\longrightarrow} \mathbb{R} \times N = N'$$

over a domain $N' \subset \mathbb{R}^{n+1}$ whose fibres are diffeomorphic to the toroidal cylinder (7.7.11). In accordance with Theorem 7.3.3, the open neighborhood $\zeta^{-1}(U_M)$ of $h(M)$ is a trivial principal bundle

$$\zeta^{-1}(U_M) = N' \times (\mathbb{R}^{1+n-r} \times T^r) \to N' \tag{7.7.12}$$

with the structure group (7.7.11) whose bundle coordinates are the generalized action-angle coordinates

$$(I_0, I_1, \ldots, I_n, t, z^1, \ldots, z^n) \tag{7.7.13}$$

such that:

(i) (t, z^a) are coordinates on the toroidal cylinder (7.7.11),

(ii) the symplectic form Ω_T on $\zeta^{-1}(U)$ reads

$$\Omega_T = dI_0 \wedge dt + dI_a \wedge dz^a,$$

(iii) $\mathcal{H}^* = I_0$,

(iv) the integrals of motion $\zeta^* F_1, \ldots, \zeta^* F_n$ depend only on the action coordinates $I_1, \ldots, I_n$.

Provided with the coordinates (7.7.13),

$$\zeta^{-1}(U_M) = U_M \times \mathbb{R}$$

is a trivial bundle possessing the fibre coordinate I_0 (3.3.4). Consequently, the non-autonomous open neighborhood U_M of an invariant submanifold M of the completely integrable Hamiltonian system (7.7.6) is diffeomorphic to the Poisson annulus

$$U_M = N \times (\mathbb{R}^{1+n-r} \times T^r) \tag{7.7.14}$$

endowed with the *generalized action-angle coordinates*

$$(I_1, \ldots, I_n, t, z^1, \ldots, z^n) \tag{7.7.15}$$

such that:

(i) the Poisson structure (3.3.7) on U_M takes the form

$$\{f, g\}_V = \partial^a f \partial_a g - \partial^a g \partial_a f,$$

(ii) the Hamiltonian (3.3.13) reads $\mathcal{H} = 0$,

(iii) the integrals of motion $F_1, \ldots, F_n$ depend only on the action coordinates $I_1, \ldots, I_n$.

The Hamilton equation (3.3.22) – (3.3.23) relative to the generalized action-angle coordinates (7.7.15) takes the form

$$z^a_t = 0, \qquad I_{ta} = 0.$$

It follows that the generalized action-angle coordinates (7.7.15) are the initial date coordinates.

Note that the generalized action-angle coordinates (7.7.15) by no means are unique. Given a smooth function $\mathcal{H}'$ on $\mathbb{R}^n$, one can provide $\zeta^{-1}(U_M)$ with the generalized action-angle coordinates

$$t, \qquad z'^a = z^a - t\partial^a \mathcal{H}', \qquad I'_0 = I_0 + \mathcal{H}'(I_b), \qquad I'_a = I_a. \tag{7.7.16}$$

With respect to these coordinates, a Hamiltonian of the autonomous Hamiltonian system on $\zeta^{-1}(U_M)$ reads $\mathcal{H}'^* = I'_0 - \mathcal{H}'$. A Hamiltonian of the non-autonomous Hamiltonian system on U endowed with the generalized action-angle coordinates (I_a, t, z'^a) is $\mathcal{H}'$.

Thus, the following has been proved.

Theorem 7.7.3. *Let $(\gamma_H, F_1, \ldots, F_n)$ be a non-autonomous completely integrable Hamiltonian system. Let M be its invariant submanifold such that the vector fields (7.7.6) on M are complete and that there exists an open neighborhood U_M of M which is a fibred manifold in mutually diffeomorphic invariant submanifolds. Then U_M is diffeomorphic to the Poisson annulus (7.7.14), and it can be provided with the generalized action-angle coordinates (7.7.15) such that the integrals of motion $(F_1, \ldots, F_n)$ and the Hamiltonian $\mathcal{H}$ depend only on the action coordinates $I_1, \ldots, I_n$.*

Let now $(\gamma_H, \Phi_1, \ldots, \Phi_k)$ be a non-autonomous superintegrable Hamiltonian system in accordance with Definition 7.7.1. The associated autonomous Hamiltonian system on T^*Q possesses $k+1$ integrals of motion

$$(\mathcal{H}^*, \zeta^*\Phi_1, \ldots, \zeta^*\Phi_k) \tag{7.7.17}$$

with the following properties.

(i) The functions (7.7.17) are mutually independent, and the map

$$\begin{aligned} \widetilde{\Phi} : T^*Q &\to (\mathcal{H}^*(T^*Q), \zeta^*\Phi_1(T^*Q), \ldots, \zeta^*\Phi_k(T^*Q)) \\ &= (I_0, \Phi_1(V^*Q), \ldots, \Phi_k(V^*Q)) = \mathbb{R} \times N = N' \end{aligned} \tag{7.7.18}$$

is a fibred manifold.

(ii) The functions (7.7.17) obey the relations

$$\{\zeta^*\Phi_\alpha, \zeta^*\Phi_\beta\} = s_{\alpha\beta} \circ \zeta^*\Phi, \qquad \{\mathcal{H}^*, \zeta^*\Phi_\alpha\} = s_{0\alpha} = 0$$

so that the matrix function with the entries $(s_{0\alpha}, s_{\alpha\beta})$ on N' is of constant corank $2n+1-k$.

Referring to Definition 7.3.1 of an autonomous superintegrable system, we come to the following.

Theorem 7.7.4. *Given a non-autonomous superintegrable Hamiltonian system (γ_H, Φ_α) on V^*Q, the associated autonomous Hamiltonian system (7.7.17) on T^*Q is superintegrable.*

There is the commutative diagram

$$\begin{array}{ccc} T^*Q & \xrightarrow{\zeta} & V^*Q \\ {\scriptstyle\widetilde{\Phi}}\big\downarrow & & \big\downarrow{\scriptstyle\Phi} \\ N' & \xrightarrow{\xi} & N \end{array}$$

where ζ (7.7.1) and

$$\xi : N' = \mathbb{R} \times N \to N$$

are trivial bundles. It follows that the fibred manifold (7.7.18) is the pull-back $\widetilde{\Phi} = \xi^*\Phi$ of the fibred manifold Φ (7.7.2) onto N'.

Let the conditions of Theorem 7.3.1 hold. If the Hamiltonian vector fields

$$(\gamma_H, \vartheta_{\Phi_1}, \ldots, \vartheta_{\Phi_k}), \qquad \vartheta_{\Phi_\alpha} = \partial^i\Phi_\alpha\partial_i - \partial_i\Phi_\alpha\partial^i,$$

of integrals of motion Φ_α on V^*Q are complete, the Hamiltonian vector fields

$$(u_{\mathcal{H}^*}, u_{\zeta^*\Phi_1}, \ldots, u_{\zeta^*\Phi_k}), \qquad u_{\zeta^*\Phi_\alpha} = \partial^i\Phi_\alpha\partial_i - \partial_i\Phi_\alpha\partial^i,$$

on T^*Q are complete. If fibres of the fibred manifold Φ (7.7.2) are connected and mutually diffeomorphic, the fibres of the fibred manifold $\widetilde{\Phi}$ (7.7.18) also are well.

Let M be a fibre of Φ (7.7.2) and $h(M)$ the corresponding fibre of $\widetilde{\Phi}$ (7.7.18). In accordance Theorem 7.3.1, there exists an open neighborhood U' of $h(M)$ which is a trivial principal bundle with the structure group (7.7.4) whose bundle coordinates are the generalized action-angle coordinates

$$(I_0, I_\lambda, t, y^\lambda, p_A, q^A), \qquad A = 1, \ldots, n-m, \qquad \lambda = 1, \ldots, k, \tag{7.7.19}$$

such that:

(i) (t, y^λ) are coordinates on the toroidal cylinder (7.7.4),

(ii) the symplectic form Ω_T on U' reads

$$\Omega_T = dI_0 \wedge dt + dI_\alpha \wedge dy^\alpha + dp_A \wedge dq^A,$$

(iii) the action coordinates (I_0, I_α) are expressed in the values of the Casimir functions $C_0 = I_0$, C_α of the coinduced Poisson structure

$$w = \partial^A \wedge \partial_A$$

on N',

(iv) a homogeneous Hamiltonian $\mathcal{H}^*$ depends on the action coordinates, namely, $\mathcal{H}^* = I_0$,

(iv) the integrals of motion $\zeta^*\Phi_1, \ldots \zeta^*\Phi_k$ are independent of the coordinates (t, y^λ).

Provided with the generalized action-angle coordinates (7.7.19), the above mentioned neighborhood U' is a trivial bundle $U' = \mathbb{R} \times U_M$ where $U_M = \zeta(U')$ is an open neighborhood of the fibre M of the fibre bundle Φ (7.7.2). As a result, we come to Theorem 7.7.1.

7.8 Quantization of superintegrable systems

In accordance with Theorem 7.3.1, any superintegrable Hamiltonian system (7.3.3) on a symplectic manifold (Z, Ω) restricted to some open neighborhood U_M (7.3.6) of its invariant submanifold M is characterized by generalized action-angle coordinates $(I_\lambda, p_A, q^A, y^\lambda)$, $\lambda = 1, \ldots, m$, $A = 1, \ldots, n-m$. They are canonical for the symplectic form Ω (7.3.7) on U_M. Then one can treat the coordinates (I_λ, p_A) as n independent functions in involution on a symplectic annulus (U_M, Ω) which constitute a completely integrable system in accordance with Definition 7.3.2. Strictly speaking, its quantization fails to be a quantization of the original superintegrable system (7.3.3) because $F_i(I_\lambda, q^A, p_A)$ are not linear functions and, consequently, the algebra (7.3.3) and the algebra

$$\{I_\lambda, p_A\} = \{I_\lambda, q^A\} = 0, \qquad \{p_A, q^B\} = \delta^B_A \tag{7.8.1}$$

are not isomorphic in general. However, one can obtain the Hamilton operator $\widehat{\mathcal{H}}$ and the Casimir operators $\widehat{C}_\lambda$ of an original superintegrable system and their spectra.

There are different approaches to quantization of completely integrable systems [69; 80]. It should be emphasized that action-angle coordinates need not be globally defined on a phase space of a completely integrable system, but form an algebra of the Poisson canonical commutation relations (7.8.1) on an open neighborhood U_M of an invariant submanifold M. Therefore, quantization of a completely integrable system with respect to the action-angle variables is a quantization of the Poisson algebra $C^\infty(U_M)$ of real smooth functions on U_M. Since there is no morphism $C^\infty(U_M) \to C^\infty(Z)$, this quantization is not equivalent to quantization of an original completely integrable system on Z and, from on a physical level, is interpreted as quantization around an invariant submanifold M. A key point is that, since U_M is not a contractible manifold, the geometric quantization technique should be called into play in order to quantize a completely integrable system around its invariant submanifold. A peculiarity of the geometric quantization procedure is that it remains equivalent under symplectic isomorphisms, but essentially depends on the choice of a polarization [11; 131].

Geometric quantization of completely integrable systems has been studied at first with respect to the polarization spanned by Hamiltonian vector fields of integrals of motion [121]. For example, the well-known Simms quantization of a harmonic oscillator is of this type [38]. However, one meets a problem that the associated quantum algebra contains affine functions of angle coordinates on a torus which are ill defined. As a consequence, elements of the carrier space of this quantization fail to be smooth, but are tempered distributions. We have developed a different variant of geometric quantization of completely integrable systems [43; 60; 65]. Since a Hamiltonian of a completely integrable system depends only on action variables, it seems natural to provide the Schrödinger representation of action variables by first order differential operators on functions of angle coordinates. For this purpose, one should choose the angle polarization of a symplectic manifold spanned by almost-Hamiltonian vector fields of angle variables.

Given an open neighborhood U_M (7.3.6) in Theorem 7.3.1, us consider its fibration

$$U_M = N_M \times \mathbb{R}^{m-r} \times T^r \to V \times \mathbb{R}^{m-r} \times T^r = \mathcal{M}, \qquad (7.8.2)$$

$$(I_\lambda, p_A, q^A, y^\lambda) \to (q^A, y^\lambda). \qquad (7.8.3)$$

Then one can think of a symplectic annulus (U_M, Ω) as being an open subbundle of the cotangent bundle $T^*\mathcal{M}$ endowed with the canonical symplectic form $\Omega_T = \Omega$ (7.3.7). This fact enables us to provide quantization of any superintegrable system on a neighborhood of its invariant submanifold as geometric quantization of the cotangent bundle $T^*\mathcal{M}$ over the toroidal cylinder $\mathcal{M}$ (7.8.2) [66]. Note that this quantization however differs from that in Section 5.2 because $\mathcal{M}$ (7.8.2) is not simply connected in general.

Let (q^A, r^a, α^i) be coordinates on the toroidal cylinder $\mathcal{M}$ (7.8.2), where $(\alpha^1, \ldots, \alpha^r)$ are angle coordinates on a torus T^r, and let (p_A, I_a, I_i) be the corresponding action coordinates (i.e., the holonomic fibre coordinates on $T^*\mathcal{M}$). Since the symplectic form Ω (7.3.7) is exact, the quantum bundle is defined as a trivial complex line bundle C over $T^*\mathcal{M}$. Let its trivialization hold fixed. Any other trivialization leads to an equivalent quantization of $T^*\mathcal{M}$. Given the associated fibre coordinate $c \in \mathbb{C}$ on $C \to T^*\mathcal{M}$, one can treat its sections as smooth complex functions on $T^*\mathcal{M}$.

The Kostant–Souriau prequantization formula (5.1.11) associates to every smooth real function f on $T^*\mathcal{M}$ the first order differential operator

$$\widehat{f} = -i\vartheta_f \rfloor D^A - fc\partial_c$$

on sections of $C \to T^*\mathcal{M}$, where ϑ_f is the Hamiltonian vector field of f and D^A is the covariant differential (5.1.3) with respect to an admissible $U(1)$-principal connection A on C. This connection preserves the Hermitian fibre metric $g(c, c') = c\overline{c}'$ in C, and its curvature obeys the prequantization condition (5.1.9). Such a connection reads

$$A = A_0 - ic(p_A dq^A + I_a dr^a + I_i d\alpha^i) \otimes \partial_c, \qquad (7.8.4)$$

where A_0 is a flat $U(1)$-principal connection on $C \to T^*\mathcal{M}$.

The classes of gauge non-conjugate flat principal connections on C are indexed by the set $\mathbb{R}^r/\mathbb{Z}^r$ of homomorphisms of the de Rham cohomology group

$$H^1_{\rm DR}(T^*\mathcal{M}) = H^1_{\rm DR}(\mathcal{M}) = H^1_{\rm DR}(T^r) = \mathbb{R}^r$$

of $T^*\mathcal{M}$ to $U(1)$. We choose their representatives of the form

$$A_0[(\lambda_i)] = dp_A \otimes \partial^A + dI_a \otimes \partial^a + dI_j \otimes \partial^j + dq^A \otimes \partial_A + dr^a \otimes \partial_a$$
$$+ d\alpha^j \otimes (\partial_j - i\lambda_j c\partial_c), \qquad \lambda_i \in [0, 1).$$

Accordingly, the relevant connection (7.8.4) on C reads

$$A[(\lambda_i)] = dp_A \otimes \partial^A + dI_a \otimes \partial^a + dI_j \otimes \partial^j \qquad (7.8.5)$$
$$+ dq^A \otimes (\partial_A - ip_A c\partial_c) + dr^a \otimes (\partial_a - iI_a c\partial_c)$$
$$+ d\alpha^j \otimes (\partial_j - i(I_j + \lambda_j)c\partial_c).$$

For the sake of simplicity, we further assume that the numbers λ_i in the expression (7.8.5) belong to $\mathbb{R}$, but bear in mind that connections $A[(\lambda_i)]$ and $A[(\lambda'_i)]$ with $\lambda_i - \lambda'_i \in \mathbb{Z}$ are gauge conjugate.

Let us choose the above mentioned *angle polarization* coinciding with the vertical polarization $VT^*\mathcal{M}$. Then the corresponding quantum algebra $\mathcal{A}$ of $T^*\mathcal{M}$ consists of affine functions

$$f = a^A(q^B, r^b, \alpha^j)p_A + a^b(q^B, r^a, \alpha^j)I_b + a^i(q^B, r^a, \alpha^j)I_i + b(q^B, r^a, \alpha^j)$$

in action coordinates (p_A, I_a, I_i). Given a connection (7.8.5), the corresponding Schrödinger operators (5.2.10) read

$$\widehat{f} = \left(-ia^A\partial_A - \frac{i}{2}\partial_A a^A\right) + \left(-ia^b\partial_i - \frac{i}{2}\partial_b a^b\right) \qquad (7.8.6)$$
$$+ \left(-ia^i\partial_i - \frac{i}{2}\partial_i a^i + a^i\lambda_i\right) - b.$$

They are Hermitian operators in the pre-Hilbert space $\mathfrak{E}_\mathcal{M}$ of complex half-densities ψ of compact support on $\mathcal{M}$ endowed with the Hermitian form

$$\langle\psi|\psi'\rangle = \int_\mathcal{M} \psi\overline{\psi}' d^{n-m}q d^{m-r}r d^r\alpha.$$

Note that, being a complex function on a toroidal cylinder $\mathbb{R}^{m-r}\times T^r$, any half-density $\psi \in \mathfrak{E}_\mathcal{M}$ is expanded into the series

$$\psi = \sum_{(n_\mu)} \phi(q^B, r^a)_{(n_j)} \exp[in_j\alpha^j], \qquad (n_j) = (n_1, \ldots, n_r) \in \mathbb{Z}^r, \qquad (7.8.7)$$

where $\phi(q^B, r^a)_{(n_\mu)}$ are half-densities of compact support on $\mathbb{R}^{n-r}$. In particular, the action operators (7.8.6) read

$$\widehat{p}_A = -i\partial_A, \qquad \widehat{I}_a = -i\partial_a, \qquad \widehat{I}_j = -i\partial_j + \lambda_j. \qquad (7.8.8)$$

It should be emphasized that

$$\widehat{a}\widehat{p}_A \neq \widehat{ap_A}, \qquad \widehat{a}\widehat{I}_b \neq \widehat{aI_b}, \qquad \widehat{a}\widehat{I}_j \neq \widehat{aI_j}, \qquad a \in C^\infty(\mathcal{M}). \qquad (7.8.9)$$

The operators (7.8.6) provide a desired quantization of a superintegrable Hamiltonian system written with respect to the action-angle coordinates. They satisfy Dirac's condition (0.0.4). However, both a Hamiltonian $\mathcal{H}$ and original integrals of motion F_i do not belong to the quantum algebra $\mathcal{A}$, unless they are affine functions in the action coordinates (p_A, I_a, I_i). In some particular cases, integrals of motion F_i can be represented by differential operators, but this representation fails to be unique because of inequalities (7.8.9), and Dirac's condition need not be satisfied. At the same time, both the Casimir functions C_λ and a Hamiltonian $\mathcal{H}$ (Proposition 7.5.2) depend only on action variables I_a, I_i. If they are polynomial in I_a, one can associate to them the operators $\widehat{C}_\lambda = C_\lambda(\widehat{I}_a, \widehat{I}_j)$, $\widehat{\mathcal{H}} = \mathcal{H}(\widehat{I}_a, \widehat{I}_j)$, acting in the space $\mathfrak{E}_{\mathcal{M}}$ by the law

$$\widehat{\mathcal{H}}\psi = \sum_{(n_j)} \mathcal{H}(\widehat{I}_a, n_j + \lambda_j)\phi(q^A, r^a)_{(n_j)} \exp[in_j\alpha^j],$$

$$\widehat{C}_\lambda\psi = \sum_{(n_j)} C_\lambda(\widehat{I}_a, n_j + \lambda_j)\phi(q^A, r^a)_{(n_j)} \exp[in_j\alpha^j].$$

Example 7.8.1. Let us consider a superintegrable system with the Lie algebra $\mathfrak{g} = so(3)$ of integrals of motion $\{F_1, F_2, F_3\}$ on a four-dimensional symplectic manifold (Z, Ω), namely,

$$\{F_1, F_2\} = F_3, \qquad \{F_2, F_3\} = F_1, \qquad \{F_3, F_1\} = F_2$$

(see Section 7.6). Since it is compact, an invariant submanifold of a superintegrable system in question is a circle $M = S^1$. We have a fibred manifold $F : Z \to N$ (7.6.22) onto an open subset $N \subset \mathfrak{g}^*$ of the Lie coalgebra $\mathfrak{g}^*$. This fibred manifold is a fibre bundle since its fibres are compact (Theorem 11.2.4). Its base N is endowed with the coordinates (x_1, x_2, x_3) such that integrals of motion $\{F_1, F_2, F_3\}$ on Z read

$$F_1 = x_1, \qquad F_2 = x_2, \qquad F_3 = x_3.$$

The coinduced Poisson structure on N is the Lie–Poisson structure (7.6.23). The coadjoint action of $so(3)$ is given by the expression (7.6.24). An orbit of the coadjoint action of dimension 2 is given by the equality (7.6.25). Let M be an invariant submanifold such that the point $F(M) \in \mathfrak{g}^*$ belongs to the orbit (7.6.25). Let us consider an open fibred neighborhood $U_M = N_M \times S^1$ of M which is a trivial bundle over an open contractible neighborhood N_M of $F(M)$ endowed with the coordinates (I, x_1, γ) defined by the equalities (7.6.27). Here, I is the Casimir function (7.6.28) on $\mathfrak{g}^*$. These coordinates are the Darboux coordinates of the Lie–Poisson structure (7.6.29) on N_M. Let ϑ_I be the Hamiltonian vector field of the Casimir function I (7.6.28). Its flows are invariant submanifolds. Let α be a parameter (7.6.30) along the flows of this vector field. Then U_M is provided with the action-angle coordinates (I, x_1, γ, α) such that the Poisson bivector on U_M takes the form (7.6.31). The action-angle variables $\{I, H_1 = x_1, \gamma\}$ constitute a superintegrable system

$$\{I, F_1\} = 0, \qquad \{I, \gamma\} = 0, \qquad \{F_1, \gamma\} = 1, \tag{7.8.10}$$

on U_M. It is related to the original one by the transformations

$$I = -\frac{1}{2}(F_1^2 + F_2^2 + F_3^2)^{1/2},$$

$$F_2 = \left(-\frac{1}{2I} - F_1^2\right)^{1/2} \sin\gamma, \qquad F_3 = \left(-\frac{1}{2I} - H_1^2\right)^{1/2} \cos\gamma.$$

Its Hamiltonian is expressed only in the action variable I. Let us quantize the superintegrable system (7.8.10). We obtain the algebra of operators

$$\widehat{f} = a\left(-i\frac{\partial}{\partial\alpha} - \lambda\right) - ib\frac{\partial}{\partial\gamma} - \frac{i}{2}\left(\frac{\partial a}{\partial\alpha} + \frac{\partial b}{\partial\gamma}\right) - c,$$

where a, b, c are smooth functions of angle coordinates (γ, α) on the cylinder $\mathbb{R} \times S^1$. In particular, the action operators read

$$\widehat{I} = -i\frac{\partial}{\partial\alpha} - \lambda, \qquad \widehat{F}_1 = -i\frac{\partial}{\partial\gamma}.$$

These operators act in the space of smooth complex functions

$$\psi(\gamma, \alpha) = \sum_k \phi(\gamma)_k \exp[ik\alpha]$$

on T^2. A Hamiltonian $\mathcal{H}(I)$ of a classical superintegrable system also can be represented by the operator

$$\widehat{\mathcal{H}}(I)\psi = \sum_k \mathcal{H}(I - \lambda)\phi(\gamma)_k \exp[ik\alpha]$$

on this space.

Chapter 8

Jacobi fields

Given a mechanical system on a configuration space $Q \to \mathbb{R}$, its extension onto the vertical tangent bundle $VQ \to \mathbb{R}$ of $Q \to \mathbb{R}$ describes the Jacobi fields of the Lagrange and Hamilton equations [54; 65; 106].

In particular, we show that Jacobi fields of a completely integrable Hamiltonian system of m degrees of freedom make up an extended completely integrable system of $2m$ degrees of freedom, where m additional integrals of motion characterize a relative motion [61].

In this Chapter, we follow the compact notation (11.2.30).

8.1 The vertical extension of Lagrangian mechanics

Given Lagrangian mechanics on a configuration bundle $Q \to \mathbb{R}$, let us consider its extension on a configuration bundle $VT \to \mathbb{R}$ equipped with the holonomic coordinates $(t, q^i, \dot{q}^i)$. [65; 106].

Remark 8.1.1. Let $Y \to X$ be a fibre bundle and VY and V^*Y its vertical tangent and cotangent bundles coordinated by $(x^\lambda, y^i, v^i = \dot{y}^i)$ and $(x^\lambda, y^i, p_i = \dot{y}_i)$, respectively. There is the canonical isomorphism (11.2.23):

$$VV^*Y \underset{VY}{=} V^*VY, \quad p_i \longleftrightarrow \dot{v}_i, \quad \dot{p}_i \longleftrightarrow \dot{y}_i. \tag{8.1.1}$$

Accordingly, any exterior form ϕ on Y gives rise to the exterior form

$$\begin{aligned} \phi_V &= \partial_V \phi = \dot{y}^i \partial_i \phi, \\ \partial_V dx^\lambda &= 0, \qquad \partial_V dy^i = d\dot{y}^i, \end{aligned} \tag{8.1.2}$$

called the *vertical extension* of ϕ onto VY so that

$$(\phi \wedge \sigma)_V = \phi_V \wedge \sigma + \phi \wedge \sigma_V, \qquad d\phi_V = (d\phi)_V$$

[106; 109]. There also is the canonical isomorphism (11.3.9):

$$J^1VY \underset{J^1Y}{=} VJ^1Y, \qquad \dot y^i_\lambda = (\dot y^i)_\lambda. \tag{8.1.3}$$

As a consequence, given a connection $\Gamma : Y \to J^1Y$ on a fibre bundle $Y \to X$, the vertical tangent map

$$V\Gamma : VY \to J^1VY = VJ^1Y$$

to Γ defines the connection

$$V\Gamma = dx^\lambda \otimes (\partial_\lambda + \Gamma^i_\lambda \partial_i + \partial_j \Gamma^i_\lambda \dot y^j \dot\partial_i) \tag{8.1.4}$$

on the vertical tangent bundle $VY \to X$. It is called the *vertical connection* to Γ. Accordingly, we have the connection

$$V^*\Gamma = dx^\lambda \otimes (\partial_\lambda + \Gamma^i_\lambda \partial_i - \partial_i \Gamma^j_\lambda \dot y_j \dot\partial^i) \tag{8.1.5}$$

on the vertical cotangent bundle $V^*Y \to X$. It is called the *covertical connection* to Γ.

Given an extended configuration space VQ, the corresponding velocity space is the jet manifold J^1VQ of $VQ \to \mathbb{R}$. Due to the canonical isomorphism (8.1.3), this velocity space

$$J^1VQ \underset{J^1Q}{=} VJ^1Q \tag{8.1.6}$$

is provided with the coordinates $(t, q^i, q^i_t, \dot q^i, \dot q^i_t)$. First order Lagrangian formalism on the velocity space (8.1.6) can be developed as the vertical extension of Lagrangian formalism on J^1Q as follows.

Let L be a Lagrangian (2.1.22) on J^1Q. Its *vertical extension* (8.1.2) onto VJ^1Q is

$$L_V = \partial_V L = \partial_V \mathcal{L} dt = (\dot q^i \partial_i + \dot q^i_t \partial^t_i)\mathcal{L} dt = \mathcal{L}_V dt. \tag{8.1.7}$$

The corresponding Lagrange equation read

$$\dot\delta_i \mathcal{L}_V = (\dot\partial_i - d_t \dot\partial^t_i)\mathcal{L} = \delta_i \mathcal{L} = 0, \tag{8.1.8}$$

$$\delta_i \mathcal{L}_V = \partial_V \delta_i \mathcal{L} = 0, \tag{8.1.9}$$

$$\partial_V = \dot q^i \partial_i + \dot q^i_t \partial^t_i + \dot q^i_{tt} \partial^{tt}_i.$$

The equation (8.1.8) is exactly the Lagrange equation for an original Lagrangian L, while the equation (8.1.9) is the well-known *variation equation* of the equation (8.1.8) [32; 106]. Substituting a solution s^i of the Lagrange equation (8.1.8) into (8.1.9), one obtains a linear differential equation whose solutions $\widetilde{s}^i$ are *Jacobi fields* of a solution s. Indeed, if $Q \to \mathbb{R}$ is a vector bundle, there is the canonical splitting $VQ = Q \oplus Q$ over $\mathbb{R}$, and $s + \overline{s}$ is a solution of the Lagrange equation (8.1.8) modulo the terms of order exceeding 1 in $\overline{s}$.

Let us consider the regular quadratic Lagrangian (2.3.17) in Example 2.3.1. The corresponding Lagrange equation takes the form (2.3.18). By virtue of Corollary 1.5.1, the second order dynamic equation (2.3.18) is equivalent to the non-relativistic geodesic equation (1.5.9) on the tangent bundle TQ with respect to the symmetric linear connection $\widetilde{K}$ (1.5.10) on $TQ \to Q$ possessing the components

$$K_\lambda{}^0{}_\nu = 0, \qquad K_\lambda{}^i{}_\nu = -(m^{-1})^{ik}\{_{\lambda k\nu}\}. \tag{8.1.10}$$

Then one can write the well-known equation for *Jacobi fields* u^λ along the geodesics of this connection [93]. Since the curvature R (11.4.22) of the connection $\widetilde{K}$ (8.1.10) has the temporal component

$$R_{\lambda\mu}{}^0{}_\beta = 0, \tag{8.1.11}$$

this equation reads

$$\dot q^\beta \dot q^\mu(\nabla_\beta(\nabla_\mu u^\alpha) - R_{\lambda\mu}{}^\alpha{}_\beta u^\lambda) = 0, \qquad \nabla_\beta \dot q^\alpha = 0, \tag{8.1.12}$$

where ∇_μ denote the covariant derivatives relative to the connection $\widetilde{K}$. Due to the equality (8.1.11), the equation (8.1.12) for the temporal component u^0 of a Jacobi field takes the form

$$\dot q^\beta \dot q^\mu(\partial_\mu\partial_\beta u^0 + K_\mu{}^\gamma{}_\beta\partial_\gamma u^0) = 0.$$

We chose its solution $u^0 = 0$ because all non-relativistic geodesics obey the constraint $\dot q^0 = 0$. Then the equation (8.1.12) coincides with the Lagrange equation (8.1.9) for the vertical extension L_V (8.1.7) of the original quadratic Lagrangian L (2.3.17) [106; 107].

8.2 The vertical extension of Hamiltonian mechanics

A phase space of a mechanical system on the extended configuration bundle VQ is the vertical cotangent bundle V^*VQ of $VQ \to \mathbb{R}$. Due to the canonical isomorphism (8.1.1), this phase space

$$V^*VQ \underset{VQ}{=} VV^*Q \tag{8.2.1}$$

is coordinated by $(t, q^i, p_i, \dot q^i = v^i, \dot p_i)$. Hamiltonian formalism on the phase space (8.2.1) can be developed as the vertical extension of Hamiltonian formalism on V^*Q, where the canonical conjugate pairs are $(q^i, \dot{p}_i)$ and $(\dot{q}^i, p_i)$.

Note that the vertical extension L_V (8.1.7) of any Lagrangian L on J^1Q yields the vertical tangent map

$$\widehat{L}_V = V\widehat{L} : VJ^1Q \underset{VQ}{\longrightarrow} VV^*Q, \tag{8.2.2}$$

$$p_i = \dot{\partial}_i^t \mathcal{L}_V = \partial_i^t \mathcal{L}, \qquad \dot{p}_i = \partial_V(\partial_i^t \mathcal{L}),$$

$$\partial_V = \dot{q}^i \partial_i + \dot{y}_t^i \partial_i^t,$$

to the Legendre map $\widehat{L}$ (2.1.30). It is called the *vertical Legendre map*. Accordingly, the phase space (8.2.1) is the *vertical Legendre bundle*. The corresponding *vertical homogeneous Legendre bundle* is the cotangent bundle T^*VQ of VQ which is coordinated by $(t, q^i, p, p_i, \dot{q}^i = v^i, \dot{p}_i)$. It is provided with the canonical Liouville form (2.2.12):

$$\overline{\Xi} = pdt + \dot{y}_i dy^i + \dot{v}_i dv^i = pdt + \dot{p}_i dq^i + p_i dv^i. \tag{8.2.3}$$

Let VT^*Q be the vertical tangent bundle of the fibre bundle $T^*Q \to \mathbb{R}$. It is equipped with the coordinates $(t, q^i, p, p_i, \dot{q}^i, \dot{p}, \dot{p}_i)$. We have the composite bundle

$$V\zeta : VT^*Q \underset{VQ}{\xrightarrow{\chi}} T^*VQ \underset{VQ}{\xrightarrow{\zeta_V}} V^*VQ = VV^*Q, \tag{8.2.4}$$

$$(t, q^i, p, p_i, \dot{q}^i, \dot{p}, \dot{p}_i) \to (t, q^i, v^i = \dot{q}^i, p = \dot{p}, \dot{q}_i = \dot{p}_i, \dot{v}_i = p_i)$$

$$\to (t, q^i, v^i = \dot{q}^i, \dot{q}_i = \dot{p}_i, \dot{v}_i = p_i),$$

where $V\zeta$ (8.2.4) is the vertical tangent map of the fibration ζ (2.2.5). With the canonical Liouville form $\overline{\Xi}$ (8.2.3) on T^*VQ, the fibre bundle VT^*Q is provided with the pull-back form

$$\chi^*\overline{\Xi} = \dot{p}dt + \dot{p}_i dq^i + p_i d\dot{q}^i = \Xi_V = \partial_V \Xi, \tag{8.2.5}$$

$$\partial_V = \dot{p}\partial_p + \dot{q}^i \partial_i + \dot{p}_i \partial^i,$$

which coincides with the vertical extension $\partial_V \Xi$ (8.1.2) of the canonical Liouville form Ξ (2.2.12) on the cotangent bundle T^*X.

Hamiltonian formalism on the vertical Legendre bundle V^*VQ is formulated similarly to that on an original phase space V^*Q in Section 3.3. Given the canonical symplectic form $d\overline{\Xi}$ on T^*VQ, the vertical Legendre bundle V^*VQ is endowed with the coinduced Poisson structure

$$\{f, g\}_{VV} = \partial^i f \dot{\partial}_i g - \dot{\partial}_i f \partial^i g + \dot{\partial}^i f \partial_i g - \partial_i f \dot{\partial}^i g.$$

Due to the isomorphism (8.1.1), the canonical three-form (3.3.11) on V^*VQ can be obtained as the vertical extension

$$\mathbf{\Omega}_V = \partial_V \mathbf{\Omega} = (d\dot{p}_i \wedge dq^i + dp_i \wedge d\dot{q}^i) \wedge dt \tag{8.2.6}$$

of the canonical three-form $\boldsymbol{\Omega}$ (3.3.11) on V^*Q.

Given a section $\overline{h}$ of the affine bundle ζ_V (8.2.4), the pull-back

$$\overline{H} = (-\overline{h})^*\Xi = \dot{p}_i dq^i + p_i d\dot{q}^i - \overline{\mathcal{H}}dt \tag{8.2.7}$$

of the canonical Liouville form Ξ (8.2.3) is a Hamiltonian form on V^*VQ. The associated Hamilton vector field (3.3.21) is

$$\gamma_{\overline{H}} = \partial_t + \dot{\partial}^i\overline{\mathcal{H}}\partial_i - \dot{\partial}_i\overline{\mathcal{H}}\partial^i + \partial^i\overline{\mathcal{H}}\dot{\partial}_i - \partial_i\overline{\mathcal{H}}\dot{\partial}^i. \tag{8.2.8}$$

It is a connection on the fibre bundle $VV^*Q \to \mathbb{R}$ which defines the corresponding Hamilton equation on V^*VQ.

Our goal is forthcoming Theorem 8.2.1 which states that any Hamiltonian system on a phase space V^*Q gives rise to a Hamiltonian system on the vertical Legendre bundle $V^*VQ = VV^*Q$.

Theorem 8.2.1. *Let γ_H be a Hamilton vector field (3.3.21) on the original phase space $V^*Q \to \mathbb{R}$ for a Hamiltonian form (3.3.14). Then the vertical connection (8.1.4):*

$$V\gamma_H = \partial_t + \partial^i\mathcal{H}\partial_i - \partial_i\mathcal{H}\partial^i + \partial_V\partial^i\mathcal{H}\dot{\partial}_i - \partial_V\partial_i\mathcal{H}\dot{\partial}^i, \tag{8.2.9}$$

to γ_H on the vertical phase space $VV^*Q \to \mathbb{R}$ is the Hamilton vector field for the Hamiltonian form

$$H_V = \partial_V H = \dot{p}_i dq^i + p_i d\dot{q}^i - \partial_V\mathcal{H}dt, \tag{8.2.10}$$
$$\partial_V\mathcal{H} = (\dot{q}^i\partial_i + \dot{p}_i\partial^i)\mathcal{H},$$

which is the vertical extension of H onto VV^*Q.

Proof. The proof follows from a direct computation. □

The Hamilton vector field $V\gamma_H$ (8.2.9) defines the Hamilton equation

$$q^i_t = \dot{\partial}^i\mathcal{H}_V = \partial^i\mathcal{H}, \tag{8.2.11}$$
$$p_{ti} = -\dot{\partial}_i\mathcal{H}_V = -\partial_i\mathcal{H}, \tag{8.2.12}$$
$$\dot{q}^i_t = \partial^i\mathcal{H}_V = \partial_V\partial^i\mathcal{H}, \tag{8.2.13}$$
$$\dot{p}_{ti} = -\partial_i\mathcal{H}_V = -\partial_V\partial_i\mathcal{H}. \tag{8.2.14}$$

The equations (8.2.11) – (8.2.12) coincide with the Hamilton equation (3.3.22) – (3.3.23) for an original Hamiltonian form H, while the equations (8.2.13) – (8.2.14) are their variation equation. Substituting a solution r of the Hamilton equation (8.2.11) – (8.2.12) into (8.2.13) – (8.2.14), one obtains a linear dynamic equations whose solutions $\overline{r}$ are *Jacobi fields* of the solution r.

The Hamiltonian form H_V (8.2.10) defines the Lagrangian L_{H_V} (3.5.1) on $J^1(V^*VQ)$ which reads

$$L_{H_V} = h_0(H_V) = [\dot{p}_i(q^i_t - \partial^i\mathcal{H}) + p_i(\dot{q}^i_t - \partial_i\mathcal{H})]dt. \qquad (8.2.15)$$

Owing to the isomorphism (8.1.3), this Lagrangian is exactly the vertical extension $(L_H)_V$ (8.1.7) of the Lagrangian L_H (3.5.1) on J^1V^*Q. Accordingly, the Hamilton equation (8.2.11) – (8.2.14) is the Lagrange equation of the Lagrangian (8.2.15), and Jacobi fields of the Hamilton equation for H are Jacobi fields of the Lagrange equation for L_H.

In conclusion, lets us describe the relationship between the vertical extensions of Lagrangian and Hamiltonian formalisms [106; 109]. The Hamiltonian form H_V (8.2.10) yields the *vertical Hamiltonian map*

$$\widehat{H}_V = V\widehat{H} : VV^*Q \underset{VQ}{\longrightarrow} VJ^1Q = J^1VQ,$$

$$q^i_t = \dot{\partial}^i(\partial_V\mathcal{H}) = \partial^i\mathcal{H}, \qquad \dot{q}^i_t = \partial_V\partial^i\mathcal{H}.$$

Proposition 8.2.1. *Let H be a Hamiltonian form on V^*Q associated with a Lagrangian L on J^1Q. Then its vertical extension H_V (8.2.10) is weakly associated with the Lagrangian L_V (8.1.7).*

Proof. If the morphisms $\widehat{H}$ and $\widehat{L}$ satisfy the relation (3.6.3), then the corresponding vertical tangent morphisms obey the relation

$$V\widehat{L}\circ V\widehat{H}\circ V\widehat{L} = V\widehat{L}.$$

The condition (3.6.4) reduces to the equality (3.6.7) which is fulfilled if H is associated with L. □

8.3 Jacobi fields of completely integrable systems

Given a completely integrable autonomous Hamiltonian system, derivatives of its integrals of motion need not be constant on trajectories of a motion. We show that Jacobi fields of a completely integrable system provide linear combinations of derivatives of integrals of motion which are integrals of motion of an extended Hamiltonian system and can characterize a relative motion.

Let us consider an autonomous Hamiltonian system on a $2m$-dimensional symplectic manifold M, coordinated by (x^λ) and endowed with a symplectic form

$$\Omega = \frac{1}{2}\Omega_{\mu\nu}dx^\mu \wedge dx^\nu. \qquad (8.3.1)$$

The corresponding Poisson bracket reads

$$\{f,f'\} = w^{\alpha\beta}\partial_\alpha f\partial_\beta f', \qquad f,f'\in C^\infty(M), \tag{8.3.2}$$

where

$$w = \frac{1}{2}w^{\alpha\beta}\partial_\alpha\wedge\partial_\beta, \qquad \Omega_{\mu\nu}w^{\mu\beta} = \delta^\beta_\nu, \tag{8.3.3}$$

is the Poisson bivector associated to Ω. Let a function $\mathcal{H}\in C^\infty(M)$ on M be a Hamiltonian of a system in question. Its Hamiltonian vector field

$$\vartheta_\mathcal{H} = -w\lfloor d\mathcal{H} = w^{\mu\nu}\partial_\mu\mathcal{H}\partial_\nu \tag{8.3.4}$$

defines the autonomous first order Hamilton equation

$$\dot x^\nu = \vartheta^\nu_\mathcal{H} = w^{\mu\nu}\partial_\mu\mathcal{H} \tag{8.3.5}$$

on M. With respect to the local Darboux coordinates (q^i,p_i), the expressions (8.3.1) – (8.3.4) read

$$\Omega = dp_i\wedge dq^i, \qquad w = \partial^i\wedge\partial_i,$$
$$\{f,f'\} = \partial^i f\partial_i f' - \partial_i f\partial^i f',$$
$$\vartheta_\mathcal{H} = \partial^i\mathcal{H}\partial_i - \partial_i\mathcal{H}\partial^i.$$

The Hamilton equation (8.3.5) takes the form

$$\dot q^i = \partial^i\mathcal{H}, \qquad \dot p_i = -\partial_i\mathcal{H}. \tag{8.3.6}$$

Let a Hamiltonian system $(M,\Omega,\mathcal{H})$ be completely integrable, i.e., there exist m independent integrals of motion F_a in involution with respect to the Poisson bracket (8.3.2). Of course, a Hamiltonian $\mathcal{H}$ itself is a first integral, but it is not independent of F_a. Moreover, one often put $F_1=\mathcal{H}$.

Let us consider Jacobi fields of the completely integrable system

$$(M,\Omega,\mathcal{H},F_a). \tag{8.3.7}$$

They obey the variation equation of the equation (8.3.6) and make up an autonomous Hamiltonian system as follows [61].

Let TM be the tangent bundle of a manifold M provided with the holonomic bundle coordinates $(x^\lambda,\dot x^\lambda)$. The symplectic form Ω (8.3.1) on M gives rise to the two-form (11.2.46):

$$\widetilde\Omega = \frac{1}{2}(\dot x^\lambda\partial_\lambda\Omega_{\mu\nu}dx^\mu\wedge dx^\nu + \Omega_{\mu\nu}d\dot x^\mu\wedge dx^\nu + \Omega_{\mu\nu}dx^\mu\wedge d\dot x^\nu), \tag{8.3.8}$$

on TM. Due to the condition (11.2.47), it is a closed form. Written with respect to the local Darboux coordinates (q^i,p_i) on M and the holonomic bundle coordinates $(q^i,p_i,\dot q^i,\dot p_i)$ on TM, the two-form (8.3.8) reads

$$\widetilde\Omega = dp_i\wedge d\dot q^i + d\dot p_i\wedge dq^i. \tag{8.3.9}$$

A glance at this expression shows that $\widetilde{\Omega}$ is a non-degenerate two-form, i.e., it is a symplectic form. Note that the conjugate pairs of coordinates and momenta with respect to this symplectic form are $(q^i, \dot{p}_i)$ and $(\dot{q}^i, p_i)$. The associated Poisson bracket on TM is

$$\{g, g'\}_{TM} = \partial^i g \dot{\partial}_i g' - \dot{\partial}_i g \partial^i g' + \dot{\partial}^i g \partial_i g' - \partial_i g \dot{\partial}^i g'. \tag{8.3.10}$$

With the tangent lift

$$\widetilde{\mathcal{H}} = \partial_T \mathcal{H}, \qquad \partial_T = (\dot{q}^j \partial_j + \dot{p}_j \partial^j), \tag{8.3.11}$$

of a Hamiltonian $\mathcal{H}$, we obtain the autonomous Hamiltonian system $(TM, \widetilde{\Omega}, \widetilde{\mathcal{H}})$ on the tangent bundle TM of M. Computing the Hamiltonian vector field

$$\vartheta_{\widetilde{\mathcal{H}}} = \dot{\partial}^i \mathcal{H} \partial_i - \dot{\partial}_i \mathcal{H} \partial^i + \partial^i \mathcal{H} \dot{\partial}_i - \partial_i \mathcal{H} \dot{\partial}^i$$

of the Hamiltonian (8.3.11) with respect to the Poisson bracket (8.3.10), we obtain the corresponding Hamilton equation

$$\mathrm{\dot{q}}^i = \dot{\partial}^i \widetilde{\mathcal{H}} = \partial^i \mathcal{H}, \qquad \mathrm{\dot{p}}_i = -\dot{\partial}_i \widetilde{\mathcal{H}} = -\partial_i \mathcal{H}, \tag{8.3.12}$$

$$\ddot{q}^i = \partial^i \widetilde{\mathcal{H}} = \partial_T \partial^i \mathcal{H}, \qquad \ddot{p}_i = -\partial_i \widetilde{\mathcal{H}} = -\partial_T \partial_i \mathcal{H}, \tag{8.3.13}$$

where $(q^i, p_i, \dot{q}^i, p_i, \mathrm{\dot{q}}^i, \mathrm{\dot{p}}_i, \ddot{q}^i, \ddot{p}_i)$ are coordinates on the double tangent bundle TTM. The equation (8.3.12) coincides with the Hamilton equation (8.3.6) of the original Hamiltonian system on M, while the equation (8.3.13) is the variation equation of the equation (8.3.12). Substituting a solution r of the Hamilton equation (8.3.12) into (8.3.13), one obtains a linear dynamic equation whose solutions $\overline{r}$ are the Jacobi fields of the solution r.

Turn now to integrals of motion of the Hamiltonian system $(\widetilde{\Omega}, \widetilde{\mathcal{H}})$ on TM. We will denote the pull-back onto TM of a function f on M by the same symbol f. The Poisson bracket $\{.,.\}_{TM}$ (8.3.10) possesses the following property. Given arbitrary functions f and f' on M and their tangent lifts $\partial_T f$ and $\partial_T f'$ on TM, we have the relations

$$\{f, f'\}_{TM} = 0, \quad \{\partial_T f, f'\}_{TM} = \{f, \partial_T f'\}_{TM} = \{f, f'\}, \tag{8.3.14}$$
$$\{\partial_T f, \partial_T f'\}_{TM} = \partial_T \{f, f'\}.$$

Let us consider the tangent lifts $\partial_T F_a$ of integrals of motion F_a of the original completely integrable system (8.3.7) on M. By virtue of the relations (8.3.14), the functions $(F_a, \partial_T F_a)$ make up a collection of $2m$ integrals of motion in involution of the Hamiltonian system $(\widetilde{\Omega}, \widetilde{\mathcal{H}})$ on TM, i.e., they are constant on solutions of the Hamilton equation (8.3.12) – (8.3.13). It is readily observed that these integrals of motion are independent on TM. Consequently, we have a completely integrable system

$$(TM, \widetilde{\Omega}, \widetilde{\mathcal{H}}, F_a, \partial_T F_a) \tag{8.3.15}$$

on the tangent bundle TM. We agree to call it the *tangent completely integrable system.*

Since integrals of motion $\partial_T F_a$ of the completely integrable system (8.3.15) depend on Jacobi fields, one may hope that they characterize a relative motion. Given a solution $r(t)$ of the Hamilton equation (8.3.6), other solutions $r'(t)$ with initial data $r'(0)$ close to $r(0)$ could be approximated $r' \approx r + \overline{r}$ by solutions $(r, \overline{r})$ of the Hamilton equation (8.3.12) – (8.3.13). However, such an approximation need not be well. Namely, if M is a vector space and $r'(0) = r(0) + \overline{s}(0)$ are the above mentioned solutions, the difference $r'(t) - (r(t) + \overline{r}(t))$, $t \in \mathbb{R}$, fails to be zero and, moreover, need not be bounded on M. Of course, if F_a is an integral of motion, then

$$F_a(r'(t)) - F_a(r(t)) = \text{const.},$$

whenever r and r' are solutions of the Hamilton equation (8.3.6). We aim to show that, under a certain condition, there exists a Jacobi field $\overline{r}$ of a solution r such that

$$F_a(r') = F_a(r) + \partial_T F_a(r, \overline{r}) \tag{8.3.16}$$

for all integrals of motion F_a of the completely integrable system (8.3.7). It follows that, given a trajectory r of the original completely integrable system (8.3.7) and the values of its integrals of motion F_a on r, one can restore the values of F_a on other trajectories r' from $F_a(r)$ and the values of integrals of motion $\partial_T F_a$ for different Jacobi fields of the solution r. Therefore, one may say that the integrals of motion $\partial_T F_a$ of the tangent completely integrable system (8.3.15) characterize a relative motion.

In accordance with Theorem 7.3.3, let

$$U = V \times \mathbb{R}^{m-k} \times T^k \tag{8.3.17}$$

be an open submanifold of M endowed with generalized action-angle coordinates (I_i, y^i), $i = 1, \ldots, m$, where (y^i) are coordinates on a toroidal cylinder $\mathbb{R}^{m-k} \times T^k$. Written with respect to these coordinates, the symplectic form on U reads

$$\Omega = dI_i \wedge dy^i,$$

while a Hamiltonian $\mathcal{H}$ and the integrals of motion F_a depend only on action coordinates I_i. The Hamilton equation (8.3.6) on U (8.3.17) takes the form

$$\dot{y}^i = \partial^i \mathcal{H}(I_j), \qquad \dot{I}_i = 0. \tag{8.3.18}$$

Let us consider the tangent completely integrable system on the tangent bundle TU of U. It is the restriction to

$$TU = V \times \mathbb{R}^{3m-k} \times T^k$$

of the tangent completely integrable system (8.3.15) on TM. Given action-angle coordinates (I_i, y^i) on U, the tangent bundle TU is provided with the holonomic coordinates

$$(I_i, y^i, \dot{I}_i, y^i). \tag{8.3.19}$$

Relative to these coordinates, the tangent symplectic form $\widetilde{\Omega}$ (8.3.8) on TU reads

$$\widetilde{\Omega} = dI_i \wedge d\dot{y}^i + d\dot{I}_i \wedge dy^i.$$

The Hamiltonian (8.3.11):

$$\widetilde{\mathcal{H}} = \partial_T \mathcal{H} = \dot{I}_i \partial^i \mathcal{H},$$

and integrals of motion $(F_a, \partial_T F_a)$ of the tangent completely integrable system on TU depend only on the coordinates $(I_j, \dot{I}_j)$. Thus, the coordinates (8.3.19) are the action-angle coordinates on TU.

The Hamilton equation (8.3.12) – (8.3.13) on TU read

$$\dot{\mathrm{I}}_i = 0, \qquad \ddot{I}_i = 0, \tag{8.3.20}$$

$$\dot{\mathrm{q}}^i = \partial^i \mathcal{H}(I_j), \qquad \ddot{q}^i = \dot{I}_k \partial^k \partial^i H(I_j), \tag{8.3.21}$$

where $(I_i, y^i, \dot{I}_i, \dot{y}^i, \dot{\mathrm{I}}_i, \dot{\mathrm{y}}^i, \ddot{I}_i, \ddot{y}^i)$ are holonomic coordinates on the double tangent bundle TTU.

Let r and r' be solutions of the Hamilton equation (8.3.6) which live in U. Consequently, they are solutions of the Hamilton equation (8.3.18) on U. Hence, their action components r_i and r'_i are constant. Let us consider the system of algebraic equations

$$F_a(r'_j) - F_a(r_j) = c_i \partial^i F_a(r_j), \qquad a = 1, \ldots, m,$$

for real numbers c_i, $i = 1, \ldots m$. Since the integrals of motion F_a have no critical points on U, this system always has a unique solution. Then let us choose a solution $(r, \overline{r})$ of the Hamilton equation (8.3.20) – (8.3.21), where the Jacobi field $\overline{r}$ of the solution r possess the action components $\overline{r}_i = c_i$. It fulfils the relations (8.3.16) for all first integrals F_a. In other words, first integrals $(F_a, \partial_T F_a)$ on TU can be replaced by the action variables $(I_i, \dot{I}_i)$. Given a solution I of the Hamilton equation (8.3.18), its another solution I' is approximated well by the solution $(I, \dot{I} = I' - I)$ of the Hamilton equation (8.3.20).

Example 8.3.1. Let us consider a one-dimensional harmonic oscillator

$$M = \mathbb{R}^2, \qquad \Omega = dp \wedge dq, \qquad \mathcal{H} = \frac{1}{2}(p^2 + q^2). \tag{8.3.22}$$

It is a completely integrable system whose integral of motion is $H(q,p)$. The action-angle coordinates (I, y) are defined on $U = \mathbb{R}^2 \setminus \{0\}$ by the relations

$$q = (2I)^{1/2} \sin y, \qquad p = (2I)^{1/2} \cos y. \tag{8.3.23}$$

Since $\mathcal{H} = I$, the corresponding Hamilton equation reads

$$\dot{I} = 0, \qquad \dot{y} = 1. \tag{8.3.24}$$

The tangent extension of the Hamiltonian system (8.3.22) is the Hamiltonian system

$$\widetilde{\Omega} = dp \wedge d\dot{q} + d\dot{p} \wedge dq, \qquad \widetilde{\mathcal{H}} = \dot{p}p + \dot{q}q.$$

It is a completely integrable system whose integrals of motion are $\mathcal{H}$ and $\widetilde{\mathcal{H}}$. The action-angle coordinates $(I, \dot{I}, y, \dot{y})$ are defined on TU by the relations

$$\dot{q} = (2I)^{-1/2}\dot{I}\sin y + (2I)^{1/2}\dot{y}\cos y,$$
$$\dot{p} = (2I)^{-1/2}\dot{I}\cos y - (2I)^{1/2}\dot{y}\sin y,$$

together with the relations (8.3.23). Since $\widetilde{\mathcal{H}} = \dot{I}$, the corresponding Hamilton equation (8.3.20) – (8.3.21) read

$$\dot{\mathrm{I}} = 0, \qquad \ddot{I} = 0, \qquad \dot{\mathrm{y}} = 1, \qquad \ddot{y} = 0. \tag{8.3.25}$$

Let

$$r = (y = t \bmod 2\pi, \ I = \text{const} \neq 0)$$

be a solution of the Hamilton equation (8.3.24). Then, for any different solution

$$r' = (y = t \bmod 2\pi, \ I' = \text{const} \neq 0)$$

of this equation, there exists a solution

$$(r, \overline{r}), \quad \overline{r} = (\dot{\mathrm{y}} = 0, \ \dot{\mathrm{I}} = I' - I)$$

of the Hamilton equation (8.3.25) such that the equality (8.3.16):

$$\mathcal{H}(I') = I' = I + \dot{\mathrm{I}} = \mathcal{H}(I) + \widetilde{\mathcal{H}}(\dot{\mathrm{I}}), \tag{8.3.26}$$

holds. Relative to the original coordinates $(q, p, \dot{q}, \dot{p})$, the above mentioned solutions r, r' and $\overline{r}$ read

$$q(t) = (2I)^{1/2} \sin t, \qquad p(t) = (2I)^{1/2} \cos t,$$
$$q'(t) = (2I')^{1/2} \sin t, \qquad p'(t) = (2I')^{1/2} \cos t,$$
$$\dot{q}(t) = (I' - I)(2I)^{-1/2} \sin t, \qquad \dot{p}(t) = (I' - I)(2I)^{-1/2} \cos t.$$

Then the equality (8.3.26) takes the form

$$\frac{1}{2}(p'(t)^2 + q'(t)^2) = \frac{1}{2}(p(t)^2 + q(t)^2) + \dot{p}(t)p(t) + \dot{q}(t)q(t).$$

It should be emphasized that

$$q'(t) \neq q(t) + \dot{q}(t), \qquad p'(t) \neq p(t) + \dot{p}(t).$$

Chapter 9

Mechanics with time-dependent parameters

At present, quantum systems with classical parameters attract special attention in connection with holonomic quantum computation.

This Chapter addresses mechanical systems with time-dependent parameters. These parameters can be seen as sections of some smooth fibre bundle $\Sigma \to \mathbb{R}$ called the *parameter bundle.* Then a configuration space of a mechanical system with time-dependent parameters is a composite fibre bundle

$$Q \xrightarrow{\pi_{Q\Sigma}} \Sigma \longrightarrow \mathbb{R} \tag{9.0.27}$$

[65; 106; 140]. Indeed, given a section $\varsigma(t)$ of a parameters bundle $\Sigma \to \mathbb{R}$, the pull-back bundle

$$Q_\varsigma = \varsigma^* Q \to \mathbb{R} \tag{9.0.28}$$

is a subbundle $i_\varsigma : Q_\varsigma \to Q$ of a fibre bundle $Q \to \mathbb{R}$ which is a configuration space of a mechanical system with a fixed *parameter function* $\varsigma(t)$.

Sections 9.1 and 9.2 are devoted to Lagrangian and Hamiltonian classical mechanics with parameters. In order to obtain the Lagrange and Hamilton equations, we treat parameters on the same level as dynamic variables. The corresponding total velocity and phase spaces are the first order jet manifold J^1Q and the vertical cotangent bundle V^*Q of the configuration bundle $Q \to \mathbb{R}$, respectively.

Section 9.3 addresses quantization of mechanical systems with time-dependent parameters. Since parameters remain classical, a phase space, that we quantize, is the vertical cotangent bundle $V_\Sigma^* Q$ of a fibre bundle $Q \to \Sigma$. We apply to $V_\Sigma^* Q \to \Sigma$ the technique of leafwise geometric quantization [58; 65].

Berry's phase factor is a phenomenon peculiar to quantum systems depending on classical time-dependent parameters [3; 15; 91; 117; 166]. It is described by driving a carrier Hilbert space of a Hamilton operator over a parameter manifold. Berry's phase factor depending only on the geometry of a path in a parameter manifold is called geometric (Section 9.4). It is characterized by a holonomy operator. A problem lies in separation of a geometric phase factor from the total evolution operator without using an adiabatic assumption.

In Section 9.5, we address the Berry phase phenomena in completely integrable systems. The reason is that, being constant under an internal dynamic evolution, action variables of a completely integrable system are driven only by a perturbation holonomy operator without any adiabatic approximation [63; 65].

9.1 Lagrangian mechanics with parameters

Let the composite bundle (9.0.27), treated as a configuration space of a mechanical system with parameters, be equipped with bundle coordinates (t, σ^m, q^i) where (t, σ^m) are coordinates on a fibre bundle $\Sigma \to \mathbb{R}$.

Remark 9.1.1. Though $Q \to \mathbb{R}$ is a trivial bundle, a fibre bundle $Q \to \Sigma$ need not be trivial.

For a time, it is convenient to regard parameters as dynamic variables. Then a total velocity space of a mechanical system with parameters is the first order jet manifold J^1Q of the fibre bundle $Q \to \mathbb{R}$. It is equipped with the adapted coordinates $(t, \sigma^m, q^i, \sigma^m_t, q^i_t)$ (see Section 11.4.4).

Let a fibre bundle $Q \to \Sigma$ be provided with a connection

$$A_\Sigma = dt \otimes (\partial_t + A^i_t \partial_i) + d\sigma^m \otimes (\partial_m + A^i_m \partial_i). \tag{9.1.1}$$

Then the corresponding vertical covariant differential (11.4.36):

$$\widetilde{D} : J^1Q \to V_\Sigma Q, \qquad \widetilde{D} = (q^i_t - A^i_t - A^i_m \sigma^m_t)\partial_i, \tag{9.1.2}$$

is defined on a configuration bundle $Q \to \mathbb{R}$.

Given a section ς of a parameter bundle $\Sigma \to \mathbb{R}$, the restriction of $\widetilde{D}$ to $J^1 i_\varsigma (J^1 Q_\varsigma) \subset J^1Q$ is the familiar covariant differential on a fibre bundle Q_ς (9.0.28) corresponding to the pull-back (11.4.37):

$$A_\varsigma = \partial_t + [(A^i_m \circ \varsigma)\partial_t \varsigma^m + (A \circ \varsigma)^i_t]\partial_i, \tag{9.1.3}$$

of the connection A_Σ (9.1.1) onto $Q_\varsigma \to \mathbb{R}$. Therefore, one can use the vertical covariant differential $\widetilde{D}$ (9.1.2) in order to construct a Lagrangian for a mechanical system with parameters on the configuration space Q (9.0.27).

We suppose that such a Lagrangian L depends on derivatives of parameters σ^m_t only via the vertical covariant differential $\widetilde{D}$ (9.1.2), i.e.,

$$L = \mathcal{L}(t, \sigma^m, q^i, \widetilde{D}^i = q^i_t - A^i_t - A^i_m \sigma^m_t)dt. \tag{9.1.4}$$

Obviously, this Lagrangian is non-regular because of the Lagrangian constraint

$$\partial^t_m \mathcal{L} + A^i_m \partial^t_i \mathcal{L} = 0.$$

As a consequence, the corresponding Lagrange equation

$$(\partial_i - d_t \partial^t_i)\mathcal{L} = 0, \tag{9.1.5}$$

$$(\partial_m - d_t \partial^t_m)\mathcal{L} = 0 \tag{9.1.6}$$

is overdefined, and it admits a solution only if a rather particular relation

$$(\partial_m + A^i_m \partial_i)\mathcal{L} + \partial^t_i \mathcal{L} d_t A^i_m = 0$$

is satisfied.

However, if a parameter function ς holds fixed, the equation (9.1.6) is replaced with the condition

$$\sigma^m = \varsigma^m(t), \tag{9.1.7}$$

and the Lagrange equation (9.1.5) only should be considered One can think of this equation under the condition (9.1.7) as being the Lagrange equation for the Lagrangian

$$L_\varsigma = J^1\varsigma^* L = \mathcal{L}(t, \varsigma^m, q^i, \widetilde{D}^i = q^i_t - A^i_t - A^i_m \partial_t \varsigma^m)dt \tag{9.1.8}$$

on a velocity space $J^1 Q_\varsigma$.

Example 9.1.1. Let us consider a one-dimensional motion of a point mass in the presence of a potential field whose center moves. A configuration space of this system is a composite fibre bundle

$$Q = \mathbb{R}^3 \to \mathbb{R}^2 \to \mathbb{R}, \tag{9.1.9}$$

coordinated by (t, σ, q) where σ, treated as a parameter, is a coordinate of the field center with respect to an inertial reference frame and q is a coordinate of a point mass relative to a field center. There is the natural inclusion

$$Q \underset{\Sigma}{\times} T\Sigma \ni (t, \sigma, q, \dot{t}, \dot{\sigma}) \to (t, \sigma, \dot{t}, \dot{\sigma}, \dot{y} = -\dot{\sigma}) \in TQ$$

which defines the connection

$$A_\Sigma = dt \otimes \partial_t + d\sigma \otimes (\partial_\sigma - \partial_q) \tag{9.1.10}$$

on a fibre bundle $Q \to \Sigma$. The corresponding vertical covariant differential (9.1.2) reads

$$\widetilde{D} = (q_t + \sigma_t)\partial_q.$$

This is a relative velocity of a point mass with respect to an inertial reference frame. Then a Lagrangian of this point mass takes the form

$$L = \left[\frac{1}{2}(q_t + \sigma_t)^2 - V(q)\right] dt. \tag{9.1.11}$$

Given a parameter function $\sigma = \varsigma(t)$, the corresponding Lagrange equation (9.1.5) reads

$$d_t(q_t + \varsigma) + \partial_q V = 0. \tag{9.1.12}$$

9.2 Hamiltonian mechanics with parameters

A total phase space of a mechanical system with time-dependent parameters on the composite bundle (9.0.27) is the vertical cotangent bundle V^*Q of $Q \to \mathbb{R}$. It is coordinated by $(t, \sigma^m, q^i, p_m, p_i)$.

Let us consider Hamiltonian forms on a phase space V^*Q which are associated with the Lagrangian L (9.1.4). The Lagrangian constraint space $N_L \subset V^*Q$ defined by this Lagrangian is given by the equalities

$$p_i = \partial_i^t \mathcal{L}, \qquad p_m + A_m^i p_i = 0, \tag{9.2.1}$$

where A_Σ is the connection (9.1.1) on a fibre bundle $Q \to \Sigma$.

Let

$$\Gamma = \partial_t + \Gamma^m(t, \sigma^r)\partial_m \tag{9.2.2}$$

be some connection on a parameter bundle $\Sigma \to \mathbb{R}$, and let

$$\gamma = \partial_t + \Gamma^m \partial_m + (A_t^i + A_m^i \Gamma^m)\partial_i \tag{9.2.3}$$

be the composite connection (11.4.29) on a fibre bundle $Q \to \mathbb{R}$ which is defined by the connection A_Σ (9.1.1) on $Q \to \Sigma$ and the connection Γ (9.2.2) on $\Sigma \to \mathbb{R}$. Then a desired L-associated Hamiltonian form reads

$$\begin{aligned} H = {} & (p_m d\sigma^m + p_i dq^i) \\ & - [p_m \Gamma^m + p_i(A_t^i + A_m^i \Gamma^m) + \mathcal{E}_\gamma(t, \sigma^m, q^i, p_i)]dt, \end{aligned} \tag{9.2.4}$$

where a Hamiltonian function $\mathcal{E}_\gamma$ satisfies the relations

$$\partial^t_i\mathcal{L}(t,\sigma^m,q^i,\widetilde{D}^i = \partial^i\mathcal{E}_\gamma(t,\sigma^m,q^i,\partial^t_i\mathcal{L}) = \partial^t_i\mathcal{L}, \tag{9.2.5}$$

$$p_i\partial^i\mathcal{E}_\gamma - \mathcal{E}_\gamma = \mathcal{L}(t,\sigma^m,q^i,\widetilde{D}^i = \partial^i\mathcal{E}_\gamma). \tag{9.2.6}$$

They are obtained by substitution of the expression (9.2.4) in the conditions (3.6.3) – (3.6.4). A key point is that the Hamiltonian form (9.2.4) is affine in momenta p_m and that the relations (9.2.5) – (9.2.6) are independent of the connection Γ (9.2.2).

The Hamilton equation (3.3.22) – (3.3.23) for the Hamiltonian form H (9.2.4) reads

$$q^i_t = A^i_t + A^i_m\Gamma^m + \partial^i\mathcal{E}_\gamma, \tag{9.2.7}$$

$$p_{ti} = -p_j(\partial_i A^j_t + \partial_i A^j_m\Gamma^m) - \partial_i\mathcal{E}_\gamma, \tag{9.2.8}$$

$$\sigma^m_t = \Gamma^m, \tag{9.2.9}$$

$$p_{tm} = -p_i(\partial_m A^i_t + \Gamma^n\partial_m A^i_n) - \partial_m\mathcal{E}_\gamma, \tag{9.2.10}$$

whereas the Lagrangian constraint (9.2.1) takes the form

$$p_i = \partial^t_i\mathcal{L}(t,q^i,\sigma^m,\partial^i\mathcal{E}_\gamma(t,\sigma^m,q^i,p_i)), \tag{9.2.11}$$

$$p_m + A^i_m p_i = 0. \tag{9.2.12}$$

If a parameter function $\varsigma(t)$ holds fixed, we ignore the equation (9.2.10) and treat the rest ones as follows.

Given $\varsigma(t)$, the equations (9.1.7) and (9.2.12) define a subbundle

$$P_\varsigma \to Q_\varsigma \to \mathbb{R} \tag{9.2.13}$$

over $\mathbb{R}$ of a total phase space $V^*Q \to \mathbb{R}$. With the connection (9.1.1), we have the splitting (11.4.35) of V^*Q which reads

$$V^*Q = A_\Sigma(V^*_\Sigma Q)\underset{Q}{\oplus}(Q\underset{Q}{\times}V^*\Sigma),$$

$$p_i\overline{d}q^i + p_m\overline{d}\sigma^m = p_i(\overline{d}q^i - A^i_m\overline{d}\sigma^m) + (p_m + A^i_m p_i)\overline{d}\sigma^m,$$

where $V^*_\Sigma Q$ is the vertical cotangent bundle of $Q\to\Sigma$. Then $V^*Q\to Q$ can be provided with the bundle coordinates

$$\overline{p}_i = p_i, \qquad \overline{p}_m = p_m + A^i_m p_i$$

compatible with this splitting. Relative to these coordinates, the equation (9.2.12) takes the form $\overline{p}_m = 0$. It follows that the subbundle

$$i_P : P_\varsigma = i^*_\varsigma(A_\Sigma(V^*_\Sigma Q)) \to V^*Q, \tag{9.2.14}$$

coordinated by (t,q^i,p_i), is isomorphic to the vertical cotangent bundle

$$V^*Q_\varsigma = i^*_\varsigma V^*_\Sigma Q$$

of the configuration space $Q_\varsigma \to \mathbb{R}$ (9.0.28) of a mechanical system with a parameter function $\varsigma(t)$. Consequently, the fibre bundle P_ς (9.2.13) is a phase space of this system.

Given a parameter function ς, there exists a connection Γ on a parameter bundle $\Sigma \to \mathbb{R}$ such that $\varsigma(t)$ is its integral section, i.e., the equation (9.2.9) takes the form

$$\partial_t\varsigma^m(t) = \Gamma^m(t,\varsigma(t)). \tag{9.2.15}$$

Then a system of equations (9.2.7), (9.2.8) and (9.2.11) under the conditions (9.1.7) and (9.2.15) describes a mechanical system with a given parameter function $\varsigma(t)$ on a phase space P_ς. Moreover, this system is the Hamilton equation for the pull-back Hamiltonian form

$$H_\varsigma = i_P^* H = p_i dq^i - [p_i(A_t^i + A_m^i\partial_t\varsigma^m) + \varsigma^*\mathcal{E}_\gamma]dt \tag{9.2.16}$$

on P_ς where

$$A_t^i + A_m^i\partial_t\varsigma^m = (i_\varsigma^*\gamma)_t^i$$

is the pull-back connection (11.4.37) on $Q_\varsigma \to \mathbb{R}$.

It is readily observed that the Hamiltonian form H_ς (9.2.16) is associated with the Lagrangian L_ς (9.1.8) on J^1Q_ς, and the equations (9.2.7), (9.2.8) and (9.2.11) are corresponded to the Lagrange equation (9.1.5).

Example 9.2.1. Let us consider a Lagrangian mechanical system on the configuration space (9.1.9) in Example 9.1.1 which is described by the Lagrangian (9.1.11). The corresponding Lagrangian constraint space is

$$p_q + p_\sigma = 0, \tag{9.2.17}$$

where $(t,\sigma,q,p_\sigma,p_q)$ are coordinates on a phase space V^*Q. Let

$$\Gamma = \partial_t + \Gamma(t,\sigma)\partial_\sigma$$

be a connection on a parameter bundle $\Sigma = \mathbb{R}^2 \to \mathbb{R}$. Given the connection A_Σ (9.1.10) on $Q \to \Sigma$, the composite connection γ (9.2.3) on a configuration bundle $Q \to \mathbb{R}$ reads

$$\gamma = \partial_t + \Gamma\partial_\sigma - \Gamma\partial_i.$$

Then the L-associated Hamiltonian form (9.2.4) reads

$$H = p_q dq + p_\sigma d\sigma - \left[-p_q\Gamma - p_\sigma\Gamma + \frac{1}{2}p_q^2 + V(q)\right] dt.$$

Given a parameter function $\sigma = \varsigma(t)$, the corresponding Hamilton equation (9.2.7) – (9.2.9) take the form

$$q_t = -\Gamma + p_q,$$
$$p_{tq} = -\partial_q V(q),$$
$$\partial_t\varsigma = \Gamma.$$

This Hamilton equation is equivalent to the Lagrange equation (9.1.12).

9.3 Quantum mechanics with classical parameters

In Sections 9.1 and 9.2 we have formulated Lagrangian and Hamiltonian classical mechanics with parameters on the composite bundle Q (9.0.27). In order to obtain Lagrange and Hamilton equations, we treat parameters on the same level as dynamic variables so that their total velocity and phase spaces are the first order jet manifold J^1Q and the vertical cotangent bundle V^*Q of a fibre bundle $Q \to \mathbb{R}$, respectively.

This Section is devoted to quantization of mechanical systems with time-dependent parameters on the composite bundle Q (9.0.27). Since parameters remain classical, a phase space that we quantize is the vertical cotangent bundle $V^*_\Sigma Q$ of a fibre bundle $Q \to \Sigma$. This phase space is equipped with holonomic coordinates (t, σ^m, q^i, p_i). It is provided with the following canonical Poisson structure. Let T^*Q be the cotangent bundle of Q equipped with the holonomic coordinates $(t, \sigma^m, q^i, p_0, p_m, p_i)$. It is endowed with the canonical Poisson structure $\{,\}_T$ (3.3.2). There is the canonical fibration

$$\zeta_\Sigma : T^*Q \xrightarrow{\zeta} V^*Q \longrightarrow V^*_\Sigma Q \tag{9.3.1}$$

(see the exact sequence (11.4.31)). Then the Poisson bracket $\{,\}_\Sigma$ on the space $C^\infty(V^*_\Sigma Q)$ of smooth real functions on $V^*_\Sigma Q$ is defined by the relation

$$\zeta^*_\Sigma \{f, f'\}_\Sigma = \{\zeta^*_\Sigma f, \zeta^*_\Sigma f'\}_T, \tag{9.3.2}$$

$$\{f, f'\}_\Sigma = \partial^k f \partial_k f' - \partial_k f \partial^k f', \qquad f, f' \in C^\infty(V^*_\Sigma Q). \tag{9.3.3}$$

The corresponding characteristic symplectic foliation $\mathcal{F}$ coincides with the fibration $V^*_\Sigma Q \to \Sigma$. Therefore, we can apply to a phase space $V^*_\Sigma Q \to \Sigma$ the technique of leafwise geometric quantization in Section 5.3 [58].

Let us assume that a manifold Q is oriented, that fibres of $V^*_\Sigma Q \to \Sigma$ are simply connected, and that

$$H^2(Q; \mathbb{Z}_2) = H^2(V^*_\Sigma Q; \mathbb{Z}_2) = 0.$$

Being the characteristic symplectic foliation of the Poisson structure (9.3.3), the fibration $V^*_\Sigma Q \to \Sigma$ is endowed with the symplectic leafwise form (3.1.31):

$$\Omega_\mathcal{F} = \widetilde{d}p_i \wedge \widetilde{d}q^i.$$

Since this form is $\widetilde{d}$-exact, its leafwise de Rham cohomology class equals zero and, consequently, it is the image of the zero de Rham cohomology class with respect to the morphism $[i^*_\mathcal{F}]$ (3.1.23). Then, in accordance with Proposition 5.3.1, the symplectic foliation $(V^*_\Sigma Q \to \Sigma, \Omega_{\mathcal{F}})$ admits prequantization.

Since the leafwise form $\Omega_{\mathcal{F}}$ is $\widetilde{d}$-exact, the prequantization bundle $C \to V^*_\Sigma Q$ is trivial. Let its trivialization

$$C = V^*_\Sigma Q \times \mathbb{C} \tag{9.3.4}$$

hold fixed, and let $(t, \sigma^m, q^k, p_k, c)$ be the corresponding bundle coordinates. Then $C \to V^*_\Sigma Q$ admits a leafwise connection

$$A_{\mathcal{F}} = \widetilde{d}p_k \otimes \partial^k + \widetilde{d}q^k \otimes (\partial_k - ip_k c\partial_c).$$

This connection preserves the Hermitian fibre metric g (3.5.3) in C, and its curvature fulfils the prequantization condition (5.3.3):

$$\widetilde{R} = -i\Omega_{\mathcal{F}} \otimes u_C.$$

The corresponding prequantization operators (5.3.2) read

$$\widehat{f} = -i\vartheta_f + (p_k \partial^k f - f), \qquad f \in C^\infty(V^*_\Sigma Q),$$
$$\vartheta_f = \partial^k f \partial_k - \partial_k f \partial^k.$$

Let us choose the canonical vertical polarization of the symplectic foliation $(V^*_\Sigma Q \to \Sigma, \Omega_{\mathcal{F}})$ which is the vertical tangent bundle $\mathbf{T} = VV^*_\Sigma Q$ of a fibre bundle

$$\pi_{VQ} : V^*_\Sigma Q \to Q.$$

It is readily observed that the corresponding quantum algebra $\mathcal{A}_{\mathcal{F}}$ consists of functions

$$f = a^i(t, \sigma^m, q^k)p_i + b(t, \sigma^m, q^k) \tag{9.3.5}$$

on $V^*_\Sigma Q$ which are affine in momenta p_k.

Following the quantization procedure in Section 5.3.3, one should consider the quantization bundle (5.3.23) which is isomorphic to the prequantization bundle C (9.3.4) because the metalinear bundle $\mathcal{D}_{1/2}[\mathcal{F}]$ of complex *fibrewise half-densities* on $V^*_\Sigma Q \to \Sigma$ is trivial owing to the identity transition functions $J_{\mathcal{F}} = 1$ (5.3.21). Then we define the representation (5.3.24) of the quantum algebra $\mathcal{A}_{\mathcal{F}}$ of functions f (9.3.5) in the space $\mathfrak{E}_{\mathcal{F}}$ of sections ρ of the prequantization bundle $C \to V^*_\Sigma Q$ which obey the condition (5.3.25) and whose restriction to each fibre of $V^*_\Sigma Q \to \Sigma$ is of compact support. Since the trivialization (9.3.4) of C holds fixed, its sections are complex functions on $V^*_\Sigma Q$, and the above mentioned condition (5.3.25) reads

$$\partial_k f \partial^k \rho = 0, \qquad f \in C^\infty(Q),$$

i.e., elements of $\mathfrak{E}_{\mathcal{F}}$ are constant on fibres of $V^*_\Sigma Q \to Q$. Consequently, $\mathfrak{E}_{\mathcal{F}}$ reduces to zero $\rho = 0$.

Therefore, we modify the leafwise quantization procedure as follows. Given a fibration

$$\pi_{Q\Sigma} : Q \to \Sigma,$$

let us consider the corresponding metalinear bundle $\mathcal{D}_{1/2}[\pi_{Q\Sigma}] \to Q$ of leafwise half-densities on $Q \to \Sigma$ and the tensor product

$$Y_Q = C_Q \otimes \mathcal{D}_{1/2}[\pi_{Q\Sigma}] = \mathcal{D}_{1/2}[\pi_{Q\Sigma}] \to Q,$$

where $C_Q = \mathbb{C} \times Q$ is the trivial complex line bundle over Q. It is readily observed that the Hamiltonian vector fields

$$\vartheta_f = a^k \partial_k - (p_j \partial_k a^j + \partial_k b)\partial^k$$

of elements $f \in \mathcal{A}_{\mathcal{F}}$ (9.3.5) are projectable onto Q. Then one can associate to each element f of the quantum algebra $\mathcal{A}_{\mathcal{F}}$ the first order differential operator

$$\begin{aligned} \widehat{f} &= (-i\nabla_{\pi_{VQ}(\vartheta_f)} + f) \otimes \mathrm{Id} + \mathrm{Id} \otimes \mathbf{L}_{\pi_{VQ}(\vartheta_f)} \\ &= -ia^k \partial_k - \frac{i}{2}\partial_k a^k - b \end{aligned} \tag{9.3.6}$$

in the space $\mathfrak{E}_Q$ of sections of the fibre bundle $Y_Q \to Q$ whose restriction to each fibre of $Q \to \Sigma$ is of compact support. Since the pull-back of $\mathcal{D}_{1/2}[\pi_{Q\Sigma}]$ onto each fibre Q_σ of $Q \to \Sigma$ is the metalinear bundle of half-densities on Q_σ, the restrictions ρ_σ of elements of $\rho \in \mathfrak{E}_Q$ to Q_σ constitute a pre-Hilbert space with respect to the non-degenerate Hermitian form

$$\langle \rho_\sigma | \rho'_\sigma \rangle_\sigma = \int_{Q_\sigma} \rho_\sigma \overline{\rho'_\sigma}.$$

Then the Schrödinger operators (9.3.6) are Hermitian operators in the pre-Hilbert $C^\infty(\Sigma)$-module $\mathfrak{E}_Q$, and provide the desired geometric quantization of the symplectic foliation $(V^*_\Sigma Q \to \Sigma, \Omega_{\mathcal{F}})$.

In order to quantize the evolution equation of a mechanical system on a phase space $V^*_\Sigma Q$, one should bear in mind that this equation is not reduced to the Poisson bracket $\{,\}_\Sigma$ on $V^*_\Sigma Q$, but is expressed in the Poisson bracket $\{,\}_T$ on the cotangent bundle T^*Q [58]. Therefore, let us start with the classical evolution equation.

Given the Hamiltonian form H (9.2.4) on a total phase space V^*Q, let $(T^*Q, \mathcal{H}^*)$ be an equivalent homogeneous Hamiltonian system with the homogeneous Hamiltonian $\mathcal{H}^*$ (3.4.1):

$$\mathcal{H}^* = p_0 + [p_m \Gamma^m + p_i(A^i_t + A^i_m \Gamma^m) + \mathcal{E}_\gamma(t, \sigma^m, q^i, p_i)]. \tag{9.3.7}$$

Let us consider the homogeneous evolution equation (3.8.3) where F are functions on a phase space $V^*_\Sigma Q$. It reads

$$\begin{aligned}&\{\mathcal{H}^*, \zeta^*_\Sigma F\}_T = 0, \qquad F\in C^\infty(V^*_\Sigma Q), \\ &\partial_t F + \Gamma^m\partial_m F + (A^i_t + A^i_m\Gamma^m + \partial^i\mathcal{E}_\gamma)\partial_i F \\ &\qquad - [p_j(\partial_i A^j_t + \partial_i A^j_m\Gamma^m) + \partial_i\mathcal{E}_\gamma]\partial^i F = 0.\end{aligned} \tag{9.3.8}$$

It is readily observed that a function $F\in C^\infty(V^*_\Sigma Q)$ obeys the equality (9.3.8) if and only if it is constant on solutions of the Hamilton equation (9.2.7) – (9.2.9). Therefore, one can think of the relation (9.3.8) as being a classical evolution equation on $C^\infty(V^*_\Sigma Q)$.

In order to quantize the evolution equation (9.3.8), one should quantize a symplectic manifold $(T^*Q, \{,\}_T)$ so that its quantum algebra $\mathcal{A}_T$ contains the pull-back $\zeta^*_\Sigma\mathcal{A}_\mathcal{F}$ of the quantum algebra $\mathcal{A}_\mathcal{F}$ of the functions (9.3.5). For this purpose, we choose the vertical polarization VT^*Q on the cotangent bundle T^*Q. The corresponding quantum algebra $\mathcal{A}_T$ consists of functions on T^*Q which are affine in momenta (p_0, p_m, p_i) (see Section 5.2). Clearly, $\zeta^*_\Sigma\mathcal{A}_\mathcal{F}$ is a subalgebra of the quantum algebra $\mathcal{A}_T$ of T^*Q.

Let us restrict our consideration to the subalgebra $\mathcal{A}'_T\subset\mathcal{A}_T$ of functions

$$f = a(t,\sigma^r)p_0 + a^m(t,\sigma^r)p_m + a^i(t,\sigma^m,q^j)p_i + b(t,\sigma^m,q^j),$$

where a and a^λ are the pull-back onto T^*Q of functions on a parameter space Σ. Of course, $\zeta^*_\Sigma\mathcal{A}_\mathcal{F}\subset\mathcal{A}'_T$. Moreover, $\mathcal{A}'_T$ admits a representation by the Hermitian operators

$$\widehat{f} = -i(a\partial_t + a^m\partial_m + a^i\partial_i) - \frac{i}{2}\partial_k a^k - b \tag{9.3.9}$$

in the carrier space $\mathfrak{C}_Q$ of the representation (9.3.6) of $\mathcal{A}_\mathcal{F}$. Then, if $\mathcal{H}^*\in\mathcal{A}'_T$, the evolution equation (9.3.8) is quantized as the Heisenberg equation (5.4.27):

$$i[\widehat{\mathcal{H}}^*, \widehat{f}] = 0, \qquad f\in\mathcal{A}_\mathcal{F}. \tag{9.3.10}$$

A problem is that the function $\mathcal{H}^*$ (9.3.7) fails to belong to the algebra $\mathcal{A}'_T$, unless the Hamiltonian function $\mathcal{E}_\gamma$ (9.2.4) is affine in momenta p_i. Let us assume that $\mathcal{E}_\gamma$ is polynomial in momenta. This is the case of almost all physically relevant models.

Lemma 9.3.1. *Any smooth function f on $V^*_\Sigma Q$ which is a polynomial of momenta p_k is decomposed in a finite sum of products of elements of the algebra $\mathcal{A}_\mathcal{F}$.*

Proof. The proof follows that of a similar statement in Section 5.4.4. □

By virtue of Lemma 9.3.1, one can associate to a polynomial Hamiltonian function $\mathcal{E}_\gamma$ an element of the enveloping algebra $\overline{\mathcal{A}}_\mathcal{F}$ of the Lie algebra $\mathcal{A}_\mathcal{F}$ (though it by no means is unique). Accordingly, the homogeneous Hamiltonian $\mathcal{H}^*$ (9.3.7) is represented by an element of the enveloping algebra $\overline{\mathcal{A}}'_T$ of the Lie algebra $\mathcal{A}'_T$. Then the Schrödinger representation (9.3.6) and (9.3.9) of the Lie algebras $\mathcal{A}_\mathcal{F}$ and $\mathcal{A}'_T$ is naturally extended to their enveloping algebras $\overline{\mathcal{A}}_\mathcal{F}$ and $\overline{\mathcal{A}}'_T$ that provides quantization

$$\widehat{\mathcal{H}}^* = -i[\partial_t + \Gamma^m\partial_m + (A_t^k + A_m^k\Gamma^m)\partial_k] - \frac{i}{2}\partial_k(A_t^k + A_m^k\Gamma^m) + \widehat{\mathcal{E}}_\gamma \quad (9.3.11)$$

of the homogeneous Hamiltonian $\mathcal{H}^*$ (9.3.7).

It is readily observed that the operator $i\widehat{\mathcal{H}}^*$ (9.3.11) obeys the Leibniz rule

$$i\widehat{\mathcal{H}}^*(r\rho) = \partial_t r\rho + r(i\widehat{\mathcal{H}}^*\rho), \qquad r \in C^\infty(\mathbb{R}), \qquad \rho \in \mathfrak{E}_Q. \quad (9.3.12)$$

Therefore, it is a connection on pre-Hilbert $C^\infty(\mathbb{R})$-module $\mathfrak{E}_Q$. The corresponding Schrödinger equation (4.6.7) reads

$$i\widehat{\mathcal{H}}^*\rho = 0, \qquad \rho \in \mathfrak{E}_Q.$$

Given a trivialization

$$Q = \mathbb{R} \times M, \quad (9.3.13)$$

there is the corresponding global decomposition

$$\widehat{\mathcal{H}}^* = -i\partial_t + \widehat{\mathcal{H}},$$

where $\widehat{\mathcal{H}}$ plays a role of the Hamilton operator. Then we can introduce the evolution operator U which obeys the equation (4.6.9):

$$\partial_t U(t) = -i\widehat{\mathcal{H}}^* \circ U(t), \qquad U(0) = \mathbf{1}.$$

It can be written as the formal time-ordered exponent

$$U = T\exp\left[-i\int_0^t \widehat{\mathcal{H}}dt'\right].$$

Given the quantum operator $\widehat{\mathcal{H}}^*$ (9.3.11), the bracket

$$\nabla\widehat{f} = i[\widehat{\mathcal{H}}^*, \widehat{f}] \quad (9.3.14)$$

defines a derivation of the quantum algebra $\overline{\mathcal{A}}_\mathcal{F}$. Since $\widehat{p}_0 = -i\partial_t$, the derivation (9.3.14) obeys the Leibniz rule

$$\nabla(r\widehat{f}) = \partial_t r\widehat{f} + r\nabla\widehat{f}, \qquad r \in C^\infty(\mathbb{R}).$$

Therefore, it is a connection on the $C^\infty(\mathbb{R})$-algebra $\overline{\mathcal{A}}_{\mathcal{F}}$, which enables one to treat quantum evolution of $\overline{\mathcal{A}}_{\mathcal{F}}$ as a parallel displacement along time (see Section 4.6). In particular, $\widehat{f}$ is parallel with respect to the connection (9.3.14) if it obeys the Heisenberg equation (9.3.10).

Now let us consider a mechanical system depending on a given parameter function $\varsigma : \mathbb{R} \to \Sigma$. Its configuration space is the pull-back bundle Q_ς (9.0.28). The corresponding phase space is the fibre bundle P_ς (9.2.14). The pull-back H_ς of the Hamiltonian form H (9.2.4) onto P_ς takes the form (9.2.16).

The homogeneous phase space of a mechanical system with a parameter function ς is the pull-back

$$\overline{P}_\varsigma = i_P^* T^* Q \tag{9.3.15}$$

onto P_ς of the fibre bundle $T^*Q \to V^*Q$ (3.3.3). The homogeneous phase space $\overline{P}_\varsigma$ (9.3.15) is coordinated by (t, q^i, p_0, p_i), and it isomorphic to the cotangent bundle T^*Q_ς. The associated homogeneous Hamiltonian on $\overline{P}_\varsigma$ reads

$$\mathcal{H}_\varsigma^* = p_0 + [p_i(A_t^i + A_m^i \partial_t \varsigma^m) + \varsigma^* \mathcal{E}_\gamma]. \tag{9.3.16}$$

It characterizes the dynamics of a mechanical system with a given parameter function ς.

In order to quantize this system, let us consider the pull-back bundle

$$\mathcal{D}_{1/2}[Q_\varsigma] = i_\varsigma^* \mathcal{D}_{1/2}[\pi_{Q\Sigma}]$$

over Q_ς and its pull-back sections $\rho_\varsigma = i_\varsigma^* \rho$, $\rho \in \mathfrak{E}_Q$. It is easily justified that these are fibrewise half-densities on a fibre bundle $Q_\varsigma \to \mathbb{R}$ whose restrictions to each fibre $i_t : Q_t \to Q_\varsigma$ are of compact support. These sections constitute a pre-Hilbert $C^\infty(\mathbb{R})$-module $\mathfrak{E}_\varsigma$ with respect to the Hermitian forms

$$\langle i_t^* \rho_\varsigma | i_t^* \rho'_\varsigma \rangle_t = \int_{Q_t} i_t^* \rho_\varsigma \overline{i_t^* \rho'_\varsigma}.$$

Then the pull-back operators

$$(\varsigma^* \widehat{f})\rho_\varsigma = (\widehat{f}\rho)_\varsigma,$$
$$\varsigma^* \widehat{f} = -ia^k(t, \varsigma^m(t), q^j)\partial_k - \frac{i}{2}\partial_k a^k(t, \varsigma^m(t), q^j) - b(t, \varsigma^m(t), q^j),$$

in $\mathfrak{E}_\varsigma$ provide the representation of the pull-back functions

$$i_\varsigma^* f = a^k(t, \varsigma^m(t), q^j)p_k + b(t, \varsigma^m(t), q^j), \qquad f \in \mathcal{A}_{\mathcal{F}},$$

on V^*Q_ς. Accordingly, the quantum operator

$$\widehat{\mathcal{H}}^*_\varsigma = -i\partial_t - i(A^i_t + A^i_m\partial_t\varsigma^m)\partial_i - \frac{i}{2}\partial_i(A^i_t + A^i_m\partial_t\varsigma^m) - \widehat{\varsigma^*\mathcal{E}_\gamma} \qquad (9.3.17)$$

coincides with the pull-back operator $\varsigma^*\widehat{\mathcal{H}}^*$, and it yields the Heisenberg equation

$$i[\widehat{\mathcal{H}}^*_\varsigma, \varsigma^*\widehat{f}] = 0$$

of a quantum system with a parameter function ς.

The operator $\widehat{\mathcal{H}}^*_\varsigma$ (9.3.17) acting in the pre-Hilbert $C^\infty(\mathbb{R})$-module $\mathfrak{E}_\varsigma$ obeys the Leibniz rule

$$i\widehat{\mathcal{H}}^*_\varsigma(r\rho_\varsigma) = \partial_t r\rho_\varsigma + r(i\widehat{\mathcal{H}}^*_\varsigma\rho_\varsigma), \qquad r \in C^\infty(\mathbb{R}), \qquad \rho_\varsigma \in \mathfrak{E}_Q, \qquad (9.3.18)$$

and, therefore, it is a connection on $\mathfrak{E}_\varsigma$. The corresponding Schrödinger equation reads

$$i\widehat{\mathcal{H}}^*_\varsigma\rho_\varsigma = 0, \qquad \rho_\varsigma \in \mathfrak{E}_\varsigma, \qquad (9.3.19)$$
$$\left[\partial_t + (A^i_t + A^i_m\partial_t\varsigma^m)\partial_i + \frac{1}{2}\partial_i(A^i_t + A^i_m\partial_t\varsigma^m) - i\widehat{\varsigma^*\mathcal{E}_\gamma}\right]\rho_\varsigma = 0.$$

With the trivialization (9.3.13) of Q, we have a trivialization of $Q_\varsigma \to \mathbb{R}$ and the corresponding global decomposition

$$\widehat{\mathcal{H}}^*_\varsigma = -i\partial_t + \widehat{\mathcal{H}}_\varsigma,$$

where

$$\widehat{\mathcal{H}}_\varsigma = -i(A^i_t + A^i_m\partial_t\varsigma^m)\partial_i - \frac{i}{2}\partial_i(A^i_t + A^i_m\partial_t\varsigma^m) + \widehat{\varsigma^*\mathcal{E}_\gamma} \qquad (9.3.20)$$

is a Hamilton operator. Then we can introduce an evolution operator U_ς which obeys the equation

$$\partial_t U_\varsigma(t) = -i\widehat{\mathcal{H}}^*_\varsigma \circ U_\varsigma(t), \qquad U_\varsigma(0) = \mathbf{1}.$$

It can be written as the formal time-ordered exponent

$$U_\varsigma(t) = T\exp\left[-i\int_0^t \widehat{\mathcal{H}}_\varsigma dt'\right]. \qquad (9.3.21)$$

9.4 Berry geometric factor

As was mentioned above, the Berry phase factor is a standard attribute of quantum mechanical systems with time-dependent classical parameters [15; 109].

Let us remind that the quantum adiabatic Berry phase has been discovered as a phase shift in the eigenfunctions of a parameter-dependent Hamiltonian when parameters traverse along a closed curve [8]. J.Hannay in [82] found a classical analogue of this phase associated to completely integrable systems and called the Hannay angles (see [9] for its non-adiabatic generalization). B.Simon in [147] has recognized that the Berry phase arises from a particular connection, called the Berry connection, on a Hermitian line bundle over a parameter space (see [117] for an analogous geometric framework of Hannay angles, determined by a parameter-dependent Hamiltonian action of a Lie group on a symplectic manifold). F.Wilczek and A.Zee in [164] generalized a notion of the adiabatic phase to the non-Abelian case corresponding to adiabatically transporting an n-fold degenerate state over the parameter manifold. They considered a vector bundle over a parameter space as a unitary bundle. E.Kiritsis in [91] has studied this bundle using homotopy theory. The reader is addressed to [166] for the case of a Hamiltonian G-space of parameters and to [151] for a homogeneous Kähler parameter manifold. The adiabatic assumption was subsequently removed by Aharonov and Anandan in [3] who suggested to considered a loop in a projective Hilbert space instead of a parameter space [2] (see [14] for the relation between the Berry and Aharonov–Anandan connections).

The Berry phase factor is described by driving a carrier Hilbert space of a Hamilton operator over cycles in a parameter manifold. The Berry geometric factor depends only on the geometry of a path in a parameter manifold and, therefore, provides a possibility to perform quantum gate operations in an intrinsically fault-tolerant way. A problem lies in separation of the Berry geometric factor from the total evolution operator without using an adiabatic assumption. Firstly, holonomy quantum computation implies exact cyclic evolution, but exact adiabatic cyclic evolution almost never exists. Secondly, an adiabatic condition requires that the evolution time must be long enough.

In a general setting, let us consider a linear (not necessarily finite-dimensional) dynamical system

$$\partial_t \psi = \widehat{S}\psi$$

whose linear (time-dependent) dynamic operator $\widehat{S}$ falls into the sum

$$\widehat{S} = \widehat{S}_0 + \Delta = \widehat{S}_0 + \partial_t\varsigma^m \Delta_m, \tag{9.4.1}$$

where $\varsigma(t)$ is a parameter function given by a section of some smooth fibre bundle $\Sigma \to \mathbb{R}$ coordinated by (t, σ^m). Let assume the following:

(i) the operators $\widehat{S}_0(t)$ and $\Delta(t')$ commute for all instants t and t',

(ii) the operator Δ depends on time only through a parameter function $\varsigma(t)$.

Then the corresponding evolution operator $U(t)$ can be represented by the product of time-ordered exponentials

$$U(t) = U_0(t) \circ U_1(t) = T\exp\left[\int_0^t \Delta dt'\right] \circ T\exp\left[\int_0^t \widehat{S}_0 dt'\right], \tag{9.4.2}$$

where the first one is brought into the ordered exponential

$$U_1(t) = T\exp\left[\int_0^t \Delta_m(\varsigma(t'))\partial_t\varsigma^m(t')dt'\right] \tag{9.4.3}$$

$$= T\exp\left[\int_{\varsigma[0,t]} \Delta_m(\sigma)d\sigma^m\right]$$

along the curve $\varsigma[0,t]$ in a parameter bundle Σ. It is the *Berry geometric factor* depending only on a trajectory of a parameter function ς. Therefore, one can think of this factor as being a displacement operator along a curve $\varsigma[0,t] \subset \Sigma$. Accordingly,

$$\Delta = \Delta_m \partial_t\varsigma^m \tag{9.4.4}$$

is called the *holonomy operator*.

However, a problem is that the above mentioned commutativity condition (i) is rather restrictive.

Turn now to the quantum Hamiltonian system with classical parameters in Section 9.3. The Hamilton operator $\widehat{\mathcal{H}}_\varsigma$ (9.3.20) in the evolution operator U (9.3.21) takes the form (9.4.1):

$$\widehat{\mathcal{H}}_\varsigma = -i\left[A^k_m\partial_k + \frac{1}{2}\partial_k A^k_m\right]\partial_t\varsigma^m + \widehat{\mathcal{H}}'(\varsigma). \tag{9.4.5}$$

Its second term $\widehat{\mathcal{H}}'$ can be regarded as a dynamic Hamilton operator of a quantum system, while the first one is responsible for the Berry geometric factor as follows.

Bearing in mind possible applications to holonomic quantum computations, let us simplify the quantum system in question. The above mentioned trivialization (9.3.13) of Q implies a trivialization of a parameter bundle $\Sigma = \mathbb{R} \times W$ such that a fibration $Q \to \Sigma$ reads

$$\mathbb{R} \times M \overset{\mathrm{Id}\times\pi_M}{\longrightarrow} \mathbb{R} \times W,$$

where $\pi_M : M \to W$ is a fibre bundle. Let us suppose that components A^k_m of the connection A_Σ (9.1.1) are independent of time. Then one can regard the second term in this connection as a connection on a fibre bundle $M \to W$. It also follows that the first term in the Hamilton operator (9.4.5) depends on time only through parameter functions $\varsigma^m(t)$. Furthermore, let the two terms in the Hamilton operator (9.4.5) mutually commute on $[0,t]$. Then the evolution operator U (9.3.21) takes the form

$$U = T\exp\left[-\int\limits_{\varsigma([0,t])}\left(A^k_m\partial_k + \frac{1}{2}\partial_k A^k_m\right)d\sigma^m\right] \tag{9.4.6}$$
$$\circ\, T\exp\left[-i\int\limits_0^t \widehat{\mathcal{H}}'dt'\right].$$

One can think of its first factor as being the parallel displacement operator along the curve $\varsigma([0,t]) \subset W$ with respect to the connection

$$\nabla_m\rho = \left(\partial_m + A^k_m\partial_k + \frac{1}{2}\partial_k A^k_m\right)\rho, \qquad \rho \in \mathfrak{E}_Q, \tag{9.4.7}$$

called the , *Berry connection* on a $C^\infty(W)$-module $\mathfrak{E}_Q$. A peculiarity of this factor in comparison with the second one lies in the fact that integration over time through a parameter function $\varsigma(t)$ depends only on a trajectory of this function in a parameter space, but not on parametrization of this trajectory by time. Therefore, the first term of the evolution operator U (9.4.6) is the Berry geometric factor. The corresponding holonomy operator (9.4.4) reads

$$\Delta = \left(A^k_m\partial_k + \frac{1}{2}\partial_k A^k_m\right)\partial_t\varsigma^m.$$

9.5 Non-adiabatic holonomy operator

We address the Berry phase phenomena (Section 9.4) in a completely integrable system of m degrees of freedom around its invariant torus T^m. The reason is that, being constant under an internal evolution, its action variables are driven only by a perturbation holonomy operator Δ. We construct such an operator for an arbitrary connection on a fibre bundle

$$W \times T^m \to W, \tag{9.5.1}$$

without any adiabatic approximation [63; 65]. In order that a holonomy operator and a dynamic Hamiltonian mutually commute, we first define a holonomy operator with respect to initial data action-angle coordinates and, afterwards, return to the original ones. A key point is that both classical evolution of action variables and mean values of quantum action operators relative to original action-angle coordinates are determined by the dynamics of initial data action and angle variables.

A generic phase space of a Hamiltonian system with time-dependent parameters is a composite fibre bundle

$$P \to \Sigma \to \mathbb{R},$$

where $\Pi \to \Sigma$ is a symplectic bundle (i.e., a symplectic foliation whose leaves are fibres of $\Pi \to \Sigma$), and

$$\Sigma = \mathbb{R} \times W \to \mathbb{R}$$

is a parameter bundle whose sections are parameter functions. In the case of a completely integrable system with time-dependent parameters, we have the product

$$P = \Sigma \times U = \Sigma \times (V \times T^m) \to \Sigma \to \mathbb{R},$$

equipped with the coordinates $(t, \sigma^\alpha, I_k, \varphi^k)$. Let us suppose for a time that parameters also are dynamic variables. The total phase space of such a system is the product

$$\Pi = V^*\Sigma \times U$$

coordinated by $(t, \sigma^\alpha, p_\alpha = \dot\sigma_\alpha, I_k, \varphi^k)$. Its dynamics is characterized by the Hamiltonian form (9.2.4):

$$\begin{aligned} H_\Sigma &= p_\alpha d\sigma^\alpha + I_k d\varphi^k - \mathcal{H}_\Sigma(t, \sigma^\beta, p_\beta, I_j, \varphi^j)dt, \\ \mathcal{H}_\Sigma &= p_\alpha \Gamma^\alpha + I_k(\Lambda^k_t + \Lambda^k_\alpha \Gamma^\alpha_t) + \widetilde{\mathcal{H}}, \end{aligned} \tag{9.5.2}$$

where $\widetilde{\mathcal{H}}$ is a function, $\partial_t + \Gamma^\alpha\partial_\alpha$ is the connection (9.2.2) on the parameter bundle $\Sigma \to \mathbb{R}$, and

$$\Lambda = dt \otimes (\partial_t + \Lambda^k_t \partial_k) + d\sigma^\alpha \otimes (\partial_\alpha + \Lambda^k_\alpha \partial_k) \tag{9.5.3}$$

is the connection (9.1.1) on the fibre bundle

$$\Sigma \times T^m \to \Sigma.$$

Bearing in mind that σ^α are parameters, one should choose the Hamiltonian $\mathcal{H}_\Sigma$ (9.5.2) to be affine in their momenta p_α. Then a Hamiltonian system with a fixed parameter function $\sigma^\alpha = \varsigma^\alpha(t)$ is described by the pull-back Hamiltonian form (9.2.16):

$$H_\varsigma = I_k d\varphi^k - \{I_k[\Lambda_t^k(t,\varphi^j) + \Lambda_\alpha^k(t,\varsigma^\beta,\varphi^j)\partial_t\varsigma^\alpha] + \widetilde{\mathcal{H}}(t,\varsigma^\beta,I_j,\varphi^j)\}dt \tag{9.5.4}$$

on a Poisson manifold

$$\mathbb{R}\times U = \mathbb{R}\times(V\times T^m). \tag{9.5.5}$$

Let $\widetilde{\mathcal{H}} = \mathcal{H}(I_i)$ be a Hamiltonian of an original autonomous completely integrable system on the toroidal domain U (9.5.5) equipped with the action-angle coordinates (I_k,φ^k). We introduce a desired holonomy operator by the appropriate choice of the connection Λ (9.5.3).

For this purpose, let us choose the initial data action-angle coordinates (I_k,ϕ^k) by the converse to the canonical transformation (7.7.16):

$$\varphi^k = \phi^k - t\partial^k\mathcal{H}. \tag{9.5.6}$$

With respect to these coordinates, the Hamiltonian of an original completely integrable system vanishes and the Hamiltonian form (9.5.4) reads

$$H_\varsigma = I_k d\phi^k - I_k[\Lambda_t^k(t,\phi^j) + \Lambda_\alpha^k(t,\varsigma^\beta,\phi^j)\partial_t\varsigma^\alpha]dt. \tag{9.5.7}$$

Let us put $\Lambda_t^k = 0$ by the choice of a reference frame associated to the initial data coordinates ϕ^k, and let us assume that coefficients Λ_α^k are independent of time, i.e., the part

$$\Lambda_W = d\sigma^\alpha \otimes (\partial_\alpha + \Lambda_\alpha^k\partial_k) \tag{9.5.8}$$

of the connection Λ (9.5.3) is a connection on the fibre bundle (9.5.1). Then the Hamiltonian form (9.5.7) reads

$$H_\varsigma = I_k d\phi^k - I_k\Lambda_\alpha^k(\varsigma^\beta,\phi^j)\partial_t\varsigma^\alpha dt. \tag{9.5.9}$$

Its Hamilton vector field (3.3.21) is

$$\gamma_H = \partial_t + \Lambda_\alpha^i\partial_t\varsigma^\alpha\partial_i - I_k\partial_i\Lambda_\alpha^k\partial_t\varsigma^\alpha\partial^i, \tag{9.5.10}$$

and it leads to the Hamilton equation

$$d_t\phi^i = \Lambda_\alpha^i(\varsigma(t),\phi^l)\partial_t\varsigma^\alpha, \tag{9.5.11}$$

$$d_tI_i = -I_k\partial_i\Lambda_\alpha^k(\varsigma(t),\phi^l)\partial_t\varsigma^\alpha. \tag{9.5.12}$$

Let us note that

$$V^*\Lambda_W = d\sigma^\alpha \otimes (\partial_\alpha + \Lambda^i_\alpha\partial_i - I_k\partial_i\Lambda^k_\alpha\partial^i) \tag{9.5.13}$$

is the lift (8.1.5) of the connection Λ_W (9.5.8) onto the fibre bundle

$$W \times (V \times T^m) \to W,$$

seen as a subbundle of the vertical cotangent bundle

$$V^*(W \times T^m) = W \times T^*T^m$$

of the fibre bundle (9.5.1). It follows that any solution $I_i(t)$, $\phi^i(t)$ of the Hamilton equation (9.5.11) – (9.5.12) (i.e., an integral curve of the Hamilton vector field (9.5.10)) is a horizontal lift of the curve $\varsigma(t) \subset W$ with respect to the connection $V^*\Lambda_W$ (9.5.13), i.e.,

$$I_i(t) = I_i(\varsigma(t)), \qquad \phi^i(t) = \phi^i(\varsigma(t)).$$

Thus, the right-hand side of the Hamilton equation (9.5.11) – (9.5.12) is the holonomy operator

$$\Delta = (\Lambda^i_\alpha\partial_t\varsigma^\alpha, -I_k\partial_i\Lambda^k_\alpha\partial_t\varsigma^\alpha). \tag{9.5.14}$$

It is not a linear operator, but the substitution of a solution $\phi(\varsigma(t))$ of the equation (9.5.11) into the Hamilton equation (9.5.12) results in a linear holonomy operator on the action variables I_i.

Let us show that the holonomy operator (9.5.14) is well defined. Since any vector field ϑ on $\mathbb{R}\times T^m$ such that $\vartheta\rfloor dt = 1$ is complete, the Hamilton equation (9.5.11) has solutions for any parameter function $\varsigma(t)$. It follows that any connection Λ_W (9.5.8) on the fibre bundle (9.5.1) is an Ehresmann connection, and so is its lift (9.5.13). Because $V^*\Lambda_W$ (9.5.13) is an Ehresmann connection, any curve $\varsigma([0,1]) \subset W$ can play a role of the parameter function in the holonomy operator Δ (9.5.14).

Now, let us return to the original action-angle coordinates (I_k, φ^k) by means of the canonical transformation (9.5.6). The perturbed Hamiltonian reads

$$\mathcal{H}' = I_k\Lambda^k_\alpha(\varsigma(t), \varphi^i - t\partial^i\mathcal{H}(I_j))\partial_t\varsigma^\alpha(t) + \mathcal{H}(I_j),$$

while the Hamilton equation (9.5.11) – (9.5.12) takes the form

$$\begin{aligned}\partial_t\varphi^i &= \partial^i\mathcal{H}(I_j) + \Lambda^i_\alpha(\varsigma(t), \varphi^l - t\partial^l\mathcal{H}(I_j))\partial_t\varsigma^\alpha(t)\\ &\quad -tI_k\partial^i\partial^s\mathcal{H}(I_j)\partial_s\Lambda^k_\alpha(\varsigma(t), \varphi^l - t\partial^l\mathcal{H}(I_j))\partial_t\varsigma^\alpha(t),\\ \partial_t I_i &= -I_k\partial_i\Lambda^k_\alpha(\varsigma(t), \varphi^l - t\partial^l\mathcal{H}(I_j))\partial_t\varsigma^\alpha(t).\end{aligned}$$

Its solution is

$$I_i(\varsigma(t)), \qquad \varphi^i(t) = \phi^i(\varsigma(t)) + t\partial^i\mathcal{H}(I_j(\varsigma(t))),$$

where $I_i(\varsigma(t))$, $\phi^i(\varsigma(t))$ is a solution of the Hamilton equation (9.5.11) – (9.5.12). We observe that the action variables I_k are driven only by the holonomy operator, while the angle variables φ^i have a non-geometric summand.

Let us emphasize that, in the construction of the holonomy operator (9.5.14), we do not impose any restriction on the connection Λ_W (9.5.8). Therefore, any connection on the fibre bundle (9.5.1) yields a holonomy operator of a completely integrable system. However, a glance at the expression (9.5.14) shows that this operator becomes zero on action variables if all coefficients Λ^k_λ of the connection Λ_W (9.5.8) are constant, i.e., Λ_W is a principal connection on the fibre bundle (9.5.1) seen as a principal bundle with the structure group T^m.

In order to quantize a non-autonomous completely integrable system on the Poisson toroidal domain $(U, \{,\}_V)$ (9.5.5) equipped with action-angle coordinates (I_i, φ^i), one may follow the instantwise geometric quantization of non-autonomous mechanics (Section 4.6). As a result, we can simply replace functions on T^m with those on $\mathbb{R}\times T^m$ [43]. Namely, the corresponding quantum algebra $\mathcal{A}\subset C^\infty(U)$ consists of affine functions

$$f = a^k(t,\varphi^j)I_k + b(t,\varphi^j) \tag{9.5.15}$$

of action coordinates I_k represented by the operators (9.3.6) in the space

$$\mathfrak{E} = \mathbb{C}^\infty(\mathbb{R}\times T^m) \tag{9.5.16}$$

of smooth complex functions $\psi(t,\varphi)$ on $\mathbb{R}\times T^m$. This space is provided with the structure of the pre-Hilbert $\mathbb{C}^\infty(\mathbb{R})$-module endowed with the non-degenerate $\mathbb{C}^\infty(\mathbb{R})$-bilinear form

$$\langle\psi|\psi'\rangle = \left(\frac{1}{2\pi}\right)^m \int_{T^m} \psi\overline{\psi}' d^m\varphi, \qquad \psi,\psi'\in\mathfrak{E}.$$

Its basis consists of the pull-back onto $\mathbb{R}\times T^m$ of the functions

$$\psi_{(n_r)} = \exp[i(n_r\phi^r)], \qquad (n_r) = (n_1,\ldots,n_m)\in\mathbb{Z}^m. \tag{9.5.17}$$

Furthermore, this quantization of a non-autonomous completely integrable system on the Poisson manifold $(U, \{,\}_V)$ is extended to the associated homogeneous completely integrable system on the symplectic annulus (7.7.12):

$$U' = \zeta^{-1}(U) = N'\times T^m \to N'$$

by means of the operator $\widehat{I}_0 = -i\partial_t$ in the pre-Hilbert module $\mathfrak{E}$ (9.5.16). Accordingly, the homogeneous Hamiltonian $\widehat{\mathcal{H}}^*$ is quantized as

$$\widehat{\mathcal{H}}^* = -i\partial_t + \widehat{\mathcal{H}}.$$

It is a Hamiltonian of a quantum non-autonomous completely integrable system. The corresponding Schrödinger equation is

$$\widehat{\mathcal{H}}^*\psi = -i\partial_t\psi + \widehat{\mathcal{H}}\psi = 0, \qquad \psi \in \mathfrak{E}. \tag{9.5.18}$$

For instance, a quantum Hamiltonian of an original autonomous completely integrable system seen as the non-autonomous one is

$$\widehat{\mathcal{H}}^* = -i\partial_t + \mathcal{H}(\widehat{I}_j).$$

Its spectrum

$$\widehat{\mathcal{H}}^*\psi_{(n_r)} = E_{(n_r)}\psi_{(n_r)}$$

on the basis $\{\psi_{(n_r)}\}$ (9.5.17) for $\mathfrak{E}$ (9.5.17) coincides with that of the autonomous Hamiltonian $\widehat{\mathcal{H}}(I_k) = \mathcal{H}(\widehat{I}_k)$. The Schrödinger equation (9.5.18) reads

$$\widehat{\mathcal{H}}^*\psi = -i\partial_t\psi + \mathcal{H}(-i\partial_k + \lambda_k)\psi = 0, \qquad \psi \in E.$$

Its solutions are the Fourier series

$$\psi = \sum_{(n_r)} B_{(n_r)} \exp[-itE_{(n_r)}]\psi_{(n_r)}, \qquad B_{(n_r)} \in \mathbb{C}.$$

Now, let us quantize this completely integrable system with respect to the initial data action-angle coordinates (I_i, ϕ^i). As was mentioned above, it is given on a toroidal domain U (9.5.5) provided with another fibration over $\mathbb{R}$. Its quantum algebra $\mathcal{A}_0 \subset C^\infty(U)$ consists of affine functions

$$f = a^k(t, \phi^j)I_k + b(t, \phi^j). \tag{9.5.19}$$

The canonical transformation (7.7.16) ensures an isomorphism of Poisson algebras $\mathcal{A}$ and $\mathcal{A}_0$. Functions f (9.5.19) are represented by the operators $\widehat{f}$ (9.3.6) in the pre-Hilbert module $\mathfrak{E}_0$ of smooth complex functions $\Psi(t, \phi)$ on $\mathbb{R} \times T^m$. Given its basis

$$\Psi_{(n_r)}(\phi) = [in_r\phi^r],$$

the operators $\widehat{I}_k$ and $\widehat{\psi}_{(n_r)}$ take the form

$$\begin{aligned} \widehat{I}_k\psi_{(n_r)} &= (n_k + \lambda_k)\psi_{(n_r)}, \\ \psi_{(n_r)}\psi_{(n'_r)} &= \psi_{(n_r)}\psi_{(n'_r)} = \psi_{(n_r+n'_r)}. \end{aligned} \tag{9.5.20}$$

The Hamiltonian of a quantum completely integrable system with respect to the initial data variables is $\widehat{\mathcal{H}}_0^* = -i\partial_t$. Then one easily obtains the isometric isomorphism

$$R(\psi_{(n_r)}) = \exp[itE_{(n_r)}]\Psi_{(n_r)}, \qquad \langle R(\psi)|R(\psi')\rangle = \langle\psi|\psi'\rangle, \qquad (9.5.21)$$

of the pre-Hilbert modules $\mathfrak{E}$ and $\mathfrak{E}_0$ which provides the equivalence

$$\widehat{I}_i = R^{-1}\widehat{I}_i R, \qquad \widehat{\psi}_{(n_r)} = R^{-1}\widehat{\Psi}_{(n_r)}R, \qquad \widehat{\mathcal{H}}^* = R^{-1}\widehat{\mathcal{H}}_0^* R \qquad (9.5.22)$$

of the quantizations of a completely integrable system with respect to the original and initial data action-angle variables.

In view of the isomorphism (9.5.22), let us first construct a holonomy operator of a quantum completely integrable system $(\mathcal{A}_0, \widehat{\mathcal{H}}_0^*)$ with respect to the initial data action-angle coordinates. Let us consider the perturbed homogeneous Hamiltonian

$$\mathbf{H}_\varsigma = \mathcal{H}_0^* + \mathbf{H}_1 = I_0 + \partial_t\varsigma^\alpha(t)\Lambda_\alpha^k(\varsigma(t), \phi^j)I_k$$

of the classical perturbed completely integrable system (9.5.9). Its perturbation term $\mathbf{H}_1$ is of the form (9.5.15) and, therefore, is quantized by the operator

$$\widehat{\mathbf{H}}_1 = -i\partial_t\varsigma^\alpha\widehat{\Delta}_\alpha = -i\partial_t\varsigma^\alpha\left[\Lambda_\alpha^k\partial_k + \frac{1}{2}\partial_k(\Lambda_\alpha^k) + i\lambda_k\Lambda_\alpha^k\right].$$

The quantum Hamiltonian $\widehat{\mathbf{H}}_\varsigma = \widehat{\mathcal{H}}_0^* + \widehat{\mathbf{H}}_1$ defines the Schrödinger equation

$$\partial_t\Psi + \partial_t\varsigma^\alpha\left[\Lambda_\alpha^k\partial_k + \frac{1}{2}\partial_k(\Lambda_\alpha^k) + i\lambda_k\Lambda_\alpha^k\right]\Psi = 0. \qquad (9.5.23)$$

If its solution exists, it can be written by means of the evolution operator $U(t)$ which is reduced to the geometric factor

$$U_1(t) = T\exp\left[i\int_0^t \partial_{t'}\varsigma^\alpha(t')\widehat{\Delta}_\alpha(t')dt'\right].$$

The latter can be viewed as a displacement operator along the curve $\varsigma[0,1] \subset W$ with respect to the connection

$$\widehat{\Lambda}_W = d\sigma^\alpha(\partial_\alpha + \widehat{\Delta}_\alpha) \qquad (9.5.24)$$

on the $\mathbb{C}^\infty(W)$-module $\mathbb{C}^\infty(W \times T^m)$ of smooth complex functions on $W \times T^m$ (see Section 4.6). Let us study weather this displacement operator exists.

Given a connection Λ_W (9.5.8), let $\Phi^i(t,\phi)$ denote the flow of the complete vector field

$$\partial_t + \Lambda_\alpha(\varsigma,\phi)\partial_t\varsigma^\alpha$$

on $\mathbb{R}\times T^m$. It is a solution of the Hamilton equation (9.5.11) with the initial data ϕ. We need the inverse flow $(\Phi^{-1})^i(t,\phi)$ which obeys the equation

$$\begin{aligned}\partial_t(\Phi^{-1})^i(t,\phi) &= -\partial_t\varsigma^\alpha\Lambda^i_\alpha(\varsigma,(\Phi^{-1})^i(t,\phi))\\ &= -\partial_t\varsigma^\alpha\Lambda^k_\alpha(\varsigma,\phi)\partial_k(\Phi^{-1})^i(t,\phi).\end{aligned}$$

Let Ψ_0 be an arbitrary complex half-form Ψ_0 on T^m possessing identical transition functions, and let the same symbol stand for its pull-back onto $\mathbb{R}\times T^m$. Given its pull-back

$$(\Phi^{-1})^*\Psi_0 = \det\left(\frac{\partial(\Phi^{-1})^i}{\partial\phi^k}\right)^{1/2}\Psi_0(\Phi^{-1}(t,\phi)), \qquad (9.5.25)$$

it is readily observed that

$$\Psi = (\Phi^{-1})^*\Psi_0\exp[i\lambda_k\phi^k] \qquad (9.5.26)$$

obeys the Schrödinger equation (9.5.23) with the initial data Ψ_0. Because of the multiplier $\exp[i\lambda_k\phi^k]$, the function Ψ (9.5.26) however is ill defined, unless all numbers λ_k equal 0 or $\pm1/2$. Let us note that, if some numbers λ_k are equal to $\pm1/2$, then $\Psi_0\exp[i\lambda_k\phi^k]$ is a half-density on T^m whose transition functions equal ±1, i.e., it is a section of a non-trivial metalinear bundle over T^m.

Thus, we observe that, if λ_k equal 0 or $\pm1/2$, then the displacement operator always exists and $\Delta = i\mathbf{H}_1$ is a holonomy operator. Because of the action law (9.5.20), it is essentially infinite-dimensional.

For instance, let Λ_W (9.5.8) be the above mentioned principal connection, i.e., Λ^k_α =const. Then the Schrödinger equation (9.5.23) where $\lambda_k = 0$ takes the form

$$\partial_t\Psi(t,\phi^j) + \partial_t\varsigma^\alpha(t)\Lambda^k_\alpha\partial_k\Psi(t,\phi^j) = 0,$$

and its solution (9.5.25) is

$$\Psi(t,\phi^j) = \Psi_0(\phi^j - (\varsigma^\alpha(t) - \varsigma^\alpha(0))\Lambda^j_\alpha).$$

The corresponding evolution operator $U(t)$ reduces to Berry's phase multiplier

$$U_1\Psi_{(n_r)} = \exp[-in_j(\varsigma^\alpha(t) - \varsigma^\alpha(0))\Lambda^j_\alpha]\Psi_{(n_r)}, \qquad n_j\in(n_r).$$

It keeps the eigenvectors of the action operators $\widehat{I}_i$.

In order to return to the original action-angle variables, one can employ the morphism R (9.5.21). The corresponding Hamiltonian reads

$$\mathbf{H} = R^{-1}\mathbf{H}_\varsigma R.$$

The key point is that, due to the relation (9.5.22), the action operators $\widehat{I}_i$ have the same mean values

$$\langle I_k\psi|\psi\rangle = \langle I_k\Psi|\Psi\rangle, \qquad \Psi = R(\psi),$$

with respect both to the original and the initial data action-angle variables. Therefore, these mean values are defined only by the holonomy operator.

In conclusion, let us note that, since action variables are driven only by a holonomy operator, one can use this operator in order to perform a dynamic transition between classical solutions or quantum states of an unperturbed completely integrable system by an appropriate choice of a parameter function ς. A key point is that this transition can take an arbitrary short time because we are entirely free with time parametrization of ς and can choose it quickly changing, in contrast with slowly varying parameter functions in adiabatic models. This fact makes non-adiabatic holonomy operators in completely integrable systems promising for several applications, e.g., quantum control and quantum computation.

Chapter 10

Relativistic mechanics

If a configuration space of a mechanical system has no preferable fibration $Q \to \mathbb{R}$, we obtain a general formulation of relativistic mechanics. A velocity space of relativistic mechanics is the first order jet manifold $J_1^1 Q$ of one-dimensional submanifolds of a configuration space Q [106; 139]. This notion of jets generalizes that of jets of sections of fibre bundles which are utilized in field theory and non-relativistic mechanics [68; 106]. The jet bundle $J_1^1 Q \to Q$ is projective, and one can think of its fibres as being spaces of the three-velocities of relativistic mechanics (Section 10.2).

The four-velocities of a relativistic system are represented by elements of the tangent bundle TQ of the configuration space Q, while the cotangent bundle T^*Q, endowed with the canonical symplectic form, plays a role of the phase space of relativistic theory. As a result, Hamiltonian relativistic mechanics can be seen as a constraint Dirac system on the hyperboloids of relativistic momenta in the phase space T^*Q.

10.1 Jets of submanifolds

Jets of sections of fibre bundles are particular jets of submanifolds of a manifold [53; 68; 95].

Given an m-dimensional smooth real manifold Z, a *k-order jet of n-dimensional submanifolds* of Z at a point $z \in Z$ is defined as an equivalence class $j_z^k S$ of n-dimensional imbedded submanifolds of Z through z which are tangent to each other at z with order $k \geq 0$. Namely, two submanifolds

$$i_S : S \to Z, \qquad i_{S'} : S' \to Z$$

through a point $z \in Z$ belong to the same equivalence class $j_z^k S$ if and only if the images of the k-tangent morphisms

$$T^k i_S : T^k S \to T^k Z, \qquad T^k i_{S'} : T^k S' \to T^k Z$$

coincide with each other. The set

$$J^k_n Z = \bigcup_{z\in Z} j^k_z S$$

of k-order jets of submanifolds is a finite-dimensional real smooth manifold, called the *k-order jet manifold of submanifolds*. For the sake of convenience, we put $J^0_n Z = Z$.

If $k > 0$, let $Y \to X$ be an m-dimensional fibre bundle over an n-dimensional base X and $J^k Y$ the k-order jet manifold of sections of $Y \to X$. Given an imbedding $\Phi : Y \to Z$, there is the natural injection

$$J^k\Phi : J^k Y \to J^k_n Z, \qquad j^k_x s \to [\Phi\circ s]^k_{\Phi(s(x))}, \tag{10.1.1}$$

where s are sections of $Y \to X$. This injection defines a chart on $J^k_n Z$. These charts provide a manifold atlas of $J^k_n Z$.

Let us restrict our consideration to first order jets of submanifolds. There is obvious one-to-one correspondence

$$\lambda_{(1)} : j^1_z S \to V_{j^1_z S} \subset T_z Z \tag{10.1.2}$$

between the jets $j^1_z S$ at a point $z \in Z$ and the n-dimensional vector subspaces of the tangent space $T_z Z$ of Z at z. It follows that $J^1_n Z$ is a fibre bundle

$$\rho : J^1_n Z \to Z \tag{10.1.3}$$

with the structure group $GL(n, m-n;\mathbb{R})$ of linear transformations of the vector space $\mathbb{R}^m$ which preserve its subspace $\mathbb{R}^n$. The typical fibre of the fibre bundle (10.1.3) is the *Grassmann manifold*

$$\mathfrak{G}(n, m-n;\mathbb{R}) = GL(m;\mathbb{R})/GL(n, m-n;\mathbb{R}).$$

This fibre bundle possesses the following coordinate atlas.

Let $\{(U; z^A)\}$ be a coordinate atlas of Z. Though $J^0_n Z = Z$, let us provide $J^0_n Z$ with an atlas where every chart $(U; z^A)$ on a domain $U \subset Z$ is replaced with the

$$\binom{m}{n} = \frac{m!}{n!(m-n)!}$$

charts on the same domain U which correspond to different partitions of the collection $(z^1\cdots z^A)$ in the collections of n and $m-n$ coordinates

$$(U; x^\lambda, y^i), \qquad \lambda = 1,\ldots,n, \qquad i = 1,\ldots,m-n. \tag{10.1.4}$$

The transition functions between the coordinate charts (10.1.4) of $J^0_n Z$ associated with a coordinate chart (U, z^A) of Z are reduced to exchange between coordinates x^λ and y^i. Transition functions between arbitrary coordinate charts of the manifold $J^0_n Z$ take the form

$$x'^\lambda = x'^\lambda(x^\mu, y^k), \qquad y'^i = y'^i(x^\mu, y^k). \tag{10.1.5}$$

Given the coordinate atlas (10.1.4) – (10.1.5) of a manifold $J^0_n Z$, the first order jet manifold $J^1_n Z$ is endowed with an atlas of adapted coordinates

$$(\rho^{-1}(U) = U \times \mathbb{R}^{(m-n)n}; x^\lambda, y^i, y^i_\lambda), \tag{10.1.6}$$

possessing transition functions

$$y'^i_\lambda = \left(\frac{\partial y'^i}{\partial y^j} y^j_\alpha + \frac{\partial y'^i}{\partial x^\alpha}\right)\left(\frac{\partial x^\alpha}{\partial y'^k} y'^k_\lambda + \frac{\partial x^\alpha}{\partial x'^\lambda}\right). \tag{10.1.7}$$

It is readily observed that the affine transition functions (11.3.1) are a particular case of the coordinate transformations (10.1.7) when the transition functions x^α (10.1.5) are independent of coordinates y'^i.

10.2 Lagrangian relativistic mechanics

As was mentioned above, a velocity space of relativistic mechanics is the first order jet manifold $J^1_1 Q$ of one-dimensional submanifolds of a configuration space Q [106; 139].

Given an m-dimensional manifold Q coordinated by (q^λ), let us consider the jet manifold $J^1_1 Q$ of its one-dimensional submanifolds. Let us provide $Q = J^0_1 Q$ with the coordinates (10.1.4):

$$(U; x^0 = q^0, y^i = q^i) = (U; q^\lambda). \tag{10.2.1}$$

Then the jet manifold

$$\rho : J^1_1 Q \to Q$$

is endowed with coordinates (10.1.6):

$$(\rho^{-1}(U); q^0, q^i, q^i_0), \tag{10.2.2}$$

possessing transition functions (10.1.5), (10.1.7) which read

$$q'^0 = q'^0(q^0, q^k), \qquad q'^0 = q'^0(q^0, q^k), \tag{10.2.3}$$

$$q'^i_0 = \left(\frac{\partial q'^i}{\partial q^j} q^j_0 + \frac{\partial q'^i}{\partial q^0}\right)\left(\frac{\partial q'^0}{\partial q^j} q^j_0 + \frac{\partial q'^0}{\partial q^0}\right)^{-1}. \tag{10.2.4}$$

A glance at the transformation law (10.2.4) shows that $J^1_1Q \to Q$ is a fibre bundle in projective spaces.

Example 10.2.1. Let $Q = M^4 = \mathbb{R}^4$ be a *Minkowski space* whose Cartesian coordinates (q^λ), $\lambda = 0, 1, 2, 3$, are subject to the *Lorentz transformations* (10.2.3):

$$q'^0 = q^0\text{ch}\alpha - q^1\text{sh}\alpha, \quad q'^1 = -q^0\text{sh}\alpha + q^1\text{ch}\alpha, \quad q'^{2,3} = q^{2,3}. \qquad (10.2.5)$$

Then q'^i (10.2.4) are exactly the Lorentz transformations

$$q'^1_0 = \frac{q^1_0\text{ch}\alpha - \text{sh}\alpha}{-q^1_0\text{sh}\alpha + \text{ch}\alpha} \qquad q'^{2,3}_0 = \frac{q^{2,3}_0}{-q^1_0\text{sh}\alpha + \text{ch}\alpha}$$

of three-velocities in relativistic mechanics [106; 139].

In view of Example 10.2.1, one can think of the velocity space J^1_1Q of relativistic mechanics as being a space of *three-velocities*. For the sake of convenience, we agree to call J^1_1Q the *three-velocity space* and its coordinate transformations (10.2.3) – (10.2.4) the *relativistic transformations*, though a dimension of Q need not equal $3+1$.

Given the coordinate chart (10.2.2) of J^1_1Q, one can regard $\rho^{-1}(U) \subset J^1_1Q$ as the first order jet manifold J^1U of sections of the fibre bundle

$$\pi : U \ni (q^0, q^i) \to (q^0) \in \pi(U) \subset \mathbb{R}. \qquad (10.2.6)$$

Then three-velocities $(q^i_0) \in \rho^{-1}(U)$ of a relativistic system on U can be treated as absolute velocities of a local non-relativistic system on the configuration space U (10.2.6). However, this treatment is broken under the relativistic transformations $q^i_0 \to q'^i_0$ (10.2.3) since they are not affine. One can develop first order Lagrangian formalism with a Lagrangian

$$L = \mathcal{L}dq^0 \in \mathcal{O}^{0,1}(\rho^{-1}(U))$$

on a coordinate chart $\rho^{-1}(U)$, but this Lagrangian fails to be globally defined on J^1_1Q (see Remark 10.2.1 below). The graded differential algebra $\mathcal{O}^*(\rho^{-1}(U))$ of exterior forms on $\rho^{-1}(U)$ is generated by horizontal forms dq^0 and contact forms $dq^i - q^i_0dq^0$. Coordinate transformations (10.2.3) preserve the ideal of contact forms, but horizontal forms are not transformed into horizontal forms, unless coordinate transition functions q^0 (10.2.3) are independent of coordinates q'^i.

In order to overcome this difficulty, let us consider a trivial fibre bundle

$$Q_R = \mathbb{R} \times Q \to \mathbb{R}, \qquad (10.2.7)$$

whose base $\mathbb{R}$ is endowed with a Cartesian coordinate τ [68]. This fibre bundle is provided with an atlas of coordinate charts

$$(\mathbb{R} \times U; \tau, q^\lambda), \tag{10.2.8}$$

where $(U; q^0, q^i)$ are the coordinate charts (10.2.1) of the manifold $J_1^0 Q$. The coordinate charts (10.2.8) possess transition functions (10.2.3). Let $J^1 Q_R$ be the first order jet manifold of the fibre bundle (10.2.7). Since the trivialization (10.2.7) is fixed, there is the canonical isomorphism (1.1.4) of $J^1 Q_R$ to the vertical tangent bundle

$$J^1 Q_R = V Q_R = \mathbb{R} \times TQ \tag{10.2.9}$$

of $Q_R \to \mathbb{R}$.

Given the coordinate atlas (10.2.8) of Q_R, the jet manifold $J^1 Q_R$ is endowed with the coordinate charts

$$((\pi^1)^{-1}(\mathbb{R} \times U) = \mathbb{R} \times U \times \mathbb{R}^m; \tau, q^\lambda, q^\lambda_\tau), \tag{10.2.10}$$

possessing transition functions

$$q'^\lambda_\tau = \frac{\partial q'^\lambda}{\partial q^\mu} q^\mu_\tau. \tag{10.2.11}$$

Relative to the coordinates (10.2.10), the isomorphism (10.2.9) takes the form

$$(\tau, q^\mu, q^\mu_\tau) \to (\tau, q^\mu, \dot q^\mu = q^\mu_\tau). \tag{10.2.12}$$

Example 10.2.2. Let $Q = M^4$ be a Minkowski space in Example 10.2.1 whose Cartesian coordinates (q^0, q^i) are subject to the Lorentz transformations (10.2.5). Then the corresponding transformations (10.2.11) take the form

$$q'^0_\tau = q^0_\tau \mathrm{ch}\alpha - q^1_\tau \mathrm{sh}\alpha, \quad q'^1_\tau = -q^0_\tau \mathrm{sh}\alpha + q^1_\tau \mathrm{ch}\alpha, \quad q'^{2,3}_\tau = q^{2,3}_\tau$$

of transformations of four-velocities in relativistic mechanics.

In view of Example 10.2.2, we agree to call fibre elements of $J^1 Q_R \to Q_R$ the *four-velocities* though the dimension of Q need not equal 4. Due to the canonical isomorphism $q^\lambda_\tau \to \dot q^\lambda$ (10.2.9), by four-velocities also are meant the elements of the tangent bundle TQ, which is called the *space of four-velocities*.

Obviously, the non-zero jet (10.2.12) of sections of the fibre bundle (10.2.7) defines some jet of one-dimensional subbundles of the manifold $\{\tau\} \times Q$ through a point $(q^0, q^i) \in Q$, but this is not one-to-one correspondence.

Since non-zero elements of J^1Q_R characterize jets of one-dimensional submanifolds of Q, one hopes to describe the dynamics of one-dimensional submanifolds of a manifold Q as that of sections of the fibre bundle (10.2.7). For this purpose, let us refine the relation between elements of the jet manifolds J^1_1Q and J^1Q_R.

Let us consider the manifold product $\mathbb{R}\times J^1_1Q$. It is a fibre bundle over Q_R. Given a coordinate atlas (10.2.8) of Q_R, this product is endowed with the coordinate charts

$$(U_R\times\rho^{-1}(U)=U_R\times U\times\mathbb{R}^{m-1};\tau,q^0,q^i,q^i_0), \tag{10.2.13}$$

possessing transition functions (10.2.3) – (10.2.4). Let us assign to an element (τ,q^0,q^i,q^i_0) of the chart (10.2.13) the elements $(\tau,q^0,q^i,q^0_\tau,q^i_\tau)$ of the chart (10.2.10) whose coordinates obey the relations

$$q^i_0q^0_\tau=q^i_\tau. \tag{10.2.14}$$

These elements make up a one-dimensional vector space. The relations (10.2.14) are maintained under coordinate transformations (10.2.4) and (10.2.11) [68]. Thus, one can associate:

$$(\tau,q^0,q^i,q^i_0)\to\{(\tau,q^0,q^i,q^0_\tau,q^i_\tau)\,|\,q^i_0q^0_\tau=q^i_\tau\}, \tag{10.2.15}$$

to each element of the manifold $\mathbb{R}\times J^1_1Q$ a one-dimensional vector space in the jet manifold J^1Q_R. This is a subspace of elements

$$q^0_\tau(\partial_0+q^i_0\partial_i)$$

of a fibre of the vertical tangent bundle (10.2.9) at a point (τ,q^0,q^i). Conversely, given a non-zero element (10.2.12) of J^1Q_R, there is a coordinate chart (10.2.10) such that this element defines a unique element of $\mathbb{R}\times J^1_1Q$ by the relations

$$q^i_0=\frac{q^i_\tau}{q^0_\tau}. \tag{10.2.16}$$

Thus, we have shown the following. Let (τ,q^λ) further be arbitrary coordinates on the product Q_R (10.2.7) and $(\tau,q^\lambda,q^\lambda_\tau)$ the corresponding coordinates on the jet manifold J^1Q_R.

Theorem 10.2.1. *(i) Any jet of submanifolds through a point $q\in Q$ defines some (but not unique) jet of sections of the fibre bundle Q_R (10.2.7) through a point $\tau\times q$ for any $\tau\in\mathbb{R}$ in accordance with the relations (10.2.14).*

(ii) Any non-zero element of J^1Q_R defines a unique element of the jet manifold J^1_1Q by means of the relations (10.2.16). However, non-zero elements of J^1Q_R can correspond to different jets of submanifolds.

(iii) Two elements $(\tau,q^\lambda,q^\lambda_\tau)$ and $(\tau,q^\lambda,q'^\lambda_\tau)$ of J^1Q_R correspond to the same jet of submanifolds if $q'^\lambda_\tau=rq^\lambda_\tau$, $r\in\mathbb{R}\setminus\{0\}$.

In the case of a Minkowski space $Q = M^4$ in Examples 10.2.1 and 10.2.2, the equalities (10.2.14) and (10.2.16) are the familiar relations between three- and four-velocities.

Based on Theorem 10.2.1, we can develop Lagrangian theory of one-dimensional submanifolds of a manifold Q as that of sections of the fibre bundle Q_R (10.2.7). Let

$$L = \mathcal{L}(\tau, q^\lambda, q^\lambda_\tau)d\tau \tag{10.2.17}$$

be a first order Lagrangian on the jet manifold J^1Q_R. The corresponding Lagrange operator (2.1.23) reads

$$\delta L = \mathcal{E}_\lambda dq^\lambda \wedge d\tau, \qquad \mathcal{E}_\lambda = \partial_\lambda \mathcal{L} - d_\tau \partial^\tau_\lambda \mathcal{L}. \tag{10.2.18}$$

It yields the Lagrange equation

$$\mathcal{E}_\lambda = \partial_\lambda \mathcal{L} - d_\tau \partial^\tau_\lambda \mathcal{L} = 0. \tag{10.2.19}$$

In accordance with Theorem 10.2.1, it seems reasonable to require that, in order to describe jets of one-dimensional submanifolds of Q, the Lagrangian L (10.2.17) on J^1Q_R possesses a gauge symmetry given by vector fields $u = \chi(\tau)\partial_\tau$ on Q_R or, equivalently, their vertical part (2.5.6):

$$u_V = -\chi q^\lambda_\tau \partial_\lambda, \tag{10.2.20}$$

which are generalized vector fields on Q_R. Then the variational derivatives of this Lagrangian obey the Noether identity

$$q^\lambda_\tau \mathcal{E}_\lambda = 0 \tag{10.2.21}$$

(see the relations (2.6.7) – (2.6.8)). We call such a Lagrangian the *relativistic Lagrangian.*

In order to obtain a generic form of a relativistic Lagrangian L, let us regard the Noether identity (10.2.21) as an equation for L. It admits the following solution. Let

$$\frac{1}{2N!}G_{\alpha_1\ldots\alpha_{2N}}(q^\nu)dq^{\alpha_1} \vee \cdots \vee dq^{\alpha_{2N}}$$

be a symmetric tensor field on Q such that the function

$$G = G_{\alpha_1\ldots\alpha_{2N}}(q^\nu)\dot q^{\alpha_1} \cdots \dot q^{\alpha_{2N}} \tag{10.2.22}$$

is positive:

$$G > 0, \tag{10.2.23}$$

everywhere on $TQ \setminus \widehat{0}(Q)$. Let $A = A_\mu(q^\nu)dq^\mu$ be a one-form on Q. Given the pull-back of G and A onto J^1Q_R due to the canonical isomorphism (10.2.9), we define a Lagrangian

$$L = (G^{1/2N} + q^\mu_\tau A_\mu)d\tau, \qquad G = G_{\alpha_1\ldots\alpha_{2N}} q^{\alpha_1}_\tau \cdots q^{\alpha_{2N}}_\tau, \tag{10.2.24}$$

on $J^1Q_R \setminus (\mathbb{R} \times \widehat{0}(Q))$ where $\widehat{0}$ is the global zero section of $TQ \to Q$. The corresponding Lagrange equation reads

$$\mathcal{E}_\lambda = \frac{\partial_\lambda G}{2NG^{1-1/2N}} - d_\tau\left(\frac{\partial^\tau_\lambda G}{2NG^{1-1/2N}}\right) + F_{\lambda\mu}q^\mu_\tau \tag{10.2.25}$$

$$= E_\beta[\delta^\beta_\lambda - q^\beta_\tau G_{\lambda\nu_2\ldots\nu_{2N}} q^{\nu_2}_\tau \cdots q^{\nu_{2N}}_\tau G^{-1}]G^{1/2N-1} = 0,$$

$$E_\beta = \left(\frac{\partial_\beta G_{\mu\alpha_2\ldots\alpha_{2N}}}{2N} - \partial_\mu G_{\beta\alpha_2\ldots\alpha_{2N}}\right) q^\mu_\tau q^{\alpha_2}_\tau \cdots q^{\alpha_{2N}}_\tau \tag{10.2.26}$$

$$- (2N-1)G_{\beta\mu\alpha_3\ldots\alpha_{2N}} q^\mu_{\tau\tau} q^{\alpha_3}_\tau \cdots q^{\alpha_{2N}}_\tau + G^{1-1/2N}F_{\beta\mu}q^\mu_\tau,$$

$$F_{\lambda\mu} = \partial_\lambda A_\mu - \partial_\mu A_\lambda.$$

It is readily observed that the variational derivatives $\mathcal{E}_\lambda$ (10.2.25) satisfy the Noether identity (10.2.21). Moreover, any relativistic Lagrangian obeying the Noether identity (10.2.21) is of type (10.2.24).

A glance at the Lagrange equation (10.2.25) shows that it holds if

$$E_\beta = \Phi G_{\beta\nu_2\ldots\nu_{2N}} q^{\nu_2}_\tau \cdots q^{\nu_{2N}}_\tau G^{-1}, \tag{10.2.27}$$

where Φ is some function on J^1Q_R. In particular, we consider the equation

$$E_\beta = 0. \tag{10.2.28}$$

Because of the Noether identity (10.2.21), the system of equations (10.2.25) is underdetermined. To overcome this difficulty, one can complete it with some additional equation. Given the function G (10.2.24), let us choose the condition

$$G = 1. \tag{10.2.29}$$

Owing to the property (10.2.23), the function G (10.2.24) possesses a nowhere vanishing differential. Therefore, its level surface W_G defined by the condition (10.2.29) is a submanifold of J^1Q_R.

Our choice of the equation (10.2.28) and the condition (10.2.29) is motivated by the following facts.

Lemma 10.2.1. *Any solution of the Lagrange equation (10.2.25) living in the submanifold W_G is a solution of the equation (10.2.28).*

Proof. A solution of the Lagrange equation (10.2.25) living in the submanifold W_G obeys the system of equations

$$\mathcal{E}_\lambda = 0, \qquad G = 1. \tag{10.2.30}$$

Therefore, it satisfies the equality

$$d_\tau G = 0. \tag{10.2.31}$$

Then a glance at the expression (10.2.25) shows that the equations (10.2.30) are equivalent to the equations

$$\begin{aligned} E_\lambda = & \left(\frac{\partial_\lambda G_{\mu\alpha_2\ldots\alpha_{2N}}}{2N} - \partial_\mu G_{\lambda\alpha_2\ldots\alpha_{2N}}\right) q^\mu_\tau q^{\alpha_2}_\tau \cdots q^{\alpha_{2N}}_\tau \\ & - (2N-1) G_{\beta\mu\alpha_3\ldots\alpha_{2N}} q^\mu_{\tau\tau} q^{\alpha_3}_\tau \cdots q^{\alpha_{2N}}_\tau + F_{\beta\mu} q^\mu_\tau = 0, \qquad (10.2.32) \\ G = & \, G_{\alpha_1\ldots\alpha_{2N}} q^{\alpha_1}_\tau \cdots q^{\alpha_{2N}}_\tau = 1. \end{aligned}$$

□

Lemma 10.2.2. *Solutions of the equation (10.2.28) do not leave the submanifold W_G (10.2.29).*

Proof. Since

$$d_\tau G = -\frac{2N}{2N-1} q^\beta_\tau E_\beta,$$

any solution of the equation (10.2.28) intersecting the submanifold W_G (10.2.29) obeys the equality (10.2.31) and, consequently, lives in W_G. □

The system of equations (10.2.32) is called the *relativistic equation.* Its components E_λ (10.2.26) are not independent, but obeys the relation

$$q^\beta_\tau E_\beta = -\frac{2N-1}{2N} d_\tau G = 0, \qquad G = 1,$$

similar to the Noether identity (10.2.21). The condition (10.2.29) is called the *relativistic constraint.*

Though the equation (10.2.25) for sections of a fibre bundle $Q_R \to \mathbb{R}$ is underdetermined, it is determined if, given a coordinate chart $(U; q^0, q^i)$ (10.2.1) of Q and the corresponding coordinate chart (10.2.8) of Q_R, we rewrite it in the terms of three-velocities q^i_0 (10.2.16) as an equation for sections of a fibre bundle $U \to \pi(U)$ (10.2.6).

Let us denote

$$\overline{G}(q^\lambda, q^i_0) = (q^0_\tau)^{-2N} G(q^\lambda, q^\lambda_\tau), \qquad q^0_\tau \neq 0. \tag{10.2.33}$$

Then we have

$$\mathcal{E}_i = q^0_\tau \left[\frac{\partial_i \overline{G}}{2N\overline{G}^{1-1/2N}} - (q^0_\tau)^{-1} d_\tau \left(\frac{\partial^0_i \overline{G}}{2N\overline{G}^{1-1/2N}} \right) + F_{ij} q^j_0 + F_{i0} \right].$$

Let us consider a solution $\{s^\lambda(\tau)\}$ of the equation (10.2.25) such that $\partial_\tau s^0$ does not vanish and there exists an inverse function $\tau(q^0)$. Then this solution can be represented by sections

$$s^i(\tau) = (\overline{s}^i \circ s^0)(\tau) \tag{10.2.34}$$

of the composite bundle

$$\mathbb{R} \times U \to \mathbb{R} \times \pi(U) \to \mathbb{R}$$

where $\overline{s}^i(q^0) = s^i(\tau(q^0))$ are sections of $U \to \pi(U)$ and $s^0(\tau)$ are sections of $\mathbb{R} \times \pi(U) \to \mathbb{R}$. Restricted to such solutions, the equation (10.2.25) is equivalent to the equation

$$\begin{aligned}\overline{\mathcal{E}}_i &= \frac{\partial_i \overline{G}}{2N\overline{G}^{1-1/2N}} - d_0\left(\frac{\partial_i^0 \overline{G}}{2N\overline{G}^{1-1/2N}}\right) \\ &\quad + F_{ij}q_0^j + F_{i0} = 0, \\ \overline{\mathcal{E}}_0 &= -q_0^i \overline{\mathcal{E}}_i.\end{aligned} \tag{10.2.35}$$

for sections $\overline{s}^i(q^0)$ of a fibre bundle $U \to \pi(U)$.

It is readily observed that the equation (10.2.35) is the Lagrange equation of the Lagrangian

$$\overline{L} = (\overline{G}^{1/2N} + q_0^i A_i + A_0)dq^0 \tag{10.2.36}$$

on the jet manifold J^1U of a fibre bundle $U \to \pi(U)$.

Remark 10.2.1. Both the equation (10.2.35) and the Lagrangian (10.2.36) are defined only on a coordinate chart (10.2.1) of Q since they are not maintained by transition functions (10.2.3) – (10.2.4).

A solution $\overline{s}^i(q^0)$ of the equation (10.2.35) defines a solution $s^\lambda(\tau)$ (10.2.34) of the equation (10.2.25) up to an arbitrary function $s^0(\tau)$. The relativistic constraint (10.2.29) enables one to overcome this ambiguity as follows.

Let us assume that, restricted to the coordinate chart $(U; q^0, q^i)$ (10.2.1) of Q, the relativistic constraint (10.2.29) has no solution $q_\tau^0 = 0$. Then it is brought into the form

$$(q_\tau^0)^{2N}\overline{G}(q^\lambda, q_0^i) = 1, \tag{10.2.37}$$

where $\overline{G}$ is the function (10.2.33). With the condition (10.2.37), every three-velocity (q_0^i) defines a unique pair of four-velocities

$$q_\tau^0 = \pm(\overline{G}(q^\lambda, q_0^i))^{1/2N}, \qquad q_\tau^i = q_\tau^0 q_0^i. \tag{10.2.38}$$

Accordingly, any solution $\overline{s}^i(q^0)$ of the equation (10.2.35) leads to solutions

$$\tau(q^0) = \pm \int (\overline{G}(q^0, \overline{s}^i(q^0), \partial_0 \overline{s}^i(q_0))^{-1/2N} dq^0,$$

$$s^i(\tau) = s^0(\tau)(\partial_i \overline{s}^i)(s^0(\tau))$$

of the equation (10.2.30) and, equivalently, the relativistic equation (10.2.32).

Example 10.2.3. Let $Q = M^4$ be a Minkowski space provided with the Minkowski metric $\eta_{\mu\nu}$ of signature $(+,---)$. This is the case of Special Relativity. Let $\mathcal{A}_\lambda dq^\lambda$ be a one-form on Q. Then

$$L = [m(\eta_{\mu\nu} q^\mu_\tau q^\nu_\tau)^{1/2} + e\mathcal{A}_\mu q^\mu_\tau] d\tau, \qquad m, e \in \mathbb{R}, \tag{10.2.39}$$

is a relativistic Lagrangian on $J^1 Q_R$ which satisfies the Noether identity (10.2.21). The corresponding relativistic equation (10.2.32) reads

$$m\eta_{\mu\nu} q^\nu_{\tau\tau} - eF_{\mu\nu} q^\nu_\tau = 0, \tag{10.2.40}$$

$$\eta_{\mu\nu} q^\mu_\tau q^\nu_\tau = 1. \tag{10.2.41}$$

This describes a relativistic massive charge in the presence of an electromagnetic field $\mathcal{A}$. It follows from the relativistic constraint (10.2.41) that $(q^0_\tau)^2 \geq 1$. Therefore, passing to three-velocities, we obtain the Lagrangian (10.2.36):

$$\overline{L} = \left[m(1 - \sum_i (q^i_0)^2)^{1/2} + e(\mathcal{A}_i q^i_0 + \mathcal{A}_0) \right] dq^0,$$

and the Lagrange equation (10.2.35):

$$d_0 \left(\frac{m q^i_0}{(1 - \sum_i (q^i_0)^2)^{1/2}} \right) + e(F_{ij} q^j_0 + F_{i0}) = 0.$$

Example 10.2.4. Let $Q = \mathbb{R}^4$ be an Euclidean space provided with the Euclidean metric ϵ. This is the case of Euclidean Special Relativity. Let $\mathcal{A}_\lambda dq^\lambda$ be a one-form on Q. Then

$$L = [(\epsilon_{\mu\nu} q^\mu_\tau q^\nu_\tau)^{1/2} + \mathcal{A}_\mu q^\mu_\tau] d\tau$$

is a relativistic Lagrangian on $J^1 Q_R$ which satisfies the Noether identity (10.2.21). The corresponding relativistic equation (10.2.32) reads

$$m\epsilon_{\mu\nu} q^\nu_{\tau\tau} - eF_{\mu\nu} q^\nu_\tau = 0, \tag{10.2.42}$$

$$\epsilon_{\mu\nu} q^\mu_\tau q^\nu_\tau = 1. \tag{10.2.43}$$

It follows from the relativistic constraint (10.2.43) that $0 \leq (q^0_\tau)^2 \leq 1$. Passing to three-velocities, one therefore meets a problem.

10.3 Relativistic geodesic equations

A glance at the relativistic Lagrangian (10.2.24) shows that, because of the gauge symmetry (10.2.20), this Lagrangian is independent of τ and, therefore, it describes an autonomous mechanical system. Accordingly, the relativistic equation (10.2.32) on Q_R is conservative and, therefore, it is equivalent to an autonomous second order equation on Q whose solutions are parameterized by the coordinate τ on a base $\mathbb{R}$ of Q_R. Given holonomic coordinates $(q^\lambda, \dot{q}^\lambda, \ddot{q}^\lambda)$ of the second tangent bundle T^2Q (see Remark 1.2.1), this autonomous second order equation (called the autonomous relativistic equation) reads

$$\begin{aligned}
&\left(\frac{\partial_\lambda G_{\mu\alpha_2\ldots\alpha_{2N}}}{2N} - \partial_\mu G_{\lambda\alpha_2\ldots\alpha_{2N}}\right)\dot{q}^\mu\dot{q}^{\alpha_2}\cdots\dot{q}^{\alpha_{2N}} \\
&\qquad - (2N-1)G_{\beta\mu\alpha_3\ldots\alpha_{2N}}\ddot{q}^\mu\dot{q}^{\alpha_3}\cdots\dot{q}^{\alpha_{2N}} + F_{\beta\mu}\dot{q}^\mu = 0, \qquad (10.3.1)\\
&G = G_{\alpha_1\ldots\alpha_{2N}}\dot{q}^{\alpha_1}\cdots\dot{q}^{\alpha_{2N}} = 1.
\end{aligned}$$

Due to the canonical isomorphism $q^\lambda_\tau \to \dot{q}^\lambda$ (10.2.9), the tangent bundle TQ is regarded as a space of four-velocities.

Generalizing Example 10.2.3, let us investigate relativistic mechanics on a pseudo-Riemannian oriented four-dimensional manifold $Q = X$, coordinated by (x^λ) and provided with a pseudo-Riemannian metric g of signature $(+,---)$. We agree to call X a *world manifold*. Let $A = A_\lambda dx^\lambda$ be a one-form on X. Let us consider the relativistic Lagrangian (10.2.24):

$$L = [(g_{\alpha\beta}x^\alpha_\tau x^\beta_\tau)^{1/2} + A_\mu x^\mu_\tau]d\tau,$$

and the relativistic constraint (10.2.29):

$$g_{\alpha\beta}x^\alpha_\tau x^\beta_\tau = 1.$$

The corresponding autonomous relativistic equation (10.2.32) on X takes the form

$$\ddot{x}^\lambda - \{_\mu{}^\lambda{}_\nu\}\dot{x}^\mu\dot{x}^\nu - g^{\lambda\beta}F_{\beta\nu}\dot{x}^\nu = 0, \qquad (10.3.2)$$

$$g = g_{\alpha\beta}\dot{x}^\alpha\dot{x}^\beta = 1, \qquad (10.3.3)$$

where $\{_\mu{}^\lambda{}_\nu\}$ is the Levi–Civita connection (11.4.24). A glance at the equality (10.3.2) shows that it is the geodesic equation (1.2.7) on TX with respect to an affine connection

$$K^\lambda_\mu = \{_\mu{}^\lambda{}_\nu\}\dot{x}^\nu + g^{\lambda\nu}F_{\nu\mu} \qquad (10.3.4)$$

on TX.

A particular form of this connection follows from the fact that the geodesic equation (10.3.2) is derived from a Lagrange equation, i.e., we are in the case of Lagrangian relativistic mechanics. In a general setting, relativistic mechanics on a pseudo-Riemannian manifold (X, g) can be formulated as follows.

Definition 10.3.1. The geodesic equation (1.2.7):

$$\ddot{x}^\mu = K^\mu_\lambda(x^\nu, \dot{x}^\nu)\dot{x}^\lambda, \tag{10.3.5}$$

on the tangent bundle TX with respect to a connection

$$K = dx^\lambda \otimes (\partial_\lambda + K^\mu_\lambda \dot{\partial}_\mu) \tag{10.3.6}$$

on $TX \to X$ is called a *relativistic geodesic equation* if a geodesic vector field of K lives in the *subbundle of hyperboloids*

$$W_g = \{\dot{x}^\lambda \in TX \mid g_{\lambda\mu}\dot{x}^\lambda \dot{x}^\mu = 1\} \subset TX \tag{10.3.7}$$

defined by the relativistic constraint (10.3.3).

Since a geodesic vector field is an integral curve of the holonomic vector field $\widehat{K}(TQ)$ (1.2.8), the equation (10.3.5) is a relativistic geodesic equation if the condition

$$\widehat{K}(TQ)\rfloor dg = (\partial_\lambda g_{\mu\nu}\dot{x}^\mu + 2g_{\mu\nu}K^\mu_\lambda)\dot{x}^\lambda \dot{x}^\nu = 0 \tag{10.3.8}$$

holds.

Obviously, the connection (10.3.4) fulfils the condition (10.3.8). Any metric connection, e.g., the Levi–Civita connection $\{_\lambda{}^\mu{}_\nu\}$ (11.4.24) on TX satisfies the condition (10.3.8).

Given a Levi–Civita connection $\{_\lambda{}^\mu{}_\nu\}$, any connection K on $TX \to X$ can be written as

$$K^\mu_\lambda = \{_\lambda{}^\mu{}_\nu\}\dot{x}^\nu + \sigma^\mu_\lambda(x^\lambda, \dot{x}^\lambda), \tag{10.3.9}$$

where

$$\sigma = \sigma^\mu_\lambda dx^\lambda \otimes \dot{\partial}_\mu \tag{10.3.10}$$

is some soldering form (11.2.61) on TX. Then the condition (10.3.8) takes the form

$$g_{\mu\nu}\sigma^\mu_\lambda \dot{x}^\lambda \dot{x}^\nu = 0. \tag{10.3.11}$$

With the decomposition (10.3.9), one can think of the relativistic geodesic equation (10.3.5):

$$\ddot{x}^\mu = \{_\lambda{}^\mu{}_\nu\}\dot{x}^\nu \dot{x}^\lambda + \sigma^\mu_\lambda(x^\nu, \dot{x}^\nu)\dot{x}^\lambda, \tag{10.3.12}$$

as describing a relativistic particle in the presence of a gravitational field g and a non-gravitational external force σ.

In order to compare relativistic and non-relativistic dynamics, let us assume that a pseudo-Riemannian world manifold (X,g) is globally hyperbolic, i.e., it admits a fibration $X \to \mathbb{R}$ over the time axis $\mathbb{R}$ such that its fibres are spatial. One can think of the bundle $X \to \mathbb{R}$ as being a configuration space of a non-relativistic mechanical system. It is provided with the adapted bundle coordinates (x^0, x^i), where the transition functions of the temporal one are $x'^0 = x^0$+const. The velocity space of this non-relativistic mechanical system is the first order jet manifold J^1X of $X \to \mathbb{R}$, coordinated by (x^λ, x^i_0).

By virtue of the canonical imbedding $J^1X \to J^1_1X$ (10.1.1), one also can treat the velocities x^i_0 of a non-relativistic system as three-velocities of a relativistic system on X restricted to an open subbundle $J^1X \subset J^1_1X$ of the bundle $J^1_1X \to X$ of three-velocities. Due to the canonical isomorphism $q^\lambda_\tau \to \dot q^\lambda$ (10.2.9), a four-velocity space of this relativistic system is the tangent bundle TX so that the relation (10.2.16) between four- and three-velocities reads

$$x^i_0 = \frac{\dot x^i}{\dot x^0}. \tag{10.3.13}$$

The relativistic constraint (10.3.3) restricts the space of four-velocities of a relativistic system to the bundle W_g (10.3.7) of hyperboloids which is the disjoint union of two connected imbedded subbundles of W^+ and W^- of TX. The relation (10.3.13) yields bundle monomorphisms of each of the subbundles $W^\pm$ to J^1X.

At the same time, there is the canonical imbedding (1.1.6) of J^1X onto the affine subbundle

$$\dot x^0 = 1, \qquad \dot x^i = x^i_0. \tag{10.3.14}$$

of the tangent bundle TX. Then one can think of elements of the subbundle (10.3.14) as being the *four-velocities of a non-relativistic system*. The relation (10.3.14) differs from the relation (10.3.13) between three- and four-velocities of a relativistic system. It follows that the four-velocities of relativistic and non-relativistic systems occupy different subbundles (10.3.7) and (10.3.14) of the tangent bundle TX.

By virtue of Theorem 1.5.1, every second order dynamic equation (1.3.3):

$$x^i_{00} = \xi^i(x^0, x^j, x^j_0), \tag{10.3.15}$$

of non-relativistic mechanics on $X \to \mathbb{R}$ is equivalent to the non-relativistic geodesic equation (1.5.7):

$$\ddot{x}^0 = 0, \qquad \dot{x}^0 = 1, \qquad \ddot{x}^i = \overline{K}{}_0^i \dot{x}^0 + \overline{K}{}_j^i \dot{x}^j \tag{10.3.16}$$

with respect to the connection

$$\overline{K} = dx^\lambda \otimes (\partial_\lambda + K_\lambda^i \dot{\partial}_i) \tag{10.3.17}$$

possessing the components

$$\overline{K}{}_\lambda^0 = 0, \qquad \xi^i = \overline{K}{}_0^i + x_0^j \overline{K}{}_j^i \mid_{\dot{x}^0=1, \dot{x}^i = x_0^i} . \tag{10.3.18}$$

Note that, written relative to bundle coordinates (x^0, x^i) adapted to a given fibration $X \to \mathbb{R}$, the connection $\overline{K}$ (10.3.18) and the non-relativistic geodesic equation (10.3.16) are well defined with respect to any coordinates on X. It also should be emphasized that the connection $\overline{K}$ (10.3.18) is not determined uniquely.

Thus, we observe that both relativistic and non-relativistic equations of motion can be seen as the geodesic equations on the same tangent bundle TX. The difference between them lies in the fact that their solutions live in the different subbundles (10.3.7) and (10.3.14) of TX.

There is the following relationship between relativistic and non-relativistic equations of motion.

Recall that, by a reference frame in non-relativistic mechanics is meant an atlas of local constant trivializations of the fibre bundle $X \to \mathbb{R}$ such that the transition functions of the spatial coordinates x^i are independent of the temporal one x^0 (Definition 1.6.2). Given a reference frame (x^0, x^i), any connection $K(x^\lambda, \dot{x}^\lambda)$ (10.3.6) on the tangent bundle $TX \to X$ defines the connection $\overline{K}$ (10.3.17) on $TX \to X$ with the components

$$\overline{K}{}_\lambda^0 = 0, \qquad \overline{K}{}_\lambda^i = K_\lambda^i. \tag{10.3.19}$$

It follows that, given a fibration $X \to \mathbb{R}$, every relativistic geodesic equation (10.3.5) yields the non-relativistic geodesic equation (10.3.16) and, consequently, the second order dynamic equation (10.3.15):

$$x_{00}^i = K_0^i(x^\lambda, 1, x_0^k) + K_j^i(x^\lambda, 1, x_0^k) x_0^j, \tag{10.3.20}$$

of non-relativistic mechanics. We agree to call this equation the *non-relativistic approximation* of the relativistic equation (10.3.5).

Note that, written with respect to a reference frame (x^0, x^i), the connection $\overline{K}$ (10.3.19) and the corresponding non-relativistic equation (10.3.20) are well defined relative to any coordinates on X. A key point is that, for another reference frame (x^0, x'^i) with time-dependent transition functions

$x^i \to x'^i$, the connection K (10.3.6) on TX yields another connection $\overline{K}'$ (10.3.17) on $TX \to X$ with the components

$$K'^0_\lambda = 0, \qquad K'^i_\lambda = \left(\frac{\partial x'^i}{\partial x^j} K^j_\mu + \frac{\partial x'^i}{\partial x^\mu}\right)\frac{\partial x^\mu}{\partial x'^\lambda} + \frac{\partial x'^i}{\partial x^0} K^{\;0}_\mu \frac{\partial x^\mu}{\partial x'^\lambda}$$

with respect to the reference frame (x^0, x'^i). It is easy to see that the connection $\overline{K}$ (10.3.19) has the components

$$K'^0_\lambda = 0, \qquad K'^i_\lambda = \left(\frac{\partial x'^i}{\partial x^j} K^j_\mu + \frac{\partial x'^i}{\partial x^\mu}\right)\frac{\partial x^\mu}{\partial x'^\lambda}$$

relative to the same reference frame. This illustrates the fact that a non-relativistic approximation is not relativistic invariant.

The converse procedure is more intricate. Firstly, a non-relativistic dynamic equation (10.3.15) is brought into the non-relativistic geodesic equation (10.3.16) with respect to the connection $\overline{K}$ (10.3.18) which is not unique defined. Secondly, one should find a pair (g, K) of a pseudo-Riemannian metric g and a connection K on $TX \to X$ such that $K^i_\lambda = \overline{K}^i_\lambda$ and the condition (10.3.8) is fulfilled.

From the physical viewpoint, the most interesting second order dynamic equations are the quadratic ones (1.5.8):

$$\xi^i = a^i_{jk}(x^\mu)x^j_0 x^k_0 + b^i_j(x^\mu)x^j_0 + f^i(x^\mu). \tag{10.3.21}$$

By virtue of Corollary 1.5.1, such an equation is equivalent to the non-relativistic geodesic equation

$$\ddot{x}^0 = 0, \qquad \dot{x}^0 = 1,$$
$$\ddot{x}^i = a^i_{jk}(x^\mu)\dot{x}^j\dot{x}^k + b^i_j(x^\mu)\dot{x}^j\dot{x}^0 + f^i(x^\mu)\dot{x}^0\dot{x}^0 \tag{10.3.22}$$

with respect to the symmetric linear connection

$$K_\lambda{}^0{}_\nu = 0, \quad K_0{}^i{}_0 = f^i, \quad K_0{}^i{}_j = \frac{1}{2}b^i_j, \quad K_k{}^i{}_j = a^i_{kj} \tag{10.3.23}$$

on the tangent bundle TX.

In particular, let the equation (10.3.21) be the Lagrange equation for a non-degenerate quadratic Lagrangian. We show that there is a reference frame such that this Lagrange equation coincides with the non-relativistic approximation of some relativistic geodesic equation with respect to a pseudo-Riemannian metric, whose spatial part is a mass tensor of a Lagrangian.

Given a coordinate systems (x^0, x^i), compatible with the fibration $X \to \mathbb{R}$, let us consider a non-degenerate quadratic Lagrangian (2.3.17):

$$L = \frac{1}{2}m_{ij}(x^\mu)x^i_0 x^j_0 + k_i(x^\mu)x^i_0 + \phi(x^\mu), \tag{10.3.24}$$

in Example 2.3.1 where m_{ij} is a Riemannian mass tensor. Similarly to Lemma 1.5.2, one can show that any quadratic polynomial on $J^1X \subset TX$ is extended to a bilinear form in TX. Then the Lagrangian L (10.3.24) can be written as

$$L = -\frac{1}{2} g_{\alpha\mu} x_0^\alpha x_0^\mu, \qquad x_0^0 = 1, \tag{10.3.25}$$

where g is the fibre metric (2.3.19):

$$g_{00} = -2\phi, \qquad g_{0i} = -k_i, \qquad g_{ij} = -m_{ij}, \tag{10.3.26}$$

in TX. The corresponding Lagrange equation takes the form (2.3.18):

$$x_{00}^i = -(m^{-1})^{ik}\{_{\lambda k\nu}\} x_0^\lambda x_0^\nu, \qquad x_0^0 = 1, \tag{10.3.27}$$

where

$$\{_{\lambda\mu\nu}\} = -\frac{1}{2}(\partial_\lambda g_{\mu\nu} + \partial_\nu g_{\mu\lambda} - \partial_\mu g_{\lambda\nu})$$

are the Christoffel symbols of the metric (10.3.26). Let us assume that this metric is non-degenerate. By virtue of Corollary 1.5.1, the second order dynamic equation (10.3.27) is equivalent to the non-relativistic geodesic equation (10.3.22) on TX which reads

$$\ddot{x}^0 = 0, \qquad \dot{x}^0 = 1,$$
$$\ddot{x}^i = \{_\lambda{}^i{}_\nu\} \dot{x}^\lambda \dot{x}^\nu - g^{i0}\{_{\lambda 0\nu}\} \dot{x}^\lambda \dot{x}^\nu. \tag{10.3.28}$$

Let us now bring the Lagrangian (2.3.17) into the form

$$L = \frac{1}{2} m_{ij}(x^\mu)(x_0^i - \Gamma^i)(x_0^j - \Gamma^j) + \phi'(x^\mu), \tag{10.3.29}$$

where Γ is a Lagrangian frame connection on $X \to \mathbb{R}$. This connection Γ defines an atlas of local constant trivializations of the bundle $X \to \mathbb{R}$ and the corresponding coordinates $(x^0, \overline{x}^i)$ on X such that the transition functions $\overline{x}^i \to \overline{x}'^i$ are independent of x^0, and $\Gamma^i = 0$ with respect to $(x^0, \overline{x}^i)$ (see Section 1.6). In this coordinates, the Lagrangian L (10.3.29) reads

$$L = \frac{1}{2} \overline{m}_{ij} \overline{x}_0^i \overline{x}_0^j + \phi'(x^\mu).$$

One can think of its first term as the kinetic energy of a non-relativistic system with the mass tensor $\overline{m}_{ij}$ relative to the reference frame Γ, while $(-\phi')$ is a potential. Let us assume that ϕ' is a nowhere vanishing function on X. Then the Lagrange equation (10.3.27) takes the form

$$\overline{x}_{00}^i = \{_\lambda{}^i{}_\nu\} \overline{x}_0^\lambda \overline{x}_0^\nu, \qquad \overline{x}_0^0 = 1, \tag{10.3.30}$$

where $\{_\lambda{}^i{}_\nu\}$ are the Christoffel symbols of the metric (10.3.26) whose components with respect to the coordinates $(x^0, \overline{x}^i)$ read

$$g_{00} = -2\phi', \qquad g_{0i} = 0, \qquad g_{ij} = -\overline{m}_{ij}. \tag{10.3.31}$$

This metric is Riemannian if $f' > 0$ and pseudo-Riemannian if $f' < 0$. Then the spatial part of the corresponding non-relativistic geodesic equation

$$\ddot{\overline{x}}^0 = 0, \qquad \dot{\overline{x}}^0 = 1, \qquad \ddot{\overline{x}}^i = \{_\lambda{}^i{}_\nu\}\dot{\overline{x}}^\lambda \dot{\overline{x}}^\nu$$

is exactly the spatial part of the relativistic geodesic equation with respect to the Levi–Civita connection of the metric (10.3.31) on TX. It follows that, as was declared above, the non-relativistic dynamic equation (10.3.30) is the non-relativistic approximation (10.3.20) of the relativistic geodesic equation (10.3.5) where K is the Levi–Civita connection of the (Riemannian or pseudo-Riemannian) metric (10.3.31).

Conversely, let us consider a relativistic geodesic equation

$$\ddot{x}^\mu = \{_\lambda{}^\mu{}_\nu\}\dot{x}^\lambda \dot{x}^\nu \tag{10.3.32}$$

with respect to a pseudo-Riemannian metric g on a world manifold X. Let $(x^0, \overline{x}^i)$ be local hyperbolic coordinates such that $g_{00} = 1$, $g_{0i} = 0$. This coordinates are associated to a non-relativistic reference frame for a local fibration $X \to \mathbb{R}$. Then the equation (10.3.32) admits the non-relativistic approximation (10.3.20):

$$x^i_{00} = \{_\lambda{}^i{}_\mu\}x^\lambda_0 x^\mu_0, \qquad x^0_0 = 1, \tag{10.3.33}$$

which is the Lagrange equation (10.3.27) for the Lagrangian (10.3.25):

$$L = \frac{1}{2}\overline{m}_{ij}\overline{x}^i_0 \overline{x}^j_0, \tag{10.3.34}$$

where $g_{00} = 1$, $g_{0i} = 0$. It describes a free non-relativistic mechanical system with the mass tensor $\overline{m}_{ij} = -g_{ij}$. Relative to another reference frame $(x^0, x^i(x^0, \overline{x}^j))$ associated with the same local splitting $X \to \mathbb{R}$, the non-relativistic approximation of the equation (10.3.32) is brought into the Lagrange equation for the Lagrangian (10.3.29):

$$L = \frac{1}{2}m_{ij}(x^\mu)(x^i_0 - \Gamma^i)(x^j_0 - \Gamma^j). \tag{10.3.35}$$

This Lagrangian describes a mechanical system in the presence of the inertial force associated with the reference frame Γ. The difference between Lagrangians (10.3.34) and (10.3.35) shows that a gravitational force can not model an inertial force in general.

10.4 Hamiltonian relativistic mechanics

We are in the case of relativistic mechanics on a pseudo-Riemannian world manifold (X, g). Given the coordinate chart (10.2.6) of its configuration space X, the homogeneous Legendre bundle corresponding to the local non-relativistic system on U is the cotangent bundle T^*U of U. This fact motivate us to think of the cotangent bundle T^*X as being the *phase space of relativistic mechanics* on X. It is provided with the canonical symplectic form

$$\Omega_T = dp_\lambda \wedge dx^\lambda \tag{10.4.1}$$

and the corresponding Poisson bracket $\{,\}$.

Remark 10.4.1. Let us note that one also considers another symplectic form $\Omega_T + F$ where F is the strength of an electromagnetic field [148].

A relativistic Hamiltonian is defined as follows [106; 136; 139]. Let $\mathcal{H}$ be a smooth real function on T^*X such that the morphism

$$\widetilde{\mathcal{H}} : T^*X \to TX, \qquad \dot{x}^\mu \circ \widetilde{\mathcal{H}} = \partial^\mu \mathcal{H}, \tag{10.4.2}$$

is a bundle isomorphism. Then the inverse image

$$N = \widetilde{\mathcal{H}}^{-1}(W_g)$$

of the subbundle of hyperboloids W_g (10.3.7) is a one-codimensional (consequently, coisotropic) closed imbedded subbundle N of T^*X given by the condition

$$\mathcal{H}_T = g_{\mu\nu}\partial^\mu \mathcal{H} \partial^\nu \mathcal{H} - 1 = 0. \tag{10.4.3}$$

We say that $\mathcal{H}$ is a *relativistic Hamiltonian* if the Poisson bracket $\{\mathcal{H}, \mathcal{H}_T\}$ vanishes on N. This means that the Hamiltonian vector field

$$\gamma = \partial^\lambda \mathcal{H} \partial_\lambda - \partial_\lambda \mathcal{H} \partial^\lambda \tag{10.4.4}$$

of H preserves the constraint N and, restricted to N, it obeys the equation (6.2.1):

$$\gamma \rfloor \Omega_N + i_N^* d\mathcal{H} = 0, \tag{10.4.5}$$

which is the Hamilton equation of a Dirac constrained system on N with a Hamiltonian $\mathcal{H}$.

The morphism (10.4.2) sends the vector field γ (10.4.4) onto the vector field

$$\gamma_T = \dot{x}^\lambda \partial_\lambda + (\partial^\mu \mathcal{H} \partial^\lambda \partial_\mu \mathcal{H} - \partial_\mu \mathcal{H} \partial^\lambda \partial^\mu \mathcal{H}) \dot{\partial}_\lambda$$

on TX. This vector field defines the autonomous second order dynamic equation

$$\ddot{x}^\lambda = \partial^\mu \mathcal{H} \partial^\lambda \partial_\mu \mathcal{H} - \partial_\mu \mathcal{H} \partial^\lambda \partial^\mu \mathcal{H} \tag{10.4.6}$$

on X which preserves the subbundle of hyperboloids (10.3.7), i.e., it is the autonomous relativistic equation (10.3.1).

Example 10.4.1. The following is a basic example of relativistic Hamiltonian mechanics. Given a one-form $A = A_\mu dq^\mu$ on X, let us put

$$\mathcal{H} = \frac{1}{2} g^{\mu\nu}(p_\mu - A_\mu)(p_\nu - A_\nu). \tag{10.4.7}$$

Then $\mathcal{H}_T = 2\mathcal{H} - 1$ and, hence, $\{\mathcal{H}, \mathcal{H}_T\} = 0$. The constraint $\mathcal{H}_T = 0$ (10.4.3) defines a one-codimensional closed imbedded subbundle N of T^*X. The Hamilton equation (10.4.5) takes the form $\gamma\rfloor\Omega_N = 0$. Its solution (10.4.4) reads

$$\dot{x}^\alpha = g^{\alpha\nu}(p_\nu - A_\nu),$$
$$\dot{p}_\alpha = -\frac{1}{2}\partial_\alpha g^{\mu\nu}(p_\mu - A_\mu)(p_\nu - A_\nu) + g^{\mu\nu}(p_\mu - A_\mu)\partial_\alpha A_\nu.$$

The corresponding autonomous second order dynamic equation (10.4.6) on X is

$$\ddot{x}^\lambda - \{{}_\mu{}^\lambda{}_\nu\}\dot{x}^\mu \dot{x}^\nu - g^{\lambda\nu} F_{\nu\mu}\dot{x}^\mu = 0, \tag{10.4.8}$$
$$\{{}_\mu{}^\lambda{}_\nu\} = -\frac{1}{2} g^{\lambda\beta}(\partial_\mu g_{\beta\nu} + \partial_\nu g_{\beta\mu} - \partial_\beta g_{\mu\nu}),$$
$$F_{\mu\nu} = \partial_\mu A_\nu - \partial_\nu A_\mu.$$

It is a relativistic geodesic equation with respect to the affine connection (10.3.4).

Since the equation (10.4.8) coincides with the generic Lagrange equation (10.3.2) on a world manifold X, one can think of H (10.4.7) as being a generic Hamiltonian of relativistic mechanics on X.

10.5 Geometric quantization of relativistic mechanics

Let us consider geometric quantization of relativistic mechanics on a pseudo-Riemannian simply connected world manifold (X, g), [65; 141]. We follow the standard geometric quantization of a cotangent bundle (Section 5.2). Let the cohomology group $H^2(X; \mathbb{Z}_2)$ be trivial.

Because the canonical symplectic form Ω_T (10.4.1) on T^*X is exact, the prequantum bundle is defined as a trivial complex line bundle C over T^*X. Let its trivialization (5.2.1):

$$C \cong T^*X \times \mathbb{C}, \tag{10.5.1}$$

hold fixed, and let $(x^\lambda, p_\lambda, c)$, $c \in \mathbb{C}$, be the associated bundle coordinates. Then one can treat sections of C (10.5.1) as smooth complex functions on T^*X. Let us note that another trivialization of C leads to an equivalent quantization of T^*X.

The Kostant–Souriau prequantization formula (5.1.11) associates to each smooth real function $f \in C^\infty(T^*X)$ on T^*X the first order differential operator

$$\widehat{f} = -i\nabla_{\vartheta_f} - f \tag{10.5.2}$$

on sections of C, where

$$\vartheta_f = \partial^\lambda f \partial_\lambda - \partial_\lambda f \partial^\lambda$$

is the Hamiltonian vector field of f and ∇ is the covariant differential (5.2.3) with respect to the admissible $U(1)$-principal connection A (5.2.2):

$$A = dp_\lambda \otimes \partial^\lambda + dx^\lambda \otimes (\partial_\lambda - icp_\lambda \partial_c), \tag{10.5.3}$$

on C. This connection preserves the Hermitian metric $g(c,c')$ (5.1.1) on C, and its curvature form obeys the prequantization condition (5.1.9). The prequantization operators (10.5.2) read

$$\widehat{f} = -i\vartheta_f + (p_\lambda \partial^\lambda f - f). \tag{10.5.4}$$

Let us choose the vertical polarization VT^*X of T^*X. The corresponding quantum algebra $\mathcal{A}_T \subset C^\infty(T^*X)$ consists of affine functions of momenta

$$f = a^\lambda(x^\mu)p_\lambda + b(x^\mu) \tag{10.5.5}$$

on T^*X. They are represented by the Schrödinger operators (5.2.10):

$$\widehat{f} = -ia^\lambda \partial_\lambda - \frac{i}{2}\partial_\lambda a^\lambda - b, \tag{10.5.6}$$

in the space E of complex half-densities ρ of compact support on X.

For the sake of simplicity, let us choose a trivial metalinear bundle $\mathcal{D}_{1/2} \to X$ associated to the orientation of X. Its sections can be written in the form

$$\rho = (-g)^{1/4}\psi,$$

where ψ are smooth complex functions on X. Then the quantum algebra $\mathcal{A}_T$ can be represented by the operators

$$\widehat{f} = -ia^\lambda\partial_\lambda - \frac{i}{2}\partial_\lambda a^\lambda - \frac{i}{4}a^\lambda\partial_\lambda \ln(-g) - b, \qquad g = \det(g_{\alpha\beta}),$$

in the space $\mathbb{C}^\infty(X)$ of these functions. It is easily justified that these operators obey Dirac's condition.

Remark 10.5.1. One usually considers the subspace $\mathfrak{E}$ of complex functions of compact support on X. It is a pre-Hilbert space with respect to the non-degenerate Hermitian form

$$\langle\psi|\psi'\rangle = \int\limits_X \psi\overline{\psi}'(-g)^{1/2}d^4x.$$

It is readily observed that $\widehat{f}$ (10.5.6) are symmetric operators $\widehat{f} = \widehat{f}^*$ on $\mathfrak{E}$, i.e.,

$$\langle\widehat{f}\psi|\psi'\rangle = \langle\psi|\widehat{f}\psi'\rangle.$$

In relativistic mechanics, the space $\mathfrak{E}$ however gets no physical meaning.

Let us note that the function $\mathcal{H}_T$ (10.4.3) need not belong to the quantum algebra $\mathcal{A}_T$. Nevertheless, one can show that, if $\mathcal{H}_T$ is a polynomial of momenta of degree k, it can be represented as a finite composition

$$\mathcal{H}_T = \sum_i f_{1i}\cdots f_{ki} \tag{10.5.7}$$

of products of affine functions (10.5.5), i.e., as an element of the enveloping algebra $\overline{\mathcal{A}}_T$ of the quantum algebra $\mathcal{A}_T$ [57]. Then it is quantized

$$\mathcal{H}_T \to \widehat{\mathcal{H}}_T = \sum_i \widehat{f}_{1i}\cdots\widehat{f}_{ki} \tag{10.5.8}$$

as an element of $\overline{\mathcal{A}}_T$. However, the representation (10.5.7) and, consequently, the quantization (10.5.8) fail to be unique.

The quantum constraint

$$\widehat{\mathcal{H}}_T\psi = 0.$$

serves as a *relativistic quantum equation.*

Example 10.5.1. Let us consider a massive relativistic charge in Example 10.4.1 whose relativistic Hamiltonian is

$$\mathcal{H} = \frac{1}{2m}g^{\mu\nu}(p_\mu - eA_\mu)(p_\nu - eA_\nu).$$

It defines the constraint

$$\mathcal{H}_T = \frac{1}{m^2} g^{\mu\nu}(p_\mu - eA_\mu)(p_\nu - eA_\nu) - 1 = 0. \tag{10.5.9}$$

Let us represent the function $\mathcal{H}_T$ (10.5.9) as the symmetric product

$$\mathcal{H}_T = \frac{(-g)^{-1/4}}{m} \cdot (p_\mu - eA_\mu) \cdot (-g)^{1/4} \cdot g^{\mu\nu} \cdot (-g)^{1/4} \cdot (p_\nu - eA_\nu) \cdot \frac{(-g)^{-1/4}}{m} - 1$$

of affine functions of momenta. It is quantized by the rule (10.5.8), where

$$(-g)^{1/4} \circ \widehat{\partial}_\alpha \circ (-g)^{-1/4} = -i\partial_\alpha.$$

Then the well-known relativistic quantum equation

$$(-g)^{-1/2}[(\partial_\mu - ieA_\mu)g^{\mu\nu}(-g)^{1/2}(\partial_\nu - ieA_\nu) + m^2]\psi = 0$$

is reproduced up to the factor $(-g)^{-1/2}$.

Chapter 11

Appendices

For the sake of convenience of the reader, this Chapter summarizes the relevant material on differential geometry of fibre bundles and modules over commutative rings [68; 76; 109; 150].

11.1 Commutative algebra

In this Section, the relevant basics on modules over commutative algebras is summarized [99; 105].

An *algebra* $\mathcal{A}$ is an additive group which is additionally provided with distributive multiplication. All algebras throughout the book are associative, unless they are Lie algebras. A *ring* is a *unital algebra*, i.e., it contains the unit element $\mathbf{1} \neq 0$. Non-zero elements of a ring form a multiplicative monoid. If this multiplicative monoid is a multiplicative group, one says that the ring has a multiplicative inverse. A *field* is a commutative ring whose non-zero elements make up a multiplicative group.

A subset $\mathcal{I}$ of an algebra $\mathcal{A}$ is called a left (resp. right) *ideal* if it is a subgroup of the additive group $\mathcal{A}$ and $ab \in \mathcal{I}$ (resp. $ba \in \mathcal{I}$) for all $a \in \mathcal{A}$, $b \in \mathcal{I}$. If $\mathcal{I}$ is both a left and right ideal, it is called a *two-sided ideal*. An ideal is a subalgebra, but a *proper* ideal (i.e., $\mathcal{I} \neq \mathcal{A}$) of a ring is not a subring because it does not contain a unit element.

Let $\mathcal{A}$ be a commutative ring. Of course, its ideals are two-sided. Its proper ideal is said to be *maximal* if it does not belong to another proper ideal. A commutative ring $\mathcal{A}$ is called *local* if it has a unique maximal ideal. This ideal consists of all non-invertible elements of $\mathcal{A}$. A proper two-sided ideal $\mathcal{I}$ of a commutative ring is called *prime* if $ab \in \mathcal{I}$ implies either $a \in \mathcal{I}$ or $b \in \mathcal{I}$. Any maximal two-sided ideal is prime. Given a two-sided ideal $\mathcal{I} \subset \mathcal{A}$, the additive factor group $\mathcal{A}/\mathcal{I}$ is an algebra, called the *factor algebra*. If $\mathcal{A}$ is a ring, then $\mathcal{A}/\mathcal{I}$ is so.

Given an algebra $\mathcal{A}$, an additive group P is said to be a left (resp. right) $\mathcal{A}$-*module* if it is provided with distributive multiplication $\mathcal{A} \times P \to P$ by elements of $\mathcal{A}$ such that $(ab)p = a(bp)$ (resp. $(ab)p = b(ap)$) for all $a, b \in \mathcal{A}$ and $p \in P$. If $\mathcal{A}$ is a ring, one additionally assumes that $\mathbf{1}p = p = p\mathbf{1}$ for all $p \in P$. Left and right module structures are usually written by means of left and right multiplications $(a, p) \to ap$ and $(a, p) \to pa$, respectively. If P is both a left module over an algebra $\mathcal{A}$ and a right module over an algebra $\mathcal{A}'$, it is called an $(\mathcal{A} - \mathcal{A}')$-bimodule (an $\mathcal{A}$-bimodule if $\mathcal{A} = \mathcal{A}'$). If $\mathcal{A}$ is a commutative algebra, an $(\mathcal{A} - \mathcal{A})$-bimodule P is said to be *commutative* if $ap = pa$ for all $a \in \mathcal{A}$ and $p \in P$. Any left or right module over a commutative algebra $\mathcal{A}$ can be brought into a commutative bimodule. Therefore, unless otherwise stated, any module over a commutative algebra $\mathcal{A}$ is called an $\mathcal{A}$-module.

A module over a field is called a *vector space*. If an algebra $\mathcal{A}$ is a module over a ring $\mathcal{K}$, it is said to be a $\mathcal{K}$-algebra. Any algebra can be seen as a $\mathbb{Z}$-algebra.

Remark 11.1.1. Any $\mathcal{K}$-algebra $\mathcal{A}$ can be extended to a unital algebra $\widetilde{\mathcal{A}}$ by the adjunction of the identity $\mathbf{1}$ to $\mathcal{A}$. The algebra $\widetilde{\mathcal{A}}$, called the *unital extension* of $\mathcal{A}$, is defined as the direct sum of $\mathcal{K}$-modules $\mathcal{K} \oplus \mathcal{A}$ provided with the multiplication

$$(\lambda_1, a_1)(\lambda_2, a_2) = (\lambda_1\lambda_2, \lambda_1 a_2 + \lambda_2 a_1 + a_1 a_2), \quad \lambda_1, \lambda_2 \in \mathcal{K}, \quad a_1, a_2 \in \mathcal{A}.$$

Elements of $\widetilde{\mathcal{A}}$ can be written as $(\lambda, a) = \lambda\mathbf{1} + a$, $\lambda \in \mathcal{K}$, $a \in \mathcal{A}$. Let us note that, if $\mathcal{A}$ is a unital algebra, the identity $\mathbf{1}_\mathcal{A}$ in $\mathcal{A}$ fails to be that in $\widetilde{\mathcal{A}}$. In this case, the algebra $\widetilde{\mathcal{A}}$ is isomorphic to the product of $\mathcal{A}$ and the algebra $\mathcal{K}(\mathbf{1} - \mathbf{1}_\mathcal{A})$.

From now on, $\mathcal{A}$ is a commutative algebra.

The following are standard constructions of new $\mathcal{A}$-modules from old ones.

• The *direct sum* $P_1 \oplus P_2$ of $\mathcal{A}$-modules P_1 and P_2 is the additive group $P_1 \times P_2$ provided with the $\mathcal{A}$-module structure

$$a(p_1, p_2) = (ap_1, ap_2), \qquad p_{1,2} \in P_{1,2}, \qquad a \in \mathcal{A}.$$

Let $\{P_i\}_{i \in I}$ be a set of modules. Their direct sum $\oplus P_i$ consists of elements $(\ldots, p_i, \ldots)$ of the Cartesian product $\prod P_i$ such that $p_i \neq 0$ at most for a finite number of indices $i \in I$.

• The *tensor product* $P\otimes Q$ of $\mathcal{A}$-modules P and Q is an additive group which is generated by elements $p\otimes q$, $p\in P$, $q\in Q$, obeying the relations

$$(p+p')\otimes q = p\otimes q + p'\otimes q, \quad p\otimes(q+q') = p\otimes q + p\otimes q',$$
$$pa\otimes q = p\otimes aq, \qquad p\in P, \qquad q\in Q, \qquad a\in\mathcal{A},$$

and it is provided with the $\mathcal{A}$-module structure

$$a(p\otimes q) = (ap)\otimes q = p\otimes(qa) = (p\otimes q)a.$$

If the ring $\mathcal{A}$ is treated as an $\mathcal{A}$-module, the tensor product $\mathcal{A}\otimes_{\mathcal{A}} Q$ is canonically isomorphic to Q via the assignment

$$\mathcal{A}\otimes_{\mathcal{A}} Q \ni a\otimes q \leftrightarrow aq \in Q.$$

• Given a submodule Q of an $\mathcal{A}$-module P, the quotient P/Q of the additive group P with respect to its subgroup Q also is provided with an $\mathcal{A}$-module structure. It is called a *factor module*.

• The set $\mathrm{Hom}_{\mathcal{A}}(P,Q)$ of $\mathcal{A}$-linear morphisms of an $\mathcal{A}$-module P to an $\mathcal{A}$-module Q is naturally an $\mathcal{A}$-module. The $\mathcal{A}$-module $P^* = \mathrm{Hom}_{\mathcal{A}}(P,\mathcal{A})$ is called the *dual* of an $\mathcal{A}$-module P. There is a natural monomorphism $P\to P^{**}$.

An $\mathcal{A}$-module P is called *free* if it has a *basis*, i.e., a linearly independent subset $I\subset P$ spanning P such that each element of P has a unique expression as a linear combination of elements of I with a finite number of non-zero coefficients from an algebra $\mathcal{A}$. Any vector space is free. Any module is isomorphic to a quotient of a free module. A module is said to be *finitely generated* (or of *finite rank*) if it is a quotient of a free module with a finite basis.

One says that a module P is *projective* if it is a direct summand of a free module, i.e., there exists a module Q such that $P\oplus Q$ is a free module. A module P is projective if and only if $P=\mathbf{p}S$ where S is a free module and $\mathbf{p}$ is a projector of S, i.e., $\mathbf{p}^2=\mathbf{p}$. If P is a projective module of finite rank over a ring, then its dual P^* is so, and P^{**} is isomorphic to P.

Now we focus on exact sequences, direct and inverse limits of modules [105; 113].

A composition of module morphisms

$$P \xrightarrow{i} Q \xrightarrow{j} T$$

is said to be *exact* at Q if $\mathrm{Ker}\, j = \mathrm{Im}\, i$. A composition of module morphisms

$$0\to P \xrightarrow{i} Q \xrightarrow{j} T \to 0 \tag{11.1.1}$$

is called a *short exact sequence* if it is exact at all the terms P, Q, and T. This condition implies that: (i) i is a monomorphism, (ii) $\operatorname{Ker} j = \operatorname{Im} i$, and (iii) j is an epimorphism onto the quotient $T = Q/P$.

One says that the exact sequence (11.1.1) is *split* if there exists a monomorphism $s : T \to Q$ such that $j \circ s = \operatorname{Id} T$ or, equivalently,

$$Q = i(P) \oplus s(T) = P \oplus T.$$

The exact sequence (11.1.1) is always split if T is a projective module.

A *directed set* I is a set with an order relation $<$ which satisfies the following three conditions:

(i) $i < i$, for all $i \in I$;

(ii) if $i < j$ and $j < k$, then $i < k$;

(iii) for any $i, j \in I$, there exists $k \in I$ such that $i < k$ and $j < k$.

It may happen that $i \neq j$, but $i < j$ and $j < i$ simultaneously.

A family of $\mathcal{A}$-modules $\{P_i\}_{i\in I}$, indexed by a directed set I, is called a *direct system* if, for any pair $i < j$, there exists a morphism $r^i_j : P_i \to P_j$ such that

$$r^i_i = \operatorname{Id} P_i, \qquad r^i_j \circ r^j_k = r^i_k, \qquad i < j < k.$$

A direct system of modules admits a *direct limit*. This is a module P_∞ together with morphisms $r^i_\infty : P_i \to P_\infty$ such that $r^i_\infty = r^j_\infty \circ r^i_j$ for all $i < j$. The module P_∞ consists of elements of the direct sum $\bigoplus_I P_i$ modulo the identification of elements of P_i with their images in P_j for all $i < j$. An example of a direct system is a *direct sequence*

$$P_0 \longrightarrow P_1 \longrightarrow \cdots P_i \xrightarrow{r^i_{i+1}} \cdots, \qquad I = \mathbb{N}. \tag{11.1.2}$$

It should be noted that direct limits also exist in the categories of commutative algebras and rings, but not in categories whose objects are non-Abelian groups.

Theorem 11.1.1. *Direct limits commute with direct sums and tensor products of modules. Namely, let $\{P_i\}$ and $\{Q_i\}$ be two direct systems of modules over the same algebra which are indexed by the same directed set I, and let P_∞ and Q_∞ be their direct limits. Then the direct limits of the direct systems $\{P_i \oplus Q_i\}$ and $\{P_i \otimes Q_i\}$ are $P_\infty \oplus Q_\infty$ and $P_\infty \otimes Q_\infty$, respectively.*

A morphism of a direct system $\{P_i, r^i_j\}_I$ to a direct system $\{Q_{i'}, \rho^{i'}_{j'}\}_{I'}$ consists of an order preserving map $f : I \to I'$ and morphisms $F_i : P_i \to Q_{f(i)}$ which obey the compatibility conditions

$$\rho^{f(i)}_{f(j)} \circ F_i = F_j \circ r^i_j.$$

If P_∞ and Q_∞ are limits of these direct systems, there exists a unique morphism $F_\infty : P_\infty \to Q_\infty$ such that

$$\rho_\infty^{f(i)} \circ F_i = F_\infty \circ r_\infty^i.$$

Moreover, direct limits preserve monomorphisms and epimorphisms. To be precise, if all $F_i : P_i \to Q_{f(i)}$ are monomorphisms or epimorphisms, so is $\Phi_\infty : P_\infty \to Q_\infty$. As a consequence, the following holds.

Theorem 11.1.2. *Let short exact sequences*

$$0 \to P_i \xrightarrow{F_i} Q_i \xrightarrow{\Phi_i} T_i \to 0 \tag{11.1.3}$$

for all $i \in I$ define a short exact sequence of direct systems of modules $\{P_i\}_I$, $\{Q_i\}_I$, and $\{T_i\}_I$ which are indexed by the same directed set I. Then there exists a short exact sequence of their direct limits

$$0 \to P_\infty \xrightarrow{F_\infty} Q_\infty \xrightarrow{\Phi_\infty} T_\infty \to 0. \tag{11.1.4}$$

In particular, the direct limit of factor modules Q_i/P_i is the factor module Q_∞/P_∞. By virtue of Theorem 11.1.1, if all the exact sequences (11.1.3) are split, the exact sequence (11.1.4) is well.

Example 11.1.1. Let P be an $\mathcal{A}$-module. We denote $P^{\otimes k} = \overset{k}{\otimes} P$. Let us consider the direct system of $\mathcal{A}$-modules with respect to monomorphisms

$$\mathcal{A} \longrightarrow (\mathcal{A} \oplus P) \longrightarrow \cdots (\mathcal{A} \oplus P \oplus \cdots \oplus P^{\otimes k}) \longrightarrow \cdots.$$

Its direct limit

$$\otimes P = \mathcal{A} \oplus P \oplus \cdots \oplus P^{\otimes k} \oplus \cdots \tag{11.1.5}$$

is an $\mathbb{N}$-graded $\mathcal{A}$-algebra with respect to the tensor product $\otimes$. It is called the *tensor algebra* of a module P. Its quotient with respect to the ideal generated by elements $p \otimes p' + p' \otimes p$, $p, p' \in P$, is an $\mathbb{N}$-graded commutative algebra, called the *exterior algebra* of a module P.

We restrict our consideration of inverse systems of modules to *inverse sequences*

$$P^0 \longleftarrow P^1 \longleftarrow \cdots P^i \overset{\pi_i^{i+1}}{\longleftarrow} \cdots. \tag{11.1.6}$$

Its *inductive limit* (*the inverse limit*) is a module P^∞ together with morphisms $\pi_i^\infty : P^\infty \to P^i$ such that $\pi_i^\infty = \pi_i^j \circ \pi_j^\infty$ for all $i < j$. It consists of elements $(\ldots, p^i, \ldots)$, $p^i \in P^i$, of the Cartesian product $\prod P^i$ such that $p^i = \pi_i^j(p^j)$ for all $i < j$.

Theorem 11.1.3. *Inductive limits preserve monomorphisms, but not epimorphisms. If a sequence*

$$0 \to P^i \xrightarrow{F^i} Q^i \xrightarrow{\Phi^i} T^i, \qquad i \in \mathbb{N},$$

of inverse systems of modules $\{P^i\}$, $\{Q^i\}$ *and* $\{T^i\}$ *is exact, so is the sequence of the inductive limits*

$$0 \to P^\infty \xrightarrow{F^\infty} Q^\infty \xrightarrow{\Phi^\infty} T^\infty.$$

In contrast with direct limits, the inductive ones exist in the category of groups which are not necessarily commutative.

Example 11.1.2. Let $\{P_i\}$ be a direct sequence of modules. Given another module Q, the modules $\mathrm{Hom}\,(P_i, Q)$ make up an inverse system such that its inductive limit is isomorphic to $\mathrm{Hom}\,(P_\infty, Q)$.

11.2 Geometry of fibre bundles

Throughout this Section, all morphisms are smooth (i.e., of class C^∞), and manifolds are smooth real and finite-dimensional. A *smooth manifold* is customarily assumed to be Hausdorff and *second-countable* (i.e., possessing a countable base for its topology). Consequently, it is a locally compact space which is a union of a countable number of compact subsets, a *separable space* (i.e., it has a countable dense subset), a paracompact and completely regular space. Being paracompact, a smooth manifold admits a partition of unity by smooth real functions. Unless otherwise stated, manifolds are assumed to be connected (and, consequently, arcwise connected). We follow the notion of a manifold without boundary.

The standard symbols $\otimes$, $\vee$, and $\wedge$ stand for the tensor, symmetric, and exterior products, respectively. The interior product (contraction) is denoted by $\rfloor$.

Given a smooth manifold Z, by $\pi_Z : TZ \to Z$ is denoted its tangent bundle. Given manifold coordinates (z^α) on Z, the tangent bundle TZ is equipped with the *holonomic coordinates*

$$(z^\lambda, \dot z^\lambda), \qquad \dot z'^\lambda = \frac{\partial z'^\lambda}{\partial z^\mu}\dot z^\mu,$$

with respect to the *holonomic frames* $\{\partial_\lambda\}$ in the tangent spaces to Z. Any manifold morphism $f : Z \to Z'$ yields the *tangent morphism*

$$Tf : TZ \to TZ', \qquad \dot z'^\lambda \circ Tf = \frac{\partial f^\lambda}{\partial z^\mu}\dot z^\mu,$$

of their tangent bundles.

The symbol $C^\infty(Z)$ stands for the ring of smooth real functions on a manifold Z.

11.2.1 Fibred manifolds

Let M and N be smooth manifolds and $f : M \to N$ a manifold morphism. Its *rank* $\mathrm{rank}_p f$ at a point $p \in M$ is defined as the rank of the tangent map

$$T_p f : T_p M \to T_{f(p)} N, \qquad p \in M.$$

Since the function $p \to \mathrm{rank}_p f$ is lower semicontinuous, a manifold morphism f of maximal rank at a point p also is of maximal rank on some open neighborhood of p. A morphism f is said to be an *immersion* if $T_p f$, $p \in M$, is injective and a *submersion* if $T_p f$, $p \in M$, is surjective. Note that a submersion is an *open map* (i.e., an image of any open set is open).

If $f : M \to N$ is an injective immersion, its range is called a *submanifold* of N. A submanifold is said to be *imbedded* if it also is a topological subspace. In this case, f is called an *imbedding*. For the sake of simplicity, we usually identify (M, f) with $f(M)$. If $M \subset N$, its natural injection is denoted by $i_M : M \to N$. There are the following criteria for a submanifold to be imbedded.

Theorem 11.2.1. *Let (M, f) be a submanifold of N.*

(i) A map f is an imbedding if and only if, for each point $p \in M$, there exists a (cubic) coordinate chart (V, ψ) of N centered at $f(p)$ so that $f(M) \cap V$ consists of all points of V with coordinates $(x^1, \ldots, x^m, 0, \ldots, 0)$.

(ii) Suppose that $f : M \to N$ is a proper map, i.e., the inverse images of compact sets are compact. Then (M, f) is a closed imbedded submanifold of N. In particular, this occurs if M is a compact manifold.

(iii) If $\dim M = \dim N$, then (M, f) is an open imbedded submanifold of N.

If a manifold morphism

$$\pi : Y \to X, \qquad \dim X = n > 0, \tag{11.2.1}$$

is a surjective submersion, one says that: (i) its domain Y is a *fibred manifold*, (ii) X is its *base*, (iii) π is a *fibration*, and (iv) $Y_x = \pi^{-1}(x)$ is a *fibre* over $x \in X$.

By virtue of the inverse function theorem [162], the surjection (11.2.1) is a fibred manifold if and only if a manifold Y admits an atlas of *fibred coordinate* charts $(U_Y; x^\lambda, y^i)$ such that (x^λ) are coordinates on $\pi(U_Y) \subset X$ and coordinate transition functions read

$$x'^\lambda = f^\lambda(x^\mu), \qquad y'^i = f^i(x^\mu, y^j).$$

The surjection π (11.2.1) is a fibred manifold if and only if, for each point $y \in Y$, there exists a local section s of $Y \to X$ passing through y. Recall that by a *local section* of the surjection (11.2.1) is meant an injection $s : U \to Y$ of an open subset $U \subset X$ such that $\pi \circ s = \operatorname{Id} U$, i.e., a section sends any point $x \in X$ into the fibre Y_x over this point. A local section also is defined over any subset $N \in X$ as the restriction to N of a local section over an open set containing N. If $U = X$, one calls s the *global section*. A range $s(U)$ of a local section $s : U \to Y$ of a fibred manifold $Y \to X$ is an imbedded submanifold of Y. A local section is a *closed map*, sending closed subsets of U onto closed subsets of Y. If s is a global section, then $s(X)$ is a closed imbedded submanifold of Y. Global sections of a fibred manifold need not exist.

Theorem 11.2.2. *Let $Y \to X$ be a fibred manifold whose fibres are diffeomorphic to $\mathbb{R}^m$. Any its section over a closed imbedded submanifold (e.g., a point) of X is extended to a global section [150]. In particular, such a fibred manifold always has a global section.*

Given fibred coordinates $(U_Y; x^\lambda, y^i)$, a section s of a fibred manifold $Y \to X$ is represented by collections of local functions $\{s^i = y^i \circ s\}$ on $\pi(U_Y)$.

Morphisms of fibred manifolds, by definition, are *fibrewise morphisms*, sending a fibre to a fibre. Namely, a *fibred morphism* of a fibred manifold $\pi : Y \to X$ to a fibred manifold $\pi' : Y' \to X'$ is defined as a pair (Φ, f) of manifold morphisms which form a commutative diagram

$$\begin{array}{ccc} Y & \xrightarrow{\Phi} & Y' \\ {\scriptstyle \pi}\downarrow & & \downarrow{\scriptstyle \pi'} \\ X & \xrightarrow{f} & X' \end{array}, \qquad \pi' \circ \Phi = f \circ \pi.$$

Fibred injections and surjections are called monomorphisms and epimorphisms, respectively. A fibred diffeomorphism is called an isomorphism or an automorphism if it is an isomorphism to itself. For the sake of brevity, a fibred morphism over $f = \operatorname{Id} X$ usually is said to be a fibred morphism over X, and is denoted by $Y \underset{X}{\longrightarrow} Y'$. In particular, a fibred automorphism over X is called a *vertical automorphism*.

11.2.2 Fibre bundles

A fibred manifold $Y \to X$ is said to be *trivial* if Y is isomorphic to the product $X \times V$. Different trivializations of $Y \to X$ differ from each other in surjections $Y \to V$.

A fibred manifold $Y \to X$ is called a *fibre bundle* if it is locally trivial, i.e., if it admits a fibred coordinate atlas $\{(\pi^{-1}(U_\xi); x^\lambda, y^i)\}$ over a cover $\{\pi^{-1}(U_\xi)\}$ of Y which is the inverse image of a cover $\mathfrak{U} = \{U_\xi\}$ of X. In this case, there exists a manifold V, called a *typical fibre*, such that Y is locally diffeomorphic to the splittings

$$\psi_\xi : \pi^{-1}(U_\xi) \to U_\xi \times V, \tag{11.2.2}$$

glued together by means of *transition functions*

$$\varrho_{\xi\zeta} = \psi_\xi \circ \psi_\zeta^{-1} : U_\xi \cap U_\zeta \times V \to U_\xi \cap U_\zeta \times V \tag{11.2.3}$$

on overlaps $U_\xi \cap U_\zeta$. Transition functions $\varrho_{\xi\zeta}$ fulfil the *cocycle condition*

$$\varrho_{\xi\zeta} \circ \varrho_{\zeta\iota} = \varrho_{\xi\iota} \tag{11.2.4}$$

on all overlaps $U_\xi \cap U_\zeta \cap U_\iota$. Restricted to a point $x \in X$, *trivialization morphisms* ψ_ξ (11.2.2) and transition functions $\varrho_{\xi\zeta}$ (11.2.3) define diffeomorphisms of fibres

$$\psi_\xi(x) : Y_x \to V, \qquad x \in U_\xi, \tag{11.2.5}$$

$$\varrho_{\xi\zeta}(x) : V \to V, \qquad x \in U_\xi \cap U_\zeta. \tag{11.2.6}$$

Trivialization charts (U_ξ, ψ_ξ) together with transition functions $\varrho_{\xi\zeta}$ (11.2.3) constitute a *bundle atlas*

$$\Psi = \{(U_\xi, \psi_\xi), \varrho_{\xi\zeta}\} \tag{11.2.7}$$

of a fibre bundle $Y \to X$. Two bundle atlases are said to be *equivalent* if their union also is a bundle atlas, i.e., there exist transition functions between trivialization charts of different atlases. All atlases of a fibre bundle are equivalent.

Given a bundle atlas Ψ (11.2.7), a fibre bundle Y is provided with the fibred coordinates

$$x^\lambda(y) = (x^\lambda \circ \pi)(y), \qquad y^i(y) = (y^i \circ \psi_\xi)(y), \qquad y \in \pi^{-1}(U_\xi),$$

called the *bundle coordinates*, where y^i are coordinates on a typical fibre V.

A fibre bundle $Y \to X$ is uniquely defined by a bundle atlas. Given an atlas Ψ (11.2.7), there exists a unique manifold structure on Y for which $\pi : Y \to X$ is a fibre bundle with a typical fibre V and a bundle atlas Ψ.

There are the following useful criteria for a fibred manifold to be a fibre bundle.

Theorem 11.2.3. *If a fibration $\pi : Y \to X$ is a proper map, then $Y \to X$ is a fibre bundle. In particular, a compact fibred manifold is a fibre bundle.*

Theorem 11.2.4. *A fibred manifold whose fibres are diffeomorphic either to a compact manifold or $\mathbb{R}^r$ is a fibre bundle [114].*

A comprehensive relation between fibred manifolds and fibre bundles is given in Remark 11.4.1. It involves the notion of an Ehresmann connection.

Forthcoming Theorems 11.2.5 – 11.2.7 describe the particular covers which one can choose for a bundle atlas [76].

Theorem 11.2.5. *Any fibre bundle over a contractible base is trivial.*

Note that a fibred manifold over a contractible base need not be trivial. It follows from Theorem 11.2.5 that any cover of a base X by *domains* (i.e., contractible open subsets) is a bundle cover.

Theorem 11.2.6. *Every fibre bundle $Y \to X$ admits a bundle atlas over a countable cover $\mathfrak{U}$ of X where each member U_ξ of $\mathfrak{U}$ is a domain whose closure $\overline{U}_\xi$ is compact.*

If a base X is compact, there is a bundle atlas of Y over a finite cover of X which obeys the condition of Theorem 11.2.6.

Theorem 11.2.7. *Every fibre bundle $Y \to X$ admits a bundle atlas over a finite cover $\mathfrak{U}$ of X, but its members need not be contractible and connected.*

A fibred morphism of fibre bundles is called a *bundle morphism*. A bundle monomorphism $\Phi : Y \to Y'$ over X onto a submanifold $\Phi(Y)$ of Y' is called a *subbundle* of a fibre bundle $Y' \to X$. There is the following useful criterion for an image and an inverse image of a bundle morphism to be subbundles.

Theorem 11.2.8. *Let $\Phi : Y \to Y'$ be a bundle morphism over X. Given a global section s of the fibre bundle $Y' \to X$ such that $s(X) \subset \Phi(Y)$, by the* kernel *of a bundle morphism Φ with respect to a section s is meant the inverse image*

$$\mathrm{Ker}\,_s\Phi = \Phi^{-1}(s(X))$$

of $s(X)$ by Φ. If $\Phi : Y \to Y'$ is a bundle morphism of constant rank over X, then $\Phi(Y)$ and $\mathrm{Ker}\,_s\Phi$ are subbundles of Y' and Y, respectively.

The following are the standard constructions of new fibre bundles from old ones.

• Given a fibre bundle $\pi : Y \to X$ and a manifold morphism $f : X' \to X$, the *pull-back* of Y by f is called the manifold

$$f^*Y = \{(x', y) \in X' \times Y : \pi(y) = f(x')\} \tag{11.2.8}$$

together with the natural projection $(x', y) \to x'$. It is a fibre bundle over X' such that the fibre of f^*Y over a point $x' \in X'$ is that of Y over the point $f(x') \in X$. There is the canonical bundle morphism

$$f_Y : f^*Y \ni (x', y)|_{\pi(y)=f(x')} \to y \in Y. \tag{11.2.9}$$

Any section s of a fibre bundle $Y \to X$ yields the *pull-back section*

$$f^*s(x') = (x', s(f(x'))$$

of $f^*Y \to X'$.

• If $X' \subset X$ is a submanifold of X and $i_{X'}$ is the corresponding natural injection, then the pull-back bundle

$$i^*_{X'}Y = Y|_{X'}$$

is called the *restriction* of a fibre bundle Y to the submanifold $X' \subset X$. If X' is an imbedded submanifold, any section of the pull-back bundle

$$Y|_{X'} \to X'$$

is the restriction to X' of some section of $Y \to X$.

• Let $\pi : Y \to X$ and $\pi' : Y' \to X$ be fibre bundles over the same base X. Their *bundle product* $Y \times_X Y'$ over X is defined as the pull-back

$$Y \underset{X}{\times} Y' = \pi^*Y' \quad \text{or} \quad Y \underset{X}{\times} Y' = \pi'^*Y$$

together with its natural surjection onto X. Fibres of the bundle product $Y \times Y'$ are the Cartesian products $Y_x \times Y'_x$ of fibres of fibre bundles Y and Y'.

• Let us consider the *composite fibre bundle*

$$Y \to \Sigma \to X. \tag{11.2.10}$$

It is provided with bundle coordinates $(x^\lambda, \sigma^m, y^i)$, where (x^λ, σ^m) are bundle coordinates on a fibre bundle $\Sigma \to X$, i.e., transition functions of coordinates σ^m are independent of coordinates y^i. Let h be a global section of a fibre bundle $\Sigma \to X$. Then the restriction $Y_h = h^*Y$ of a fibre bundle $Y \to \Sigma$ to $h(X) \subset \Sigma$ is a subbundle of a fibre bundle $Y \to X$.

11.2.3 *Vector bundles*

A fibre bundle $\pi : Y \to X$ is called a *vector bundle* if both its typical fibre and fibres are finite-dimensional real vector spaces, and if it admits a bundle atlas whose trivialization morphisms and transition functions are linear isomorphisms. Then the corresponding bundle coordinates on Y are *linear bundle coordinates* (y^i) possessing linear transition functions $y'^i = A^i_j(x)y^j$. We have

$$y = y^i e_i(\pi(y)) = y^i \psi_\xi(\pi(y))^{-1}(e_i), \qquad \pi(y) \in U_\xi, \tag{11.2.11}$$

where $\{e_i\}$ is a fixed basis for a typical fibre V of Y and $\{e_i(x)\}$ are the fibre bases (or the *frames*) for the fibres Y_x of Y associated to a bundle atlas Ψ.

By virtue of Theorem 11.2.2, any vector bundle has a global section, e.g., the canonical global *zero-valued section* $\widehat{0}(x) = 0$.

Theorem 11.2.9. *Let a vector bundle $Y \to X$ admit $m = \dim V$ nowhere vanishing global sections s_i which are linearly independent, i.e., $\overset{m}{\wedge} s_i \neq 0$. Then Y is trivial.*

Global sections of a vector bundle $Y \to X$ constitute a projective $C^\infty(X)$-module $Y(X)$ of finite rank. It is called the *structure module of a vector bundle*. Serre–Swan Theorem 11.5.2 states the categorial equivalence between the vector bundles over a smooth manifold X and projective $C^\infty(X)$-modules of finite rank.

There are the following particular constructions of new vector bundles from the old ones.

• Let $Y \to X$ be a vector bundle with a typical fibre V. By $Y^* \to X$ is denoted the *dual vector bundle* with the typical fibre V^*, dual of V. The *interior product* of Y and Y^* is defined as a fibred morphism

$$\rfloor : Y \otimes Y^* \underset{X}{\longrightarrow} X \times \mathbb{R}.$$

• Let $Y \to X$ and $Y' \to X$ be vector bundles with typical fibres V and V', respectively. Their *Whitney sum* $Y \oplus_X Y'$ is a vector bundle over X with the typical fibre $V \oplus V'$.

• Let $Y \to X$ and $Y' \to X$ be vector bundles with typical fibres V and V', respectively. Their *tensor product* $Y \otimes_X Y'$ is a vector bundle over X with the typical fibre $V \otimes V'$. Similarly, the *exterior product* of vector bundles $Y \wedge_X Y'$ is defined. The exterior product

$$\wedge Y = X \times \mathbb{R} \underset{X}{\oplus} Y \underset{X}{\oplus} \overset{2}{\wedge} Y \underset{X}{\oplus} \cdots \oplus \overset{k}{\wedge} Y, \qquad k = \dim Y - \dim X, \tag{11.2.12}$$

is called the *exterior bundle.*

• If Y' is a subbundle of a vector bundle $Y \to X$, the *factor bundle* Y/Y' over X is defined as a vector bundle whose fibres are the quotients Y_x/Y'_x, $x \in X$.

By a morphism of vector bundles is meant a *linear bundle morphism,* which is a linear fibrewise map whose restriction to each fibre is a linear map.

Given a linear bundle morphism $\Phi : Y' \to Y$ of vector bundles over X, its *kernel* $\operatorname{Ker}\Phi$ is defined as the inverse image $\Phi^{-1}(\widehat{0}(X))$ of the canonical zero-valued section $\widehat{0}(X)$ of Y. By virtue of Theorem 11.2.8, if Φ is of constant rank, its kernel and its range are vector subbundles of the vector bundles Y' and Y, respectively. For instance, monomorphisms and epimorphisms of vector bundles fulfil this condition.

Remark 11.2.1. Given vector bundles Y and Y' over the same base X, every linear bundle morphism

$$\Phi : Y_x \ni \{e_i(x)\} \to \{\Phi_i^k(x)e'_k(x)\} \in Y'_x$$

over X defines a global section

$$\Phi : x \to \Phi_i^k(x)e^i(x) \otimes e'_k(x)$$

of the tensor product $Y \otimes Y'^*$, and *vice versa.*

A sequence

$$Y' \xrightarrow{i} Y \xrightarrow{j} Y''$$

of vector bundles over the same base X is called *exact* at Y if $\operatorname{Ker} j = \operatorname{Im} i$. A sequence of vector bundles

$$0 \to Y' \xrightarrow{i} Y \xrightarrow{j} Y'' \to 0 \tag{11.2.13}$$

over X is said to be a *short exact sequence* if it is exact at all terms Y', Y, and Y''. This means that i is a bundle monomorphism, j is a bundle epimorphism, and $\operatorname{Ker} j = \operatorname{Im} i$. Then Y'' is isomorphic to a factor bundle Y/Y'. Given an exact sequence of vector bundles (11.2.13), there is the exact sequence of their duals

$$0 \to Y''^* \xrightarrow{j^*} Y^* \xrightarrow{i^*} Y'^* \to 0.$$

One says that the exact sequence (11.2.13) is *split* if there exists a bundle monomorphism $s : Y'' \to Y$ such that $j \circ s = \operatorname{Id} Y''$ or, equivalently,

$$Y = i(Y') \oplus s(Y'') = Y' \oplus Y''.$$

Theorem 11.2.10. *Every exact sequence of vector bundles (11.2.13) is split [85].*

The tangent bundle TZ and the cotangent bundle T^*Z of a manifold Z exemplify vector bundles.

Given an atlas $\Psi_Z = \{(U_\iota, \phi_\iota)\}$ of a manifold Z, the tangent bundle is provided with the *holonomic bundle atlas*

$$\Psi_T = \{(U_\iota, \psi_\iota = T\phi_\iota)\}. \tag{11.2.14}$$

The associated linear bundle coordinates are holonomic coordinates $(\dot z^\lambda)$.

The *cotangent bundle* of a manifold Z is the dual $T^*Z \to Z$ of the tangent bundle $TZ \to Z$. It is equipped with the holonomic coordinates

$$(z^\lambda, \dot z_\lambda). \qquad \dot z'_\lambda = \frac{\partial z^\mu}{\partial z'^\lambda}\dot z_\mu,$$

with respect to the *coframes* $\{dz^\lambda\}$ for T^*Z which are the duals of $\{\partial_\lambda\}$.

The tensor product of tangent and cotangent bundles

$$T = (\overset{m}{\otimes} TZ)\otimes(\overset{k}{\otimes} T^*Z), \qquad m, k \in \mathbb{N}, \tag{11.2.15}$$

is called a *tensor bundle*, provided with holonomic bundle coordinates $\dot z^{\alpha_1\cdots\alpha_m}_{\beta_1\cdots\beta_k}$ possessing transition functions

$$\dot z'^{\alpha_1\cdots\alpha_m}_{\beta_1\cdots\beta_k} = \frac{\partial z'^{\alpha_1}}{\partial z^{\mu_1}}\cdots\frac{\partial z'^{\alpha_m}}{\partial z^{\mu_m}}\frac{\partial z^{\nu_1}}{\partial z'^{\beta_1}}\cdots\frac{\partial z^{\nu_k}}{\partial z'^{\beta_k}}\dot z^{\mu_1\cdots\mu_m}_{\nu_1\cdots\nu_k}.$$

Let $\pi_Y : TY \to Y$ be the tangent bundle of a fibred manifold $\pi : Y \to X$. Given fibred coordinates (x^λ, y^i) on Y, it is equipped with the holonomic coordinates $(x^\lambda, y^i, \dot x^\lambda, \dot y^i)$. The tangent bundle $TY \to Y$ has the subbundle $VY = \mathrm{Ker}\,(T\pi)$, which consists of the vectors tangent to fibres of Y. It is called the *vertical tangent bundle* of Y, and it is provided with the holonomic coordinates $(x^\lambda, y^i, \dot y^i)$ with respect to the *vertical frames* $\{\partial_i\}$. Every fibred morphism $\Phi : Y \to Y'$ yields the linear bundle morphism over Φ of the vertical tangent bundles

$$V\Phi : VY \to VY', \qquad \dot y'^i \circ V\Phi = \frac{\partial \Phi^i}{\partial y^j}\dot y^j. \tag{11.2.16}$$

It is called the *vertical tangent morphism*.

In many important cases, the vertical tangent bundle $VY \to Y$ of a fibre bundle $Y \to X$ is trivial, and it is isomorphic to the bundle product

$$VY = Y \underset{X}{\times} \overline{Y}, \tag{11.2.17}$$

where $\overline{Y} \to X$ is some vector bundle. One calls (11.2.17) the *vertical splitting*. For instance, every vector bundle $Y \to X$ admits the *canonical vertical splitting*

$$VY = Y \underset{X}{\oplus} Y. \tag{11.2.18}$$

The *vertical cotangent bundle* $V^*Y \to Y$ of a fibred manifold $Y \to X$ is defined as the dual of the vertical tangent bundle $VY \to Y$. It is not a subbundle of the cotangent bundle T^*Y, but there is the canonical surjection

$$\zeta : T^*Y \ni \dot{x}_\lambda dx^\lambda + \dot{y}_i dy^i \to \dot{y}_i \overline{d}y^i \in V^*Y, \tag{11.2.19}$$

where the bases $\{\overline{d}y^i\}$, possessing transition functions

$$\overline{d}y'^i = \frac{\partial y'^i}{\partial y^j}\overline{d}y^j,$$

are the duals of the vertical frames $\{\partial_i\}$ of the vertical tangent bundle VY.

For any fibred manifold Y, there exist the exact sequences of vector bundles

$$0 \to VY \longrightarrow TY \xrightarrow{\pi_T} Y \underset{X}{\times} TX \to 0, \tag{11.2.20}$$

$$0 \to Y \underset{X}{\times} T^*X \to T^*Y \to V^*Y \to 0. \tag{11.2.21}$$

Their splitting, by definition, is a connection on $Y \to X$ (Section 11.4.1).

Let us consider the tangent bundle TT^*X of T^*X and the cotangent bundle T^*TX of TX. Relative to coordinates $(x^\lambda, p_\lambda = \dot{x}_\lambda)$ on T^*X and $(x^\lambda, \dot{x}^\lambda)$ on TX, these fibre bundles are provided with the coordinates $(x^\lambda, p_\lambda, \dot{x}^\lambda, \dot{p}_\lambda)$ and $(x^\lambda, \dot{x}^\lambda, \dot{x}_\lambda, \ddot{x}_\lambda)$, respectively. By inspection of the coordinate transformation laws, one can show that there is an isomorphism

$$\alpha : TT^*X = T^*TX, \qquad p_\lambda \longleftrightarrow \ddot{x}_\lambda, \quad \dot{p}_\lambda \longleftrightarrow \dot{x}_\lambda, \tag{11.2.22}$$

of these bundles over TX. Given a fibred manifold $Y \to X$, there is the similar isomorphism

$$\alpha_V : VV^*Y = V^*VY, \qquad p_i \longleftrightarrow \ddot{y}_i, \quad \dot{p}_i \longleftrightarrow \dot{y}_i, \tag{11.2.23}$$

over VY, where $(x^\lambda, y^i, p_i, \dot{y}^i, \dot{p}_i)$ and $(x^\lambda, y^i, \dot{y}^i, \dot{y}_i, \ddot{y}_i)$ are coordinates on VV^*Y and V^*VY, respectively.

11.2.4 *Affine bundles*

Let $\overline{\pi} : \overline{Y} \to X$ be a vector bundle with a typical fibre $\overline{V}$. An *affine bundle* modelled over the vector bundle $\overline{Y} \to X$ is a fibre bundle $\pi : Y \to X$ whose typical fibre V is an affine space modelled over $\overline{V}$, all the fibres Y_x of Y are affine spaces modelled over the corresponding fibres $\overline{Y}_x$ of the vector bundle $\overline{Y}$, and there is an affine bundle atlas

$$\Psi = \{(U_\alpha, \psi_\chi), \varrho_{\chi\zeta}\}$$

of $Y \to X$ whose local trivializations morphisms ψ_χ (11.2.5) and transition functions $\varrho_{\chi\zeta}$ (11.2.6) are affine isomorphisms.

Dealing with affine bundles, we use only *affine bundle coordinates* (y^i) associated to an affine bundle atlas Ψ. There are the bundle morphisms

$$Y \underset{X}{\times} \overline{Y} \underset{X}{\longrightarrow} Y, \qquad (y^i, \overline{y}^i) \to y^i + \overline{y}^i,$$
$$Y \underset{X}{\times} Y \underset{X}{\longrightarrow} \overline{Y}, \qquad (y^i, y'^i) \to y^i - y'^i,$$

where $(\overline{y}^i)$ are linear coordinates on a vector bundle $\overline{Y}$.

By virtue of Theorem 11.2.2, affine bundles have global sections, but in contrast with vector bundles, there is no canonical global section of an affine bundle. Let $\pi : Y \to X$ be an affine bundle. Every global section s of an affine bundle $Y \to X$ modelled over a vector bundle $\overline{Y} \to X$ yields the bundle morphisms

$$Y \ni y \to y - s(\pi(y)) \in \overline{Y}, \tag{11.2.24}$$
$$\overline{Y} \ni \overline{y} \to s(\pi(y)) + \overline{y} \in Y. \tag{11.2.25}$$

In particular, every vector bundle Y has a natural structure of an affine bundle due to the morphisms (11.2.25) where $s = \widehat{0}$ is the canonical zero-valued section of Y.

Theorem 11.2.11. *Any affine bundle $Y \to X$ admits bundle coordinates $(x^\lambda, \widetilde{y}^i)$ possessing linear transition functions $\widetilde{y}'^i = A^i_j(x)\widetilde{y}^j$ [68].*

By a morphism of affine bundles is meant a bundle morphism $\Phi : Y \to Y'$ whose restriction to each fibre of Y is an affine map. It is called an *affine bundle morphism*. Every affine bundle morphism $\Phi : Y \to Y'$ of an affine bundle Y modelled over a vector bundle $\overline{Y}$ to an affine bundle Y' modelled over a vector bundle $\overline{Y}'$ yields an unique linear bundle morphism

$$\overline{\Phi} : \overline{Y} \to \overline{Y}', \qquad \overline{y}'^i \circ \overline{\Phi} = \frac{\partial \Phi^i}{\partial y^j}\overline{y}^j, \tag{11.2.26}$$

called the *linear derivative* of Φ.

Every affine bundle $Y \to X$ modelled over a vector bundle $\overline{Y} \to X$ admits the *canonical vertical splitting*

$$VY = Y \underset{X}{\times} \overline{Y}. \tag{11.2.27}$$

11.2.5 Vector fields

Vector fields on a manifold Z are global sections of the tangent bundle $TZ \to Z$.

The set $\mathcal{T}_1(Z)$ of vector fields on Z is both a $C^\infty(Z)$-module and a real Lie algebra with respect to the *Lie bracket*

$$u = u^\lambda \partial_\lambda, \qquad v = v^\lambda \partial_\lambda,$$
$$[v, u] = (v^\lambda \partial_\lambda u^\mu - u^\lambda \partial_\lambda v^\mu)\partial_\mu.$$

Remark 11.2.2. A vector field u on an imbedded submanifold $N \subset Z$ is said to be a section of the tangent bundle $TZ \to Z$ over N. It should be emphasized that this is not a vector field on a manifold N since $u(N)$ does not belong to $TN \subset TX$ in general. A vector field on a submanifold $N \subset Z$ is called tangent to N if $u(N) \subset TN$.

Given a vector field u on X, a *curve*

$$c : \mathbb{R} \supset (,) \to Z$$

in Z is said to be an *integral curve* of u if $Tc = u(c)$. Every vector field u on a manifold Z can be seen as an *infinitesimal generator* of a local one-parameter group of local diffeomorphisms (a *flow*), and *vice versa* [93]. One-dimensional orbits of this group are integral curves of u.

Remark 11.2.3. Let $U \subset Z$ be an open subset and $\epsilon > 0$. Recall that by a local one-parameter group of local diffeomorphisms of Z defined on $(-\epsilon, \epsilon) \times U$ is meant a map

$$G : (-\epsilon, \epsilon) \times U \ni (t, z) \to G_t(z) \in Z$$

which possesses the following properties:

- for each $t \in (-\epsilon, \epsilon)$, the mapping G_t is a diffeomorphism of U onto the open subset $G_t(U) \subset Z$;
- $G_{t+t'}(z) = (G_t \circ G_{t'})(z)$ if $t + t' \in (-\epsilon, \epsilon)$.

If such a map G is defined on $\mathbb{R} \times Z$, it is called the one-parameter group of diffeomorphisms of Z. If a local one-parameter group of local diffeomorphisms of Z is defined on $(-\epsilon, \epsilon) \times Z$, it is uniquely prolonged onto $\mathbb{R} \times Z$ to a one-parameter group of diffeomorphisms of Z [93]. As was mentioned above, a local one-parameter group of local diffeomorphisms G on $U \subset Z$ defines a local vector field u on U by setting $u(z)$ to be the tangent vector to the curve $s(t) = G_t(z)$ at $t = 0$. Conversely, let u be a vector field on a manifold Z. For each $z \in Z$, there exist a number $\epsilon > 0$, a neighborhood U of z and a unique local one-parameter group of local diffeomorphisms on $(-\epsilon, \epsilon) \times U$, which determines u.

A vector field is called *complete* if its flow is a one-parameter group of diffeomorphisms of Z.

Theorem 11.2.12. *Any vector field on a compact manifold is complete.*

A vector field u on a fibred manifold $Y \to X$ is called *projectable* if it is projected onto a vector field on X, i.e., there exists a vector field τ on X such that

$$\tau \circ \pi = T\pi \circ u.$$

A projectable vector field takes the coordinate form

$$u = u^\lambda(x^\mu)\partial_\lambda + u^i(x^\mu, y^j)\partial_i, \qquad \tau = u^\lambda \partial_\lambda. \tag{11.2.28}$$

A projectable vector field is called *vertical* if its projection onto X vanishes, i.e., if it lives in the vertical tangent bundle VY.

A vector field $\tau = \tau^\lambda \partial_\lambda$ on a base X of a fibred manifold $Y \to X$ gives rise to a vector field on Y by means of a connection on this fibre bundle (see the formula (11.4.3) below). Nevertheless, every tensor bundle (11.2.15) admits the *functorial lift* of vector fields

$$\widetilde{\tau} = \tau^\mu \partial_\mu + [\partial_\nu \tau^{\alpha_1} \dot{x}^{\nu\alpha_2\cdots\alpha_m}_{\beta_1\cdots\beta_k} + \ldots - \partial_{\beta_1}\tau^\nu \dot{x}^{\alpha_1\cdots\alpha_m}_{\nu\beta_2\cdots\beta_k} - \ldots]\dot{\partial}^{\beta_1\cdots\beta_k}_{\alpha_1\cdots\alpha_m}, \tag{11.2.29}$$

where we employ the compact notation

$$\dot{\partial}_\lambda = \frac{\partial}{\partial \dot{x}^\lambda}. \tag{11.2.30}$$

This lift is an $\mathbb{R}$-linear monomorphism of the Lie algebra $\mathcal{T}_1(X)$ of vector fields on X to the Lie algebra $\mathcal{T}_1(Y)$ of vector fields on Y. In particular, we have the functorial lift

$$\widetilde{\tau} = \tau^\mu \partial_\mu + \partial_\nu \tau^\alpha \dot{x}^\nu \frac{\partial}{\partial \dot{x}^\alpha} \tag{11.2.31}$$

of vector fields on X onto the tangent bundle TX and their functorial lift

$$\widetilde{\tau} = \tau^\mu \partial_\mu - \partial_\beta \tau^\nu \dot{x}_\nu \frac{\partial}{\partial \dot{x}_\beta} \tag{11.2.32}$$

onto the cotangent bundle T^*X.

Let $Y \to X$ be a vector bundle. Using the canonical vertical splitting (11.2.18), we obtain the canonical vertical vector field

$$u_Y = y^i \partial_i \tag{11.2.33}$$

on Y, called the *Liouville vector field*. For instance, the Liouville vector field on the tangent bundle TX reads

$$u_{TX} = \dot{x}^\lambda \dot{\partial}_\lambda. \tag{11.2.34}$$

Accordingly, any vector field $\tau = \tau^\lambda \partial_\lambda$ on a manifold X has the canonical *vertical lift*

$$\tau_V = \tau^\lambda \dot{\partial}_\lambda \tag{11.2.35}$$

onto the tangent bundle TX.

11.2.6 *Multivector fields*

A *multivector field* ϑ of degree $|\vartheta| = r$ (or, simply, an r-vector field) on a manifold Z is a section

$$\vartheta = \frac{1}{r!}\vartheta^{\lambda_1\ldots\lambda_r}\partial_{\lambda_1}\wedge\cdots\wedge\partial_{\lambda_r} \tag{11.2.36}$$

of the exterior product $\overset{r}{\wedge}TZ \to Z$. Let $\mathcal{T}_r(Z)$ denote the $C^\infty(Z)$-module space of r-vector fields on Z. All multivector fields on a manifold Z make up the graded commutative algebra $\mathcal{T}_*(Z)$ of global sections of the exterior bundle $\wedge TZ$ (11.2.12) with respect to the exterior product $\wedge$.

The graded commutative algebra $\mathcal{T}_*(Z)$ is endowed with the *Schouten–Nijenhuis* bracket

$$\begin{aligned}
&[.,.]_{\rm SN} : \mathcal{T}_r(Z)\times\mathcal{T}_s(Z)\to\mathcal{T}_{r+s-1}(Z), \qquad (11.2.37)\\
&[\vartheta,\upsilon]_{\rm SN} = \vartheta\bullet\upsilon + (-1)^{rs}\upsilon\bullet\vartheta,\\
&\vartheta\bullet\upsilon = \frac{r}{r!s!}(\vartheta^{\mu\lambda_2\ldots\lambda_r}\partial_\mu\upsilon^{\alpha_1\ldots\alpha_s}\partial_{\lambda_2}\wedge\cdots\wedge\partial_{\lambda_r}\wedge\partial_{\alpha_1}\wedge\cdots\wedge\partial_{\alpha_s}).
\end{aligned}$$

This generalizes the Lie bracket of vector fields. It obeys the relations

$$[\vartheta,\upsilon]_{\rm SN} = (-1)^{|\vartheta||\upsilon|}[\upsilon,\vartheta]_{\rm SN}, \tag{11.2.38}$$

$$[\nu,\vartheta\wedge\upsilon]_{\rm SN} = [\nu,\vartheta]_{\rm SN}\wedge\upsilon + (-1)^{(|\nu|-1)|\vartheta|}\vartheta\wedge[\nu,\upsilon]_{\rm SN}, \tag{11.2.39}$$

$$\begin{aligned}
&(-1)^{|\nu|(|\upsilon|-1)}[\nu,[\vartheta,\upsilon]_{\rm SN}]_{\rm SN} + (-1)^{|\vartheta|(|\nu|-1)}[\vartheta,[\upsilon,\nu]_{\rm SN}]_{\rm SN} \qquad (11.2.40)\\
&\quad + (-1)^{|\upsilon|(|\vartheta|-1)}[\upsilon,[\nu,\vartheta]_{\rm SN}]_{\rm SN} = 0.
\end{aligned}$$

The *Lie derivative of a multivector field* ϑ along a vector field u is defined as

$$\mathbf{L}_u\upsilon = [u,\vartheta]_{\rm SN},$$

$$\mathbf{L}_u(\vartheta\wedge\upsilon) = \mathbf{L}_u\vartheta\wedge\upsilon + \vartheta\wedge\mathbf{L}_u\upsilon.$$

Given an r-vector field ϑ (11.2.36) on a manifold Z, its *tangent lift* $\widetilde{\vartheta}$ onto the tangent bundle TZ of Z is defined by the relation

$$\widetilde{\vartheta}(\widetilde{\sigma}^r,\ldots,\widetilde{\sigma}^1) = \widetilde{\vartheta(\sigma^r,\ldots,\sigma^1)} \tag{11.2.41}$$

where [75]:

- $\sigma^k = \sigma^k_\lambda dz^\lambda$ are arbitrary one-forms on a manifold Z,
- by

$$\widetilde{\sigma}^k = \dot z^\mu\partial_\mu\sigma^k_\lambda dz^\lambda + \sigma^k_\lambda d\dot z^\lambda$$

are meant their tangent lifts (11.2.46) onto the tangent bundle TZ of Z,

- the right-hand side of the equality (11.2.41) is the tangent lift (11.2.44) onto TZ of the function $\vartheta(\sigma^r,\ldots,\sigma^1)$ on Z.

The tangent lift (11.2.41) takes the coordinate form

$$\widetilde{\vartheta} = \frac{1}{r!}[\dot{z}^\mu \partial_\mu \vartheta^{\lambda_1\ldots\lambda_r} \dot{\partial}_{\lambda_1} \wedge \cdots \wedge \dot{\partial}_{\lambda_r} + \vartheta^{\lambda_1\ldots\lambda_r} \sum_{i=1}^{r} \dot{\partial}_{\lambda_1} \wedge \cdots \wedge \partial_{\lambda_i} \wedge \cdots \wedge \dot{\partial}_{\lambda_r}]. \tag{11.2.42}$$

In particular, if τ is a vector field on a manifold Z, its tangent lift (11.2.42) coincides with the functorial lift (11.2.31).

The Schouten–Nijenhuis bracket commutes with the tangent lift (11.2.42) of multivectors, i.e.,

$$[\widetilde{\vartheta}, \widetilde{\upsilon}]_{\rm SN} = \widetilde{[\vartheta, \upsilon]_{\rm SN}}. \tag{11.2.43}$$

11.2.7 *Differential forms*

An *exterior r-form* on a manifold Z is a section

$$\phi = \frac{1}{r!}\phi_{\lambda_1\ldots\lambda_r} dz^{\lambda_1} \wedge \cdots \wedge dz^{\lambda_r}$$

of the exterior product $\overset{r}{\wedge} T^*Z \to Z$, where

$$dz^{\lambda_1} \wedge \cdots \wedge dz^{\lambda_r} = \frac{1}{r!}\epsilon^{\lambda_1\ldots\lambda_r}{}_{\mu_1\ldots\mu_r} dz^{\mu_1} \otimes \cdots \otimes dz^{\mu_r},$$
$$\epsilon^{\ldots\lambda_i\ldots\lambda_j\ldots}{}_{\ldots\mu_p\ldots\mu_k\ldots} = -\epsilon^{\ldots\lambda_j\ldots\lambda_i\ldots}{}_{\ldots\mu_p\ldots\mu_k\ldots} = -\epsilon^{\ldots\lambda_i\ldots\lambda_j\ldots}{}_{\ldots\mu_k\ldots\mu_p\ldots},$$
$$\epsilon^{\lambda_1\ldots\lambda_r}{}_{\lambda_1\ldots\lambda_r} = 1.$$

Sometimes, it is convenient to write

$$\phi = \phi'_{\lambda_1\ldots\lambda_r} dz^{\lambda_1} \wedge \cdots \wedge dz^{\lambda_r}$$

without the coefficient $1/r!$.

Let $\mathcal{O}^r(Z)$ denote the $C^\infty(Z)$-module of exterior r-forms on a manifold Z. By definition, $\mathcal{O}^0(Z) = C^\infty(Z)$ is the ring of smooth real functions on Z. All exterior forms on Z constitute the graded algebra $\mathcal{O}^*(Z)$ of global sections of the exterior bundle $\wedge T^*Z$ (11.2.12) endowed with the *exterior product*

$$\phi = \frac{1}{r!}\phi_{\lambda_1\ldots\lambda_r} dz^{\lambda_1} \wedge \cdots \wedge dz^{\lambda_r}, \qquad \sigma = \frac{1}{s!}\sigma_{\mu_1\ldots\mu_s} dz^{\mu_1} \wedge \cdots \wedge dz^{\mu_s},$$
$$\phi \wedge \sigma = \frac{1}{r!s!}\phi_{\nu_1\ldots\nu_r}\sigma_{\nu_{r+1}\ldots\nu_{r+s}} dz^{\nu_1} \wedge \cdots \wedge dz^{\nu_{r+s}}$$
$$= \frac{1}{r!s!(r+s)!}\epsilon^{\nu_1\ldots\nu_{r+s}}{}_{\alpha_1\ldots\alpha_{r+s}}\phi_{\nu_1\ldots\nu_r}\sigma_{\nu_{r+1}\ldots\nu_{r+s}} dz^{\alpha_1} \wedge \cdots \wedge dz^{\alpha_{r+s}},$$

such that

$$\phi \wedge \sigma = (-1)^{|\phi||\sigma|} \sigma \wedge \phi,$$

where the symbol $|\phi|$ stands for the form degree. The algebra $\mathcal{O}^*(Z)$ also is provided with the *exterior differential*

$$d\phi = dz^\mu \wedge \partial_\mu \phi = \frac{1}{r!} \partial_\mu \phi_{\lambda_1 \dots \lambda_r} dz^\mu \wedge dz^{\lambda_1} \wedge \dots \wedge dz^{\lambda_r}$$

which obeys the relations

$$d \circ d = 0, \qquad d(\phi \wedge \sigma) = d(\phi) \wedge \sigma + (-1)^{|\phi|} \phi \wedge d(\sigma).$$

The exterior differential d makes $\mathcal{O}^*(Z)$ into a differential graded algebra, called the *exterior algebra*.

Given a manifold morphism $f : Z \to Z'$, any exterior k-form ϕ on Z' yields the *pull-back exterior form* $f^*\phi$ on Z given by the condition

$$f^*\phi(v^1, \dots, v^k)(z) = \phi(Tf(v^1), \dots, Tf(v^k))(f(z))$$

for an arbitrary collection of tangent vectors $v^1, \cdots, v^k \in T_z Z$. We have the relations

$$f^*(\phi \wedge \sigma) = f^*\phi \wedge f^*\sigma, \qquad df^*\phi = f^*(d\phi).$$

In particular, given a fibred manifold $\pi : Y \to X$, the pull-back onto Y of exterior forms on X by π provides the monomorphism of graded commutative algebras $\mathcal{O}^*(X) \to \mathcal{O}^*(Y)$. Elements of its range $\pi^*\mathcal{O}^*(X)$ are called *basic forms*. Exterior forms

$$\phi : Y \to \overset{r}{\wedge} T^*X, \qquad \phi = \frac{1}{r!} \phi_{\lambda_1 \dots \lambda_r} dx^{\lambda_1} \wedge \dots \wedge dx^{\lambda_r},$$

on Y such that $u \rfloor \phi = 0$ for an arbitrary vertical vector field u on Y are said to be *horizontal forms*. Horizontal forms of degree $n = \dim X$ are called *densities*.

In the case of the tangent bundle $TX \to X$, there is a different way to lift exterior forms on X onto TX [75; 102]. Let f be a function on X. Its *tangent lift* onto TX is defined as the function

$$\widetilde{f} = \dot{x}^\lambda \partial_\lambda f. \tag{11.2.44}$$

Let σ be an r-form on X. Its *tangent lift* onto TX is said to be the r-form $\widetilde{\sigma}$ given by the relation

$$\widetilde{\sigma}(\widetilde{\tau}_1, \dots, \widetilde{\tau}_r) = \widetilde{\sigma(\tau_1, \dots, \tau_r)}, \tag{11.2.45}$$

where τ_i are arbitrary vector fields on X and $\widetilde{\tau}_i$ are their functorial lifts (11.2.31) onto TX. We have the coordinate expression

$$\sigma = \frac{1}{r!}\sigma_{\lambda_1\cdots\lambda_r} dx^{\lambda_1} \wedge \cdots \wedge dx^{\lambda_r},$$
$$\widetilde{\sigma} = \frac{1}{r!}[\dot{x}^\mu \partial_\mu \sigma_{\lambda_1\cdots\lambda_r} dx^{\lambda_1} \wedge \cdots \wedge dx^{\lambda_r} \tag{11.2.46}$$
$$+ \sum_{i=1}^{r} \sigma_{\lambda_1\cdots\lambda_r} dx^{\lambda_1} \wedge \cdots \wedge d\dot{x}^{\lambda_i} \wedge \cdots \wedge dx^{\lambda_r}].$$

The following equality holds:

$$d\widetilde{\sigma} = \widetilde{d\sigma}. \tag{11.2.47}$$

The *interior product* (or *contraction*) of a vector field u and an exterior r-form ϕ on a manifold Z is given by the coordinate expression

$$u\rfloor\phi = \sum_{k=1}^{r} \frac{(-1)^{k-1}}{r!} u^{\lambda_k} \phi_{\lambda_1\ldots\lambda_k\ldots\lambda_r} dz^{\lambda_1} \wedge \cdots \wedge \widehat{dz}^{\lambda_k} \wedge \cdots \wedge dz^{\lambda_r}$$
$$= \frac{1}{(r-1)!} u^\mu \phi_{\mu\alpha_2\ldots\alpha_r} dz^{\alpha_2} \wedge \cdots \wedge dz^{\alpha_r},$$

where the caret $\widehat{}$ denotes omission. It obeys the relations

$$\phi(u_1, \ldots, u_r) = u_r\rfloor \cdots u_1\rfloor\phi,$$
$$u\rfloor(\phi \wedge \sigma) = u\rfloor\phi \wedge \sigma + (-1)^{|\phi|}\phi \wedge u\rfloor\sigma. \tag{11.2.48}$$

A generalization of the interior product to multivector fields is the *left interior product*

$$\vartheta\rfloor\phi = \phi(\vartheta), \qquad |\vartheta| \leq |\phi|, \qquad \phi \in \mathcal{O}^*(Z), \qquad \vartheta \in \mathcal{T}_*(Z),$$

of multivector fields and exterior forms. It is defined by the equalities

$$\phi(u_1 \wedge \cdots \wedge u_r) = \phi(u_1, \ldots, u_r), \qquad \phi \in \mathcal{O}^*(Z), \qquad u_i \in \mathcal{T}_1(Z),$$

and obeys the relation

$$\vartheta\rfloor\upsilon\rfloor\phi = (\upsilon \wedge \vartheta)\rfloor\phi = (-1)^{|\upsilon||\vartheta|}\upsilon\rfloor\vartheta\rfloor\phi, \qquad \phi \in \mathcal{O}^*(Z), \qquad \vartheta, \upsilon \in \mathcal{T}_*(Z).$$

The *Lie derivative* of an exterior form ϕ along a vector field u is

$$\mathbf{L}_u\phi = u\rfloor d\phi + d(u\rfloor\phi), \tag{11.2.49}$$
$$\mathbf{L}_u(\phi \wedge \sigma) = \mathbf{L}_u\phi \wedge \sigma + \phi \wedge \mathbf{L}_u\sigma. \tag{11.2.50}$$

In particular, if f is a function, then

$$\mathbf{L}_u f = u(f) = u\rfloor df.$$

An exterior form ϕ is invariant under a local one-parameter group of diffeomorphisms G_t of Z (i.e., $G_t^*\phi = \phi$) if and only if its Lie derivative along the infinitesimal generator u of this group vanishes, i.e.,

$$\mathbf{L}_u\phi = 0.$$

Following physical terminology (Definition 1.10.3), we say that a vector field u is a *symmetry* of an exterior form ϕ.

A *tangent-valued r-form* on a manifold Z is a section

$$\phi = \frac{1}{r!}\phi^\mu_{\lambda_1\dots\lambda_r} dz^{\lambda_1} \wedge \cdots \wedge dz^{\lambda_r} \otimes \partial_\mu \tag{11.2.51}$$

of the tensor bundle

$$\overset{r}{\wedge} T^*Z \otimes TZ \to Z.$$

Remark 11.2.4. There is one-to-one correspondence between the tangent-valued one-forms ϕ on a manifold Z and the linear bundle endomorphisms

$$\widehat{\phi} : TZ \to TZ, \quad \widehat{\phi} : T_zZ \ni v \to v\rfloor\phi(z) \in T_zZ, \tag{11.2.52}$$

$$\widehat{\phi}^* : T^*Z \to T^*Z, \quad \widehat{\phi}^* : T_z^*Z \ni v^* \to \phi(z)\rfloor v^* \in T_z^*Z, \tag{11.2.53}$$

over Z (Remark 11.2.1). For instance, the *canonical tangent-valued one-form*

$$\theta_Z = dz^\lambda \otimes \partial_\lambda \tag{11.2.54}$$

on Z corresponds to the identity morphisms (11.2.52) and (11.2.53).

Remark 11.2.5. Let $Z = TX$, and let TTX be the tangent bundle of TX. It is called the *double tangent bundle*. There is the bundle endomorphism

$$J(\partial_\lambda) = \dot\partial_\lambda, \qquad J(\dot\partial_\lambda) = 0 \tag{11.2.55}$$

of TTX over X. It corresponds to the canonical tangent-valued form

$$\theta_J = dx^\lambda \otimes \dot\partial_\lambda \tag{11.2.56}$$

on the tangent bundle TX. It is readily observed that $J \circ J = 0$.

The space $\mathcal{O}^*(Z) \otimes \mathcal{T}_1(Z)$ of tangent-valued forms is provided with the *Frölicher–Nijenhuis bracket*

$$\begin{aligned}
&[,]_{\rm FN} : \mathcal{O}^r(Z) \otimes \mathcal{T}_1(Z) \times \mathcal{O}^s(Z) \otimes \mathcal{T}_1(Z) \to \mathcal{O}^{r+s}(Z) \otimes \mathcal{T}_1(Z),\\
&[\alpha \otimes u,\, \beta \otimes v]_{\rm FN} = (\alpha \wedge \beta) \otimes [u,v] + (\alpha \wedge \mathbf{L}_u\beta) \otimes v \\
&\quad - (\mathbf{L}_v\alpha \wedge \beta) \otimes u + (-1)^r(d\alpha \wedge u\rfloor\beta) \otimes v + (-1)^r(v\rfloor\alpha \wedge d\beta) \otimes u,\\
&\alpha \in \mathcal{O}^r(Z), \qquad \beta \in \mathcal{O}^s(Z), \qquad u, v \in \mathcal{T}_1(Z).
\end{aligned} \tag{11.2.57}$$

Its coordinate expression is

$$[\phi,\sigma]_{\rm FN} = \frac{1}{r!s!}(\phi^\nu_{\lambda_1\ldots\lambda_r}\partial_\nu\sigma^\mu_{\lambda_{r+1}\ldots\lambda_{r+s}} - \sigma^\nu_{\lambda_{r+1}\ldots\lambda_{r+s}}\partial_\nu\phi^\mu_{\lambda_1\ldots\lambda_r}$$
$$- r\phi^\mu_{\lambda_1\ldots\lambda_{r-1}\nu}\partial_{\lambda_r}\sigma^\nu_{\lambda_{r+1}\ldots\lambda_{r+s}} + s\sigma^\mu_{\nu\lambda_{r+2}\ldots\lambda_{r+s}}\partial_{\lambda_{r+1}}\phi^\nu_{\lambda_1\ldots\lambda_r})$$
$$dz^{\lambda_1}\wedge\cdots\wedge dz^{\lambda_{r+s}}\otimes\partial_\mu,$$
$$\phi\in\mathcal{O}^r(Z)\otimes\mathcal{T}_1(Z), \qquad \sigma\in\mathcal{O}^s(Z)\otimes\mathcal{T}_1(Z).$$

There are the relations

$$[\phi,\sigma]_{\rm FN} = (-1)^{|\phi||\psi|+1}[\sigma,\phi]_{\rm FN}, \tag{11.2.58}$$
$$[\phi,[\sigma,\theta]_{\rm FN}]_{\rm FN} = [[\phi,\sigma]_{\rm FN},\theta]_{\rm FN} \tag{11.2.59}$$
$$+(-1)^{|\phi||\sigma|}[\sigma,[\phi,\theta]_{\rm FN}]_{\rm FN},$$
$$\phi,\sigma,\theta\in\mathcal{O}^*(Z)\otimes\mathcal{T}_1(Z).$$

Given a tangent-valued form θ, the *Nijenhuis differential* on $\mathcal{O}^*(Z)\otimes\mathcal{T}_1(Z)$ is defined as the morphism

$$d_\theta:\psi\to d_\theta\psi=[\theta,\psi]_{\rm FN}, \qquad \psi\in\mathcal{O}^*(Z)\otimes\mathcal{T}_1(Z).$$

By virtue of (11.2.59), it has the property

$$d_\phi[\psi,\theta]_{\rm FN} = [d_\phi\psi,\theta]_{\rm FN} + (-1)^{|\phi||\psi|}[\psi,d_\phi\theta]_{\rm FN}.$$

In particular, if $\theta=u$ is a vector field, the Nijenhuis differential is the *Lie derivative of tangent-valued forms*

$$\mathbf{L}_u\sigma = d_u\sigma = [u,\sigma]_{\rm FN} = \frac{1}{s!}(u^\nu\partial_\nu\sigma^\mu_{\lambda_1\ldots\lambda_s} - \sigma^\nu_{\lambda_1\ldots\lambda_s}\partial_\nu u^\mu$$
$$+ s\sigma^\mu_{\nu\lambda_2\ldots\lambda_s}\partial_{\lambda_1}u^\nu)dx^{\lambda_1}\wedge\cdots\wedge dx^{\lambda_s}\otimes\partial_\mu, \qquad \sigma\in\mathcal{O}^s(Z)\otimes\mathcal{T}_1(Z).$$

If ϕ is a tangent-valued one-form, the Nijenhuis differential

$$d_\phi\phi = [\phi,\phi]_{\rm FN} \tag{11.2.60}$$
$$= (\phi^\mu_\nu\partial_\mu\phi^\alpha_\beta - \phi^\mu_\beta\partial_\mu\phi^\alpha_\nu - \phi^\alpha_\mu\partial_\nu\phi^\mu_\beta + \phi^\alpha_\mu\partial_\beta\phi^\mu_\nu)dz^\nu\wedge dz^\beta\otimes\partial_\alpha$$

is called the *Nijenhuis torsion*.

Let $Y\to X$ be a fibred manifold. We consider the following subspaces of the space $\mathcal{O}^*(Y)\otimes\mathcal{T}_1(Y)$ of tangent-valued forms on Y:

- *horizontal tangent-valued forms*

$$\phi: Y\to\overset{r}{\wedge}T^*X\underset{Y}{\otimes}TY,$$

$$\phi = dx^{\lambda_1}\wedge\cdots\wedge dx^{\lambda_r}\otimes\frac{1}{r!}[\phi^\mu_{\lambda_1\ldots\lambda_r}(y)\partial_\mu + \phi^i_{\lambda_1\ldots\lambda_r}(y)\partial_i],$$

- *projectable horizontal tangent-valued forms*

$$\phi = dx^{\lambda_1} \wedge \cdots \wedge dx^{\lambda_r} \otimes \frac{1}{r!}[\phi^\mu_{\lambda_1\ldots\lambda_r}(x)\partial_\mu + \phi^i_{\lambda_1\ldots\lambda_r}(y)\partial_i],$$

- *vertical-valued form*

$$\phi : Y \to \overset{r}{\wedge} T^*X \underset{Y}{\otimes} VY, \quad \phi = \frac{1}{r!}\phi^i_{\lambda_1\ldots\lambda_r}(y) dx^{\lambda_1} \wedge \cdots \wedge dx^{\lambda_r} \otimes \partial_i,$$

- vertical-valued one-forms, called *soldering forms*,

$$\sigma = \sigma^i_\lambda(y) dx^\lambda \otimes \partial_i, \tag{11.2.61}$$

- *basic soldering forms*

$$\sigma = \sigma^i_\lambda(x) dx^\lambda \otimes \partial_i.$$

Remark 11.2.6. The tangent bundle TX is provided with the canonical soldering form θ_J (11.2.56). Due to the canonical vertical splitting

$$VTX = TX \underset{X}{\times} TX, \tag{11.2.62}$$

the canonical soldering form (11.2.56) on TX defines the canonical tangent-valued form θ_X (11.2.54) on X. By this reason, tangent-valued one-forms on a manifold X also are called soldering forms.

We also mention the TX-valued forms

$$\phi : Y \to \overset{r}{\wedge} T^*X \underset{Y}{\otimes} TX, \tag{11.2.63}$$
$$\phi = \frac{1}{r!}\phi^\mu_{\lambda_1\ldots\lambda_r} dx^{\lambda_1} \wedge \cdots \wedge dx^{\lambda_r} \otimes \partial_\mu,$$

and V^*Y-valued forms

$$\phi : Y \to \overset{r}{\wedge} T^*X \underset{Y}{\otimes} V^*Y, \tag{11.2.64}$$
$$\phi = \frac{1}{r!}\phi_{\lambda_1\ldots\lambda_r i} dx^{\lambda_1} \wedge \cdots \wedge dx^{\lambda_r} \otimes \overline{d}y^i.$$

It should be emphasized that (11.2.63) are not tangent-valued forms, while (11.2.64) are not exterior forms. They exemplify *vector-valued forms.* Given a vector bundle $E \to X$, by a E-valued k-form on X, is meant a section of the fibre bundle

$$(\overset{k}{\wedge} T^*X) \underset{X}{\otimes} E^* \to X.$$

11.2.8 *Distributions and foliations*

A subbundle $\mathbf{T}$ of the tangent bundle TZ of a manifold Z is called a regular *distribution* (or, simply, a distribution). A vector field u on Z is said to be *subordinate* to a distribution $\mathbf{T}$ if it lives in $\mathbf{T}$. A distribution $\mathbf{T}$ is called *involutive* if the Lie bracket of $\mathbf{T}$-subordinate vector fields also is subordinate to $\mathbf{T}$.

A subbundle of the cotangent bundle T^*Z of Z is called a *codistribution* $\mathbf{T}^*$ on a manifold Z. For instance, the *annihilator* $\operatorname{Ann}\mathbf{T}$ of a distribution $\mathbf{T}$ is a codistribution whose fibre over $z \in Z$ consists of covectors $w \in T^*_z$ such that $v\rfloor w = 0$ for all $v \in \mathbf{T}_z$.

There is the following criterion of an involutive distribution [162].

Theorem 11.2.13. *Let* $\mathbf{T}$ *be a distribution and* $\operatorname{Ann}\mathbf{T}$ *its annihilator. Let* $\wedge\operatorname{Ann}\mathbf{T}(Z)$ *be the ideal of the exterior algebra* $\mathcal{O}^*(Z)$ *which is generated by sections of* $\operatorname{Ann}\mathbf{T} \to Z$. *A distribution* $\mathbf{T}$ *is involutive if and only if the ideal* $\wedge\operatorname{Ann}\mathbf{T}(Z)$ *is a differential ideal, i.e.,*

$$d(\wedge\operatorname{Ann}\mathbf{T}(Z)) \subset \wedge\operatorname{Ann}\mathbf{T}(Z).$$

The following local coordinates can be associated to an involutive distribution [162].

Theorem 11.2.14. *Let* $\mathbf{T}$ *be an involutive* r*-dimensional distribution on a manifold* Z, $\dim Z = k$. *Every point* $z \in Z$ *has an open neighborhood* U *which is a domain of an adapted coordinate chart* $(z^1, \ldots, z^k)$ *such that, restricted to* U, *the distribution* $\mathbf{T}$ *and its annihilator* $\operatorname{Ann}\mathbf{T}$ *are spanned by the local vector fields* $\partial/\partial z^1, \cdots, \partial/\partial z^r$ *and the local one-forms* $dz^{r+1}, \ldots, dz^k$, *respectively.*

A connected submanifold N of a manifold Z is called an *integral manifold* of a distribution $\mathbf{T}$ on Z if $TN \subset \mathbf{T}$. Unless otherwise stated, by an integral manifold is meant an integral manifold of dimension of $\mathbf{T}$. An integral manifold is called *maximal* if no other integral manifold contains it. The following is the classical theorem of Frobenius [93; 162].

Theorem 11.2.15. *Let* $\mathbf{T}$ *be an involutive distribution on a manifold* Z. *For any* $z \in Z$, *there exists a unique maximal integral manifold of* $\mathbf{T}$ *through* z, *and any integral manifold through* z *is its open subset.*

Maximal integral manifolds of an involutive distribution on a manifold Z are assembled into a regular foliation $\mathcal{F}$ of Z.

A regular r-dimensional *foliation* (or, simply, a foliation) $\mathcal{F}$ of a k-dimensional manifold Z is defined as a partition of Z into connected r-dimensional submanifolds (the *leaves* of a foliation) F_ι, $\iota \in I$, which possesses the following properties [132; 154].

A manifold Z admits an adapted coordinate atlas

$$\{(U_\xi; z^\lambda, z^i)\}, \quad \lambda = 1, \ldots, k-r, \qquad i = 1, \ldots, r, \tag{11.2.65}$$

such that transition functions of coordinates z^λ are independent of the remaining coordinates z^i. For each leaf F of a foliation $\mathcal{F}$, the connected components of $F \cap U_\xi$ are given by the equations $z^\lambda =$const. These connected components and coordinates (z^i) on them make up a coordinate atlas of a leaf F. It follows that tangent spaces to leaves of a foliation $\mathcal{F}$ constitute an involutive distribution $T\mathcal{F}$ on Z, called the *tangent bundle to the foliation* $\mathcal{F}$. The factor bundle

$$V\mathcal{F} = TZ/T\mathcal{F},$$

called the *normal bundle* to $\mathcal{F}$, has transition functions independent of coordinates z^i. Let $T\mathcal{F}^* \to Z$ denote the dual of $T\mathcal{F} \to Z$. There are the exact sequences

$$0 \to T\mathcal{F} \xrightarrow{i_\mathcal{F}} TX \longrightarrow V\mathcal{F} \to 0, \tag{11.2.66}$$

$$0 \to \operatorname{Ann} T\mathcal{F} \longrightarrow T^*X \xrightarrow{i^*_\mathcal{F}} T\mathcal{F}^* \to 0 \tag{11.2.67}$$

of vector bundles over Z.

A pair $(Z, \mathcal{F})$, where $\mathcal{F}$ is a foliation of Z, is called a *foliated manifold.* It should be emphasized that leaves of a foliation need not be closed or imbedded submanifolds. Every leaf has an open *saturated neighborhood* U, i.e., if $z \in U$, then a leaf through z also belongs to U.

Any submersion $\zeta : Z \to M$ yields a foliation

$$\mathcal{F} = \{F_p = \zeta^{-1}(p)\}_{p \in \zeta(Z)}$$

of Z indexed by elements of $\zeta(Z)$, which is an open submanifold of M, i.e., $Z \to \zeta(Z)$ is a fibred manifold. Leaves of this foliation are closed imbedded submanifolds. Such a foliation is called *simple.* Any (regular) foliation is locally simple.

Example 11.2.1. Every smooth real function f on a manifold Z with nowhere vanishing differential df is a submersion $Z \to \mathbb{R}$. It defines a one-codimensional foliation whose leaves are given by the equations

$$f(z) = c, \qquad c \in f(Z) \subset \mathbb{R}.$$

This is the *foliation of level surfaces* of the function f, called a *generating function*. Every one-codimensional foliation is locally a foliation of level surfaces of some function on Z. The level surfaces of an arbitrary smooth real function f on a manifold Z define a *singular foliation* $\mathcal{F}$ on Z [89]. Its leaves are not submanifolds in general. Nevertheless if $df(z) \neq 0$, the restriction of $\mathcal{F}$ to some open neighborhood U of z is a foliation with the generating function $f|_U$.

11.2.9 *Differential geometry of Lie groups*

Let G be a real Lie group of $\dim G > 0$, and let $L_g : G \to gG$ and $R_g : G \to Gg$ denote the action of G on itself by left and right multiplications, respectively. Clearly, L_g and $R_{g'}$ for all $g, g' \in G$ mutually commute, and so do the tangent maps TL_g and $TR_{g'}$.

A vector field ξ_l (resp. ξ_r) on a group G is said to be *left-invariant* (resp. *right-invariant*) if $\xi_l \circ L_g = TL_g \circ \xi_l$ (resp. $\xi_r \circ R_g = TR_g \circ \xi_r$). Left-invariant (resp. right-invariant) vector fields make up the *left Lie algebra* $\mathfrak{g}_l$ (resp. the *right Lie algebra* $\mathfrak{g}_r$) of G.

There is one-to-one correspondence between the left-invariant vector field ξ_l (resp. right-invariant vector fields ξ_r) on G and the vectors $\xi_l(e) = TL_{g^{-1}}\xi_l(g)$ (resp. $\xi_r(e) = TR_{g^{-1}}\xi_l(g)$) of the tangent space T_eG to G at the unit element e of G. This correspondence provides T_eG with the left and the right Lie algebra structures. Accordingly, the left action L_g of a Lie group G on itself defines its *adjoint representation*

$$\xi_r \to \operatorname{Ad} g(\xi_r) = TL_g \circ \xi_r \circ L_{g^{-1}} \tag{11.2.68}$$

in the right Lie algebra $\mathfrak{g}_r$.

Let $\{\epsilon_m\}$ (resp. $\{\varepsilon_m\}$) denote the basis for the left (resp. right) Lie algebra, and let c^k_{mn} be the *right structure constants*

$$[\varepsilon_m, \varepsilon_n] = c^k_{mn}\varepsilon_k.$$

There is the morphism

$$\rho : \mathfrak{g}_l \ni \epsilon_m \to \varepsilon_m = -\epsilon_m \in \mathfrak{g}_r$$

between left and right Lie algebras such that

$$[\epsilon_m, \epsilon_n] = -c^k_{mn}\epsilon_k.$$

The tangent bundle $\pi_G : TG \to G$ of a Lie group G is trivial. There are the following two canonical isomorphisms

$$\varrho_l : TG \ni q \to (g = \pi_G(q), TL_g^{-1}(q)) \in G \times \mathfrak{g}_l,$$
$$\varrho_r : TG \ni q \to (g = \pi_G(q), TR_g^{-1}(q)) \in G \times \mathfrak{g}_r.$$

Therefore, any action

$$G \times Z \ni (g, z) \to gz \in Z$$

of a Lie group G on a manifold Z on the left yields the homomorphism

$$\mathfrak{g}_r \ni \varepsilon \to \xi_\varepsilon \in \mathcal{T}_1(Z) \tag{11.2.69}$$

of the right Lie algebra $\mathfrak{g}_r$ of G into the Lie algebra of vector fields on Z such that

$$\xi_{\mathrm{Ad}\, g(\varepsilon)} = Tg \circ \xi_\varepsilon \circ g^{-1}. \tag{11.2.70}$$

Vector fields ξ_ε are said to be the *infinitesimal generators* of a representation of the Lie group G in Z.

In particular, the adjoint representation (11.2.68) of a Lie group in its right Lie algebra yields the *adjoint representation*

$$\varepsilon' : \varepsilon \to \mathrm{ad}\, \varepsilon'(\varepsilon) = [\varepsilon', \varepsilon], \qquad \mathrm{ad}\, \varepsilon_m(\varepsilon_n) = c^k_{mn}\varepsilon_k, \tag{11.2.71}$$

of the right Lie algebra $\mathfrak{g}_r$ in itself.

The dual $\mathfrak{g}^* = T^*_e G$ of the tangent space $T_e G$ is called the *Lie coalgebra*). It is provided with the basis $\{\varepsilon^m\}$ which is the dual of the basis $\{\varepsilon_m\}$ for $T_e G$. The group G and the right Lie algebra $\mathfrak{g}_r$ act on $\mathfrak{g}^*$ by the *coadjoint representation*

$$\begin{aligned} &\langle \mathrm{Ad}^* g(\varepsilon^*), \varepsilon\rangle = \langle \varepsilon^*, \mathrm{Ad}\, g^{-1}(\varepsilon)\rangle, \quad \varepsilon^* \in \mathfrak{g}^*, \quad \varepsilon \in \mathfrak{g}_r, \\ &\langle \mathrm{ad}^* \varepsilon'(\varepsilon^*), \varepsilon\rangle = -\langle \varepsilon^*, [\varepsilon', \varepsilon]\rangle, \qquad \varepsilon' \in \mathfrak{g}_r, \\ &\mathrm{ad}^* \varepsilon_m(\varepsilon^n) = -c^n_{mk}\varepsilon^k. \end{aligned} \tag{11.2.72}$$

Remark 11.2.7. In the literature (e.g., [1]), one can meet another definition of the coadjoint representation in accordance with the relation

$$\langle \mathrm{Ad}^* g(\varepsilon^*), \varepsilon\rangle = \langle \varepsilon^*, \mathrm{Ad}\, g(\varepsilon)\rangle.$$

The Lie coalgebra $\mathfrak{g}^*$ of a Lie group G is provided with the canonical Poisson structure, called the *Lie–Poisson structure* [1; 104]. It is given by the bracket

$$\{f, g\}_{\mathrm{LP}} = \langle \varepsilon^*, [df(\varepsilon^*), dg(\varepsilon^*)]\rangle, \qquad f, g \in C^\infty(\mathfrak{g}^*), \tag{11.2.73}$$

where $df(\varepsilon^*), dg(\varepsilon^*) \in \mathfrak{g}_r$ are seen as linear mappings from $T_{\varepsilon^*}\mathfrak{g}^* = \mathfrak{g}^*$ to $\mathbb{R}$. Given coordinates z_k on $\mathfrak{g}^*$ with respect to the basis $\{\varepsilon^k\}$, the Lie–Poisson bracket (11.2.73) and the corresponding Poisson bivector field w read

$$\{f, g\}_{\mathrm{LP}} = c^k_{mn} z_k \partial^m f \partial^n g, \qquad w_{mn} = c^k_{mn} z_k.$$

One can show that symplectic leaves of the Lie–Poisson structure on the coalgebra $\mathfrak{g}^*$ of a connected Lie group G are orbits of the coadjoint representation (11.2.72) of G on $\mathfrak{g}^*$ [163].

11.3 Jet manifolds

This Section collects the relevant material on jet manifolds of sections of fibre bundles [68; 94; 109; 145].

11.3.1 *First order jet manifolds*

Given a fibre bundle $Y \to X$ with bundle coordinates (x^λ, y^i), let us consider the equivalence classes $j^1_x s$ of its sections s, which are identified by their values $s^i(x)$ and the values of their partial derivatives $\partial_\mu s^i(x)$ at a point $x \in X$. They are called the *first order jets* of sections at x. One can justify that the definition of jets is coordinate-independent. A key point is that the set J^1Y of first order jets $j^1_x s$, $x \in X$, is a smooth manifold with respect to the adapted coordinates $(x^\lambda, y^i, y^i_\lambda)$ such that

$$y^i_\lambda(j^1_x s) = \partial_\lambda s^i(x), \qquad y'^i_\lambda = \frac{\partial x^\mu}{\partial x'^\lambda}(\partial_\mu + y^j_\mu \partial_j) y'^i. \tag{11.3.1}$$

It is called the first order *jet manifold* of a fibre bundle $Y \to X$. We call (y^i_λ) the *jet coordinate*.

A jet manifold J^1Y admits the natural fibrations

$$\pi^1 : J^1Y \ni j^1_x s \to x \in X, \tag{11.3.2}$$

$$\pi^1_0 : J^1Y \ni j^1_x s \to s(x) \in Y. \tag{11.3.3}$$

A glance at the transformation law (11.3.1) shows that π^1_0 is an affine bundle modelled over the vector bundle

$$T^*X \underset{Y}{\otimes} VY \to Y. \tag{11.3.4}$$

It is convenient to call π^1 (11.3.2) the *jet bundle*, while π^1_0 (11.3.3) is said to be the *affine jet bundle*.

Let us note that, if $Y \to X$ is a vector or an affine bundle, the jet bundle π_1 (11.3.2) is so.

Jets can be expressed in terms of familiar tangent-valued forms as follows. There are the canonical imbeddings

$$\lambda_{(1)} : J^1Y \underset{Y}{\to} T^*X \underset{Y}{\otimes} TY,$$

$$\lambda_{(1)} = dx^\lambda \otimes (\partial_\lambda + y^i_\lambda \partial_i) = dx^\lambda \otimes d_\lambda, \tag{11.3.5}$$

$$\theta_{(1)} : J^1Y \underset{Y}{\to} T^*Y \underset{Y}{\otimes} VY,$$

$$\theta_{(1)} = (dy^i - y^i_\lambda dx^\lambda) \otimes \partial_i = \theta^i \otimes \partial_i, \tag{11.3.6}$$

where d_λ are said to be *total derivatives*, and θ^i are called *contact forms*.

We further identify the jet manifold J^1Y with its images under the canonical morphisms (11.3.5) and (11.3.6), and represent the jets $j^1_x s = (x^\lambda, y^i, y^i_\mu)$ by the tangent-valued forms $\lambda_{(1)}$ (11.3.5) and $\theta_{(1)}$ (11.3.6).

Sections and morphisms of fibre bundles admit prolongations to jet manifolds as follows.

Any section s of a fibre bundle $Y \to X$ has the *jet prolongation* to the section

$$(J^1 s)(x) = j^1_x s, \qquad y^i_\lambda \circ J^1 s = \partial_\lambda s^i(x),$$

of the jet bundle $J^1Y \to X$. A *section of the jet bundle* $J^1Y \to X$ is called *integrable* if it is the jet prolongation of some section of a fibre bundle $Y \to X$.

Any bundle morphism $\Phi : Y \to Y'$ over a diffeomorphism f admits a *jet prolongation* to a bundle morphism of affine jet bundles

$$J^1\Phi : J^1Y \underset{\Phi}{\longrightarrow} J^1Y', \qquad {y'}^i_\lambda \circ J^1\Phi = \frac{\partial (f^{-1})^\mu}{\partial x'^\lambda} d_\mu \Phi^i. \tag{11.3.7}$$

Any projectable vector field u (11.2.28) on a fibre bundle $Y \to X$ has a *jet prolongation* to the projectable vector field

$$\begin{aligned} J^1 u &= r_1 \circ J^1 u : J^1Y \to J^1TY \to TJ^1Y, \\ J^1 u &= u^\lambda \partial_\lambda + u^i \partial_i + (d_\lambda u^i - y^i_\mu \partial_\lambda u^\mu) \partial_i^\lambda, \end{aligned} \tag{11.3.8}$$

on the jet manifold J^1Y. In order to obtain (11.3.8), the canonical bundle morphism

$$r_1 : J^1TY \to TJ^1Y, \qquad \dot y^i_\lambda \circ r_1 = (\dot y^i)_\lambda - y^i_\mu \dot x^\mu_\lambda,$$

is used. In particular, there is the canonical isomorphism

$$VJ^1Y = J^1VY, \qquad \dot y^i_\lambda = (\dot y^i)_\lambda. \tag{11.3.9}$$

11.3.2 *Second order jet manifolds*

Taking the first order jet manifold of the jet bundle $J^1Y \to X$, we obtain the *repeated jet manifold* J^1J^1Y provided with the adapted coordinates

$$(x^\lambda, y^i, y^i_\lambda, \widehat{y}^i_\mu, y^i_{\mu\lambda})$$

possessing transition functions

$$\begin{aligned} &{y'}^i_\lambda = \frac{\partial x^\alpha}{\partial x'^\lambda} d_\alpha {y'}^i, \qquad \widehat{y'}^i_\lambda = \frac{\partial x^\alpha}{\partial x'^\lambda} \widehat{d}_\alpha {y'}^i, \qquad {y'}^i_{\mu\lambda} = \frac{\partial x^\alpha}{\partial x'^\mu} \widehat{d}_\alpha {y'}^i_\lambda, \\ &d_\alpha = \partial_\alpha + y^j_\alpha \partial_j + y^j_{\nu\alpha} \partial_j^\nu, \qquad \widehat{d}_\alpha = \partial_\alpha + \widehat{y}^j_\alpha \partial_j + y^j_{\nu\alpha} \partial^\nu_j. \end{aligned}$$

There exist two different affine fibrations of J^1J^1Y over J^1Y:

- the familiar affine jet bundle (11.3.3):

$$\pi_{11} : J^1J^1Y \to J^1Y, \qquad y^i_\lambda \circ \pi_{11} = y^i_\lambda, \tag{11.3.10}$$

- the affine bundle

$$J^1\pi^1_0 : J^1J^1Y \to J^1Y, \qquad y^i_\lambda \circ J^1\pi^1_0 = \widehat{y}^i_\lambda. \tag{11.3.11}$$

In general, there is no canonical identification of these fibrations. The points $q \in J^1J^1Y$, where

$$\pi_{11}(q) = J^1\pi^1_0(q),$$

form an affine subbundle $\widehat{J}^2Y \to J^1Y$ of J^1J^1Y called the *sesquiholonomic jet manifold*. It is given by the coordinate conditions $\widehat{y}^i_\lambda = y^i_\lambda$, and is coordinated by $(x^\lambda, y^i, y^i_\lambda, y^i_{\mu\lambda})$.

The *second order* (or *holonomic*) jet manifold J^2Y of a fibre bundle $Y \to X$ can be defined as the affine subbundle of the fibre bundle $\widehat{J}^2Y \to J^1Y$ given by the coordinate conditions $y^i_{\lambda\mu} = y^i_{\mu\lambda}$. It is modelled over the vector bundle

$$\overset{2}{\vee} T^*X \underset{J^1Y}{\otimes} VY \to J^1Y,$$

and is endowed with adapted coordinates $(x^\lambda, y^i, y^i_\lambda, y^i_{\lambda\mu} = y^i_{\mu\lambda})$, possessing transition functions

$$y'^i_\lambda = \frac{\partial x^\alpha}{\partial x'^\lambda} d_\alpha y'^i, \qquad y'^i_{\mu\lambda} = \frac{\partial x^\alpha}{\partial x'^\mu} d_\alpha y'^i_\lambda. \tag{11.3.12}$$

The second order jet manifold J^2Y also can be introduced as the set of the equivalence classes $j^2_x s$ of sections s of the fibre bundle $Y \to X$, which are identified by their values and the values of their first and second order partial derivatives at points $x \in X$, i.e.,

$$y^i_\lambda(j^2_x s) = \partial_\lambda s^i(x), \qquad y^i_{\lambda\mu}(j^2_x s) = \partial_\lambda\partial_\mu s^i(x).$$

The equivalence classes $j^2_x s$ are called the *second order jets* of sections.

Let s be a section of a fibre bundle $Y \to X$, and let J^1s be its jet prolongation to a section of a jet bundle $J^1Y \to X$. The latter gives rise to the section J^1J^1s of the repeated jet bundle $J^1J^1Y \to X$. This section takes its values into the second order jet manifold J^2Y. It is called the *second order jet prolongation of a section* s, and is denoted by J^2s.

Theorem 11.3.1. *Let $\overline{s}$ be a section of the jet bundle $J^1Y \to X$, and let $J^1\overline{s}$ be its jet prolongation to a section of the repeated jet bundle $J^1J^1Y \to X$. The following three facts are equivalent:*

- *$\overline{s} = J^1s$ where s is a section of a fibre bundle $Y \to X$,*
- *$J^1\overline{s}$ takes its values into $\widehat{J}^2Y$,*
- *$J^1\overline{s}$ takes its values into J^2Y.*

11.3.3 Higher order jet manifolds

The notion of first and second order jet manifolds is naturally extended to higher order jet manifolds.

The *k-order jet manifold* J^kY of a fibre bundle $Y \to X$ comprises the equivalence classes $j^k_x s$, $x \in X$, of sections s of Y identified by the $k+1$ terms of their Tailor series at points $x \in X$. The jet manifold J^kY is provided with the adapted coordinates

$$(x^\lambda, y^i, y^i_\lambda, \ldots, y^i_{\lambda_k\cdots\lambda_1}),$$
$$y^i_{\lambda_l\cdots\lambda_1}(j^k_x s) = \partial_{\lambda_l}\cdots\partial_{\lambda_1}s^i(x), \qquad 0 \leq l \leq k.$$

Every section s of a fibre bundle $Y \to X$ gives rise to the section $J^k s$ of a fibre bundle $J^kY \to X$ such that

$$y^i_{\lambda_l\cdots\lambda_1} \circ J^k s = \partial_{\lambda_l}\cdots\partial_{\lambda_1}s^i, \qquad 0 \leq l \leq k.$$

The following operators on exterior forms on jet manifolds are utilized:

• the *total derivative* operator

$$d_\lambda = \partial_\lambda + y^i_\lambda\partial_i + y^i_{\lambda\mu}\partial^\mu_i + \cdots, \tag{11.3.13}$$

obeying the relations

$$d_\lambda(\phi \wedge \sigma) = d_\lambda(\phi)\wedge\sigma + \phi\wedge d_\lambda(\sigma),$$
$$d_\lambda(d\phi) = d(d_\lambda(\phi)),$$

in particular,

$$d_\lambda(f) = \partial_\lambda f + y^i_\lambda\partial_i f + y^i_{\lambda\mu}\partial^\mu_i f + \cdots, \qquad f \in C^\infty(J^kY),$$
$$d_\lambda(dx^\mu) = 0, \qquad d_\lambda(dy^i_{\lambda_l\cdots\lambda_1}) = dy^i_{\lambda\lambda_l\cdots\lambda_1};$$

• the *horizontal projection* h_0 given by the relations

$$h_0(dx^\lambda) = dx^\lambda, \qquad h_0(dy^i_{\lambda_k\cdots\lambda_1}) = y^i_{\mu\lambda_k\ldots\lambda_1}dx^\mu, \tag{11.3.14}$$

in particular,

$$h_0(dy^i) = y^i_\mu dx^\mu, \qquad h_0(dy^i_\lambda) = y^i_{\mu\lambda}dx^\mu;$$

• the *total differential*

$$d_H(\phi) = dx^\lambda \wedge d_\lambda(\phi), \tag{11.3.15}$$

possessing the properties

$$d_H \circ d_H = 0, \qquad h_0 \circ d = d_H \circ h_0.$$

11.3.4 *Differential operators and differential equations*

Jet manifolds provide the standard language for the theory of differential equations and differential operators [21; 53; 95].

Definition 11.3.1. Let Z be an $(m+n)$-dimensional manifold. A system of k-order partial differential equations (or, simply, a *differential equation*) in n variables on Z is defined to be a closed smooth submanifold $\mathfrak{E}$ of the k-order jet bundle $J^k_n Z$ of n-dimensional submanifolds of Z.

By its *solution* is meant an n-dimensional submanifold S of Z whose k-order jets $[S]^k_z$, $z \in S$, belong to $\mathfrak{E}$.

Definition 11.3.2. A k-order differential equation in n variables on a manifold Z is called a *dynamic equation* if it can be algebraically solved for the highest order derivatives, i.e., it is a section of the fibration $J^k_n Z \to J^{k-1}_n Z$.

In particular, a *first order dynamic equation* in n variables on a manifold Z is a section of the jet bundle $J^1_n Z \to Z$. Its image in the tangent bundle $TZ \to Z$ by the correspondence $\lambda_{(1)}$ (10.1.2) is an n-dimensional vector subbundle of TZ. If $n=1$, a dynamic equation is given by a vector field

$$\dot{z}^\lambda(t) = u^\lambda(z(t)) \tag{11.3.16}$$

on a manifold Z. Its solutions are integral curves $c(t)$ of the vector field u.

Let $Y \to X$ be a fibre bundle. There are several equivalent definitions of (non-linear) differential operators. We start with the following.

Definition 11.3.3. Let $E \to X$ be a vector bundle. A *k-order E-valued differential operator* on a fibre bundle $Y \to X$ is defined as a section $\mathcal{E}$ of the pull-back bundle

$$\mathrm{pr}_1 : E^k_Y = J^k Y \underset{X}{\times} E \to J^k Y. \tag{11.3.17}$$

Given bundle coordinates (x^λ, y^i) on Y and (x^λ, χ^a) on E, the pull-back (11.3.17) is provided with coordinates $(x^\lambda, y^j_\Sigma, \chi^a)$, $0 \le |\Sigma| \le k$. With respect to these coordinates, a differential operator $\mathcal{E}$ seen as a closed imbedded submanifold $\mathcal{E} \subset E^k_Y$ is given by the equalities

$$\chi^a = \mathcal{E}^a(x^\lambda, y^j_\Sigma). \tag{11.3.18}$$

There is obvious one-to-one correspondence between the sections $\mathcal{E}$ (11.3.18) of the fibre bundle (11.3.17) and the bundle morphisms

$$\Phi : J^k Y \underset{X}{\longrightarrow} E, \tag{11.3.19}$$

$$\Phi = \mathrm{pr}_2 \circ \mathcal{E} \iff \mathcal{E} = (\mathrm{Id}\, J^k Y, \Phi).$$

Therefore, we come to the following equivalent definition of differential operators on $Y \to X$.

Definition 11.3.4. A fibred morphism

$$\mathcal{E} : J^kY \underset{X}{\to} E \tag{11.3.20}$$

is called a k-order *differential operator* on the fibre bundle $Y \to X$. It sends each section $s(x)$ of $Y \to X$ onto the section $(\mathcal{E} \circ J^k s)(x)$ of the vector bundle $E \to X$ [21; 95].

The *kernel* of a differential operator is the subset

$$\operatorname{Ker} \mathcal{E} = \mathcal{E}^{-1}(\widehat{0}(X)) \subset J^kY, \tag{11.3.21}$$

where $\widehat{0}$ is the zero section of the vector bundle $E \to X$, and we assume that $\widehat{0}(X) \subset \mathcal{E}(J^kY)$.

Definition 11.3.5. A system of k-order partial differential equations (or, simply, a *differential equation*) on a fibre bundle $Y \to X$ is defined as a closed subbundle $\mathfrak{E}$ of the jet bundle $J^kY \to X$.

Its *solution* is a (local) section s of the fibre bundle $Y \to X$ such that its k-order jet prolongation J^ks lives in $\mathfrak{E}$.

For instance, if the kernel (11.3.21) of a differential operator $\mathcal{E}$ is a closed subbundle of the fibre bundle $J^kY \to X$, it defines a differential equation

$$\mathcal{E} \circ J^k s = 0.$$

The following condition is sufficient for a kernel of a differential operator to be a differential equation.

Theorem 11.3.2. *Let the morphism (11.3.20) be of constant rank. Its kernel (11.3.21) is a closed subbundle of the fibre bundle $J^kY \to X$ and, consequently, is a k-order differential equation.*

11.4 Connections on fibre bundles

There are different equivalent definitions of a connection on a fibre bundle $Y \to X$. We define it both as a splitting of the exact sequence (11.2.20) and a global section of the affine jet bundle $J^1Y \to$Y [68; 109; 145].

11.4.1 *Connections*

A *connection* on a fibred manifold $Y \to X$ is defined as a splitting (called the horizontal splitting)

$$\Gamma : Y \underset{X}{\times} TX \underset{Y}{\to} TY, \qquad \Gamma : \dot{x}^\lambda \partial_\lambda \to \dot{x}^\lambda(\partial_\lambda + \Gamma^i_\lambda(y)\partial_i), \tag{11.4.1}$$
$$\dot{x}^\lambda \partial_\lambda + \dot{y}^i \partial_i = \dot{x}^\lambda(\partial_\lambda + \Gamma^i_\lambda \partial_i) + (\dot{y}^i - \dot{x}^\lambda \Gamma^i_\lambda)\partial_i,$$

of the exact sequence (11.2.20). Its range is a subbundle of $TY \to Y$ called the *horizontal distribution*. By virtue of Theorem 11.2.10, a connection on a fibred manifold always exists. A connection Γ (11.4.1) is represented by the horizontal tangent-valued one-form

$$\Gamma = dx^\lambda \otimes (\partial_\lambda + \Gamma^i_\lambda \partial_i) \tag{11.4.2}$$

on Y which is projected onto the canonical tangent-valued form θ_X (11.2.54) on X.

Given a connection Γ on a fibred manifold $Y \to X$, any vector field τ on a base X gives rise to the projectable vector field

$$\Gamma\tau = \tau \rfloor \Gamma = \tau^\lambda(\partial_\lambda + \Gamma^i_\lambda \partial_i) \tag{11.4.3}$$

on Y which lives in the horizontal distribution determined by Γ. It is called the *horizontal lift* of τ by means of a connection Γ.

The splitting (11.4.1) also is given by the vertical-valued form

$$\Gamma = (dy^i - \Gamma^i_\lambda dx^\lambda) \otimes \partial_i, \tag{11.4.4}$$

which yields an epimorphism $TY \to VY$. It provides the corresponding splitting

$$\Gamma : V^*Y \ni \overline{d}y^i \to dy^i - \Gamma^i_\lambda dx^\lambda \in T^*Y, \tag{11.4.5}$$
$$\dot{x}_\lambda dx^\lambda + \dot{y}_i dy^i = (\dot{x}_\lambda + \dot{y}_i \Gamma^i_\lambda)dx^\lambda + \dot{y}_i(dy^i - \Gamma^i_\lambda dx^\lambda),$$

of the dual exact sequence (11.2.21).

In an equivalent way, connections on a fibred manifold $Y \to X$ are introduced as global sections of the affine jet bundle $J^1Y \to Y$. Indeed, any global section Γ of $J^1Y \to Y$ defines the tangent-valued form $\lambda_1 \circ \Gamma$ (11.4.2). It follows from this definition that connections on a fibred manifold $Y \to X$ constitute an affine space modelled over the vector space of soldering forms σ (11.2.61). One also deduces from (11.3.1) the coordinate transformation law of connections

$$\Gamma'^i_\lambda = \frac{\partial x^\mu}{\partial x'^\lambda}(\partial_\mu + \Gamma^j_\mu \partial_j)y'^i.$$

Remark 11.4.1. Any connection Γ on a fibred manifold $Y \to X$ yields a horizontal lift of a vector field on X onto Y, but need not defines the similar lift of a *path* in X into Y. Let

$$\mathbb{R} \supset [,] \ni t \to x(t) \in X, \qquad \mathbb{R} \ni t \to y(t) \in Y,$$

be smooth paths in X and Y, respectively. Then $t \to y(t)$ is called a *horizontal lift* of $x(t)$ if

$$\pi(y(t)) = x(t), \qquad \dot{y}(t) \in H_{y(t)}Y, \qquad t \in \mathbb{R},$$

where $HY \subset TY$ is the horizontal subbundle associated to the connection Γ. If, for each path $x(t)$ $(t_0 \leq t \leq t_1)$ and for any $y_0 \in \pi^{-1}(x(t_0))$, there exists a horizontal lift $y(t)$ $(t_0 \leq t \leq t_1)$ such that $y(t_0) = y_0$, then Γ is called the *Ehresmann connection*. A fibred manifold is a fibre bundle if and only if it admits an Ehresmann connection [76].

Hereafter, we restrict our consideration to connections on fibre bundles. The following are two standard constructions of new connections from old ones.

• Let Y and Y' be fibre bundles over the same base X. Given connections Γ on Y and Γ' on Y', the bundle product $Y \underset{X}{\times} Y'$ is provided with the *product connection*

$$\Gamma \times \Gamma' = dx^\lambda \otimes \left(\partial_\lambda + \Gamma^i_\lambda \frac{\partial}{\partial y^i} + \Gamma'^j_\lambda \frac{\partial}{\partial y'^j}\right). \tag{11.4.6}$$

• Given a fibre bundle $Y \to X$, let $f : X' \to X$ be a manifold morphism and f^*Y the pull-back of Y over X'. Any connection Γ (11.4.4) on $Y \to X$ yields the *pull-back connection*

$$f^*\Gamma = \left(dy^i - \Gamma^i_\lambda(f^\mu(x'^\nu), y^j) \frac{\partial f^\lambda}{\partial x'^\mu} dx'^\mu\right) \otimes \partial_i \tag{11.4.7}$$

on the pull-back bundle $f^*Y \to X'$.

Every connection Γ on a fibre bundle $Y \to X$ defines the first order differential operator

$$D^\Gamma : J^1Y \underset{Y}{\to} T^*X \underset{Y}{\otimes} VY, \tag{11.4.8}$$
$$D^\Gamma = \lambda_1 - \Gamma \circ \pi^1_0 = (y^i_\lambda - \Gamma^i_\lambda) dx^\lambda \otimes \partial_i,$$

on Y called the *covariant differential*. If $s : X \to Y$ is a section, its *covariant differential*

$$\nabla^\Gamma s = D^\Gamma \circ J^1 s = (\partial_\lambda s^i - \Gamma^i_\lambda \circ s) dx^\lambda \otimes \partial_i \tag{11.4.9}$$

and its *covariant derivative* $\nabla^\Gamma_\tau s = \tau\rfloor\nabla^\Gamma s$ along a vector field τ on X are introduced. In particular, a (local) section s of $Y \to X$ is called an *integral section* for a connection Γ (or *parallel* with respect to Γ) if s obeys the equivalent conditions

$$\nabla^\Gamma s = 0 \quad \text{or} \quad J^1 s = \Gamma \circ s. \tag{11.4.10}$$

Let Γ be a connection on a fibre bundle $Y \to X$. Given vector fields τ, τ' on X and their horizontal lifts $\Gamma\tau$ and $\Gamma\tau'$ (11.4.3) on Y, let us consider the vertical vector field

$$R(\tau,\tau') = \Gamma[\tau,\tau'] - [\Gamma\tau,\Gamma\tau'] = \tau^\lambda\tau'^\mu R^i_{\lambda\mu}\partial_i, \tag{11.4.11}$$

$$R^i_{\lambda\mu} = \partial_\lambda\Gamma^i_\mu - \partial_\mu\Gamma^i_\lambda + \Gamma^j_\lambda\partial_j\Gamma^i_\mu - \Gamma^j_\mu\partial_j\Gamma^i_\lambda. \tag{11.4.12}$$

It can be seen as the contraction of vector fields τ and τ' with the vertical-valued horizontal two-form

$$R = \frac{1}{2}[\Gamma,\Gamma]_{\rm FN} = \frac{1}{2}R^i_{\lambda\mu}dx^\lambda\wedge dx^\mu\otimes\partial_i \tag{11.4.13}$$

on Y called the *curvature form* of a connection Γ.

Given a connection Γ and a soldering form σ, the *torsion* of Γ with respect to σ is defined as the vertical-valued horizontal two-form

$$T = [\Gamma,\sigma]_{\rm FN} = (\partial_\lambda\sigma^i_\mu + \Gamma^j_\lambda\partial_j\sigma^i_\mu - \partial_j\Gamma^i_\lambda\sigma^j_\mu)dx^\lambda\wedge dx^\mu\otimes\partial_i. \tag{11.4.14}$$

11.4.2 *Flat connections*

A *flat* (or *curvature-free*) connection is a connection Γ on a fibre bundle $Y \to X$ which satisfies the following equivalent conditions:

• its curvature vanishes everywhere on Y;

• its horizontal distribution is involutive;

• there exists a local integral section for the connection Γ through any point $y \in Y$.

By virtue of Theorem 11.2.15, a flat connection Γ yields a foliation of Y which is transversal to the fibration $Y \to X$. It called a *horizontal foliation*. Its leaf through a point $y \in Y$ is locally defined by an integral section s_y for the connection Γ through y. Conversely, let a fibre bundle $Y \to X$ admit a horizontal foliation such that, for each point $y \in Y$, the leaf of this foliation through y is locally defined by a section s_y of $Y \to X$ through y. Then the map

$$\Gamma : Y \ni y \to j^1_{\pi(y)}s_y \in J^1Y$$

sets a flat connection on $Y \to X$. Hence, there is one-to-one correspondence between the flat connections and the horizontal foliations of a fibre bundle $Y \to X$.

Given a horizontal foliation of a fibre bundle $Y \to X$, there exists the associated atlas of bundle coordinates (x^λ, y^i) on Y such that every leaf of this foliation is locally given by the equations y^i =const., and the transition functions $y^i \to y'^i(y^j)$ are independent of the base coordinates x^λ [68]. It is called the *atlas of constant local trivializations*. Two such atlases are said to be equivalent if their union also is an atlas of the same type. They are associated to the same horizontal foliation. Thus, the following is proved.

Theorem 11.4.1. *There is one-to-one correspondence between the flat connections Γ on a fibre bundle $Y \to X$ and the equivalence classes of atlases of constant local trivializations of Y such that $\Gamma = dx^\lambda \otimes \partial_\lambda$ relative to the corresponding atlas.*

Example 11.4.1. Any trivial bundle has flat connections corresponding to its trivializations. Fibre bundles over a one-dimensional base have only flat connections.

Example 11.4.2. Let $(Z, \mathcal{F})$ be a foliated manifold endowed with the adapted coordinate atlas $\Psi_\mathcal{F} = \{(U; z^\lambda, z^i)\}$ (11.2.65). With respect to this atlas, the normal bundle $V\mathcal{F} \to Z$ to $\mathcal{F}$ is provided with coordinates $(z^\lambda, z^i, \dot{z}^\lambda)$ whose fibre coordinates $\dot{z}^\lambda$ have transition functions independent of coordinates z^i on leaves of the foliation. Therefore, restricted to a leaf F, the normal bundle $V\mathcal{F}|_F \to F$ has transition functions independent of coordinates on its base F, i.e., it is equipped with a bundle atlas of local constant trivializations. In accordance with Proposition 11.4.1, this atlas provides the fibre bundle $V\mathcal{F}|_F \to F$ with the corresponding flat connection, called *Bott's connection*. This connection is canonical in the sense that any two different adapted coordinate atlases $\Psi_\mathcal{F}$ and $\Psi'_\mathcal{F}$ on Z also form an atlas of this type and, therefore, induce equivalent bundle atlases of constant local trivializations on $V\mathcal{F}|_F$.

11.4.3 *Linear connections*

Let $Y \to X$ be a vector bundle equipped with linear bundle coordinates (x^λ, y^i). It admits a *linear connection*

$$\Gamma = dx^\lambda \otimes (\partial_\lambda + \Gamma_\lambda{}^i{}_j(x) y^j \partial_i). \tag{11.4.15}$$

There are the following standard constructions of new linear connections from old ones.

• Any linear connection Γ (11.4.15) on a vector bundle $Y \to X$ defines the *dual linear connection*

$$\Gamma^* = dx^\lambda \otimes (\partial_\lambda - \Gamma_\lambda{}^j{}_i(x) y_j \partial^i) \tag{11.4.16}$$

on the dual bundle $Y^* \to X$.

• Let Γ and Γ' be linear connections on vector bundles $Y \to X$ and $Y' \to X$, respectively. The *direct sum connection* $\Gamma \oplus \Gamma'$ on the Whitney sum $Y \oplus Y'$ of these vector bundles is defined as the product connection (11.4.6).

• Similarly, the tensor product $Y \otimes Y'$ of vector bundles possesses the *tensor product connection*

$$\Gamma \otimes \Gamma' = dx^\lambda \otimes \left[\partial_\lambda + (\Gamma_\lambda{}^i{}_j y^{ja} + \Gamma'_\lambda{}^a{}_b y^{ib}) \frac{\partial}{\partial y^{ia}}\right]. \tag{11.4.17}$$

The curvature of a linear connection Γ (11.4.15) on a vector bundle $Y \to X$ is usually written as a Y-valued two-form

$$\begin{aligned} & R = \frac{1}{2} R_{\lambda\mu}{}^i{}_j(x) y^j dx^\lambda \wedge dx^\mu \otimes e_i, \\ & R_{\lambda\mu}{}^i{}_j = \partial_\lambda \Gamma_\mu{}^i{}_j - \partial_\mu \Gamma_\lambda{}^i{}_j + \Gamma_\lambda{}^h{}_j \Gamma_\mu{}^i{}_h - \Gamma_\mu{}^h{}_j \Gamma_\lambda{}^i{}_h, \end{aligned} \tag{11.4.18}$$

due to the canonical vertical splitting $VY = Y \times Y$, where $\{\partial_i\} = \{e_i\}$. For any two vector fields τ and τ' on X, this curvature yields the zero order differential operator

$$R(\tau, \tau')s = ([\nabla^\Gamma_\tau, \nabla^\Gamma_{\tau'}] - \nabla^\Gamma_{[\tau,\tau']})s \tag{11.4.19}$$

on section s of a vector bundle $Y \to X$.

An important example of linear connections is a connection

$$K = dx^\lambda \otimes (\partial_\lambda + K_\lambda{}^\mu{}_\nu \dot{x}^\nu \dot{\partial}_\mu) \tag{11.4.20}$$

on the tangent bundle TX of a manifold X. It is called a *world connection* or, simply, a *connection on a manifold* X. The dual connection (11.4.16) on the cotangent bundle T^*X is

$$K^* = dx^\lambda \otimes (\partial_\lambda - K_\lambda{}^\mu{}_\nu \dot{x}_\mu \dot{\partial}^\nu). \tag{11.4.21}$$

The curvature of the world connection K (11.4.20) reads

$$\begin{aligned} & R = \frac{1}{2} R_{\lambda\mu}{}^\alpha{}_\beta \dot{x}^\beta dx^\lambda \wedge dx^\mu \otimes \partial_\alpha, \\ & R_{\lambda\mu}{}^\alpha{}_\beta = \partial_\lambda K_\mu{}^\alpha{}_\beta - \partial_\mu K_\lambda{}^\alpha{}_\beta + K_\lambda{}^\gamma{}_\beta K_\mu{}^\alpha{}_\gamma - K_\mu{}^\gamma{}_\beta K_\lambda{}^\alpha{}_\gamma. \end{aligned} \tag{11.4.22}$$

Its *Ricci tensor* $R_{\lambda\beta} = R_{\lambda\mu}{}^{\mu}{}_{\beta}$ is introduced.

A *torsion of a world connection* is defined as the torsion (11.4.14) of the connection K (11.4.20) on the tangent bundle TX with respect to the canonical vertical-valued form $dx^\lambda \otimes \dot\partial_\lambda$. Due to the vertical splitting of VTX, it also is written as a tangent-valued two-form

$$T = \frac{1}{2}T_\mu{}^\nu{}_\lambda dx^\lambda \wedge dx^\mu \otimes \partial_\nu, \qquad T_\mu{}^\nu{}_\lambda = K_\mu{}^\nu{}_\lambda - K_\lambda{}^\nu{}_\mu, \tag{11.4.23}$$

on X. The world connection (11.4.20) is called symmetric if its torsion (11.4.23) vanishes.

For instance, let a manifold X be provided with a non-degenerate fibre metric

$$g \in \overset{2}{\vee}\mathcal{O}^1(X), \qquad g = g_{\lambda\mu}dx^\lambda \otimes dx^\mu,$$

in the tangent bundle TX, and with the dual metric

$$g \in \overset{2}{\vee}\mathcal{T}^1(X), \qquad g = g^{\lambda\mu}\partial_\lambda \otimes \partial_\mu,$$

in the cotangent bundle T^*X. Then there exists a world connection K such that g is its integral section, i.e.,

$$\nabla_\lambda g^{\alpha\beta} = \partial_\lambda\, g^{\alpha\beta} - g^{\alpha\gamma}K_\lambda{}^\beta{}_\gamma - g^{\beta\gamma}K_\lambda{}^\alpha{}_\gamma = 0.$$

It is called the *metric connection*. There exists a unique symmetric metric connection

$$K_\lambda{}^\nu{}_\mu = \{_\lambda{}^\nu{}_\mu\} = -\frac{1}{2}g^{\nu\rho}(\partial_\lambda g_{\rho\mu} + \partial_\mu g_{\rho\lambda} - \partial_\rho g_{\lambda\mu}). \tag{11.4.24}$$

This is the *Levi–Civita connection*, whose components (11.4.24) are called *Christoffel symbols*.

A manifold X which admits a flat world connection is called *parallelizable*. However, the components $K_\lambda{}^\mu{}_\nu$ (11.4.20) of a flat world connection K need not be zero because they are written with respect to holonomic coordinates. Namely, the torsion (11.4.23) of a flat connection need not vanish. A manifold X possessing a flat symmetric connection is called *locally affine*. Such a manifold can be provided with a coordinate atlas (x^μ) with transition functions $x'^\mu = x^\mu + c^\mu$, c^μ =const. Therefore, locally affine manifolds are toroidal cylinders $\mathbb{R}^m \times T^k$.

11.4.4 *Composite connections*

Let us consider the composite bundle $Y \to \Sigma \to X$ (11.2.10), coordinated by $(x^\lambda, \sigma^m, y^i)$. Let us consider the jet manifolds $J^1\Sigma$, $J^1_\Sigma Y$, and J^1Y of the fibre bundles $\Sigma \to X$, $Y \to \Sigma$ and $Y \to X$, respectively. They are parameterized respectively by the coordinates

$$(x^\lambda, \sigma^m, \sigma^m_\lambda), \quad (x^\lambda, \sigma^m, y^i, \widetilde{y}^i_\lambda, y^i_m), \quad (x^\lambda, \sigma^m, y^i, \sigma^m_\lambda, y^i_\lambda).$$

There is the canonical map

$$\varrho : J^1\Sigma \underset{\Sigma}{\times} J^1_\Sigma Y \underset{Y}{\longrightarrow} J^1Y, \qquad y^i_\lambda \circ \varrho = y^i_m \sigma^m_\lambda + \widetilde{y}^i_\lambda. \tag{11.4.25}$$

Using the canonical map (11.4.25), we can consider the relations between connections on fibre bundles $Y \to X$, $Y \to \Sigma$ and $\Sigma \to X$ [109; 145].

Connections on fibre bundles $Y \to X$, $Y \to \Sigma$ and $\Sigma \to X$ read

$$\gamma = dx^\lambda \otimes (\partial_\lambda + \gamma^m_\lambda \partial_m + \gamma^i_\lambda \partial_i), \tag{11.4.26}$$

$$A_\Sigma = dx^\lambda \otimes (\partial_\lambda + A^i_\lambda \partial_i) + d\sigma^m \otimes (\partial_m + A^i_m \partial_i), \tag{11.4.27}$$

$$\Gamma = dx^\lambda \otimes (\partial_\lambda + \Gamma^m_\lambda \partial_m). \tag{11.4.28}$$

The canonical map ϱ (11.4.25) enables us to obtain a connection γ on $Y \to X$ in accordance with the diagram

$$\begin{array}{ccc} J^1\Sigma \underset{\Sigma}{\times} J^1_\Sigma Y & \overset{\varrho}{\longrightarrow} & J^1Y \\ {\scriptstyle (\Gamma, A)} \uparrow & & \uparrow {\scriptstyle \gamma} \\ \Sigma \underset{X}{\times} Y & \longleftarrow & Y \end{array}$$

This connection, called the *composite connection*, reads

$$\gamma = dx^\lambda \otimes [\partial_\lambda + \Gamma^m_\lambda \partial_m + (A^i_\lambda + A^i_m \Gamma^m_\lambda)\partial_i]. \tag{11.4.29}$$

It is a unique connection such that the horizontal lift $\gamma\tau$ on Y of a vector field τ on X by means of the connection γ (11.4.29) coincides with the composition $A_\Sigma(\Gamma\tau)$ of horizontal lifts of τ onto Σ by means of the connection Γ and then onto Y by means of the connection A_Σ. For the sake of brevity, let us write $\gamma = A_\Sigma \circ \Gamma$.

Given the composite bundle Y (11.2.10), there is the exact sequence

$$0 \to V_\Sigma Y \to VY \to Y \underset{\Sigma}{\times} V\Sigma \to 0, \tag{11.4.30}$$

$$0 \to Y \underset{\Sigma}{\times} V^*\Sigma \to V^*Y \to V^*_\Sigma Y \to 0, \tag{11.4.31}$$

where $V_\Sigma Y$ denotes the vertical tangent bundle of a fibre bundle $Y \to \Sigma$ coordinated by $(x^\lambda, \sigma^m, y^i, \dot{y}^i)$. Let us consider the splitting

$$\begin{aligned} B : VY \ni v = \dot{y}^i \partial_i + \dot{\sigma}^m \partial_m &\to v \rfloor B \\ &= (\dot{y}^i - \dot{\sigma}^m B^i_m)\partial_i \in V_\Sigma Y, \\ B = (\overline{d}y^i - B^i_m \overline{d}\sigma^m) &\otimes \partial_i \in V^*Y \underset{Y}{\otimes} V_\Sigma Y, \end{aligned} \tag{11.4.32}$$

of the exact sequence (11.4.30). Then the connection γ (11.4.26) on $Y \to X$ and the splitting B (11.4.32) define a connection

$$\begin{aligned} A_\Sigma &= B \circ \gamma : TY \to VY \to V_\Sigma Y, \\ A_\Sigma &= dx^\lambda \otimes (\partial_\lambda + (\gamma^i_\lambda - B^i_m \gamma^m_\lambda)\partial_i) \\ &\quad + d\sigma^m \otimes (\partial_m + B^i_m \partial_i), \end{aligned} \tag{11.4.33}$$

on the fibre bundle $Y \to \Sigma$.

Conversely, every connection A_Σ (11.4.27) on a fibre bundle $Y \to \Sigma$ provides the splittings

$$VY = V_\Sigma Y \underset{Y}{\oplus} A_\Sigma (Y \underset{\Sigma}{\times} V\Sigma), \tag{11.4.34}$$
$$\dot y^i \partial_i + \dot\sigma^m \partial_m = (\dot y^i - A^i_m \dot\sigma^m)\partial_i + \dot\sigma^m(\partial_m + A^i_m \partial_i),$$
$$V^*Y = (Y \underset{\Sigma}{\times} V^*\Sigma) \underset{Y}{\oplus} A_\Sigma(V^*_\Sigma Y), \tag{11.4.35}$$
$$\dot y_i \overline{d}y^i + \dot\sigma_m \overline{d}\sigma^m = \dot y_i(\overline{d}y^i - A^i_m \overline{d}\sigma^m) + (\dot\sigma_m + A^i_m \dot y_i)\overline{d}\sigma^m,$$

of the exact sequences (11.4.30) – (11.4.31). Using the splitting (11.4.34), one can construct the first order differential operator

$$\widetilde D : J^1Y \to T^*X \underset{Y}{\otimes} V_\Sigma Y, \quad \widetilde D = dx^\lambda \otimes (y^i_\lambda - A^i_\lambda - A^i_m \sigma^m_\lambda)\partial_i, \tag{11.4.36}$$

called the *vertical covariant differential*, on the composite fibre bundle $Y \to X$.

The vertical covariant differential (11.4.36) possesses the following important property. Let h be a section of a fibre bundle $\Sigma \to X$, and let $Y_h \to X$ be the restriction of a fibre bundle $Y \to \Sigma$ to $h(X) \subset \Sigma$. This is a subbundle $i_h : Y_h \to Y$ of a fibre bundle $Y \to X$. Every connection A_Σ (11.4.27) induces the pull-back connection (11.4.7):

$$A_h = i^*_h A_\Sigma = dx^\lambda \otimes [\partial_\lambda + ((A^i_m \circ h)\partial_\lambda h^m + (A \circ h)^i_\lambda)\partial_i] \tag{11.4.37}$$

on $Y_h \to X$. Then the restriction of the vertical covariant differential $\widetilde D$ (11.4.36) to $J^1 i_h(J^1Y_h) \subset J^1Y$ coincides with the familiar covariant differential D^{A_h} (11.4.8) on Y_h relative to the pull-back connection A_h (11.4.37).

11.5 Differential operators and connections on modules

This Section addresses the notion of a linear differential operator on a module over an arbitrary commutative ring [95; 109].

Let $\mathcal{K}$ be a commutative ring and $\mathcal{A}$ a commutative $\mathcal{K}$-ring. Let P and Q be $\mathcal{A}$-modules. The $\mathcal{K}$-module $\mathrm{Hom}_{\mathcal{K}}(P,Q)$ of $\mathcal{K}$-module homomorphisms $\Phi : P \to Q$ can be endowed with the two different $\mathcal{A}$-module structures

$$(a\Phi)(p) = a\Phi(p), \qquad (\Phi \bullet a)(p) = \Phi(ap), \qquad a \in \mathcal{A}, \quad p \in P. \tag{11.5.1}$$

For the sake of convenience, we refer to the second one as an $\mathcal{A}^\bullet$-module structure. Let us put

$$\delta_a \Phi = a\Phi - \Phi \bullet a, \qquad a \in \mathcal{A}. \tag{11.5.2}$$

Definition 11.5.1. An element $\Delta \in \mathrm{Hom}_{\mathcal{K}}(P,Q)$ is called a Q-valued *differential operator* of order s on P if

$$\delta_{a_0} \circ \cdots \circ \delta_{a_s} \Delta = 0$$

for any tuple of $s+1$ elements $a_0, \ldots, a_s$ of $\mathcal{A}$. The set $\mathrm{Diff}_s(P,Q)$ of these operators inherits the $\mathcal{A}$- and $\mathcal{A}^\bullet$-module structures (11.5.1).

In particular, zero order differential operators obey the condition

$$\delta_a \Delta(p) = a\Delta(p) - \Delta(ap) = 0, \qquad a \in \mathcal{A}, \qquad p \in P,$$

and, consequently, they coincide with $\mathcal{A}$-module morphisms $P \to Q$. A first order differential operator Δ satisfies the condition

$$\delta_b \circ \delta_a \Delta(p) = ba\Delta(p) - b\Delta(ap) - a\Delta(bp) + \Delta(abp) = 0, \quad a, b \in \mathcal{A}. \tag{11.5.3}$$

The following fact reduces the study of Q-valued differential operators on an $\mathcal{A}$-module P to that of Q-valued differential operators on a ring $\mathcal{A}$.

Theorem 11.5.1. *Let us consider the $\mathcal{A}$-module morphism*

$$h_s : \mathrm{Diff}_s(\mathcal{A}, Q) \to Q, \qquad h_s(\Delta) = \Delta(\mathbf{1}). \tag{11.5.4}$$

Any Q-valued s-order differential operator $\Delta \in \mathrm{Diff}_s(P,Q)$ on P uniquely factorizes as

$$\Delta : P \xrightarrow{\mathfrak{f}_\Delta} \mathrm{Diff}_s(\mathcal{A}, Q) \xrightarrow{h_s} Q \tag{11.5.5}$$

through the morphism h_s (11.5.4) and some homomorphism

$$\mathfrak{f}_\Delta : P \to \mathrm{Diff}_s(\mathcal{A}, Q), \qquad (\mathfrak{f}_\Delta p)(a) = \Delta(ap), \qquad a \in \mathcal{A}, \tag{11.5.6}$$

of an $\mathcal{A}$-module P to an $\mathcal{A}^\bullet$-module $\mathrm{Diff}_s(\mathcal{A}, Q)$. The assignment $\Delta \to \mathfrak{f}_\Delta$ defines the isomorphism

$$\mathrm{Diff}_s(P,Q) = \mathrm{Hom}_{\mathcal{A}-\mathcal{A}^\bullet}(P, \mathrm{Diff}_s(\mathcal{A}, Q)). \tag{11.5.7}$$

Let $P = \mathcal{A}$. Any zero order Q-valued differential operator Δ on $\mathcal{A}$ is defined by its value $\Delta(\mathbf{1})$. Then there is an isomorphism

$$\mathrm{Diff}\,_0(\mathcal{A}, Q) = Q$$

via the association

$$Q \ni q \to \Delta_q \in \mathrm{Diff}\,_0(\mathcal{A}, Q),$$

where Δ_q is given by the equality $\Delta_q(\mathbf{1}) = q$. A first order Q-valued differential operator Δ on $\mathcal{A}$ fulfils the condition

$$\Delta(ab) = b\Delta(a) + a\Delta(b) - ba\Delta(\mathbf{1}), \qquad a, b \in \mathcal{A}.$$

It is called a Q-valued *derivation* of $\mathcal{A}$ if $\Delta(\mathbf{1}) = 0$, i.e., the *Leibniz rule*

$$\Delta(ab) = \Delta(a)b + a\Delta(b), \qquad a, b \in \mathcal{A}, \tag{11.5.8}$$

holds. One obtains at once that any first order differential operator on $\mathcal{A}$ falls into the sum

$$\Delta(a) = a\Delta(\mathbf{1}) + [\Delta(a) - a\Delta(\mathbf{1})]$$

of the zero order differential operator $a\Delta(\mathbf{1})$ and the derivation $\Delta(a) - a\Delta(\mathbf{1})$. If ∂ is a Q-valued derivation of $\mathcal{A}$, then $a\partial$ is well for any $a \in \mathcal{A}$. Hence, Q-valued derivations of $\mathcal{A}$ constitute an $\mathcal{A}$-module $\mathfrak{d}(\mathcal{A}, Q)$, called the *derivation module*. There is the $\mathcal{A}$-module decomposition

$$\mathrm{Diff}\,_1(\mathcal{A}, Q) = Q \oplus \mathfrak{d}(\mathcal{A}, Q). \tag{11.5.9}$$

If $P = Q = \mathcal{A}$, the derivation module $\mathfrak{d}\mathcal{A}$ of $\mathcal{A}$ also is a Lie $\mathcal{K}$-algebra with respect to the Lie bracket

$$[u, u'] = u \circ u' - u' \circ u, \qquad u, u' \in \mathcal{A}. \tag{11.5.10}$$

Accordingly, the decomposition (11.5.9) takes the form

$$\mathrm{Diff}\,_1(\mathcal{A}) = \mathcal{A} \oplus \mathfrak{d}\mathcal{A}. \tag{11.5.11}$$

Definition 11.5.2. A *connection* on an $\mathcal{A}$-module P is an $\mathcal{A}$-module morphism

$$\mathfrak{d}\mathcal{A} \ni u \to \nabla_u \in \mathrm{Diff}\,_1(P, P) \tag{11.5.12}$$

such that the first order differential operators ∇_u obey the *Leibniz rule*

$$\nabla_u(ap) = u(a)p + a\nabla_u(p), \quad a \in \mathcal{A}, \quad p \in P. \tag{11.5.13}$$

Though ∇_u (11.5.12) is called a connection, it in fact is a *covariant differential* on a module P.

Let P be a commutative $\mathcal{A}$-ring and $\mathfrak{d}P$ the derivation module of P as a $\mathcal{K}$-ring. The $\mathfrak{d}P$ is both a P- and $\mathcal{A}$-module. Then Definition 11.5.2 is modified as follows.

Definition 11.5.3. A *connection* on an $\mathcal{A}$-ring P is an $\mathcal{A}$-module morphism

$$\mathfrak{d}\mathcal{A} \ni u \to \nabla_u \in \mathfrak{d}P \subset \mathrm{Diff}_1(P, P), \tag{11.5.14}$$

which is a connection on P as an $\mathcal{A}$-module, i.e., obeys the Leinbniz rule (11.5.13).

For instance, let $Y \to X$ be a smooth vector bundle. Its global sections form a $C^\infty(X)$-module $Y(X)$. The following *Serre–Swan theorem* shows that such modules exhaust all projective modules of finite rank over $C^\infty(X)$ [68].

Theorem 11.5.2. *Let X be a smooth manifold. A $C^\infty(X)$-module P is isomorphic to the structure module of a smooth vector bundle over X if and only if it is a projective module of finite rank.*

This theorem states the categorial equivalence between the vector bundles over a smooth manifold X and projective modules of finite rank over the ring $C^\infty(X)$ of smooth real functions on X. The following are corollaries of this equivalence

The derivation module of the real ring $C^\infty(X)$ coincides with the $C^\infty(X)$-module $\mathcal{T}(X)$ of vector fields on X. Its dual is isomorphic to the module $\mathcal{T}(X)^* = \mathcal{O}^1(X)$ of differential one-forms on X.

If P is a $C^\infty(X)$-module, one can reformulate Definition 11.5.2 of a connection on P as follows.

Definition 11.5.4. A connection on a $C^\infty(X)$-module P is a $C^\infty(X)$-module morphism

$$\nabla : P \to \mathcal{O}^1(X) \otimes P, \tag{11.5.15}$$

which satisfies the Leibniz rule

$$\nabla(fp) = df \otimes p + f\nabla(p), \qquad f \in C^\infty(X), \qquad p \in P.$$

It associates to any vector field $\tau \in \mathcal{T}(X)$ on X a first order differential operator ∇_τ on P which obeys the Leibniz rule

$$\nabla_\tau(fp) = (\tau\rfloor df)p + f\nabla_\tau p. \tag{11.5.16}$$

In particular, let $Y \to X$ be a vector bundle and $Y(X)$ its structure module. The notion of a connection on the structure module $Y(X)$ is equivalent to the standard geometric notion of a connection on a vector bundle $Y \to X$ [109].

11.6 Differential calculus over a commutative ring

Let $\mathfrak{g}$ be a Lie algebra over a commutative ring $\mathcal{K}$. Let $\mathfrak{g}$ act on a $\mathcal{K}$-module P on the left such that

$$[\varepsilon, \varepsilon']p = (\varepsilon \circ \varepsilon' - \varepsilon' \circ \varepsilon)p, \qquad \varepsilon, \varepsilon' \in \mathfrak{g}.$$

Then one calls P the Lie algebra $\mathfrak{g}$-*module*. Let us consider $\mathcal{K}$-multilinear skew-symmetric maps

$$c^k : \overset{k}{\times} \mathfrak{g} \to P.$$

They form a $\mathfrak{g}$-module $C^k[\mathfrak{g}; P]$. Let us put $C^0[\mathfrak{g}; P] = P$. We obtain the cochain complex

$$0 \to P \xrightarrow{\delta^0} C^1[\mathfrak{g}; P] \xrightarrow{\delta^1} \cdots C^k[\mathfrak{g}; P] \xrightarrow{\delta^k} \cdots \tag{11.6.1}$$

with respect to the *Chevalley–Eilenberg coboundary operators*

$$\delta^k c^k(\varepsilon_0, \ldots, \varepsilon_k) = \sum_{i=0}^{k} (-1)^i \varepsilon_i c^k(\varepsilon_0, \ldots, \widehat{\varepsilon}_i, \ldots, \varepsilon_k) \tag{11.6.2}$$

$$+ \sum_{1 \leq i < j \leq k} (-1)^{i+j} c^k([\varepsilon_i, \varepsilon_j], \varepsilon_0, \ldots, \widehat{\varepsilon}_i, \ldots, \widehat{\varepsilon}_j, \ldots, \varepsilon_k),$$

where the caret $\widehat{}$ denotes omission [50]. For instance, we have

$$\delta^0 p(\varepsilon_0) = \varepsilon_0 p, \tag{11.6.3}$$

$$\delta^1 c^1(\varepsilon_0, \varepsilon_1) = \varepsilon_0 c^1(\varepsilon_1) - \varepsilon_1 c^1(\varepsilon_0) - c^1([\varepsilon_0, \varepsilon_1]). \tag{11.6.4}$$

The complex (11.6.1) is called the *Chevalley–Eilenberg complex*, and its cohomology $H^*(\mathfrak{g}, P)$ is the *Chevalley–Eilenberg cohomology* of a Lie algebra $\mathfrak{g}$ with coefficients in P.

Let $\mathcal{A}$ be a commutative $\mathcal{K}$-ring. Since the derivation module $\mathfrak{d}\mathcal{A}$ of $\mathcal{A}$ is a Lie $\mathcal{K}$-algebra, one can associate to $\mathcal{A}$ the Chevalley–Eilenberg complex $C^*[\mathfrak{d}\mathcal{A}; \mathcal{A}]$. Its subcomplex of $\mathcal{A}$-multilinear maps is a differential graded algebra.

A *graded algebra* Ω^* over a commutative ring $\mathcal{K}$ is defined as a direct sum

$$\Omega^* = \bigoplus_k \Omega^k$$

of $\mathcal{K}$-modules Ω^k, provided with an associative multiplication law $\alpha \cdot \beta$, $\alpha, \beta \in \Omega^*$, such that $\alpha \cdot \beta \in \Omega^{|\alpha|+|\beta|}$, where $|\alpha|$ denotes the degree of an element $\alpha \in \Omega^{|\alpha|}$. In particular, it follows that Ω^0 is a $\mathcal{K}$-algebra $\mathcal{A}$, while $\Omega^{k>0}$ are $\mathcal{A}$-bimodules and Ω^* is an $(\mathcal{A}-\mathcal{A})$-algebra. A graded algebra is said to be *graded commutative*

$$\alpha \cdot \beta = (-1)^{|\alpha||\beta|}\beta \cdot \alpha, \qquad \alpha, \beta \in \Omega^*.$$

A graded algebra Ω^* is called the *differential graded algebra* or the *differential calculus* over $\mathcal{A}$ if it is a cochain complex of $\mathcal{K}$-modules

$$0 \to \mathcal{K} \longrightarrow \mathcal{A} \xrightarrow{\delta} \Omega^1 \xrightarrow{\delta} \cdots \Omega^k \xrightarrow{\delta} \cdots \tag{11.6.5}$$

with respect to a coboundary operator δ which obeys the *graded Leibniz rule*

$$\delta(\alpha \cdot \beta) = \delta\alpha \cdot \beta + (-1)^{|\alpha|}\alpha \cdot \delta\beta. \tag{11.6.6}$$

In particular, $\delta : \mathcal{A} \to \Omega^1$ is a Ω^1-valued derivation of a $\mathcal{K}$-algebra $\mathcal{A}$. The cochain complex (11.6.5) is said to be the *abstract de Rham complex* of the differential graded algebra (Ω^*, δ). Cohomology $H^*(\Omega^*)$ of the complex (11.6.5) is called the *abstract de Rham cohomology*.

One considers the minimal differential graded subalgebra $\Omega^*\mathcal{A}$ of the differential graded algebra Ω^* which contains $\mathcal{A}$. Seen as an $(\mathcal{A}-\mathcal{A})$-algebra, it is generated by the elements δa, $a \in \mathcal{A}$, and consists of monomials

$$\alpha = a_0\delta a_1 \cdots \delta a_k, \qquad a_i \in \mathcal{A},$$

whose product obeys the *juxtaposition rule*

$$(a_0\delta a_1) \cdot (b_0\delta b_1) = a_0\delta(a_1 b_0) \cdot \delta b_1 - a_0 a_1 \delta b_0 \cdot \delta b_1$$

in accordance with the equality (11.6.6). The differential graded algebra $(\Omega^*\mathcal{A}, \delta)$ is called the *minimal differential calculus* over $\mathcal{A}$.

Let now $\mathcal{A}$ be a commutative $\mathcal{K}$-ring possessing a non-trivial Lie algebra $\mathfrak{d}\mathcal{A}$ of derivations. Let us consider the extended Chevalley–Eilenberg complex

$$0 \to \mathcal{K} \xrightarrow{\text{in}} C^*[\mathfrak{d}\mathcal{A}; \mathcal{A}]$$

of the Lie algebra $\mathfrak{d}\mathcal{A}$ with coefficients in the ring $\mathcal{A}$, regarded as a $\mathfrak{d}\mathcal{A}$-module [68]. It is easily justified that this complex contains a subcomplex $\mathcal{O}^*[\mathfrak{d}\mathcal{A}]$ of $\mathcal{A}$-multilinear skew-symmetric maps

$$\phi^k : \overset{k}{\times} \mathfrak{d}\mathcal{A} \to \mathcal{A} \tag{11.6.7}$$

with respect to the Chevalley–Eilenberg coboundary operator

$$d\phi(u_0,\ldots,u_k)=\sum_{i=0}^{k}(-1)^i u_i(\phi(u_0,\ldots,\widehat{u}_i,\ldots,u_k)) \qquad (11.6.8)$$
$$+\sum_{i<j}(-1)^{i+j}\phi([u_i,u_j],u_0,\ldots,\widehat{u}_i,\ldots,\widehat{u}_j,\ldots,u_k).$$

In particular, we have

$$(da)(u)=u(a), \qquad a\in\mathcal{A}, \qquad u\in\mathfrak{d}\mathcal{A},$$
$$(d\phi)(u_0,u_1)=u_0(\phi(u_1))-u_1(\phi(u_0))-\phi([u_0,u_1]), \quad \phi\in\mathcal{O}^1[\mathfrak{d}\mathcal{A}],$$
$$\mathcal{O}^0[\mathfrak{d}\mathcal{A}]=\mathcal{A},$$
$$\mathcal{O}^1[\mathfrak{d}\mathcal{A}]=\mathrm{Hom}_{\mathcal{A}}(\mathfrak{d}\mathcal{A},\mathcal{A})=\mathfrak{d}\mathcal{A}^*.$$

It follows that $d(\mathbf{1})=0$ and d is a $\mathcal{O}^1[\mathfrak{d}\mathcal{A}]$-valued derivation of $\mathcal{A}$.

The graded module $\mathcal{O}^*[\mathfrak{d}\mathcal{A}]$ is provided with the structure of a graded $\mathcal{A}$-algebra with respect to the exterior product

$$\phi\wedge\phi'(u_1,...,u_{r+s}) \qquad (11.6.9)$$
$$=\sum_{i_1<\cdots<i_r;j_1<\cdots<j_s}\mathrm{sgn}^{i_1\cdots i_r j_1\cdots j_s}_{1\cdots r+s}\,\phi(u_{i_1},\ldots,u_{i_r})\phi'(u_{j_1},\ldots,u_{j_s}),$$
$$\phi\in\mathcal{O}^r[\mathfrak{d}\mathcal{A}], \qquad \phi'\in\mathcal{O}^s[\mathfrak{d}\mathcal{A}], \qquad u_k\in\mathfrak{d}\mathcal{A},$$

where $\mathrm{sgn}^{\cdots}_{\cdots}$ is the sign of a permutation. This product obeys the relations

$$d(\phi\wedge\phi')=d(\phi)\wedge\phi'+(-1)^{|\phi|}\phi\wedge d(\phi'), \quad \phi,\phi'\in\mathcal{O}^*[\mathfrak{d}\mathcal{A}],$$
$$\phi\wedge\phi'=(-1)^{|\phi||\phi'|}\phi'\wedge\phi. \qquad (11.6.10)$$

By virtue of the first one, $\mathcal{O}^*[\mathfrak{d}\mathcal{A}]$ is a differential graded $\mathcal{K}$-algebra, called the *Chevalley–Eilenberg differential calculus* over a $\mathcal{K}$-ring $\mathcal{A}$. The relation (11.6.10) shows that $\mathcal{O}^*[\mathfrak{d}\mathcal{A}]$ is a graded commutative algebra.

The *minimal Chevalley–Eilenberg differential calculus* $\mathcal{O}^*\mathcal{A}$ over a ring $\mathcal{A}$ consists of the monomials

$$a_0da_1\wedge\cdots\wedge da_k, \qquad a_i\in\mathcal{A}.$$

Its complex

$$0\to\mathcal{K}\longrightarrow\mathcal{A}\xrightarrow{d}\mathcal{O}^1\mathcal{A}\xrightarrow{d}\cdots\mathcal{O}^k\mathcal{A}\xrightarrow{d}\cdots \qquad (11.6.11)$$

is said to be the *de Rham complex* of a $\mathcal{K}$-ring $\mathcal{A}$, and its cohomology $H^*(\mathcal{A})$ is called the *de Rham cohomology* of $\mathcal{A}$.

For instance, the minimal Chevalley–Eilenberg differential calculus over the ring $C^\infty(Z)$ of smooth real functions on a smooth manifold Z coincides with the differential graded algebra $\mathcal{O}^*(Z)$ of exterior forms on Z.

11.7 Infinite-dimensional topological vector spaces

There are several standard topologies introduced on an infinite-dimensional (complex or real) vector space and its dual [135]. Topological vector spaces throughout the book are assumed to be locally convex. Unless otherwise stated, by the dual V' of a topological vector space V is meant its *topological dual*, i.e., the space of continuous linear maps of $V \to \mathbb{R}$.

Let us note that a topology on a vector space V is often determined by a set of seminorms. A non-negative real function p on V is called a *seminorm* if it satisfies the conditions

$$p(\lambda x) = |\lambda| p(x), \qquad p(x+y) \leq p(x) + p(y), \qquad x, y \in V, \quad \lambda \in \mathbb{R}.$$

A seminorm p for which $p(x) = 0$ implies $x = 0$ is called a *norm*. Given any set $\{p_i\}_{i \in I}$ of seminorms on a vector space V, there is the coarsest topology on V compatible with the algebraic structure such that all seminorms p_i are continuous. It is a locally convex topology whose base of closed neighborhoods consists of the set

$$\{x : \sup_{1 \leq i \leq n} p_i(x) \leq \varepsilon\}, \qquad \varepsilon > 0.$$

It is called the *topology defined by a set of seminorms*. A topology defined by a norm is called the *normed topology*. A complete normed topological space is called the *Banach space*.

Let V and W be two vector spaces whose Cartesian product $V \times W$ is provided with a bilinear form $\langle v, w \rangle$, called the *interior product*, which obeys the following conditions:

• for any element $v \neq 0$ of V, there exists an element $w \in W$ such that $\langle v, w \rangle \neq 0$;

• for any element $w \neq 0$ of W, there exists an element $v \in V$ such that $\langle v, w \rangle \neq 0$.

Then one says that (V, W) is a *dual pair*. If (V, W) is a dual pair, so is (W, V). Clearly, W is isomorphic to a vector subspace of the algebraic dual V^* of V, and V is a subspace of the algebraic dual of W.

Given a dual pair (V, W), every vector $w \in W$ defines the seminorm $p_w = |\langle v, w \rangle|$ on V. The coarsest topology $\sigma(V, W)$ on V making all these seminorms continuous is called the *weak topology determined by W on V*. It also is the coarsest topology on V such that all linear forms in $W \subset V^*$ are continuous. Moreover, W coincides with the (topological) dual V' of V provided with the weak topology $\sigma(V, W)$, and $\sigma(V, W)$ is the coarsest topology on V such that $V' = W$. Of course, the weak topology is Hausdorff.

For instance, if V is a Hausdorff topological vector space with the (topological) dual V', then (V, V') is a dual pair. The *weak topology* $\sigma(V, V')$ on V is coarser than the original topology on V. Since (V', V) also is a dual pair, the dual V' of V can be provided with the *weak** topology $\sigma(V', V)$. Then V is the dual of V', equipped with the weak* topology.

The weak* topology is the coarsest case of a topology of uniform convergence on V'. A subset M of a vector space V is said to absorb a subset $N \subset V$ if there is a number $\epsilon \geq 0$ such that $N \subset \lambda M$ for all λ with $|\lambda| \geq \epsilon$. An *absorbent set* is one which absorbs all points. A subset N of a topological vector space V is called *bounded* if it is absorbed by any neighborhood of the origin of V. Let (V, V') be a dual pair and $\mathcal{N}$ some family of weakly bounded subsets of V. Every $N \subset \mathcal{N}$ yields the seminorm

$$p_N(v') = \sup_{v \in N} |\langle v, v' \rangle|$$

on the dual V' of V. The topology on V' defined by the set of seminorms p_N, $N \in \mathcal{N}$, is called the *topology of uniform convergence on the sets* of $\mathcal{N}$. When $\mathcal{N}$ is a set of all finite subsets of V, we have the coarsest topology of uniform convergence which is the above mentioned weak* topology $\sigma(V', V)$. The finest topology of uniform convergence is obtained by taking $\mathcal{N}$ to be the set of all weakly bounded subsets of V. It is called the *strong topology*. The dual V'' of V', provided with the strong topology, is called the *bidual*. One says that V is *reflexive* if $V = V''$.

Since (V', V) is a dual pair, the vector space V also can be provided with the topology of uniform convergence on the subsets of V', e.g., the weak* and strong topologies. Moreover, any Hausdorff locally convex topology on V is a topology of uniform convergence. The coarsest and finest topologies of them are the weak* and strong topologies, respectively. There is the following chain

$$\text{weak}^* < \text{weak} < \text{original} < \text{strong}$$

of topologies on V, where $<$ means "to be finer".

For instance, let V be a normed space. The dual V' of V also is equipped with the norm

$$\|v'\|' = \sup_{\|v\|=1} |\langle v, v' \rangle|, \qquad v \in V, \qquad v' \in V'. \tag{11.7.1}$$

Let us consider the set of all balls $\{v : \|v\| \leq \epsilon, \epsilon > 0\}$ in V. The topology of uniform convergence on this set coincides with strong and normed topologies on V' because weakly bounded subsets of V also are bounded by the norm. Normed and strong topologies on V also are equivalent. Let

$\overline{V}$ denote the completion of a normed space V. Then V' is canonically identified to $(\overline{V})'$ as a normed space, though weak* topologies on V' and $(\overline{V})'$ are different. Let us note that both V' and V'' are Banach spaces. If V is a Banach space, it is closed in V'' with respect to the strong topology on V'' and dense in V'' equipped with the weak* topology. One usually considers the weak*, weak and normed (equivalently, strong) topologies on a Banach space.

It should be recalled that topology on a finite-dimensional vector space is locally convex and Hausdorff if and only if it is determined by the Euclidean norm.

In conclusion, let us say a few words about morphisms of topological vector spaces.

A linear morphism between two topological vector spaces is called *weakly continuous* if it is continuous with respect to the weak topologies on these vector spaces. In particular, any continuous morphism between topological vector spaces is weakly continuous [135].

A linear morphism between two topological vector spaces is called *bounded* if the image of a bounded set is bounded. Any continuous morphism is bounded. A topological vector space is called the *Mackey space* if any bounded endomorphism of this space is continuous. Metrizable and, consequently, normed spaces are of this type.

Any linear morphism $\gamma : V \to W$ of topological vector spaces yields the *dual morphism* $\gamma' : W' \to V'$ of the their topological duals such that

$$\langle v, \gamma'(w)\rangle = \langle \gamma(v), w\rangle, \qquad v \in V, \qquad w \in W.$$

If γ is weakly continuous, then γ' is weakly* continuous. If V and W are normed spaces, then any weakly continuous morphism $\gamma : V \to W$ is continuous and strongly continuous. Given normed topologies on V' and W', the dual morphism $\gamma' : W' \to V'$ is continuous if and only if γ is continuous.

Bibliography

[1] Abraham, R. and Marsden, J. (1978). *Foundations of Mechanics* (Benjamin Cummings Publ. Comp., London).

[2] Anandan, J. and Aharonov, Y. (1987). Phase change during a cyclic quantum evolution, *Phys. Rev. Lett* **58**, 1593.

[3] Anandan, J. and Aharonov, Y. (1988). Geometric quantum phase and angles, *Phys. Rev. D* **38**, 1863.

[4] Arnold, V. (Ed.) (1990). *Dynamical Systems III, IV* (Springer, Berlin).

[5] Ashtekar, A. and Stillerman, M. (1986). Geometric quantization and constrained system, *J. Math. Phys.* **27**, 1319.

[6] Asorey, M., Cariñena, J. and Paramion, M. (1982). Quantum evolution as a parallel transport, *J. Math. Phys.* **23**, 1451.

[7] Bates, L. (1988). Examples for obstructions to action-angle coordinates, *Proc. Roy. Soc. Edinburg, Sect. A* **110**, 27.

[8] Berry, M. (1985). Classical adiabatic angles and quantal adiabatic phase, *J. Phys. A* **18**, 15.

[9] Berry, M. and Hannay, J. (1988). Classical non-adiabatic angles, *J. Phys. A* **21**, L325.

[10] Bessega, C. (1966). Every infinite-dimensional Hilbert space is diffeomorphic with its unit sphere, *Bull. Acad. Polon. Sci.* **XIV**, 27.

[11] Blattner, R. (1977). The metalinear geometry of non-real polarizations, In: *Differential Geometric Methods in Mathematical Physics (Proc. Sympos. Univ. Bonn, Bonn 1975)*, Lect. Notes in Math. **570** (Springer, New York), p. 11.

[12] Blau, M. (1988). On the geometric quantization of constrained systems, *Class. Quant. Grav.* **5**, 1033.

[13] Bogoyavlenskij, O. (1998). Extended integrability and bi-Hamiltonian systems, *Commun. Math. Phys.* **196**, 19.

[14] Bohm, A. and Mostafazadeh, A. (1994). The relation between the Berry and the Anandan – Aharonov connections for $U(N)$ bundles, *J. Math. Phys.* **35**, 1463.

[15] Bohm, A., Mostafazadeh, A., Koizumi, H., Niu, Q. and Zwanziger, J. (2003). *The Geometric Phase in Quantum Systems* (Springer, Berlin).

[16] Bolsinov, A. and Jovanović, B. (2003). Noncommutative integrability, moment map and geodesic flows, *Ann. Global Anal. Geom.* **23**, 305.

[17] Bourbaki, N. (1987). *Topological vector spaces, Chap. 1-5* (Springer, Berlin).

[18] Bratteli, O. and Robinson, D. (1975). Unbounded derivations of C^*-algebras, *Commun. Math. Phys.* **42**, 253.

[19] Bratteli, O. and Robinson, D. (2002). *Operator Algebras and Quantum Statistical Mechanics, Vol.1, Second Edition* (Springer, Berlin).

[20] Broer, H., Huitema, G., Takens, F. and Braaksma, B. (1990). Unfoldings and bifurcations of quasi-periodic tori, *Mem. Amer. Math. Soc.* **421**, 1.

[21] Bryant, R., Chern, S., Gardner, R., Goldschmidt, H. and Griffiths, P. (1991). *Exterior Differential Systems* (Springer, Berlin).

[22] Brylinski, J.-L. (1993). *Loop spaces, Characteristic Classes and Geometric Quantization* (Birkhäuser, Boston).

[23] Carey, A., Crowley, D. and Murray, M. (1998). Principal bundles and the Dixmier-Douady class, *Commun. Math. Phys.* **193**, 171.

[24] Cassinelli, G., de Vito, E., Lahti, P. and Levrero, A. (1998). Symmetries of the quantum state space and group representations, *Rev. Math. Phys.* **10**, 893.

[25] Chinea, D., de León, M. and Marrero, J. (1994). The constraint algoritm for time-dependent Lagrangian, *J. Math. Phys.* **35**, 3410.

[26] Cirelli, R., Manià, A. and Pizzocchero, L. (1990). Quantum mechanics as an infinite-dimensional Hamiltonian system with uncertainty structure, *J. Math. Phys.* **31**, 2891, 2898.

[27] Crampin, M., Sarlet, W and Thompson, G. (2000). Bi-differential calculi, bi-Hamiltonian systems and conformal Killing tensors, *J. Phys. A* **33**, 8755.

[28] Cushman, R. and Bates, L. (1997). *Global Aspects of Classical Integrable Systems* (Birkhäuser, Basel).

[29] Daleckiĭ, Ju. and Kreĭn, M. (1974). *Stability of Solutions of Differential Equations in Banch Space*, Transl. Math. Monographs (AMS, Providence).

[30] Dazord, P. and Delzant, T. (1987). Le probleme general des variables actions-angles, *J. Diff. Geom.* **26**, 223.

[31] Dewisme, A. and Bouquet, S. (1993). First integrals and symmetries of time-dependent Hamiltonian systems, *J. Math. Phys* **34**, 997.

[32] Dittrich, W. and Reuter, M. (1994). *Classical and Quantum Dynamics* (Springer, Berlin).

[33] Dixmier, J. (1977). *C^*-Algebras* (North-Holland, Amsterdam).

[34] Doplicher, S., Kastler, D. and Robinson, D. (1966). Covariance algebras in field theory and statistical mechanics, *Commun. Math. Phys.* **3**, 1.

[35] Dustermaat, J. (1980). On global action-angle coordinates, *Commun. Pure Appl. Math.* **33**, 687.

[36] Echeverría Enríquez, A., Muñoz Lecanda, M. and Román Roy, N. (1995). Non-standard connections in classical mechanics, *J. Phys. A* **28**, 5553.

[37] Echeverría Enríquez, A., Muñoz Lecanda, M. and Román Roy, N. (1991). Geometrical setting of time-dependent regular systems. Alternative models, *Rev. Math. Phys.* **3**, 301.

[38] Echeverría Enríquez, A., Muñoz Lecanda, M., Román Roy, N. and Victoria-Monge, C. (1998). Mathematical foundations of geometric quantization, *Extracta Math.* **13**, 135.

[39] Emch, G. (1972). *Algebraic Methods in Statistical Mechanics and Quantum Field Theory* (Wiley–Interscience, New York).

[40] Fassó, F. (1998). Quasi-periodicity of motions and completely integrability of Hamiltonian systems, *Ergod. Theor. Dynam. Sys.* **18**, 1349.

[41] Fassó, F. (2005). Superintegrable Hamiltonian systems: geometry and applications, *Acta Appl. Math.* **87**, 93.

[42] Fatibene, L., Ferraris, M., Francaviglia, M. and McLenaghan, R. (2002). Generalized symmetries in mechanics and field theories, *J. Math. Phys.* **43**, 3147.

[43] Fiorani, E., Giachetta, G. and Sardanashvily, G. (2002). Geometric quantization of time-dependent completely integrable Hamiltonian systems, *J. Math. Phys.* **43**, 5013.

[44] Fiorani, E., Giachetta, G. and Sardanashvily, G. (2003). The Liouville – Arnold – Nekhoroshev theorem for noncompact invariant manifolds, *J. Phys. A* **36**, L101.

[45] Fiorani, E. (2004). Completely and partially integrable systems in the noncompact case, *Int. J. Geom. Methods Mod. Phys.* **1**, 167.

[46] Fiorani, E. and Sardanashvily, G. (2006). Noncommutative integrability on noncompact invariant manifold, *J. Phys. A* **39**, 14035.

[47] Fiorani, E. and Sardanashvily, G. (2007). Global action-angle coordinates for completely integrable systems with noncompact invariant manifolds, *J. Math. Phys.* **48**, 032001.

[48] Fiorani, E. (2008). Geometrical aspects of integrable systems, *Int. J. Geom. Methods Mod. Phys.* **5**, 457.

[49] Fiorani, E. (2009). Momentum maps, independent first integrals and integrability for central potentials *Int. J. Geom. Methods Mod. Phys.* **6**, 1323.

[50] Fuks, D. (1986). *Cohomology of Infinite-Dimensional Lie Algebras* (Consultants Bureau, New York).

[51] Gaeta, G. (2002). The Poincaré – Lyapounov – Nekhoroshev theorem, *Ann. Phys.* **297**, 157.

[52] Gaeta, G. (2003). The Poincaré–Nekhoroshev map, *J. Nonlin. Math. Phys.* **10**, 51.

[53] Giachetta, G., Mangiarotti, L. and Sardanashvily, G. (1997). *New Lagrangian and Hamiltonian Methods in Field Theory* (World Scientific, Singapore).

[54] Giachetta, G., Mangiarotti, L. and Sardanashvily, G. (1999). Nonholonomic constraints in time-dependent mechanics, *J. Math. Phys.* **40**, 1376.

[55] Giachetta, G., Mangiarotti, L. and Sardanashvily, G. (1999). Covariant Hamilton equations for field theory, *J. Phys. A* **32**, 6629.

[56] Giachetta, G., Mangiarotti, L. and Sardanashvily, G. (1999). Nonrelativistic geodesic motion, *Int. J. Theor. Phys.* **38** 2703.

[57] Giachetta, G., Mangiarotti, L. and Sardanashvily, G. (2002). Covariant geometric quantization of nonrelativistic time-dependent mechanics, *J. Math. Phys* **43**, 56.

[58] Giachetta, G., Mangiarotti, L. and Sardanashvily, G. (2002). Geometric quantization of mechanical systems with time-dependent parameters, *J. Math. Phys* **43**, 2882.

[59] Giachetta, G., Mangiarotti, L. and Sardanashvily, G. (2002). Action-angle coordinates for time-dependent completely integrable Hamiltonian systems, *J. Phys. A* **35**, L439.

[60] Giachetta, G., Mangiarotti, L. and Sardanashvily, G. (2002). Geometric quantization of completely integrable systems in action-angle variables, *Phys. Lett. A* **301**, 53.

[61] Giachetta, G., Mangiarotti, L. and Sardanashvily, G. (2003). Jacobi fields of completely integrable Hamiltonian systems, *Phys. Lett. A* **309**, 382.

[62] Giachetta, G., Mangiarotti, L. and Sardanashvily, G. (2003). Bi-Hamiltonian partially integrable systems, *J. Math. Phys.* **44**, 1984.

[63] Giachetta, G., Mangiarotti, L. and Sardanashvily, G. (2004). Nonadiabatic holonomy operators in classical and quantum completely integrable systems, *J. Math. Phys* **45**, 76.

[64] Giachetta, G., Mangiarotti, L. and Sardanashvily, G. (2005). Lagrangian supersymmetries depending on derivatives. Global analysis and cohomology, *Commun. Math. Phys.* **259**, 103.

[65] Giachetta, G., Mangiarotti, L. and Sardanashvily, G. (2005). *Geometric and Topological Algebraic Methods in Quantum Mechanics* (World Scientific, Singapore).

[66] Giachetta, G., Mangiarotti, L. and Sardanashvily, G. (2007). Quantization of noncommutative completely integrable Hamiltonian systems, *Phys. Lett. A* **362**, 138.

[67] Giachetta, G., Mangiarotti, L. and Sardanashvily, G. (2009). On the notion of gauge symmetries of generic Lagrangian field theory, *J. Math. Phys.* **50**, 012903.

[68] Giachetta, G., Mangiarotti, L. and Sardanashvily, G. (2009). *Advanced Classical Field Theory* (World Scientific, Singapore).

[69] De Gosson, M. (2001). The symplectic camel and phase space quantization, *J. Phys. A* **34**, 10085.

[70] Gotay, M., Nester, J. and Hinds, G. (1978). Presymplectic manifolds and the Dirac–Bergman theory of constraints, *J. Math.Phys.* **19**, 2388.

[71] Gotay, M. and Śniatycki, J. (1981). On the quantization of presymplectic dynamical systems via coisotropic imbeddings, *Commun. Math. Phys.* **82**, 377.

[72] Gotay, M. (1982). On coisotropic imbeddings of presymplectic manifolds, *Proc. Amer. Math. Soc.* **84**, 111.

[73] Gotay, M. (1986). Constraints, reduction and quantization, *J. Math. Phys.* **27**, 2051.

[74] Gotay, M. (1991). A multisymplectic framework for classical field theory and the calculus of variations. I. Covariant Hamiltonian formalism, In: *Mechanics, Analysis and Geometry: 200 Years after Lagrange* (North Holland, Amsterdam) p. 203.

[75] Grabowski, J. and Urbański, P. (1995). Tangent lifts of Poisson and related structures, *J. Phys. A* **28**, 6743.

[76] Greub, W., Halperin, S. and Vanstone, R. (1972). *Connections, Curvature and Cohomology* (Academic Press, New York).

[77] Grundling, H. and Hurst, A. (1988). The quantum theory of second class constraints, *Commun. Math. Phys.* **119**, 75.

[78] Grundling, H. and Lledo, F. (2000). Local quantum constraints, *Rev. Math. Phys.* **12**, 1159.

[79] Guillemin, V. and Sternberg, S. (1984). *Symplectic Techniques in Physics* (Cambr. Univ. Press, Cambridge).

[80] Gutzwiller, M. (1990). *Chaos in Classical and Quantum Mechanics* (Springer, Berlin).

[81] Hamoui, A. and Lichnerowicz, A. (1984). Geometry of dynamical systems with time-dependent constraints and time-dependent Hamiltonians: An approach towards quantization, *J. Math. Phys.* **25**, 923.

[82] Hannay, J. (1985). Angle variable in adiabatic excursion of an integrable Hamiltonian, *J. Phys. A* **18**, 221.

[83] Harlet, J. (1996). Time and time functions in parametrized non-relativistic quantum mechanics, *Class. Quant. Grav.* **13**, 361.

[84] De la Harpe, P. (1972). *Classical Banach-Lie Algebras and Banach-Lie Groups of Operators in Hilbert Space*, Lect. Notes in Math. **285** (Springer, Berlin).

[85] Hirzebruch, F. (1966). *Topological Methods in Algebraic Geometry* (Springer, Berlin).

[86] Horuzhy, S. (1990). *Introduction to Algebraic Quantum Field Theory*, Mathematics and its Applications (Soviet Series) **19** (Kluwer, Dordrecht).

[87] Ibragimov, N. (1985). *Transformation Groups Applied to Mathematical Physics* (Riedel, Boston).

[88] Iliev, B. (2001). Fibre bundle formulation of nonrelativistic quantum mechanics: I-III, *J. Phys. A* **34**, 4887, 4919, 4935.

[89] Kamber, F. and Tondeur, P. (1975). *Foliated Bundles and Characteristic Classes*, Lect. Notes in Mathematics **493** (Springer, Berlin).

[90] Kimura, T. (1993). Generalized classical BRST cohomology and reduction of Poisson manifolds, *Commun. Math. Phys.* **151**, 155.

[91] Kiritsis, E. (1987). A topological investigation of the quantum adiabatc phase, *Commun. Math. Phys.* **111**, 417.

[92] Kishimoto, A. (1976). Dissipations and derivations, *Commun. Math. Phys.* **47**, 25.

[93] Kobayashi, S. and Nomizu, K. (1963). *Foundations of Differential Geometry, Vol. 1* (John Wiley, New York - Singapore).

[94] Kolář, I., Michor, P. and Slovák, J. (1993). *Natural Operations in Differential Geometry* (Springer, Berlin).

[95] Krasil'shchik, I., Lychagin, V. and Vinogradov, A. (1985). *Geometry of Jet Spaces and Nonlinear Partial Differential Equations* (Gordon and Breach, Glasgow).

[96] Kriegl, A. and Michor, P. (1997). *The Convenient Setting for Global Analysis*, AMS Math. Surveys and Monographs, **53** (AMS, Providence).

[97] Kuiper, N. (1965). Contractibility of the unitary group of a Hilbert space, *Topology* **3**, 19.

[98] Lam, C. (1998). Decomposition of time-ordered products and path-ordered exponentials, *J. Math. Phys.* **39**, 5543.

[99] Lang, S. (1993). *Algebra* (Addison–Wisley, New York).

[100] Lang, S. (1995). *Differential and Riemannian Manifolds*, Gradutate Texts in Mathematics **160** (Springer, New York).

[101] Lazutkin, V. (1993). *KAM Theory and Semiclassical Approximations to Eigenfunctions* (Springer, Berlin).

[102] De León, M. and Rodrigues, P. (1989). *Methods of Differential Geometry in Analytical Mechanics* (North-Holland, Amsterdam).

[103] De León, M., Marrero J. and Martín de Diego, D. (1996). Time-dependent conctrained Hamiltonian systems and Dirac bracket, *J. Phys. A* **29**, 6843.

[104] Libermann, P. and Marle, C-M. (1987). *Symplectic Geometry and Analytical Mechanics* (D.Reidel Publishing Company, Dordrecht).

[105] Mac Lane, S. (1967). *Homology* (Springer, Berlin).

[106] Mangiarotti, L. and Sardanashvily, G. (1998). *Gauge Mechanics* (World Scientific, Singapore).

[107] Mangiarotti, L. and Sardanashvily, G. (2000). On the geodesic form of second order dynamic equations, *J. Math. Phys.* **41**, 835.

[108] Mangiarotti, L. and Sardanashvily, G. (2000). Constraints in Hamiltonian time-dependent mechanics, *J. Math. Phys.* **41**, 2858.

[109] Mangiarotti, L. and Sardanashvily, G. (2000). *Connections in Classical and Quantum Field Theory* (World Scientific, Singapore).

[110] Mangiarotti, L. and Sardanashvily, G. (2007). Quantum mechanics with respect to different reference frames, *J. Math. Phys.* **48**, 082104.

[111] Marle, C.-M. (1997). The Schouten–Nijenhuis bracket and interior products, *J. Geom. Phys.* **23**, 350.

[112] Massa, E. and Pagani, E. (1994). Jet bundle geometry, dynamical connections and the inverse problem of Lagrangian mechanics, *Ann. Inst. Henri Poincaré* **61**, 17.

[113] Massey, W. (1978). *Homology and Cohomology Theory* (Marcel Dekker, Inc., New York).

[114] Meigniez, G. (2002). Submersions, fibrations and bundles, *Trans. Amer. Math. Soc.* **354**, 3771.

[115] Mishchenko, A. and Fomenko, A. (1978). Generalized Liouville method of integration of Hamiltonian systems, *Funct. Anal. Appl.* **12**, 113.

[116] Molino, P. (1988). *Riemannian Foliations* (Birkhäuser, Boston).

[117] Montgomery, R. (1988). The connection whose holonomy is the classical adiabatic angles of Hannay and Berry and its generalization to the non-integrable case, *Commun. Math. Phys.* **120**, 269.

[118] Morandi, G., Ferrario, C., Lo Vecchio, G., Marmo, G. and Rubano, C. (1990). The inverse problem in the calculus of variations and the geometry of the tangent bundle, *Phys. Rep.* **188**, 147.

[119] Mostov, M. (1976). Continuous cohomology of spaces with two topologies, *Mem. Amer. Math. Soc.* **7**, No.175.

[120] Muñoz-Lecanda, M. (1989). Hamiltonian systems with constraints: A geometric approach, *Int. J. Theor. Phys.* **28**, 1405.

[121] Mykytiuk, I., Prykarpatsky, A., Andrushkiw, R. and Samoilenko, V. (1994). Geometric quantization of Neumann-type completely integrable Hamiltonian systems, *J. Math. Phys.* **35**, 1532.

[122] Nekhoroshev, N. (1994). The Poincaré – Lyapounov – Liouville – Arnold theorem, *Funct. Anal. Appl.* **28**, 128.

[123] Nunes da Costa, J. and Petalidou, F. (2002). Reduction of Jacobi-Nijenhuis manifolds, *J. Geom. Phys.* **41**, 181.

[124] Olver, P. (1986). *Applications of Lie Groups to Differential Equations* (Springer, Berlin).

[125] Onishchik, A. (Ed.) (1993). *Lie Groups and Lie Algebras I. Foundations of Lie Theory, Lie Transformation Groups* (Springer, Berlin).

[126] Oteo, J. and Ros, J. (2000). From time-ordered products to Magnus expansion, *J. Math. Phys.* **41**, 3268.

[127] Palais, R. (1957). A global formulation of Lie theory of transformation groups, *Mem. Am. Math. Soc.* **22**, 1.

[128] Pedersen, G. (1979). *C^*-Algebras and Their Automorphism Groups* (Academic Press, London).

[129] Powers, R. (1971). Self-adjoint algebras of unbounded operators, I, *Commun. Math. Phys.* **21**, 85.

[130] Powers, R. (1974). Self-adjoint algebras of unbounded operators, II, *Trans. Amer. Math. Soc.* **187**, 261.

[131] Rawnsley, J. (1977). On the cohomology groups of a polarisation and diagonal quantisation, *Trans. Amer. Math. Soc.* **230**, 235.

[132] Reinhart, B. (1983). *Differential Geometry and Foliations* (Springer, Berlin).

[133] Riewe, F. (1996). Nonconservative Lagrangian and Hamiltonian mechanics, *Phys. Rev. E* **53**, 1890.

[134] Robart, T. (1997). Sur l'intégrabilité des sous-algèbres de Lie en dimension infinie, *Can. J. Math.* **49**, 820.

[135] Robertson, A. and Robertson, W. (1973). *Topological Vector Spaces* (Cambridge Univ. Press., Cambridge).

[136] Rovelli, C. (1991). Time in quantum gravity: A hypothesis, *Phys. Rev.* **D43**, 442.

[137] Sakai, S. (1971). *C^*-algebras and W^*-algebras* (Springer, Berlin).

[138] Sardanashvily, G. (1995). *Generalized Hamiltonian Formalism for Field Theory. Constraint Systems.* (World Scientific, Singapore).

[139] Sardanashvily, G. (1998). Hamiltonian time-dependent mechanics, *J. Math. Phys.* **39**, 2714.

[140] Sardanashvily, G. (2000). Classical and quantum mechanics with time-dependent parameters, *J. Math. Phys.* **41**, 5245.

[141] Sardanashvily, G. (2003). Geometric quantization of relativistic Hamiltonian mechanics, *Int. J. Theor. Phys.* **44**, 697.

[142] Sardanashvily, G. (2008). Classical field theory. Advanced mathematical formulation, *Int. J. Geom. Methods Mod. Phys.* **5**, 1163.

[143] Sardanashvily, G. (2009). Gauge conservation laws in a general setting. Superpotential, *Int. J. Geom. Methods Mod. Phys.* **6**, 1046.

[144] Sardanashvily, G. (2009). Superintegrable Hamiltonian systems with non-compact invariant submanifolds. Kepler system, *Int. J. Geom. Methods Mod. Phys.* **6**, 1391.

[145] Saunders, D. (1989). *The Geometry of Jet Bundles* (Cambridge Univ. Press, Cambridge).

[146] Schmüdgen, K. (1990). *Unbounded Operator Algebras and Representation Theory* (Birkhäuser, Berlin).

[147] Simon, B. (1983). Holonomy, the quantum adiabatic theorem and Berry's phase, *Phys. Rev. Lett.* **51**, 2167.

[148] Śniatycki, J. (1980). *Geometric Quantization and Quantum Mechanics* (Springer, Berlin).

[149] Souriau, J. (1970). *Structures des Systemes Dynamiques* (Dunod, Paris, 1970).

[150] Steenrod, N. (1972). *The Topology of Fibre Bundles* (Princeton Univ. Press, Princeton).

[151] Strahov, E. (2001). Berry's phase for compact Lie groups, *J. Math. Phys.* **42**, 2008.

[152] Sundermeyer, K. (1982). *Constrained Dynamics* (Springer, Berlin).

[153] Sussmann, H. (1973). Orbits of families of vector fields and integrability of distributions, *Trans. Amer. Math. Soc.* **180**, 171.

[154] Tamura, I. (1992). *Topology of Foliations: An Introduction*, Transl. Math. Monographs **97** (AMS, Providence).

[155] Vaisman, I. (1973). *Cohomology and Differential Forms* (Marcel Dekker, Inc., New York).

[156] Vaisman, I. (1991). On the geometric quantization of Poisson manifolds, *J. Math. Phys.* **32**, 3339.

[157] Vaisman, I. (1994). *Lectures on the Geometry of Poisson Manifolds* (Birkhäuser, Basel).

[158] Vaisman, I. (1997). On the geometric quantization of the symplectic leaves of Poisson manifolds, *Diff. Geom. Appl.* **7**, 265.

[159] Varadarjan, V. (1985). *Geometry of Quantum Theory* (Springer, Berlin).

[160] Vassiliou, E.(1978). On the infinite dimensional holonomy theorem, *Bull. Soc. Roy. Sc. Liège* **9**, 223.

[161] Vinogradov, A. and Kupershmidt, B. (1977). The structure of Hamiltonian mechanics, *Russian Math. Surveys* **32** (4), 177.

[162] Warner, F. (1983). *Foundations of Differential Manifolds and Lie Groups* (Springer, Berlin).

[163] Weinstein, A. (1983). The local structure of Poisson manifolds, *J. Diff. Geom.* **18**, 523.

[164] Wilczek, F. and Zee, A. (1984). Appearance of gauge structure in simple dynamical systems, *Phys. Rev. Lett.* **52**, 2111.

[165] Woodhouse, N. (1992). *Geometric Quantization* (Clarendon Press, Oxford).

[166] Wu, Y. (1990). Classical non-abelian Berry's phase, *J. Math. Phys.* **31**, 294.

Index